Studies in Computational Intelligence

Volume 1026

Series Editor
Janusz Kacprzyk, Polish Academy of Sciences, Warsaw, Poland

The series "Studies in Computational Intelligence" (SCI) publishes new developments and advances in the various areas of computational intelligence—quickly and with a high quality. The intent is to cover the theory, applications, and design methods of computational intelligence, as embedded in the fields of engineering, computer science, physics and life sciences, as well as the methodologies behind them. The series contains monographs, lecture notes and edited volumes in computational intelligence spanning the areas of neural networks, connectionist systems, genetic algorithms, evolutionary computation, artificial intelligence, cellular automata, self-organizing systems, soft computing, fuzzy systems, and hybrid intelligent systems. Of particular value to both the contributors and the readership are the short publication timeframe and the world-wide distribution, which enable both wide and rapid dissemination of research output.

Indexed by SCOPUS, DBLP, WTI Frankfurt eG, zbMATH, SCImago.

All books published in the series are submitted for consideration in Web of Science.

More information about this series at https://link.springer.com/bookseries/7092

David Camacho · Domenico Rosaci ·
Giuseppe M. L. Sarné · Mario Versaci
Editors

Intelligent Distributed Computing XIV

Springer

Editors
David Camacho
Computer Systems Engineering Department
Universidad Politécnica de Madrid
Madrid, Spain

Giuseppe M. L. Sarné
Department of Psychology
University of Milan Bicocca
Milan, Italy

Domenico Rosaci
DIIES Department
University of Reggio Calabria
Reggio Calabria, Italy

Mario Versaci
DICEAM Department
University of Reggio Calabria
Reggio Calabria, Italy

ISSN 1860-949X ISSN 1860-9503 (electronic)
Studies in Computational Intelligence
ISBN 978-3-030-96626-3 ISBN 978-3-030-96627-0 (eBook)
https://doi.org/10.1007/978-3-030-96627-0

This Springer imprint is published by the registered company Springer Nature Switzerland AG
The registered company address is: Gewerbestrasse 11, 6330 Cham, Switzerland

Organization

General Chair

David Camacho, Universidad Politécnica de Madrid, Spain
Domenico Rosaci, University Mediterranea of Reggio Calabria, Italy
Giuseppe M. L. Sarné, University of Milan Bicocca, Italy
Mario Versaci, University Mediterranea of Reggio Calabria, Italy

Steering Committee

Janusz Kacprzyk, Polish Academy of Sciences, Poland
Costin Badica, University of Craiova, Romania
David Camacho, Universidad Politécnica de Madrid, Spain
Paulo Novais, University of Minho, Portugal
Filip Zavoral, Charles University Prague, Czech Republic
Frances Brazier, Delft University of Technology, The Netherlands
George A. Papadopoulos, University of Cyprus, Cyprus
Giancarlo Fortino, University of Calabria, Italy
Kees Nieuwenhuis, Thales Research and Technology, The Netherlands
Marcin Paprzycki, Polish Academy of Sciences, Poland
Michele Malgeri, University of Catania, Italy
Mohammad Essaaidi, Abdelmalek Essaadi University in Tetuan
Mirjana Ivanovic, University of Novi Sad, Serbia
Amal El Fallah Seghrouchni, LIP6/University Pierre and Marie Curie, France

Program Chairs

Amal El Fallah, University of Pierre and Marie Curie, Paris, France
Igor Kotenko, St. Petersburg Federal Research Center of the Russian Academy of Sciences, Russia
Mirjana Ivanovic, University of Novi Sad, Serbia
Jason J. Jung, Chung-Ang University, South Korea
Paolo Novais, University of Minho, Portogallo
Andrea Omicini, University "Alma Mater Studiorum" of Bologna, Italy

Program Committee

Jemal Abawajy, Deakin University
Amelia Badica, University of Craiova, Romania
Fabrizio Baiardi, University of Pisa
Doina Bein, California State University
Diana Gratiela Berbecaru, Polytechnic di Torino
Cristina Bianca, Technical University of Cluj-Napoca
Giacomo Cabri, University of Modena and Reggio Emilia
Vincenza Carchiolo, University of Catania
Jesus Carretero, University Carlos III of Madrid
Dmitry Chaly, Yaroslavl State University
Andrey Chechulin, St. Petersburg Institute for Informatics and Automation of the Russian Academy of Sciences
Sergei Chernyi, Admiral Makarov State University of Maritime and Inland Shipping
Angelo Ciaramella, University of Naples "Parthenope"
Franco Cicirelli, ICAR, CNR
Giuseppe Coviello, NEC Laboratories America
Claudio De Farias, PPGI-IM/NCE-UFRJ
Vasily Desnitsky, St. Petersburg Institute for Informatics and Automation of the Russian Academy of Sciences
Giuseppe Di Fatta, University of Reading
Marios Dikaiakos, University of Cyprus
Chiara Ditta, Alma Mater Studiorum—University of Bologna
Elena Doynikova, St. Petersburg Institute for Informatics and Automation of the Russian Academy of Sciences
Sisi Duan, Tsinghua University
Lidia Fotia, University of Salerno
Maria Ganzha, Warsaw University of Technology
Javier Garcia Blas, University Carlos III of Madrid
Dimitrios Georgakopoulos, Swinburne University of Technology
Sandra Gesing, University of Notre Dame

Daniela Gifu, “Alexandru Ioan Cuza” University
Raffaele Gravina, University of Calabria
Alexander Grusho, Moscow State University
Cheol-Ho Hong, Chung-Ang University
Bahman Javadi, Western Sydney University
X Jiang, Delft University of Technology
Qi Kang, Tongji University
Dimitrios Katsaros, University of Thessaly
Igor Kotenko, St. Petersburg Federal Research Center of the Russian Academy of Sciences
Georgia Kougka, Aristotle University of Thessaloniki
Giuliano Laccetti, University of Naples Federico II and INFN
Marco Lapegna, University of Naples Federico II
Bin Li, Fujian University of Technology
Wenfeng Li, Wuhan University of Technology
Antonio Liotta, University of Bozen
Carmelo Fabio Longo, University of Catania
Valeria Loscri, INRIA
Douglas Macedo, Federal University of Santa Catarina
Michele Malgeri, University of Catania
Stefano Mariani, University of Modena and Reggio Emilia
Maria-Cristina Marinescu, Barcelona supercomputing Center
Carlo Mastroianni, ICAR, CNR
Nouredine Melab, University of Lille/INRIA Lille/CNRS CRIStAL
Fabrizio Messina, University of Catania
Manuela Montangero, University of Modena and Reggio Emilia
Raffaele Montella, University of Naples Parthenope
Andea Francesco Morabito, University Mediterranea of Reggio Calabria
Evgenia Novikova, Saint-Petersburg Electrotechnical University “LETI”
Sergio Ochoa, University of Chile
Mariacristina Onete, Université de Limoges
George Pallis, University of Cyprus
Marcin Paprzycki, IBS PAN and WSM
Marianadia Postorino, Alma Mater Studiorum—University of Bologna
Shahram Rahimi, Mississippi State University
Alessandro Ricci, Alma Mater Studiorum—University of Bologna
Roland Rieke, Fraunhofer Institute for Secure Information Technology SIT
Igor Saenko, St. Petersburg Institute for Information and Automation of RAS
Alessandro Sapienza, ISTC-CNR
Ilias Savvas, University of Thessaly
Riaz Ahmed Shaikh, King Abdul Aziz University
Yulia Shichkina, Saint Petersburg Electrotechnical University "LETI"
Federico Silla, Universitat Politècnica de València
Ivor Spence, Queen’s University Belfast
Giandomenico Spezzano, ICAR, CNR

Stanimir Stoyanov, University of Plovdiv "Paisii Hilendarski"
Yusak Susilo, University of Natural Resources and Life Sciences
Parimala Thulasiram, University of Manitoba
Paolo Trunfio, University of Calabria
Salvatore Venticinque, University Vanvitelli
Ilya Viksnin, ITMO University
Norihiko Yoshida, Rissho University

Publicity Chair

Giancarlo Fortino, University of Calabria, Italy
Costin Badica, University of Craiova, Romania

Publications Chair

Fabrizio Messina, University of Catania, Italy

Special Sessions Co-Chairs

Antonio Liotta, University of Bolzano, Italy
Claudio Savaglio, ICAR-CNR, Italy

Finance Chair

Antonio Guerrieri, ICAR-CNR, Italy

Industry Chair

Giuseppe Vizzari, University of Milano Bicocca, Italy

Local Organizing Committee

Daniela Streva, University Mediterranea of Reggio Calabria, Italy

Lidia Fotia, University of Salerno, Italy
Maurizio Campolo, University Mediterranea of Reggio Calabria, Italy

Preface

Distributed intelligent computing has emerged from the synergistic application of different approaches and techniques of intelligent and distributed computing as an extension of studies and researches in the field of artificial intelligence combined with computational intelligence, distributed systems and multi-agent systems.

This book collects some recent scientific contributions in the field of intelligent and distributed computing, including new architectures and algorithms. Significant attention has also been paid to new techniques and applications for the Internet of Things, intelligent transportation, cybersecurity, and trust and reputation systems. The book includes the peer-reviewed proceedings of the 14th International Symposium on Intelligent Distributed Computing (IDC 2021), which was held in Scilla, Italy, from September 16–18, 2021 in online mode due to the problematics inherited by the COVSARS2 pandemic that had already forced to move to 2021 the 14th edition scheduled to take place in 2020.

This 2021 edition of the IDC continues the legacy of the previous IDC Symposium Series as an initiative of two research groups: (i) Systems Research Institute, Polish Academy of Sciences, Warsaw, Poland, and (ii) Department of Software Engineering, University of Craiova, Craiova, Romania.

After a rigorous peer review process, the International Program Committee selected 43 papers, which are published in these conference proceedings. The selection of these contributions was rigorous with the intention of maintaining the high quality of the conference. We would like to thank the Program Committee for their hard work during the review process. This peer review phase is very important to achieve the high-quality standards of the IDC conference. We would also like to thank Janusz Kacprzyk, editor of the Studies in Computational Intelligence series and a member of the steering committee, for his support and encouragement for the development of the IDC symposium series. The organization also feels indebted to Thomas Ditzinger and Ramamoorthy Rajangam of Springer, for their kind support during the production of the symposium proceedings.

The 43 papers, from 18 countries, published in this volume address a wide variety of topics related to the theory and applications of distributed intelligent computing. We would also like to highlight the organization of three different special sessions: Intelligent Distributed Computing for Cyber-Physical Security and Safety (IDCCPSS), Future Intelligent Transport Solutions (FITS), and Trust and Reputation in intelligent Environments (TRIE).

IDC 2021 had the privilege of two outstanding Keynotes from distinguished guest speakers, Prof. Costin Badica, Full Professor, University of Craiova, Romania and Prof. Antonio Liotta, Full Professor, University of Bozen, Italy.

Special thanks to all the members of the local organizing committee who contributed to the success of the event.

Madrid, Spain — David Camacho
Reggio Calabria, Italy — Domenico Rosaci
Milan, Italy — Giuseppe M. L. Sarné
Reggio Calabria, Italy — Mario Versaci
September 2021

Contents

TRIE (Special Session) & COMPUTING—Chair: Lidia Fotia

INTERNET OF THINGS (Main Track)—Chair: David Camacho

EBASI: IoT-Based Emotion and Behaviour Recognition System Against Elderly People Social Isolation

Clara I. Valero, Saverio Gravina, Alessio Ciuffetti, Marco Di Gioia, Raffaele Gravina, Fernando Boronat, and Carlos E. Palau

Abstract In recent decades, great strides have been made in the prevention of chronic diseases by setting healthier lifestyles for seniors trying to set effective health models. However, despite this great effort, there is still a lack of solutions on what we can call social care. Nowadays, Social isolation is a big health and social problem that communities are facing. Older adults face many challenges to their independence, often relating the feeling of loneliness and a decline in mobility or cognition. Active and healthy aging is still an aspiration not only critical in making public welfare systems sustainable but also for the mental and social well-being of people. Access to assistive technologies is being identified as an important facilitator for successful ageing that can improve the independent living of seniors as well as the quality of life in general. This paper proposes the application of Internet of Things (IoT) to detect, avoid and prevent elderly social isolation. This can be achieved by improving the social inclusion and facilitate the uptake of evidence-based active healthy ageing interventions.

C. I. Valero (✉) · F. Boronat · C. E. Palau
Communications Department, Universitat Politècnica de València, Valencia, Spain
e-mail: clavalpe@upv.es

F. Boronat
e-mail: fboronat@dcom.upv.es

C. E. Palau
e-mail: cpalau@dcom.upv.es

S. Gravina · A. Ciuffetti · M. Di Gioia
DS TECH, Rome, Italy
e-mail: saverio.gravina@dstech.it

A. Ciuffetti
e-mail: a.ciuffetti@dstech.it

M. Di Gioia
e-mail: m.digioia@pharmawizard.com

R. Gravina
University of Calabria, Rende, Italy
e-mail: r.gravina@dimes.unical.it

D. Camacho et al. (eds.), *Intelligent Distributed Computing XIV*, Studies in Computational Intelligence 1026, https://doi.org/10.1007/978-3-030-96627-0_1

1 Introduction

In the last decades life expectancy has increased significantly and this trend is expected to continue. According to World Health Organization (WHO) [1], the number of older persons worldwide (2021) is slightly more than 1 billion—about 13.5% of the global population. By 2030, 1 in 6 people will be 60 years of age or older. On the other hand, according to the results of the European Social Survey (ESS) [2], around 7% of adults in Europe frequently feel lonely. In contrast with feelings of loneliness, many more adults in Europe (18%) are socially isolated. This is a problem that worsens with age and can have significant negative consequences. Loneliness and isolation not only pose a major threat to the physical health of the elderly but can also have serious effects on mental health.

In this paper an non-invasive and low-cost IoT system for living environment capable of detection, avoidance and prevention of elderly social isolation is presented. The IoT-Based Emotion and Behaviour Recognition System Against Elderly People Social Isolation (EBASI) system aims to improving social inclusion and facilitate the uptake of evidence-based active healthy ageing interventions. Its main target user are retired seniors who live alone and away from their families. The IoT system recognizes person's activities and change of emotions by the integration of IoT sensor data. To detect loneliness patterns, a multimodal sensor-based recognition method has been designed and implemented. When patterns of loneliness are detected, the system sends positive encouragement to the elderly and, if it is considered to be a worrying situation, sends a warning to the caregiver or family. This work was performed as part of the European H2020 project DIATOMIC.[1]

The remainder of the paper is structured as follows. Section 2 describes the motivation of this work in addition to present previous related works. Section 3 introduces the multimodal sensor-based recognition method, the functional and technical system architecture as well as the validation methodology. Section 4 presents the results obtained from installing the solution in two houses. Finally, concluding remarks are given in Sect. 5.

2 Motivation and Related Work

2.1 Motivation

This work aims to detect and face elder social isolation and loneliness. Social isolation is a multidimensional concept, to which both aspects of a "structural" nature (such as living alone and the scarcity of relationships) and aspects of a "functional" nature (such as the material and emotional support conveyed by existing relationships) contribute. The main cause of social isolation is the absence or relative scarcity of

[1] https://diatomic.eu/.

social relationships, this represents a major health risk factor comparable, if not superior, to well-known risk factors such as cigarette smoking, alcohol abuse and obesity. For the elderly people, social isolation has also been found to be a major cause to the decline in cognitive skills and more generally to a worse state of health, psychological, mental and physical [3, 4]. Previous studies have shown that the social participation and maintenance of healthy social networks is essential for preserving individuals from emotional distress, cognitive decline, and physical disability [5].

The most important issue for family members, specially when acting as a caregiver, is to setup and maintain strong relationships with elderly people. When older adults begin to require care, they need emotional and social supports that are different from the usual exchanges among family members [6]. Despite efforts to maintain reciprocity, a major shift in the balance of reciprocity in the caregiver-recipient relationship is inevitable. As the needs of the care recipient increase, the care recipient contributes less to the relationship while needing more from it [7]. In addition, the care recipient's physical and mental condition will worsen over time and therefore needs a great deal of emotional support from the caregiver. Caregivers must deal with unknown depressive symptoms, anxiety, irritability or anger in the care recipient.

If the cognitive or functional decline occurs slowly but progressively the changes can be very subtle and therefore very difficult to detect. In this case, they are more easily recognizable when looking back in time. Conversely, changes in relationships may occur suddenly when the person's condition seriously declines in a short period of time. For example, among caregivers of people who have suffered a stroke, the most stressful problems are in the relationship between the caregiver and the stroke survivor (including poor communication, frustration with role reversal, and intimacy issues) [8]. Emotional support is perceived to be the most time-consuming task [9]. In a study focused on the first year of care after a stroke, surveyed caregivers reported that the problems perceived as most stressful were that the recipient of care seemed sad or depressed, spoke of feeling lonely, had trouble controlling bowels, felt worthless or burdened, and/or seemed anxious or worried [10].

2.2 Related Work

Currently there are a large number of smart home solutions to monitor the well-being and health of the elderly. However, to the best of the author's knowledge, at the present time there are very few that focus on monitoring and analyzing people's state of mind. Of the existing IoT smart home solutions, some have similarities in their objective with the present work.

In [11] the authors present an IoT solution whose goal is to detect possible aging people's behavior changes by using unobtrusive monitoring technologies embedded in real environments or in objects of daily living. The system aims to detect sedentary lifestyle and health related problems at early stage. For indoor monitoring, they use motion sensors, contact sensors, and bed sensors. Outdoor monitoring is based on smartphones carried by participants and beacons deployed in places of interest of participants in the pilot city. These sensors allow the collection of raw data and objective information in real time, which are processed to detect behavior changes.

On the other hand, the authors of [12] propose a smart in-home healthcare system based on the use of IoT for smart and individualized monitoring of elderly patients in their homes. By making use of images and facial expressions, the solution monitors patients and elderly people that have special needs. The system captures facial expressions to determine whether the patient is experiencing pain and to record their emotional state. The system is composed of several sensors and a decision maker. The sensors are devices distributed in the environment that collects information on the patient's health. This information is then sent to the decision making element. It processes the information received and makes decisions aimed at preserving the health of patients.

3 Materials and Methods

3.1 *Multimodal Sensor-Based Loneliness Recognition Methodology*

The needs of an elderly person suffering from social isolation are closely linked to maintaining positive relationships and emotions.

Behavioral activities and attitude that are symptoms of social isolation and loneliness are identified in order to consequently define the parameters and indicators that need to be monitored. Moreover, identifying the causes and related indicators of the isolation help to define the type of interaction that the system needs to activate with the elders, with the family members and/or caregivers.

Loneliness depends on several things, such as mental process and physical process. The authors propose the measurement of the isolation of elderly people in three axes: self-care, social loneliness, and functional mobility. To detect these three elements, the authors have defined Social Activities of Daily Living (SADL) and simple Activities of Daily Living (ADL): SADL focus on the social activities to measure the person's sense of community, ADL focus on self-care and abilities that the senior can do by him/herself.

3.2 *Functional Architecture*

The proposed multimodal sensor-based recognition method for SADL and ADL is implemented in an IoT-based physical & emotion recognition sensors network and an AI algorithm in order to build a smart living environment. EBASI recognizes a person's activities and change of emotions by integrating the sensors' data: microwave sensor for vital data, video camera for facial expressions, microphone for speech tone, environmental sensors for temperature, and brightness.

The interfaces between EBASI and the users (senior, family member and caregiver) are a mobile application for the family member as well as a domotic assistance speaker for the senior. The mobile application allows the caregiver the ability to set the thresholds of the ADL and SADL sensors parameters. It provides periodical reports and sends notifications in case of critical situations. Moreover, the system permits to upload media content (photos or videos) so that an interaction is triggered with the elderly person, asking him/her if he/she wishes to view them on the TV. The domotic assistant is directly linked to a speaker. It starts the interaction phase, allowing the senior to watch the videos or pictures that the family member has uploaded in addition to listening to music.

Functionally, the system operation is divided into three main phases: (i) monitoring and detection, (ii) interacting and alerting and (iii) reporting and forecasting.

During the monitoring and detection phase, different sensors (see Sect. 3.3) integrated into the domotic assistant send ADL and SADL data. These behaviors are monitored in four spaces of senior's house: the bathroom, the kitchen, the living room and the bedroom.

The interaction and alerting phase starts as soon as the system detects an ADL or SADL that is over the configured threshold (see Sect. 3.1). Depending on the value of the measure received the system will send a positive interaction to the person or alert the caregiver through a mobile notification. Interaction includes encouragements, incentives and invitations.

The reporting and forecast phase is designed to send to the family member a periodic report on the ADL and SADL parameters and to show a forecast trend of them, taking into account the improvement rate of senior behavior and emotional status. The mobile application provides family members with interactive real-time reports to monitor the activities of the elderly and provide early warnings about the state of social isolation. Moreover, it allows access to historical activities logs (recorded on cloud) to determine social isolation trends and forecast evolution of the situation.

3.3 Technical Architecture

Figure 1 shows the EBASI technical system architecture. Data sources are composed of a set of IoT sensors strategically distributed in the home of the seniors and a smart bracelet for the detection of biometric data. To monitor the activities of the elders the following IoT sensors are used: (i) HC-SR04 ultrasonic distance sensor, (ii) BH1750 light sensor, (iii) MD30–60 pressure (weight) sensor and (iv) a rain detector (AZDelivery). Moreover, a smartband (Xiaomi Mi Band 3) is used to get the following raw data: accelerometer data, heart rate data and steps data. Wi-Fi Direct communication protocol is used to connect the sensor directly to a Raspberry Pi 3. For that purpose, each sensor has been equipped with ESP8266 modules in

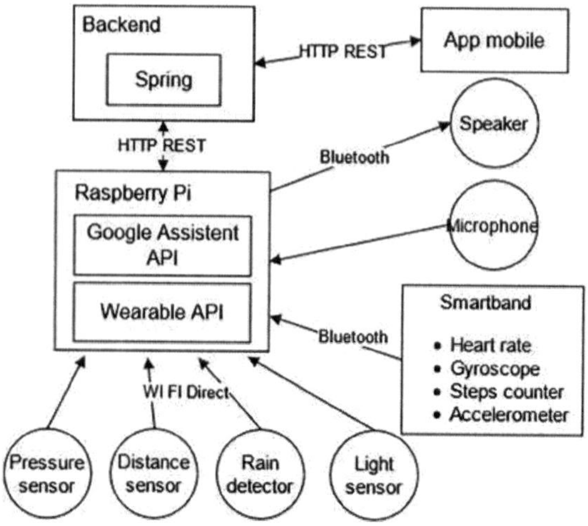

Fig. 1 System architecture

order to support the Wi-Fi Direct with an additional one connected to the Raspberry, that acts as a receiver. To avoid operational problems, especially in the case of IoT sensors intended to be activated by contact with water, 3D models of enclosures have been designed. They enclose the electronic components and the battery that powers the sensors.

The Raspberry Pi is used as both, a controller of the data collected by sensors and as a home automation assistant. It interacts with the user vocally by means of the Google assistant API, a microphone (Anker Bluetooth Speaker) and a speaker (Movo MC 1000).

The backend is the heart of the platform. It takes care of receiving the data collected by Raspberry Pi, to perform analyses on it and, based on certain parameters, initiates an interaction with the user or alert the family member with push notifications. The system has also the function of receiving multimedia content (photos, videos), saving them and sending them to the frontend for viewing on TV. A MongoDB database is used for saving thresholds and sensor measurements whereas a PostgreSQL database is used for saving utilities, album and pictures. The backend software is based on Spring framework and Java language.

The mobile application is an integral part of the system, where the relationship between caregiver, domotic assistant and user takes place. Its functionalities have been described in the previous subsection. It is native iOS mobile application written in Swift.

Every data exchange between frontend (Raspberry Pi, smartphone application) and the backend is performed on HTTP REST protocol and JSON format.

3.4 Validation Methodology

The solution was tested in the laboratory before being deployed in a pilot to verify and validate the robustness, performance, scalability and security compliance of the platform.

For pilot validation, due to the special circumstances that have occurred in relation with the COVID 19 crisis, it was not possible to install the solution in five senior apartments as initially planned. The consequent social distancing made it possible to carry out the pilot inside the apartments of some members of the technical team. It is important to note that the users mirrored the senior user as they lived alone, leaving the house very few times and working at home. The pilot phase lasted 30 days.

The users of the pilot were two and were selected on a shortlist of 75 candidates. Recruitment was made on a voluntary basis. Informed written consent was obtained prior to any data collection in order to guarantee that the senior had fully understood the aim of the project and the procedure. The inclusion criteria were the following:

- Individuals who live alone without a family or with a distant one.
- Absence of serious physical or cognitive impairment, which categorizes the elderly taking into account their physical and cognitive condition. The two users can perform the necessary tasks in daily life. Still, some assistance can needed to help you continue to function on their own.
- Score >50 on the State Trait Anxiety Inventory (STAI) Y2 Questionnaire [13].
- Score >5 on the Pittsburgh Sleep Quality Inventory (PSQI) test [14].
- Score >7 on Perceived Stress Scale (PSS) [15]; that could be useful due to the COVID19 situation.

Based on the user profile and the user's apartment layouts, sensors have been positioned to cover and identify the widest range of physical and social activities. Figure 2 shows the location of the sensors in user A (left) and user B (right) apartments. Parameter thresholds have been adjusted for each user, taking into account his or her particular lifestyle. Only one caregiver was selected for the pilot and managed the two users through the mobile application.

In order to validate the system, a questionnaire was provided to users in order to verify when the system has generated false positives in terms of: (i) accuracy in detection frequency, (ii) accuracy of detection hours, (iii) identification of ADL and SADL activities, (iv) adherence of the types of interaction with the activities detected and (v) adherence of alerting notifications with monitoring activities.

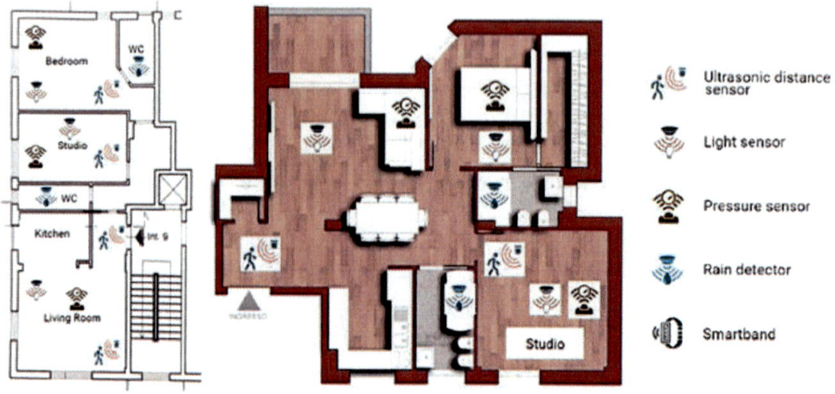

Fig. 2 Pilot

Table 1 Reported monitored data

User	Activity	Reported data	False positives
A	Showering	17	3
	Bathing	16	2
	Slepping	176 h	2
	Sitting without doing nothing	96 h	N/A
	Going outside	8	0
	Playing, watching TV, listening to music, have a call with someone	115 h	5
B	Showering	19	1
	Bathing	13	7
	Sleeping	207 h	10
	Sitting without doing nothing	31 h	~4
	Going outside	16	1
	Playing, watching TV, listening to music, have a call with someone	67h	3

4 Results and Discussion

Following the validation methodology introduced in the previously, this Section presents the results obtained after concluding the pilot phase.

Table 2 System interactions

User	Motivation events	Number of interactions	Interaction	Caregiver's alerting	False positives
A	The caregiver uploaded pictures or videos	5	Invitation	0	0
	The user slept more than what is needed	2	Encouragement	2	1
	The user slept less than what is needed	12	Encouragement	0	3
	The user had a shower	0		0	0
	Sitting without doing anything	7	Incentive 2 Encouragement 5	2	0
	Going outside	0		0	0
	Playing, watching TV, listening to music	1	Incentive	1	0
B	The caregiver uploaded pictures or videos	5	Invitation	0	0
	The user slept more than what is needed	0		0	0
	The user slept less than what is needed	2	Encouragement	0	0
	The user had a shower	1	Encouragement	0	0
	Sitting without doing anything	2	Incentive	1	0
	Going outside	2	Encouragement	2	1
	Playing, watching TV, listening to music	7	Incentive	0	0

Table 1 contains a summary of the data monitored and recorded by the system during the pilot. The system has detected false positives in both user installations. The shower sensors have detected nine erroneous events such as cleaning the shower or leaving the faucet on without turning it off correctly. The bath sensors have detected errors caused by incorrect positioning of sensors. Regarding sleeping time, due to movements and variations in heartbeat during the night the detection was inaccurate in some cases. It needs to be fine tuned. In relation to the time spent doing nothing, while working on the computer some minor movements were not detected triggering an inactivity record. It was not possible to quantify the number of times this happened. In relation to the number of times users have left the home, in user B installations an event was not detected because the smart bracelet was discharged and therefore not connected to the system. This occurred because the home's exits are detected as a combination of two events: the door motion sensor detects an event and the Bluetooth signal of the smartband goes lost. Lastly, eight false positives have been reported in the detection of playful activities. This is due to some protracted noise/sounds have been interpreted as activity causing a false positive.

Table 2 shows the system interactions performed during the pilot. For user A, only false positives related to the number of hours of sleep are observed. In some cases, the pressure sensor under the mattress continued to record the pressure without the user actually being on the mattress. On other occasions, the pressure sensor stopped recording because the user had moved too far away from the sensor position while sleeping. To correct this, smartband accelerometer and heart rate data was also taken into account to determine if the user is sleeping. For user B, only one false positive has occurred. The error was caused by the user forgetting to wear the smartband and therefore the system detected that the user was stuck inside the house. This is due to the recognition of the user outside the house was designed to be analyzed both by a motion sensor and by the loss of the smartband signal.

The system has demonstrated a good performance in terms of identification of physical activities, identifying them with more than 75% degree of accuracy. With regard to SADL interactions, some have been confused, sometimes not too close to what the user might need. The reason for this is obviously because the data collection was incorrect. In this scenario the only valid alternative is to use a Kinetic sensor or a video camera, with the permission of the elderly.

5 Conclusions

This work presented a technological solution for monitoring daily activity and emotions of elderly people at home and stimulating them to have with a more integrated social activity. It introduced a multimodal sensor based recognition methodology for social activities of daily living, integrated with a domotic system as EBASI. The sensors monitor the daily activities and emotions of the seniors, estimates their social relationships and takes care of the seniors who are isolated from the community.

Then, integrating all the collected data, the system could interact with the residents, reduces their loneliness feelings by forming a good relationship in community. The system has been successfully tested and validated in real scenarios.

References

1. World Health Organization.: Decade of healthy ageing: baseline report: summary. World Health Organization (2021)
2. Abrams, D., Swift, H.J.: Experiences and expressions of ageism: topline results (UK) from round 4 of the European Social Survey London. European Social Survey (2007)
3. O'Súilleabháin, P.S., Gallagher, S., Steptoe, A.: Loneliness, living alone, and all-cause mortality: the role of emotional and social loneliness in the elderly during 19 years of follow-up. Psychosom. Med. **81**(6), 521 (2019)
4. Holt-Lunstad, J, Smith, T.B., Baker, M., Harris, T., Stephenson, D.: Loneliness and social isolation as risk factors for mortality: a meta-analytic review. Perspect. Psychol. Sci. **10**(2), 227–237 (2015)
5. Burholt, Vanessa, Scharf, Thomas: Poor health and loneliness in later life: the role of depressive symptoms, social resources, and rural environments. J. Gerontol. B Psychol. Sci. Soc. Sci. **69**(2), 311–324 (2014)
6. Brody, E.M.: Parent care as a normative family stress. Gerontol. **25**(1), 19–29 (1985)
7. Pearlin, L.I., Mullan, J.T., Semple, S.J., Skaff, M.M.: Caregiving and the stress process: an overview of concepts and their measures. Gerontol. **30**(5), 583–594 (1990)
8. King, R.B., Ainsworth, C.R., Ronen, M., Hartke, R.J.: Stroke caregivers: pressing problems reported during the first months of caregiving. J. Neurosci. Nurs.: J. Am. Assoc. Neurosci. Nurses **42**(6), 302 (2010)
9. Bakas, T., Austin, J.K., Jessup, S.L., Williams, L.S., Oberst, M.T.: Time and difficulty of tasks provided by family caregivers of stroke survivors. J. Neurosci. Nurs. **36**(2), 95 (2004)
10. Haley, W.E., Allen, J.Y., Grant, J.S., Clay, O.J., Perkins, M., Roth, D.L.: Problems and benefits reported by stroke family caregivers: results from a prospective epidemiological study. Stroke **40**(6), 2129–2133 (2009)
11. Aloulou, Hamdi, Mokhtari, Mounir, Abdulrazak, Bessam: Pilot site deployment of an IoT solution for older adults' early behavior change detection. Sensors **20**(7), 1888 (2020)
12. Mano, L.Y., Faiçal, B.S., Nakamura, L.H., Gomes, P.H., Libralon, G.L., Meneguete, R.I., Geraldo Filho, P.R., Giancristofaro, G.T., Pessin, G., Krishnamachari, B., et al.: Exploiting IoT technologies for enhancing health smart homes through patient identification and emotion recognition. Comput. Commun. **89**, 178–190 (2016)
13. Gault, R.H.: A history of the questionnaire method of research in psychology. Pedagog. Semin. **14**(3), 366–383 (1907)
14. Buysse, D.J., Reynolds III, C.F., Monk, T.H., Berman, S.R., Kupfer, D.J.: The pittsburgh sleep quality index: a new instrument for psychiatric practice and research. Psychiatry Res. **28**(2), 193–213 (1989)
15. Cohen, S., Kamarck, T., Mermelstein, R.: A global measure of perceived stress. J. Health Soc. Behav. 385–396 (1983)

Smart Health Services Based on IoT and GIS

Aldina Avdić, Ulfeta Marovac, and Dragan Janković

Abstract Today, computer systems have the ability to access numerous user context data at any time, e.g. about its location, movement, habits, and other data that can be collected from sensors on mobile devices. This data can be processed and used for various purposes in the form of applications that can improve interaction with the user. Of particular importance are the purposes for providing health services. This paper proposes, describes, and provides an overview of the implementation of five smart health services based on the Internet of Things (IoT) and geographic information systems (GIS), namely: finding the nearest health care facilities, air quality monitoring, improving medical services, smart transport for patients, and mobile SOS system.

1 Introduction

With the advancement of information and telecommunication technologies (ICT), the number of devices connected to the network has increased, and thus the amount of data that needs to be processed in a specific way. This led to the development of Internet of Things (IoT) [1] and Big Data [2]. These are also two technologies whose development was key to the formation of smart cities [3], which are urban areas that use the benefits of ICT technologies to implement different types of services that allow their citizens an easier and more comfortable life. These services can be used for smart learning, transportation, health, energy efficiency, agriculture, administration, etc. One of the most important components of a smart city is smart health [4].

A. Avdić (✉) · U. Marovac
Department for Technical Sciences, State University of Novi Pazar, Novi Pazar, Serbia
e-mail: apljaskovic@np.ac.rs

U. Marovac
e-mail: umarovac@np.ac.rs

D. Janković
Faculty of Electronic Engineering, University of Nis, Nis, Serbia
e-mail: dragan.jankovic@elfak.ni.ac.rs

D. Camacho et al. (eds.), *Intelligent Distributed Computing XIV*, Studies in Computational Intelligence 1026, https://doi.org/10.1007/978-3-030-96627-0_2

The terms e-health [5] and mobile health [6] are often encountered. Mobile health is a type of e-health that is possible for users of mobile devices. Smart health is also a subset of e-health, which has a specific data flow. As mobile and e-health services are user-oriented, smart health services also serve in part to create collective intelligence to form information that is useful for the entire smart city, i.e. its inhabitants. Thus, these types of services may include various information obtained from individual users or from health care institutions that may be publicly available in the form of maps, boards, visualizations, instructions, and recommendations, etc., which will facilitate and make more efficient and simpler execution of the established medical procedure.

As the previously mentioned maps that contain useful information for the inhabitants of a smart city, geographic information systems (GIS) [7] and APIs intended for these purposes can be used for their realization.

The research question of this paper is whether and how IoT and GIS can be used to create smart health services. The motivation for this research is the creation of smart health services based on IoT and GIS which would be useful for patients and citizens in a smart city. For the realization of this research, the method of description was applied to the existing smart health services based on IoT and GIS. Five services are proposed and using the modeling method and use case diagrams are shown. Details of the implementation and a screenshot of the demo systems that implement these services are given.

The paper is organized as follows. The second section presents the application of IoT and GIS for medical purposes through related research. This is followed by a proposal and a description of IoT and GIS health services (finding the nearest health care facilities, air quality monitoring, improving medical services, smart transport for patients, and mobile SOS system), and the implementation of the service is given. In the end, the conclusion and directions of further research are given.

2 Definitions and Related Work

GIS (Geographic Information System) is a framework for collecting, managing and analyzing spatial data and their associated features. GIS integrates many types of data, analyzes spatial locations, and organizes layers of information into visualizations using maps and 3D scenes. This provides a deeper insight into spatial data, such as patterns, relationships and situations, thus helping users to make smarter decisions [7].

The Google Maps API is a popular Internet geographic information visualization service launched by Google in early 2005 to provide a visual interface, allowing a page to be embedded in an external website and adding a layer of specific data to the map [9].

GeoJSON is a standard for geospatial data on the Internet that is also used by the Google Maps API [10].

The following is an overview of papers describing smart health services based on IoT and GIS. The importance of GIS for public health is described in detail in the book [8]. In [11], there are application that suggests routes where air pollution is lower to users who have a problem with the respiratory organs. There are the services that are provided through various types of wearable devices and using IoT and wireless networks, often M2M communication. In [12], wearable devices within the RFID-WSN hybrid system that enable location monitoring and real-time patient monitoring are described. In [13], services for measuring blood sugar, heart electrocardiogram, blood pressure, and temperature using IoT are described. The importance of IoT in the implementation of health services in smart cities is also described in [14, 15]. The paper [16] describes the integration of IoT and cloud services in order to determine the disease based on the symptoms that are reflected in the changes in the voice of patients. The paper [17] provides an overview of challenges for the formation of smart homes for the care of the elderly and chronically ill, who use technological infrastructure (sensors, cameras, wearables, and web services) in order to respects the patients' desire to be at home.

3 Smart Health Services Based on IoT and GIS

3.1 Finding the Nearest Health Care Facilities

Centralized information system (IS) for smart health should take advantage of available GIS systems that contain information about health facilities and pharmacies. The user should ensure that in one place he can search and display outpatient clinics, hospitals, pharmacies, private practices, with associated information about them, such as ratings of institutions and a link to additional information on the website and display the path from its location to the selected institutions. Figure 1 shows a UML diagram of use cases for a smart health service to search and display the nearest health facilities and pharmacies.

These features are already provided by the Google Maps API, and the benefits of integrating into other websites with a selection of subsets of required geodata, so it could also be integrated into the proposed smart health IS. Figure 1 also shows Google Maps with a choice to display pharmacies and hospitals. Additional information offers a view of the image associated with the object, rating and comments, a link to the website if the object has it, the ability to share and the ability to display the path from the user's location to the selected object.

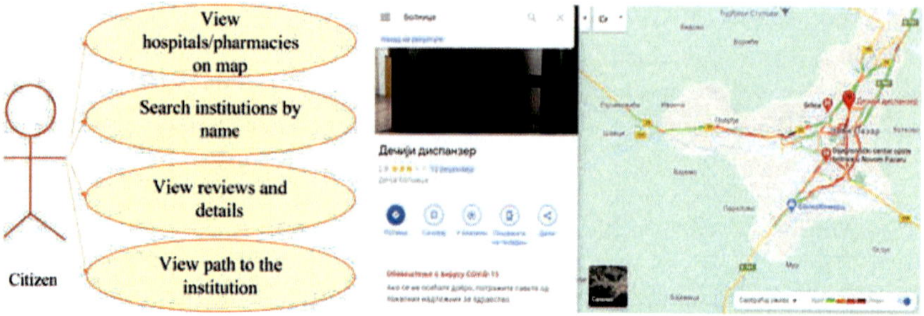

Fig. 1 Diagram of use cases for the search and display service of the nearest health care institutions and its selection on Google Maps

3.2 Air Quality Monitoring

The proposed service is capable of collecting data from different types of IoT devices and from residents and manipulating them using data analysis techniques in order to inform citizens about the air quality in their immediate vicinity and provides the possibility of providing assistance if a citizen finds himself in a crisis situation. The data is collected from air quality sensors that have a fixed location in the smart city infrastructure. The system provides data both on the website and on a mobile phone or another smart device. But it is also interactive, i.e. although it receives information from users, it also sends notifications to users. It also offers the possibility of calling for help from a given location, whereby the medical service for emergency situations is activated. Knowing the location of the user and possibly his private data enables more efficient assistance in emergencies.

The scenario of using the proposed platform is described by the UML diagram shown in Fig. 2, which describes the following scenarios: • Notifications: Air pollution in an area has risen beyond the threshold. The application contains the geolocations of its subscribers and notifies all subscribers in the contaminated area. • Asking for medical help: If a citizen, a subscriber of the service, has breathing problems, he can ask for help by pressing a special button available on the client-side of the application and this call is automatically forwarded to the health service. Using the personal data left by the citizen during the subscription, geolocation, and other data on the control panel, the health service can send ambulance assistance. • Reporting of sudden air pollution: If a citizen has noticed a critical situation (e.g. fire), this situation can be reported. After processing this notification, this information is posted on the dashboard, and alerts are sent to all subscribers in the incident area [18].

Figure 2 also shows a visualization of the map of the AQI (air quality index) values shown by the sensors located at the marked locations nearby. The color indicates the degree of pollution and its connection with the impact on the health of the inhabitants.

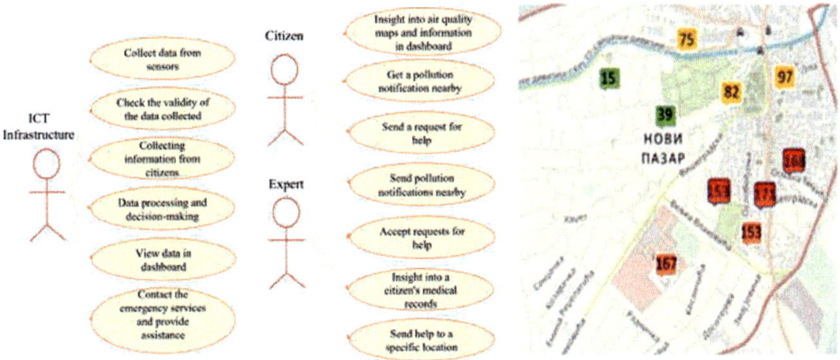

Fig. 2 A use case diagram for air quality monitoring service and the layout of the pollution map

3.3 Improving Medical Services

We are witnessing the current pandemic of Covid-19 disease caused by the new corona virus, which has affected numerous activities and increased the presence of the use of online services, both in health care and outside it. The reason for this was the advice to avoid physical contact. The forecasts are that similar pandemics can happen in the future, but also in conditions without pandemics, in fast life in urban areas, it is convenient to use the opportunities offered by ICT and do jobs electronically that would require hours of waiting in lines. The problems that patients face years ago, today in the case of overloading the health system, have only culminated. In smaller cities, doctors often instruct patients to perform specialist examinations or certain interventions in health centers in larger cities, for the performance of which there are no conditions in smaller health centers. Therefore, instructions are given that must be approved by the medical commission, often the travel costs are covered by the RHIF, and the submission and insight into the status of the request are currently done without the use of electronic services. The question is what else can be done to reduce unnecessary administrative items, long queues, unnecessary visits to the doctor to show results and get advice, submit documents to the medical commission, and the like.

Figure 3 shows cases of using services to improve the provision of medical services. Here, cases of use are classified into two entities, the first refers to clinics, triage, surgeries, laboratories, and similar institutions where queues are formed despite scheduling examinations, and sometimes scheduling is impossible (as was the case during the current COVID-19 pandemic). This service has two types of users, and they are a citizen and a professional.

The citizen has the following possibilities within the entity marked as an ambulance: selection of the ambulance for which the queue is reported, electronic queuing using the health insurance id and location to prevent abuse, insight into the number of patients in the queue, insight into the dynamics of completion of the examination,

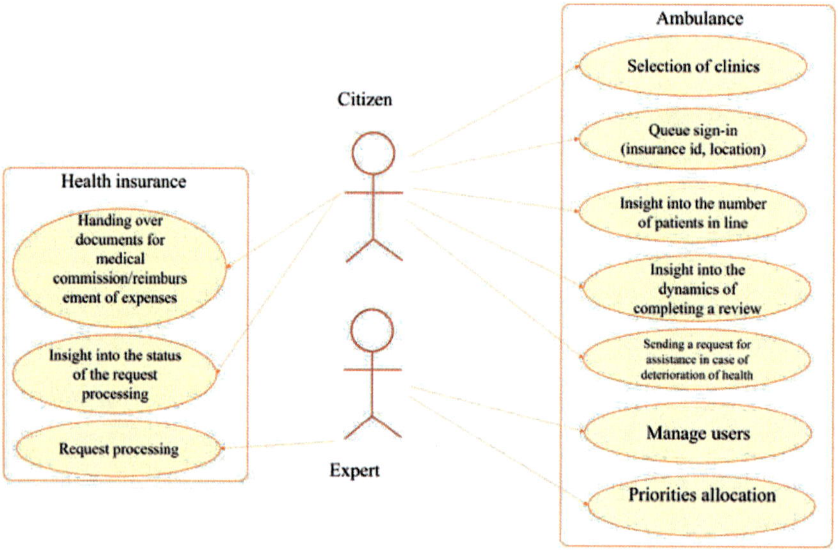

Fig. 3 A use case diagram of the improvement of medical services

staff in case of deterioration. An expert in this entity has the ability to manage registered users, and change priorities (in case of alarm by a citizen whose health has deteriorated). The professional has the health insurance id of the patient waiting in line, and therefore an insight into what diseases the patient has and can change his priority if it is a pregnant woman, an elderly citizen, or a chronic patient.

In an entity designated as health insurance, this service provides the user with online submission of a request for referral to another health facility (to be approved by a medical commission), reimbursement, and online insight into the approval of a request for a medical commission or reimbursement status. An expert in this entity processes the request i.e. changes its status.

3.4 Smart Transport for Patients

Two ways for using IoT and GIS are proposed in order to make smart transportation services for patient. First is based on vibration data measured with an accelerator in smart devices, and second is using maps and parking spots information for finding the nearest parking place for the ambulance car.

There are several scenarios of how vibration data measured with an accelerator can be used to improve patient transport conditions. The first scenario is based on data on the level of vibration at a given location. The level of comfort for a given location and the route on which the measurement is performed can be assessed on the basis of comfort. The second scenario includes, in addition to the above data, the recording of vehicle data in which the measurement is performed. In this way, it is possible to compare the recorded comfortable routes when different emergency vehicles are used. If the data recorded in one of the vehicles differs significantly from the others, the cause of discomfort maybe only the vehicle and not the route. In this case, the system suggests the most comfortable vehicle and signals that there are vehicles that may need repair. The third scenario is the recording of driver data because in the same way as in the above case, it can be the cause of vibrations (due to the rough driving style), and not the vehicle or the route. On this occasion, the system recommends the most responsible available driver based on his assessment by the vibration measuring system [19].

In order to process the collected information on driving vibrations, which are used to create publicly available comfort maps, a client-server application has been developed, described in detail in [20]. The client part is a mobile Android application that is used to record and send information about recorded comfortable, less comfortable, or uncomfortable points while driving. The server part processes this data and as an output generates comfort maps for the locations where the measurements were performed (the frequency of measurements can be adjusted). The realization of this part was done using Java web technologies and the Google Maps API for displaying GIS data. The measured data for each interval is stored in an XML or CSV file, depending on the user's choice. By sending data to the server part, the comfort map for the appropriate locations is updated (Fig. 4).

The service intended for efficient parking of ambulances accesses the data of the parking service for appropriate parking lots, and based on the time of received SMS messages for payment of parking spaces from a certain location, it is calculated how many parking spaces are available at the moment. The client part of the service communicates with the Google Maps API, in order to obtain data on parking lots near the user who activates the request. To enable this, you need to allow access to the user's location. Based on this data, the user received information about the distance of the parking lot, the time required to reach it, and the number of free spaces for that parking (Fig. 4) [21].

3.5 *Mobile SOS System*

The smart health service for the mobile SOS system is implemented in such a way that in the fastest possible way, a request for help can be sent with just one click. This click automatically calls and alerts the emergency service. In this way, the emergency service immediately has insight into the information about the patient. The existence of this service is also important as a component of the air pollution monitoring service described in the Sect. 3.2. It is imagined that on a wearable device, such as a watch or bracelet, there is an SOS button that activates the mobile SOS system. The location of the user who calls for help is determined, the nearest ambulance is sent,

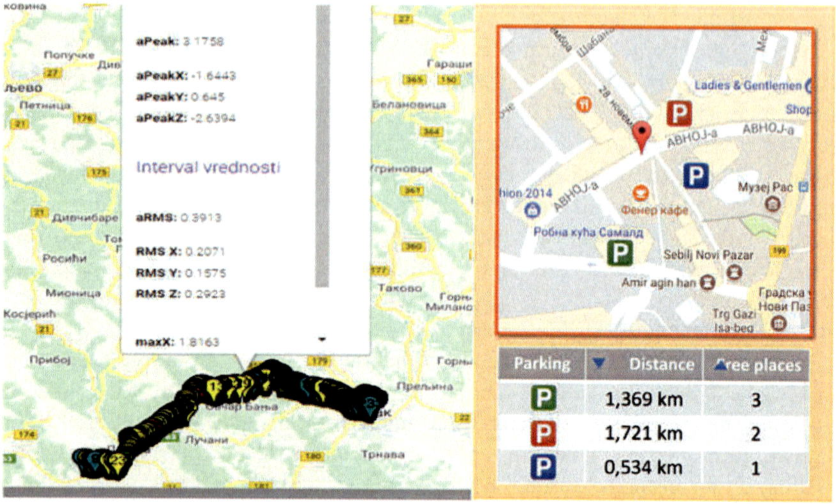

Fig. 4 Smart transportation for patients

and the condition from the patient's history is suggested to the health workers from the vehicle.

Acknowledgements This paper is partially supported by the Ministry of Education, Science and Technological Development of the Republic of Serbia under projects III44007 and ON 174026.

References

1. Tavares, P.D.S., Rodrigues, E.B.: IoT-based architecture for data analytics of arboviruses in smart cities. In: 2018 IEEE Symposium on Computers and Communications (ISCC), pp. 952–957. IEEE, Natal, Brazil (2018)
2. Batty, M.: Big data, smart cities and city planning. Dialogues Hum. Geogr. **3**(3), 274–279 (2013)
3. Mohanty, S.P., Choppali, U., Kougianos, E.: Everything you wanted to know about smart cities: the internet of things is the backbone. IEEE Consum. Electron. Mag. **5**(3), 60–70 (2016)
4. Solanas, A., Patsakis, C., Conti, M., Martinez-Balleste, A.: Smart health: a context-aware health paradigm within smart cities. IEEE Commun. Mag. **52**(8), 74–81 (2014)
5. Mettler, M.: Blockchain technology in healthcare: the revolution starts here: e-Health networking, applications and services (Healthcom). In: 18th International Conference on IEEE, pp. 1–3. IEEE (2016)
6. Fox, S., Duggan, M.: Mobile Health 2010. Pew Internet & American Life Project, Washington DC (2010)
7. Davies, F.B.: Object-based geographic information system (GIS), U.S. Patent No. 6,247,019. 12 Jun (2001)
8. Cromley, E.K., McLafferty, S.L.: GIS and public health. Guilford Press (2011)
9. Google Maps API.: https://developers.google.com/maps/documentation. Last accessed 15 July 2021

10. GEOJSON.: https://leafletjs.com/examples/geojson/. Last accessed 15 July 2021
11. Mehta, Y., Pai, M.M., Mallissery, S., Singh, S.: Cloud enabled air quality detection, analysis and prediction-a smart city application for smart health. In: Big Data and Smart City (ICBDSC), 3rd MEC International Conference on IEEE, pp. 1–7 (2016)
12. Adame, T., Bel, A., Carreras, A., Meliá-Seguí, J., Oliver, M., Pous, R.: CUIDATS: an RFID-WSN hybrid monitoring system for smart healthcare environments. Futur. Gener. Comput. Syst. **78**, 602–615 (2018)
13. Natarajan, K., Prasath, B., Kokila, P.: Smart health care system using internet of things. J. Netw. Commun. Emerg. Technol. (JNCET) **6**(3), 37–42 (2016)
14. Ullah, K., Shah, M.A., Zhang, S.: Effective ways to use Internet of Things in the field of medical and smart health care. In: Intelligent Systems Engineering (ICISE), International Conference on IEEE, pp. 372–379 (2016)
15. Gomaa, A., Zhang, C., Hasan, M., Roche, M.B., Hynes, S.: Supportive glucose sensing mobile application to improve the accuracy of continuous glucose monitors. In: International Conference on Smart Health, Springer, pp. 206–212 (2014)
16. Muhammad, G., Rahman, S.M.M., Alelaiwi, A., Alamri, A.: Smart health solution integrating IoT and cloud: a case study of voice pathology monitoring. IEEE Commun. Mag. **55**(1), 69–73 (2017)
17. Sprint, G., Cook, D.J., Fritz, R., Schmitter-Edgecombe, M.: Using smart homes to detect and analyze health events. Computer **49**(11), 29–37 (2016)
18. Avdic, A., Kajan, E., Jankovic, D., Avdic, Dz.: Towards context-aware smart healthcare platform. Int. J. Electr. Eng. Comput. **3**(1), 26–31 (2019)
19. Jovanovic, Z., Avdic, A., Jankovic, D.: Transportation in smart cities—tracking and improving driving comfort. In: Konjovic, Z., Zdravkovic, M., Trajanovic, M. (eds.) ICIST 2018 Proceedings, pp. 346–349 (2018)
20. Jovanovic, Z., Avdic, A., Jankovic, D., Vujicic, D.: Smart transportation in the service of improving healthcare in smart cities. IcETRAN (2018)
21. Avdic, A., Avdic, D., Marovac, U., Kajan, E., Ljajic, A.: A concept of efficient parking in smart cities. In: Proceedings of the 25th Telecommunition Forum Telfor, pp. 1–4. Belgrade (2017)

A Review of Internet of Things Platforms Through the IoT-A Reference Architecture

Giancarlo Fortino, Antonio Guerrieri, Claudio Savaglio, and Giandomenico Spezzano

Abstract For simplifying and speeding up the development of Internet of Things (IoT) ecosystem, there has been a proliferation of IoT Platforms, built up according to different design principles, computing paradigms, technologies, and targets. This paper proposes a review of main examples populating the wide landscape of IoT Platforms and their comparison on the basis of the IoT-A Reference Architecture. In such a way, heterogeneous IoT Platforms (both current and future) can be analyzed regardless of their low-level specifications but exclusively through the lens of those key functionalities and architectural building blocks that enable the interplay among devices, data flow, software, and stakeholders within the IoT ecosystem.

1 Introduction

The Internet of Things (IoT) [6, 7], at his highest realization, is an ecosystem of heterogeneous devices (sensors, actuators, gateways, smart gadgets, RFID/NFC tags, edge nodes, cloud servers, etc.) and software components (API, middlewares, ontologies, Digital Libraries, Digital Twins, etc.) that seamlessly cooperate and interact with human users to provide advanced and contextualized cyber-physical services [11]. Indeed, the IoT is widely considered a technological cornerstone for a plethora of

G. Fortino
University of Calabria, DIMES, Italy and National Research Council of Italy (CNR)—Institute for high performance computing and networking (ICAR), Rende, Italy
e-mail: giancarlo.fortino@unical.it

A. Guerrieri (✉) · C. Savaglio · G. Spezzano
National Research Council of Italy (CNR)—Institute for high performance computing and networking (ICAR), Rende, Italy
e-mail: guerrieri@icar.cnr.it

C. Savaglio
e-mail: savaglio@icar.cnr.it

G. Spezzano
e-mail: spezzano@icar.cnr.it

D. Camacho et al. (eds.), *Intelligent Distributed Computing XIV*, Studies in Computational Intelligence 1026, https://doi.org/10.1007/978-3-030-96627-0_3

emerging Smart-* applications (Smart Factory, Smart City, Smart Vehicles, Smart Buildings) as well as a key enabler for well-established domains like Ambient Assisted Living, domotics, industrial automation, etc.

Due to the IoT complexity, many frameworks and platforms have been recently proposed for supporting developers in architecting, programming, and exploiting such articulated bundles of hardware, data, and applications. There has been a proliferation especially of IoT Platforms because these are featured by a shorter distance to the implementation phase, good flexibility, and higher usability with respect to IoT Frameworks [6]. In more detail, IoT Platforms allow achieving more specific operational outcomes with reduced time-to-market, risks, and development costs, by delivering (in a secure way) key functionalities like applications enablement, data flow design, devices, and connectivity management. As result, instead of focusing on the lower levels of the technology stack, IoT Platforms allow for a greater concentration of resources in value-add applications, thus resulting extremely attractive both for individual developers and small/medium/large enterprises. As a matter of fact, an increasing number of IoT Platforms exist at the state-of-the-art, both with general- and domain-specific purposes (e.g., IoT Platform for COVID-19 prevention and control, IoT energy Platforms, IoT Platforms for Smart City, etc.), and they are constantly updated to support latest technology solutions (Edge Computing, Blockchain, Digital Twins, just to name a few).

Likewise, there exists a good number of survey works that analyze the scenario of IoT Platforms with different approaches and goals. A short list of key functionalities (device management, connectivity, data analytics, simulation and visualizations tools) is used to compare IoT Platforms able to support large-scale deployment [3, 10], cloud-based systems [14], do it yourself (DIY) projects [13] or Industrial IoT scenarios [12]. High-level features like scalability, stability, interoperability, pricing models etc., instead, have been recognized as discriminant factors in [2, 12, 15] to drive the adoption of an IoT Platform from an enterprise perspective. In [4], instead, a systematic and more comprehensive review of IoT Platforms based on their supported development phases (modeling, analysis, design, implementation, testing) is provided, in order to explore the aspect of the engineering life-cycle process. Only two surveys, namely [8, 9], provide a Reference Architecture (RA) as a comparison framework for systematically analyze so much different IoT Platforms with a suitable level of detail. This is a wise and effective choice because a RA allows to simultaneously consider, on the one hand, the aforementioned IoT Platforms' high-level functionalities and, on the other hand, their main building blocks as well as their design choices towards a concrete development of the IoT ecosystem, regardless of low-level specifications.

Along such lines, in this paper, we provide a review of the main IoT Platforms and we compare them through the IoT-A RA [1]. Indeed, differently from [8, 9], we do not introduce the umpteenth RA (see [5] for a detailed list) but we consider the most acknowledged one, widely referred and globally considered the baseline for the other successful RAs presented over the years, like AIOTI, IEEE P2413, WSO2, etc. The exploitation of the IoT-A RA provides generality to the provided contribution and soundness to the performed analysis, thus eventually steering the reader in the wide landscape of IoT Platforms.

The rest of the paper is organized as follows. In Sect. 2 we report the IoT-A RA that considers the main functionalities and building blocks of an IoT Platform. According to such RA, we survey the main IoT Platforms in Sect. 3. Discussion and final remarks conclude the paper.

2 IoT-A Reference Architecture

The IoT-A project [1], funded by European FP7-ICT, was aimed at creating an architectural reference model for interoperable and scalable IoT ecosystems as well as at outlining key building blocks, development principles, and guidelines for the technical design of their protocols, interfaces, and algorithms. One of the main advantages of the IoT-A architectural model is the creation of a single global ecosystem of things that communicate seamlessly with each other to support the advanced digital services necessary for the digital transformation of industrial and business processes.

The resulting IoT-A RA mainly consists of six layers:

The *Communication Layer* considers the variety of communication schemes derived from the many technologies belonging to IoT systems and it provides a common interface to the IoT Service layer. Here, a set of protocols as well as hardware and software solutions (gateway, proxies, APIs, etc.) for enabling the communication within and between IoT systems find the place, making this layer fundamental for providing the basic interconnectivity.

The *Virtual Entity Layer* provides the solutions for interacting (e.g., discovering, managing, shadowing, looking up services) with the physical IoT systems' components through cyber counterparts such as Digital Twins, Virtual Entities, Software Agents, etc. All these virtual aliases allow storing the history of their related asset, retrieving and updating information, monitoring the status and properties, by means of remote high-level interfaces.

The *IoT Service & Service Organization Layer* deals with IoT Service, a primary abstraction representing the entry point for the functionalities provided by the IoT Systems and their components. Therefore, being linked to all the other layers, it is central within the RA, also providing utilities related to services discovery, look-up, name resolution, composition, and orchestration.

The *IoT Process Layer* is in charge of integrating conventional process management systems with the IoT System, by providing the functional concepts, interfaces, and tools necessary to augment traditional (business) processes. Therefore, this layer provides an environment for the modeling of the overall IoT system as well as for analytics and visualization tools for processing and displaying the data. Other important functionalities like simulation might be placed at this layer to provide support to the IoT process execution.

The *Security Layer* ensures the security and the privacy of an IoT System by means of well-established mechanisms of authorization, authentication, key exchange and management, trust and reputation. Since security and privacy are concerns of paramount importance for all the components of IoT Systems and by-design solutions are typically necessary to achieve a satisfactory degree of protection, this layer is transversal to all the previous ones.

The *Management Layer* combines all functionalities that are needed to govern an IoT system like fault handling, configuration, accounting, performance. Solutions aimed at automatic load balancing, horizontal and vertical scalability, data backup, usage policy, etc. are addressed here and are designed for being as much automated as possible, thus minimizing the need for human intervention. As for the Security Layer, also this is a cross-layer.

Given this brief description of IoT-A RA and by extracting its essentials, concrete yet heterogeneous IoT Platforms can be effectively compared regardless of their low-level specifications or specific targets, as done in the following.

3 State-of-the-Art Analysis

This section quickly overviewes some well-known IoT Platforms and applies the IoT-A RA introduced above to make a proper comparison among them. Such overview does not want to be exhaustive about IoT Platforms, but it just wants to provide a comparison among the main IoT Platforms to date available. Table 1 summarizes the main differences, according to IoT-A RA, among the considered platforms and will be commented in the last section.

Siemens MindSphere[1] is the Siemens IoT Platform declared as the leading "industrial IoT as a service" solution. MindSphere uses advanced analytics and AI to make running IoT solutions from the edge to the cloud with data from connected products, plants, and systems to optimize operations, create better quality products, and deploy new business models. It allows to quickly collect, monitor, and analyze data in real-time and offers insights that improve efficiency and profitability. MindSphere can run on different cloud infrastructures such as AWS, Azure, Alibaba. In particular, The MindSphere Gateway and the MindConnect Integration APIs provide great support for the Communication Layer as well as the Siemens Digital Twins and the MindSphere Predictive Learning Analytics for the IoT Service and Process Layers.

IBM Watson was initially conceived as a question answering computing system developed to apply, to such a field, several advanced technologies like information retrieval, knowledge representation, advanced reasoning, and machine learning. Now, Watson comprehends a set of business-ready tools, applications, and solutions, designed to reduce both the costs and the hurdles of AI adoption while optimizing outcomes and responsible use of AI. Among the offered tools, the *Watson IoT*

[1] Siemens MindSphere. https://siemens.mindsphere.io/en.

Table 1 Analysis of main IoT platforms according to the IoT-A RA (✓ standard, ✓ ✓ advanced, ✓ ✓ ✓ peculiar feature)

	Communication layer	Virtual entity layer	IoT service layer	IoT process layer	Security layer	Management layer
MindSphere	✓ ✓	✓ ✓	✓ ✓	✓ ✓	✓	✓ ✓
IBM watson	✓ ✓	✓ ✓ ✓	✓ ✓	✓ ✓ ✓	✓	✓ ✓
Amazon AWS	✓ ✓	✓	✓ ✓ ✓	✓ ✓	✓	✓ ✓ ✓
SAP IoT	✓ ✓	✓ ✓	✓ ✓	✓ ✓ ✓	✓	✓ ✓ ✓
Microsoft azure	✓ ✓	✓ ✓ ✓	✓ ✓	✓ ✓ ✓	✓	✓ ✓
Bosch IoT suite	✓ ✓ ✓	✓ ✓ ✓	✓ ✓	✓ ✓	✓	✓ ✓
Hitachi lumada	✓ ✓	✓ ✓ ✓	✓ ✓	✓ ✓ ✓	✓ ✓	✓ ✓
Google IoT core	✓ ✓	✓	✓ ✓ ✓	✓ ✓	✓	✓ ✓
Cumulocity IoT	✓ ✓	✓	✓ ✓ ✓	✓ ✓ ✓	✓	✓ ✓
Oracle IoT cloud service	✓ ✓	✓	✓ ✓ ✓	✓ ✓	✓	✓ ✓

Platform[2] allows to use fully managed, cloud-hosted services with capabilities for device registration, connectivity, control, rapid visualization, data storage, and asset monitoring, especially through the IBM Cloud BLUEMIX, based on Cloud Foundry, and IBM Maximo. All of this, allows the Platform to provide great support to both IoT Service and IoT Process Layers. The built-in support for simulation tools is a distinctive feature of the Watson IoT Platform belonging to the Virtual Entity Layer. Moreover, it fully supports the Blockchain for data exchanging and allows users to take full advantage of the Watson cognitive APIs applied to the IoT.

Amazon Web Services (AWS) is the biggest player of the cloud market. It is a platform of web services offering different instruments for computing, storing, and networking. All of this is provided at different levels of abstraction. All the services are available through common Internet protocols although both APIs and Gateway can be customized. The most used services comprehend *EC2*, which has been developed to provide virtual machines, and *S3*, which provides storage. A very useful service offered by AWS is *AWS IoT*,[3] which allows to connect IoT devices and collect, store, and analyze their data. AWS clients can choose to use one of many data-centers scattered around the world or, through AWS IoT Greengrass, to move data processing and analysis as close to the end-point as necessary. AWS IoT relies on several components useful to comply with specific functions. Among these, the

[2] Watson IoT Platform. https://www.ibm.com/cloud/watson-iot-platform.

[3] AWS IoT. https://aws.amazon.com/iot/.

AWS IoT Device Management, which gives support for both the IoT Service and the Management Layers, permits to cleverly organize and monitor IoT devices, and the *AWS IoT Core*, which allows the device connection and characterizes the AWS IoT's Communication Layer.

SAP is one of the most important producers of software for the management of business processes. It aims to develop solutions enabling effective data processing and information flow across organizations. SAP software provides many business functions with a whole single view. This allows companies to manage complex business processes with only one instrument. The *SAP Internet of Things (SAP IoT)*[4] solution uses SAP to tackle the IoT world so offering capabilities to address industrial IoT use cases. SAP IoT enables users to re-design business processes with embedded IoT services and data, so giving high support to the IoT Process Layer. SAP IoT exploits cloud services for building IoT applications, enriches IoT data with business context, and provides analytical services with live integration to *SAP Analytics Cloud.* Moreover, it relies on scalable device management services so fully supporting the Management Layer. The SAP IoT Simulator uses sensory and machine data that is collected and analyzed on the *SAP HANA Cloud Platform.* A certified catalog of third-party connectors and APIs make SAP IoT greatly interoperable with the other main (I)IoT Platforms.

Microsoft Azure is the general brand name for Microsoft's cloud-computing created for building, testing, deploying, and controlling applications and services through Microsoft-managed data centers. Azures provides a plethora of services ranging from software as a service (SaaS), to platform as a service (PaaS), and infrastructure as a service (IaaS). Moreover, it has been built to manage different programming languages, tools, and frameworks, also from different vendors. *Azure IoT*[5] is a set of Azure cloud services that connect, monitor, and control IoT devices. In Azure IoT, an IoT solution is composed of one or more IoT devices and one or more back-end services running in the cloud that communicate with each other. The back-end services gather sensor data and determine how to process and store that data. Such functions are provided, among the others, by the *Azure IoT Central* component, that accelerates the creation of IoT solutions, and reduces the burden and cost of IoT management, operations, and development so giving full support to the IoT Process Layer. Moreover, among all, Azure IoT allows to *(i)* deploy SaaS solutions for IoT with minimal cloud expertise, *(ii)* customize industry-specific IoT solutions for common IoT scenarios, *(iii)* delocalize intelligence from the cloud to edge devices (Azure IoT Edge), *(iv)* use digital models of physical things (*Microsoft Azure Digital Twins*). This last one is a very important and well implemented characteristic of Azure IoT that allows to give full support to the Virtual Entity Layer.

---

[4] SAP Internet of Things. https://www.sap.com/products/iot-data-services.html.

[5] Azure IoT. https://azure.microsoft.com/en-us/overview/iot/.

Bosch IoT Suite[6] is an open-source-based IoT platform specialized in developing, testing, and running scalable IoT services and applications. Such services and applications can improve productivity across all industries, such as agriculture, homes & buildings, energy, manufacturing, and mobility. Great support to the Communication Layer is provided through the Bosch IoT Gateway and a rich set of local and remote APIs that allow to securely connect several kinds of heterogeneous devices, sensors, micro-controllers, and manage IoT data both at the edge and at the cloud. The Bosch IoT Hub, instead, provides a managed inventory of digital twins for IoT device assets, whose data are normalized, stored, and enriched by the Bosch IoT Insights at the IoT Service Layer.

Hitachi Vantara Lumada[7] from Hitachi Vantara is an IIoT platform that helps customers to rapidly create digital solutions. Lumada claims to provide users with, at Virtual Entity Layer, an advanced support to digital twins and digital twins modeling Furthermore, at IoT Process Layer, it equips the *Lumada Analytics* for advanced analytics and AI algorithms and the *Hitachi Visualization Suite*, which allows a versatile management of an IoT System. Lumada provides secure connectivity with industrial IoT sensors, actuators, and gateways. It is realized in modular components that can be distributed both at the edge and in the cloud so permitting a fully distributed system. Finally, Lumada is open and interoperable in the sense that it can work with any industrial system and any cloud, including with MATLAB and Simulink for modeling and simulating.

Google Cloud is a set of cloud computing services by Google providing users with computing, data storage, data analytics, and machine learning functionalities. The component thought from Google for adding IoT functionalities to its platform is the so-called *Google Cloud IoT Core*,[8] which augments the Google Cloud platform with heterogeneous devices connection and management. Google Cloud IoT Core, together with all the other components from the Google Cloud platform, allows an intelligent, versatile, and scalable management of IoT ecosystems. Among such components, fundamental for IoT are: *(i)* the *Cloud Dataflow*, which provides streaming and batch analytics functions and refers to the IoT Process Layer, *(ii) Cloud Pub/Sub*, which manages connections and communications so enhancing the Communication Layer, *(iii) Big Query*, which allows performing data warehouse and fast querying on the data acquired, *(iv)* and the *Cloud ML Engine* for training, deploying, and running machine learning models on the IoT data. The last two points excellently support the IoT Service Layer.

---

[6] Bosch IoT Suite. https://bosch.io/iot-technology/.

[7] Hitachi Vantara Lumada. https://www.hitachivantara.com/en-us/products/iot-software-solutions/lumada-software-for-iiot.html.

[8] Google Cloud IoT Core. https://cloud.google.com/solutions/iot/.

Cumulocity IoT[9] is a recent platform built from scratch with the aim to be open, rapid to deploy, and distributed. Several heterogeneous devices can be connected to the platform in minutes and transparently used. It has been implemented to control IoT data and devices in (almost) real-time and without the need to implement code. Cumulocity IoT offers, so as to support the IoT Service and the IoT Process Layers, a set of analytics and smart rules specifically implemented for business users as well as full developer tools for the more advanced users. Moreover, it has been developed to help in the connection of diverse applications and to synchronize data among them through several pre-built connectors. It works with lots of heterogeneous cloud apps, including SAP and Microsoft.

Oracle IoT[10] is a very interesting platform of public cloud services enabling customers to build and run a wide range of different applications in a scalable, secure, highly available, and high-performance environment. Among the Oracle Cloud products, the most significant ones regard the execution of advanced analytics, the unified database, and the code-less application development platform. Oracle IoT Cloud Service has the aim to simplify the IoT so allowing to rapidly assimilate IoT devices into any digital strategy and create novel IoT services, so supporting the IoT Service Layer. Oracle IoT Cloud Service wants to permit the integration of existing applications with IoT in an easy and efficient way. Moreover, Oracle IoT Cloud Service provides, for the IoT Process Layer, real-time analysis tools that allow correlations, aggregations, and filtering of any kind of IoT data. It also provides built-in integration with other Oracle services. Oracle IoT Cloud Service carefully adopts secure and reliable communications.

4 Discussion and Conclusions

IoT Platforms play a key role to simplify the development of IoT Systems, reason why many of them have been recently proposed with different goals, targets, and functionalities. In this paper, we have reviewed the main IoT Platforms currently available by exploiting, as a comparison framework, the IoT-A RA (which mainly consists of the six layers of Communication, Virtual Entity, IoT Service, IoT Process, Security, and Management). The performed analysis has provided an overview of the main IoT Platforms by also highlighting similarities and differences among the surveyed works as well as some open challenges in the field. In particular, *(i)* all the listed IoT Platforms provide advanced asset interconnectivity solutions at the Communication Layer (mostly, through MQTT, HTTP, REST, OPC and secured APIs) as well as proprietary/third-party analytic and visualisation tools at the IoT Process Layer; *(ii)* there are several Cloud-based IoT Platforms adopting, mainly for the sake of usability, IaaS, PaaS and SaaS solutions at the Management Layer

[9] Cumulocity IoT. https://www.softwareag.cloud/site/product/cumulocity-iot.html/.

[10] Oracle Cloud. https://www.oracle.com/cloud/.

to ensure horizontal/vertical scalability, automatic scaling, multi-zone backup, auto load balancing; *(iii)* there is an increasing general interest in open-source solutions, aimed at lowering cost and interoperability barriers, at all the architectural layers; *(iv)* the provided mechanisms at the Security Layer are typically limited to conventional cryptographic protocols (i.e., HTTPS, TLS) and vendor-specific policy, while trustworthiness is generally ignored and only the IIoT Platforms deserve attention to cyber-security aspects; *(v)* said that the majority of IoT Platforms are Cloud-based, few of them support hybrid Cloud or Edge Computing architectures, although the benefits of decentralized paradigms (especially for the IoT Service, Security and Management Layers) are undoubted; and *(vi)* except to few IoT Platforms, there is poor attention to the modeling tools and, as result, rarely the Digital Twins or similar virtual alias are fully supported. Especially the last three points, beside being important distinctive features for the current IoT Platforms, represent interesting research directions to be further inspected in the future.

Acknowledgements This work has been partially supported by the COGITO (A COGnItive dynamic sysTem to allOw buildings to learn and adapt) project, funded by the Italian Government (PON ARS01 00836) and by the CNR project "Industrial transition and resilience of post-Covid19 Societies - Sub-project: Energy Efficient Cognitive Buildings".

References

1. Bassi, A., Bauer, M., Fiedler, M., Kramp, T., Van Kranenburg, R., Lange, S., Meissner, S.: Enabling things to talk. Springer Nature (2013)
2. Bhatia, A., Yusuf, Z., Ritter, D., Hunke, N.: Who will win the IoT platform wars? BCG Perspectives (2017)
3. Erhan, L., Ndubuaku, M., Di Mauro, M., Song, W., Chen, M., Fortino, G., Bagdasar, O., Liotta, A.: Smart anomaly detection in sensor systems: a multi-perspective review. Inf. Fusion (2020)
4. Fahmideh, M., Zowghi, D.: An exploration of IoT platform development. Inf. Syst. **87** (2020)
5. Fernandez, E.B.: Reference architectures for the IoT: a survey
6. Fortino, G., Savaglio, C., Spezzano, G., Zhou, M.: Internet of things as system of systems: a review of methodologies, frameworks, platforms, and tools. IEEE Trans. Syst., Man, Cybern.: Syst. (2020)
7. Guerrieri, A., Loscri, V., Rovella, A., Fortino, G.: Management of cyber physical objects in the future internet of things. Springer International Publishing (2016)
8. Guth, J., Breitenbücher, U., Falkenthal, M., Fremantle, P., Kopp, O., Leymann, F., Reinfurt, L.: A detailed analysis of iot platform architectures: concepts, similarities, and differences, pp. 81–101. Springer (2018)
9. Guth, J., Breitenbücher, U., Falkenthal, M., Leymann, F., Reinfurt, L.: Comparison of IoT platform architectures: a field study based on a reference architecture. In: 2016 Cloudification of the Internet of Things (CIoT). pp. 1–6. IEEE (2016)
10. Hejazi, H., Rajab, H., Cinkler, T., Lengyel, L.: Survey of platforms for massive IoT. In: 2018 IEEE International Conference on Future IoT Technologies (Future IoT). pp. 1–8. IEEE (2018)
11. Nord, J.H., Koohang, A., Paliszkiewicz, J.: The internet of things: review and theoretical framework. Expert Syst. Appl. **133**, 97–108 (2019)
12. Perry, M.J.: Evaluating and choosing an IoT platform. O'Reilly Media (2016)
13. Singh, K.J., Kapoor, D.S.: Create your own internet of things: a survey of IoT platforms. IEEE Consum. Electron. Mag. **6**(2), 57–68 (2017)

14. Sruthi, M., Kavitha, B.: A survey on IoT platform. Int. J. Sci. Res. Mod. Educ. (IJSRME), ISSN (online), 2455–5630 (2016)
15. Ullah, M., Smolander, K.: Highlighting the key factors of an IoT platform. In: 2019 42nd International Conference on Information and Communication Technology, Electronics and Microelectronics (MIPRO), pp. 901–906. IEEE (2019)

Critical Comparison of Data Imputation Techniques at IoT Edge

Laura Erhan, Mario Di Mauro, Ovidiu Bagdasar, and Antonio Liotta

Abstract The advances within the Internet of Things and sensor systems put the focus on the improvement of the data reliability as close to the edge as possible. This work investigates how well-established techniques can be used for the imputation of contaminated data. We look at the performance of four algorithms for different contamination rates and error bursts of variable length. Furthermore, the algorithms are also evaluated on a constrained environment to showcase the behavior of data imputation methods at the edge of an IoT-based system.

1 Introduction

The Internet of Things is becoming a crucial paradigm to manage the "new oil", namely the data. Such data come from a plethora of domains including healthcare [1], urban environments [2], industry [3], green energy [4], and many others [5]. Unluckily, this type of data are often prone to being corrupted in terms of missing data or anomalous data. This is common as the sensors (data collection points) are susceptible to faults or malfunctions.

We advocate for performing essential data cleaning such as data imputation as close to the data source, in order to increase the reliability of the managed data and to avoid the transmission of faulty, incomplete and unreliable information. To this

L. Erhan (✉) · O. Bagdasar
University of Derby, Kedleston Rd, Derby DE22 1GB, UK
e-mail: L.Erhan@derby.ac.uk

O. Bagdasar
e-mail: O.Bagdasar@derby.ac.uk

M. Di Mauro
University of Salerno, Via Giovanni Paolo II, 132, 84084 Fisciano, SA, Italy
e-mail: mdimauro@unisa.it

A. Liotta
Free University of Bozen-Bolzano, Piazza Universitá, 1, 39100 Bolzano, BZ, Italy
e-mail: antonio.liotta@unibz.it

D. Camacho et al. (eds.), *Intelligent Distributed Computing XIV*, Studies in Computational Intelligence 1026, https://doi.org/10.1007/978-3-030-96627-0_4

extent, in the present work we compare four data imputation techniques from both the statistical and the machine learning world by accounting in particular the *bursty errors* (trails of missing data points of different lengths). We make use of a real world based IoT pollution dataset which we contaminate by introducing bursts of missing values in order to operate on various contamination rates. The four compared methods (mean, MICE, missForest, and kNN) are compared by means of the Root Mean Square Error (RMSE), a significant metric to evaluate the difference between predicted and observed values. The carried out experiments highlight the ability of kNN and missForest to accurately recover data in the context of contamination rates of up to 25% and burst sizes of up to 100 data points as opposed to the statistical based methods. Furthermore, they showcase the impact that the bursty errors have on the data imputation process, i.e. the higher the burst size, the less accurate all the considered methods become.

This paper is organised as follows: Sect. 2 places our work within the literature; Sect. 3 provides the reader with information about the used dataset and the considered data imputation methods; Sect. 4 describes the experiments and discusses the obtained results, while Sect. 5 provides the conclusions and directions for future work.

2 Related Work

The application of data imputation techniques is becoming even more popular due to the massive presence of sensor devices which gather a huge (and heterogeneous) amount of data. Actually, the gathered data can be corrupted for various reasons: sensors malfunctioning, hostile environment where sensors are supposed to operate, communication problems in the data transfer from the sensor to another node (e.g. a fusion center). Accordingly, the research in this field seems to be focused on two big aspects: the first one pertains to the design of new algorithms or techniques amenable to handle the data imputation problem to be solved directly on-board of each single sensor; the second one concerns the possibility of analyzing the correlations among sensors in a distributed environment so that a corrupted measurement from a sensor could be replaced by a healthy measurement from another sensor.

As regards the first aspect, many techniques take a cue from existing statistical or machine learning concepts to tackle the data imputation problem. For instance, various statistic-based methods to deal with the missing data problem across an IoT environment include: Bayesian Maximum Entropy (BME) adopted in [6]; Probabilistic Matrix Factorization (PMF) used in [7] in conjunction with a k-means algorithm to estimate the similarity degree among closest sensors; Gaussian Mixture Models (GMM) exploited in [8]; Singular Spectrum Analysis (SSA) adopted in [9].

Other works introduce more sophisticated frameworks such as: [10] relying on the combination between Successive Geometric Transformation Models (SGTMs) and General Regression Neural Networks (GRNNs); [11] where a dedicated platform (dubbed *Itr-MS-STLecImp*) allows the management of missing data in time-series data produced by IoT-based devices; [12] where a customized algorithm (*ESTICS*) operating in the domain of the so-called compressed sensing is able to reconstruct a huge amount of missing data.

As regards the second aspect, the research is particularly focused on cluster correlations among sensors. Valuable examples include: [13] where the "group opinion" concept is exploited to replace missing data coming from a specific sensor; [14] where the data imputation is performed by sensors at the edge of the network by considering only the healthy measurements coming from the internal sensors; [15] where a kNN algorithm along with cross-correlated information is used to estimate the missing measurements of specific devices.

Even if all the aforementioned works rely on solid methodologies, they exhibit two practical drawbacks: the first one is that all the proposed algorithms are supposed to run on a standard architecture (e.g. a PC or a laptop), and the second one is that the missing values are mostly generated through artificial models. Conversely, in our work we try to fill these gaps by considering: (i) a constrained architecture (Raspberry PI) to demonstrate the benefit of applying specific techniques directly on board of a sensor; (ii) a realistic environmental dataset where the data can miss not only at random, but also according to a bursty model.

3 Methodology

In this section we give an overview of the dataset chosen to assess the experiments, the contamination method based on the bursty errors, and the chosen algorithms for the data imputation process.

3.1 The Air Pollution Dataset

We choose an artificially generated dataset representative of an IoT framework exploited for the environmental monitoring: the Pollution Measurements dataset, part of the *City Pulse Smart City Datasets* [16]. The pollution data consist of observations for the air quality index. It was generated as to accompany real vehicular traffic obtained from an analysis conducted in the city of Aarhus (Denmark), during an experimental campaign. The values are generated every 5 min using a pre-selected pattern.[1]

The dataset includes 17, 568 samples, with each sample having associated a set of environmental measurements (e.g. ozone, sulfure dioxide, etc.) along with location data (latitude and longitude), and the timestamp. For the purpose of our analysis we consider missing data just in the Ozone column. A description of the whole dataset is

[1] http://iot.ee.surrey.ac.uk:8080/datasets/pollution/readme.txt.

Table 1 Description of the dataset pollution measurements

	Ozone	Particullate matter	Carbon monoxide	Sulfure dioxide	Nitrogen dioxide
Count	**17568**	17568	17568	17568	17568
Mean	**92.42**	106.12	100.54	131.66	159.18
Std	**46.18**	52.01	49.66	50.51	43.43
Min	**15**	15	15	15	18
25%	**54**	60	56	99	134
50%	**87**	107	99	131	173
75%	**127**	146	138	177	193
Max	**215**	215	215	215	215

Table 2 The bursty method used for corrupting the dataset

Adopted impairment method	Description
Bursty	We randomly select bursts of data points of a given size to be invalidated from the dataset, in order to reach the desired impairment level  20 data points, desired impairment rate of 30% with burst size = 3 ●- original data point, ●- invalidated data point (N/A)

given in Table 1, where: std stands for standard deviation, and 25, 50, 75% represent, respectively, the first quartile, the second quartile (median), and the third quartile.

In its original form, the dataset used in this work can be considered curated and with no missing values. In order to test and compare different methods for data imputation, we impaired the dataset by introducing missing data as described in Table 2. We consider errors of a bursty nature. Bursts of a given size of data points are randomly chosen to be invalidated. The impairment level is expressed in terms of corrupted data points. This type of errors are often encountered in an IoT setting.

3.2 *Techniques Chosen for the Edge Data Imputation*

In this work we analyse and compare well-known and readily available techniques for data imputation for bursts of missing data points. We choose two statistical based methods (mean and MICE), and two machine learning based techniques (kNN and missForest).

Mean imputation is maybe one of the most common approaches, where missing values are substituted with the mean of the considered variable. In the considered case, the missing data pertaining to the ozone measurements are substituted with the mean of the observed ozone values.

Multiple imputation by chained equations (MICE) data imputation is a robust multiple imputation technique which takes into account multiple predictions for each missing value. The procedure fills in the missing data through an iterative series of predictive models, as detailed in [17, 18].

missForest data imputation is an iterative technique relying on the random forest principle as illustrated in [19]. The principle is to average the data over a number of different decision trees (unpruned classification or regression trees).

kNN data impuation works by replacing the missing data according to the values of the closest k neighbours data identified by means of the euclidean distance [20]. We choose $k = 3$ so as to keep the search of neighbours to a minimum.

In this work, we use the python library functions *impyute.imputation.cs.mice*, the *missForest* method part of the *missingpy* Python library, and the *KNNImputer* method part of the *sklearn.impute* Python library. We apply the techniques on the whole dataset (comprising of the 5 columns as described in Table 1).

4 Results

In this section we give an insight of how the four aforementioned algorithms perform for the task of data imputation in different scenarios. We remark that the dataset contamination rate ranges from 1 to 25%, and the length of the bursts of missing data points ranges from 1 to 200.

Figure 1 offers an in depth perspective of the evolution of the RMSE for each algorithm considering bursts of size 1–200 points, for contamination rates between 1 and 25% with a step of 5. It is easily noticeable that the two machine learning based techniques outperform the statistical based methods in all the considered scenarios, especially for lower burst sizes (below 100 points). For an impairment rate of 1%, while kNN and missForest do outperform mean and MICE, the difference in RMSE values is not as large as in the other scenarios for burst sizes below 75 points. As the impairment rate grows, meaning that more chunks of data are missing from the dataset, the statistical based methods are not able to accurately impute missing values, resulting in higher RMSE values. MICE does perform better than mean which is to be expected given that the latter trivially imputes the mean value across all the existing data points for all the missing value slots. However, the difference is not too large since MICE suffers from the assumption that there exists a linear relationship between the involved variables. Unluckily, such a situation is rarely met in real-world cases. Conversely, kNN and missForest make use of other information such as neighboring values, and are able to more accurately predict the missing values even at higher contamination rates, especially for lower burst sizes.

By looking at the six presented scenarios, one notices that at the point of the burst size being equal to 100 points, there is no longer a clear difference in RMSE values among the four techniques. Furthermore, up to a burst size of 25 points,

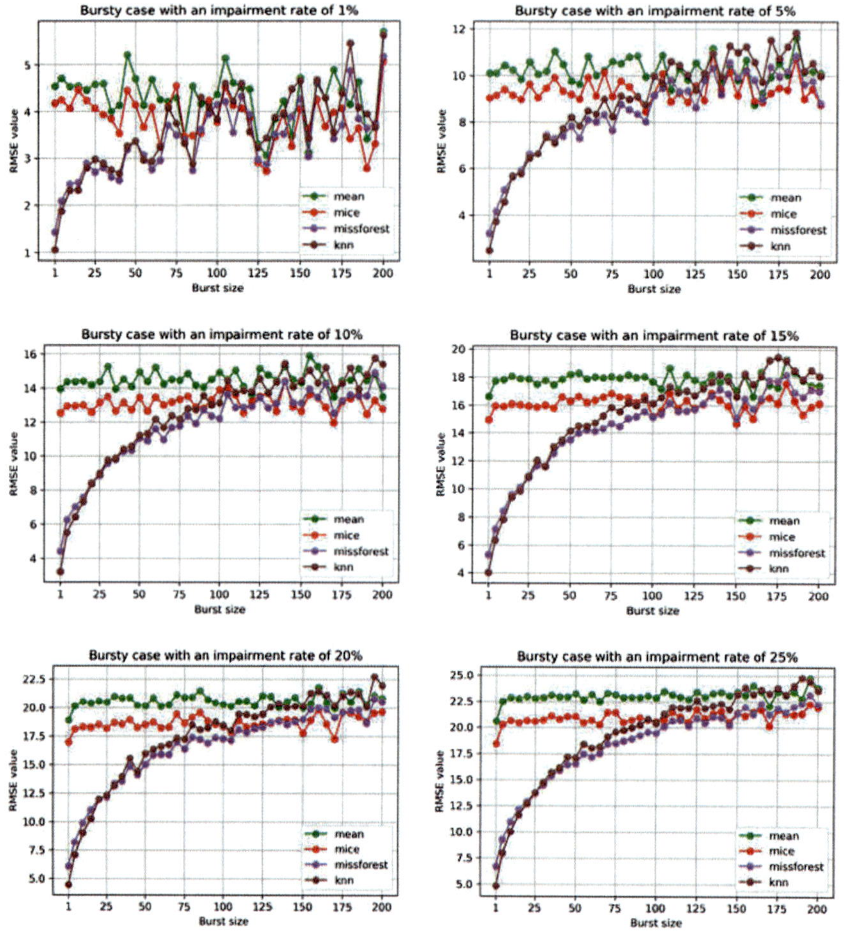

Fig. 1 Evolution of RMSE with impairment rate in the bursty case

the performance achieved by using kNN and missForest for the data imputation process is 50% higher than that of the mean and MICE methods. Cases considering a contamination rate of more than 25% were not considered as the RMSE values were worsening and the data imputation process was no longer considered reliable.

Figure 2 offers a more in depth look into the performance of the four methods for the following scenarios: a fairly low contamination rate (5%) with a burst size of 25 and 100 points, and a higher contamination rate (25%) with the same burst sizes of 25 and 100 points. Figure 2 depicts the density plots and showcases how the differently imputed datasets compare to the original and corrupted dataset. For the case of 25% contamination rate, the density plot corresponding to mean is not included. This is done as to avoid skewing the overall view caused by the decrease in performance.

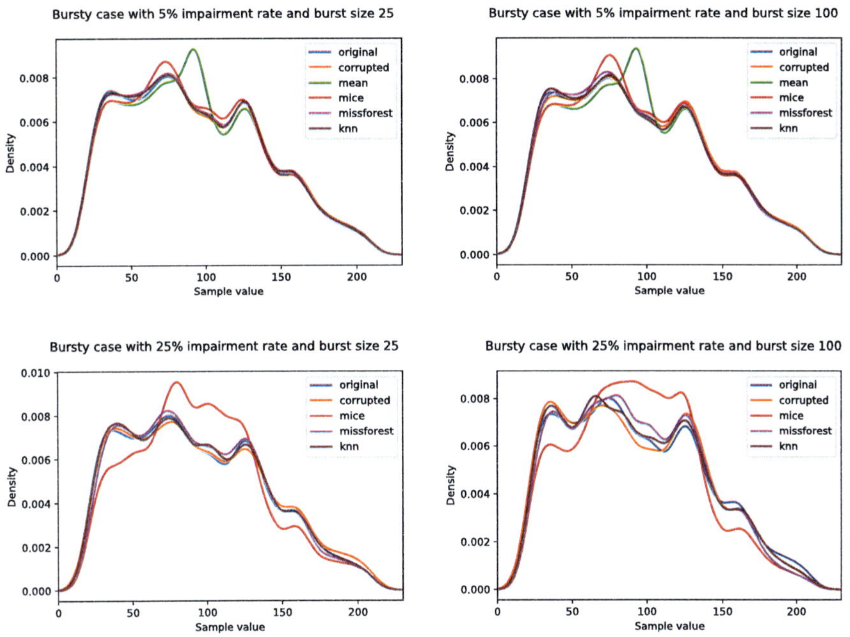

Fig. 2 Density plots for different scenarios

One interesting thing to note in Fig. 2 is how the increase in burst size affects the performance of the algorithms for the same contamination rate.

The experiments behind the presented results were carried out both on a powerful machine, namely a laptop with an Intel(R) Core(TM) i7-8850H CPU @ 2.60 GHz with 16 GB of RAM, and on a constrained IoT device, namely the Raspberry Pi 4B with 4 GB of RAM. The execution times on the RPI 4B were higher than those of a laptop but still within an acceptable range for real time edge data imputation for our considered dataset: 0.02 s for mean, up to 16 s for MICE, up to 18 s for missForest, and up to 38 s for kNN.

5 Conclusion

In this work we investigate how the issue of missing data within the IoT ecosystem can be tackled in order to improve the reliability of incoming sensor data. We compare four data imputation techniques in terms of performance in the context of bursty errors (trails of missing data points of different lengths). Starting from a clean representative IoT pollution dataset, we introduce bursts of missing values as to achieve different contamination rates. We evaluate mean, MICE, missForest, and kNN on their ability to reconstruct the original dataset by means of RMSE. This first set of experiments showcases the ability of kNN and missForest to accurately impute missing values for contamination rates of up to 25% with burst sizes of up to 100 points. Furthermore, it highlights the impact that the size of the error burst has on the recovery ability of the algorithms even when considering the same contamination rate. The higher the burst size, the more unreliable the data imputation methods become. We run the techniques within a constrained IoT device, namely the RPI 4B, in order to showcase the possibility of carrying out essential data cleaning operations (in our case this amounts to imputing missing data points) close to the data source, at the edge of the system. Future work includes a more in depth set of experiments for different scenarios, such as higher contamination rates considering lower burst sizes, and time and space complexity analysis to further analyze the ability of carrying out data imputation at the edge.

References

1. Tariq, U.U., Ali, H., Liu, L., Hardy, J., Kazim, M., Ahmed, W.: Energy-aware scheduling of streaming applications on edge-devices in IoT based healthcare. IEEE Trans. Green Commun. Netw. (in press) (2021)
2. Ferrara, E., Fragale, L., Fortino, G., Song, W., Perra, C., Di Mauro, M., Liotta, A.: An AI approach to collecting and analyzing human interactions with urban environments. IEEE Access **7**, 141476–141486 (2019)
3. Pace, P., Aloi, G., Gravina, R., Caliciuri, G., Fortino, G., Liotta, A.: An edge-based architecture to support efficient applications for healthcare industry 4.0. IEEE Trans. Ind. Inform. **15**(1), 481–489 (2019)
4. Xiang, X., Gui, J., Xiong, N.N.: An integral data gathering framework for supervisory control and data acquisition systems in green IoT. IEEE Trans. Green Commun. Netw. (in press) (2021)
5. Erhan, L., Ndubuaku, M., Di Mauro, M., Song, W., Chen, M., Fortino, G., Bagdasar, O., Liotta, A.: Smart anomaly detection in sensor systems: a multi-perspective review. Inf. Fusion **67**, 64–79 (2021)
6. González-Vidal, A., Rathore, P., Rao, A.S., Mendoza-Bernal, J., Palaniswami, M., Skarmeta-Gómez, A.F.: Missing data imputation with Bayesian maximum entropy for Internet of Things applications. IEEE Internet Things J. (in press) (2020)
7. Fekade, B., Maksymyuk, T., Kyryk, M., Jo, M.: Probabilistic recovery of incomplete sensed data in IoT. IEEE Internet Things J. **5**(4), 2282–2292 (2018)
8. Yan, X., Xiong, W., Hu, L., Wang, F., Zhao, K.: Missing value imputation based on gaussian mixture model for the internet of things. Math. Probl. Eng. **15** (2015)
9. Peixoto, M.L.M., Souza, I., Barbosa, M., Lecomte, G., Batista, B.G., Kuehne, B.T., Filho, D.M.L.: Data missing problem in smart surveillance environment. In: Proceedings of the IEEE HPCS, pp. 962–969 (2018)
10. Roman Tkachenko, Ivan Izonin, Natalia Kryvinska, Ivanna Dronyuk, and Khrystyna Zub. An approach towards increasing prediction accuracy for the recovery of missing IoT data based on the GRNN-SGTM ensemble. *Sensors*, 20(2625), 2020
11. Liu, Y., Dillon, T., Yu, W., Rahayu, W., Mostafa, F.: Missing value imputation for industrial IoT sensor data with large gaps. IEEE Internet Things J. **7**(8), 6855–6867 (2020)
12. Kong, L., Xia, M., Liu, X., Wu, M., Liu, X.: Data loss and reconstruction in sensor networks. In: Proceedings of the IEEE INFOCOM, pp. 1654–1662 (2013)
13. P. Fountas and K. Kolomvatsos. Ensemble based data imputation at the Edge. In *proc. IEEE ICTAI*, pages 961–968, 2020

14. Kolomvatsos, K., Papadopoulou, P., Anagnostopoulos, C., Hadjiefthymiades, S.: A spatio-temporal data imputation model for supporting analytics at the Edge. In: Lecture Notes in Computer Science, vol. 11701, pp. 138–150. Springer, Cham (2019)
15. Pan, L., Li, J.: K-Nearest neighbor based missing data estimation algorithm in wireless sensor networks. Wirel. Sens. Netw. **2**, 115–122 (2010)
16. Ali, M.I., Gao, F., Mileo, A.: Citybench: a configurable benchmark to evaluate RSP engines using smart city datasets. In: Proceedings of the ISWC, pp. 374–389 (2015)
17. Azur, Melissa J., Stuart, Elizabeth A., Frangakis, Constantine, Leaf, Philip J.: Multiple imputation by chained equations: what is it and how does it work?: multiple imputation by chained equations. Int. J. Methods Psychiatr. Res. **20**(1), 40–49 (2011)
18. Raghunathan, T.W., Lepkowksi, J.M., Van Hoewyk, J., Solenbeger, P.: A multivariate technique for multiply imputing missing values using a sequence of regression models. Surv. Pract. **27**, 85–95 (2001)
19. Stekhoven, Daniel J., Bühlmann, Peter: MissForest–non-parametric missing value imputation for mixed-type data. Bioinformatics **28**(1), 112–118 (2012)
20. Troyanskaya, O., Cantor, M., Sherlock, G., Brown, P., Hastie, T., Tibshirani, R., Botstein, D., Altman, R.B.: Missing value estimation methods for DNA microarrays. Bioinformatics **17**(6), 520–525 (2001). June

A Novel Approach for Calculating Real-Time Composite Indicators Relying on Internet of Things and Industrial Data Spaces

Andreu Belsa, Rafael Vaño, Ignacio Lacalle, Matilde Julián, Fernando Boronat, and Carlos E. Palau

Abstract This paper aims at devising an IoT-based system to collaboratively calculate quantitative composite indicators among Industrial actors. Requirements have been considered to build a reliable, trusted, anonymised environment within which involved stakeholders can safely share information on the indicators while keeping sovereignty over their data. Technologies such as enhanced Context Brokers, the usage of Industrial Data Spaces agents and the introduction of distributed publish-subscribe data exchange schemes lay the foundations of the proposed solution. This work entails the design of the system altogether with a proposal of technologies for its deployment. Drawing from previous public-funded projects' results and highly relying on open-source technologies, the authors have come up with a solution that aims to be proven in real environments in the near future.

1 Introduction

A "data collaborative" governance perspective is being realised as the paramount trend for effective, educated, useful decision making in an increasing number of domains [1]. The fact of bringing together several actors involved in real phenomena

A. Belsa (✉) · R. Vaño · I. Lacalle · M. Julián · F. Boronat · C. E. Palau
Communications Department, Universitat Politècnica de València, València, Spain
e-mail: anbelpel@upv.es

R. Vaño
e-mail: ravagar2@upv.es

I. Lacalle
e-mail: iglaub@upv.es

M. Julián
e-mail: majuse@upv.es

F. Boronat
e-mail: fboronat@dcom.upv.es

C. E. Palau
e-mail: cpalau@dcom.upv.es

D. Camacho et al. (eds.), *Intelligent Distributed Computing XIV*, Studies in Computational Intelligence 1026, https://doi.org/10.1007/978-3-030-96627-0_5

allows them to actually share actionable data. Only having such data available drives those actors to agree on a common view of the reality (composite indicator) getting enough insights for addressing complex problems. Although this can be easily understood from a bird's eye, it poses several challenges that now (and only now) are foreseeable to be solved thanks to technology.

First, those intervening actors must be engaged in an active way, being aware of the benefits that can be obtained from using composite indicators to represent a complex real phenomenon [2]. Second, the collection of data (from a myriad of heterogeneous sources) must be performed in a transparent, unbiased and unified way, democratising information and procedures [3]. For doing so, robust data models must be defined to which the different actors can rely on. Third, reluctance by Industrial entities to share their data must be overcome by using accepted standards thus endorsing the adoption of the technology [4]. Fourth, and most important, mechanisms must be put in place to guarantee what is known as "data sovereignty", which stands for the ability of coping with data ownership, privacy protection and origin tracking along the data process chain [5]. This becomes specially relevant in Industrial environments, where distrust is usually assumed and specific privacy and integrity safeties must be ensured.

The approach suggested in this paper is to leverage current trends in the technological state of the art to produce such a "data collaborative" solution addressed to (mainly) Industrial actors. In particular, the proposal consists of an Internet of Things (IoT) platform enhanced by joining the worlds of data spaces, data sharing techniques and added-value services for representing real phenomena thanks to composite indicators and a collaborative, multi-actor approach along the whole process chain. The hypothesis is that such a system can be devised and deployed by leveraging open-source software and building upon the results of various European Commissions' H2020-funded projects.

The following paper is going to describe the technical design of a solution to cover the aims described in this section. Section 2 provides some context on where the solution has come up from, inspired by the raise of various technologies that are being explored during the last few years by the authors. Section 3 describes the architecture of the solution as well as the overall approach while Sect. 4 relates some open source technological choices to deploy it. Finally, Sect. 5 aims at offering some future perspectives, proof of concept scenarios and lessons learnt while devising the proposed solution.

2 Context

The goal of the proposed solution is to devise the combination of various technologies and mechanisms to deliver an IoT-based trust system producing real-time composite indicators in Industrial environments. This section aims at presenting the context where the authors started from to tackle such an exercise.

Internet of Things (IoT) and Big Data infrastructure Nowadays, IoT is everywhere. It is calculated that 46 billion of devices will be connected to the network by the end of this year (2021) alone [6]. In terms of market growth, some studies suggest that the IoT integration market will reach 4,066.79 million US dollars by 2023 [9]. Technologically, IoT already enables access to large amounts of information from smart applications and sensors in real time from different domains. Along with this, Big Data Infrastructures (such as cloud data centers) take advantage of this amount of data to offer data-driven decision-making solutions. This lets more traditional approaches, such as the calculation of composite indicators, to take advantage of this entire technological stack. For addressing the IoT needs of the solution, the authors drew from the results of the H2020 projects INTER-IoT [7] and CREATE-IoT [8], as well as from well-established IoT de-facto standards such as FIWARE [22], NGSI entities and publish-subscribe data exchange schemes. The work of these projects has provided a multi-layered approach and several tools to facilitate the interoperability among heterogeneous IoT sensors, platforms, applications or services that enables different IoT scenarios.

Composite Indicators A composite indicator is the mathematical combination of individual indicators that represent different dimensions of a concept whose description is the objective of the analysis [10]. Composite indicators are much easier to interpret than trying to find a common trend in many separate indicators. However, composite indicators may send misleading or non-robust policy messages if they are poorly constructed or misinterpreted. On the one hand, statisticians may resent that work in data collection and editing is wasted or hidden behind a single number of dubious significance. On the other hand, stakeholders are tempted to summarise complex processes into a single figure in a too light fashion [11]. Previous research point out that currently there are no relevant tools or frameworks in the market focused in the design, management and visualization of the full life-cycle of the composite indicators calculation based on the integration of heterogeneous data sources, considering IoT real-time information, Artificial Intelligence (AI) and Big Data technologies. Research projects have proposed composite indicators. For example, H2020 PIXEL Project [12] provides a complete approach for maritime ports domain and focused in environmental purposes. Another example is H2020 CITYkeys project [13], which provides indicators for Smart Cities. However, none of these projects offers a tool that is independent of the domain or purpose. This paper is built on the results provided by PIXEL and the work delivered in [18] by broadening (in domain) and enhancing (in technical ambition) their scope.

Secured, trusted Industrial data sharing The creation of a secure and trusted data sharing environment was a key concern in the proposed solution. In such a system, information is an element of utmost importance, therefore the origin, validity and integrity of the raw data must be guaranteed. Thus, data sovereignty was identified as a requirement, allowing to safely integrate data from different owners to calculate the composite indicators (see Fig. 1). Here, some barriers were identified by the authors that required technological coverage: (i) Data owners are reluctant to share their data with others, being especially applicable to private entities in the Industrial context. (ii) They somehow need to be aware of the purpose and reason for sharing it. For the context of this paper, the data will be used to interpret in a quick glance real phenomena—in which those entities are involved-, analyse the evolution in time, prepare rankings or comparisons (with other entities involved). Considering the latter, the barriers might be even bigger. However, the benefits are huge and so must them be transmitted, including democratisation of information, collaborative decision making and joint growth. (iii) They also need to be sure that data are shared in a safe way. It is difficult to attract users to share data and participate in a ranking/evolution analysis if they are afraid that the rest of the participants may inspect their raw data or results. In this sense, the challenge is to build an "anonymised, secured, trusted" environment allowing users to achieve the aforementioned data sovereignty and be able to audit the exchanges of data. The International Data Spaces Association **(IDSA)** defines a reference architecture and a formal standard to be used for creating and operating virtual data spaces. The IDS Architecture [14] is based on commonly accepted data governance models, standards and technologies facilitating secure exchange and easy linkage of data within business ecosystems. The IDS Architecture ensures digital sovereignty for data owners making data available for being exchanged or shared. It thereby constitutes the basis for developing and offering smart services and for establishing innovative business processes. IDS will significantly contribute to the success of digital platforms and applications in the fields of AI, IoT, or Big Data. In addition, it is being used as reference architecture in projects such as DataPorts [15], GAIA-X [16] or i4Trust [17]. IDSA already has over one-hundred member organizations from twenty countries.

3 Proposed Solution

The aim of this paper is to define a novel approach where the calculation of composite indicators is performed in a trusted environment. The ideas behind the following chapters are the following (see Fig. 1):

- Diverse actors intervening in a phenomenon (e.g., how the automotive stakeholders contribute to the sustainability of cars manufacturing processes) collaborate to define and fulfil an agreed composite indicator (e.g., Car Production Sustainability—CPS).
- Those actors establish internal sub-indicators ("raw indicators" or just "indicators") that are calculated within the stakeholder premises under a standardised procedure.
- Indicators are shared using safe mechanisms in a trusted environment to calculate the final composite indicator (e.g., CPS), that will typically range between 0 and 1, and that will be available to all organisations involved.

The proposed solution aims to offer enablers that provide simple, close to plug and play mechanisms to integrate the shared data in an interoperability platform that follows an IoT middleware approach. The enablers can facilitate the generation

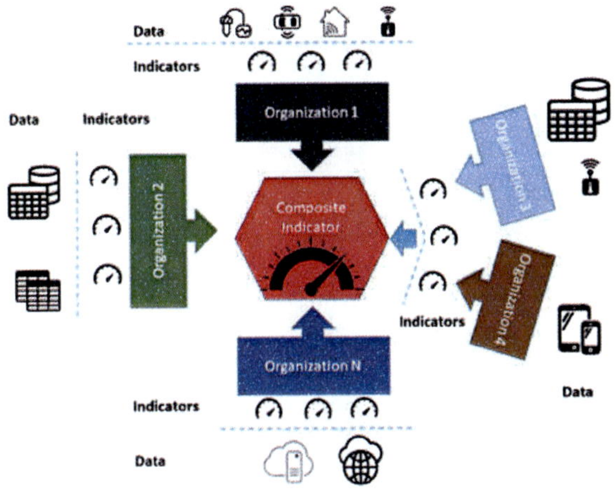

Fig. 1 Data sharing domains

of the indicators in real time based on the data collected from the data sources. In addition, internal mechanisms can perform the calculation of the composite indicator efficiently by taking advantage of Big Data analysis. The solution to be proposed must consider that: (i) the system will be dealing with data coming from different entities which have to be combined, (ii) reliability on the calculations is a paramount concern, (iii) anonymisation should be sought so that neutral comparisons can be achieved. Finally, it should offer visualisation and data management mechanisms allowing users to leverage the added value of the results of the process.

The approach proposed in this paper offers an automatic way to collect the indicators and calculate the composite index, which will save time in all the internal processes related with these tasks. Nevertheless, applying innovative IoT and Big Data techniques can offer more added value functionalities.

3.1 *Global Architecture*

The functional blocks of the proposed solution are shown in Fig. 2. The first block comprises the Data Processing Services, which obtain the data from the available data sources and translate it into the common data model to enable semantic interoperability. This block provides real-time "raw indicators" and historical values. These values are used as input by the Analytics Services and the Composite Indicators Services to actually deliver the Composite Indicator after mathematically-rationed data processing mechanisms. The Analytics Services apply Machine Learning methods to predict future values of the indicators obtained by the Data Processing services. The Composite Indicators Services obtain the indicators from the Data Processing Services and calculate the composite indicator. In addition, this block can make use

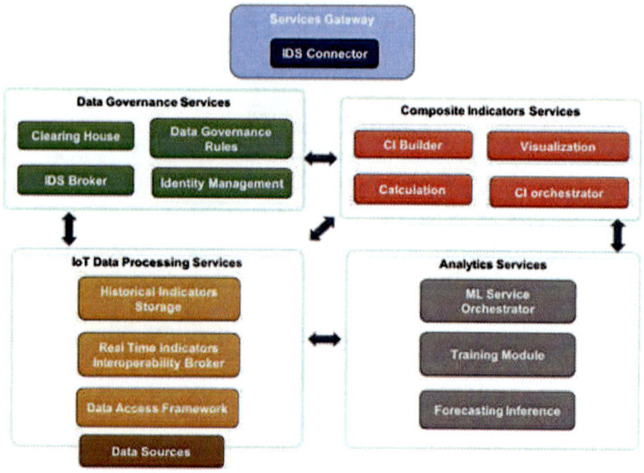

Fig. 2 Functional blocks of the global architecture proposed

of the results provided by the Analytics Services to estimate the future values of the composite indicators. This block also includes visualisation tools. Finally, the Data Governance services and the Services Gateway allow sharing data among different organisations in a trusted way following the IDSA architecture. The Data Governance Services provide a secure and trusted environment for data sharing and define the rules for data sharing. The Services Gateway allows sharing data with other organizations in a trusted way following the IDSA standard.

Figure 3 shows the architecture of the proposed solution considering that it is being deployed and used in a single organisation, suggesting as well some open-source deployment options that are further discussed in Sect. 4. The schema in a distributed environment would add certain complexity due to the fact of sharing data among Industrial contexts—check Sect. 3.3 for more details.

The lowest layer of the proposed architecture is in charge of obtaining the data from the data sources. The Historical Indicators Storage receives the values of the indicators in real time and generates the historical record. The real-time and historical values of the indicators are used by the Composite Indicators Services and the Analytics Services.

This architecture presents several advantages. From the point of view of data acquisition, the users do not need to deal with the technological issues regarding data access, they simply need to focus on the data they want to acquire. From the indicators definition point of view, the use of data models guarantees to be part of an active ecosystem in which existing models can be reused or new ones can be adopted by a community. From the composite index design point of view, the entire calculation tree can be shared with others in order to potentially convert the designed calculation methodology into an open standard. From the calculation point of view, Big Data and AI solutions can help in the imputation of missing data, improve the normalization of the data or predict future values of the indicators. From the point of

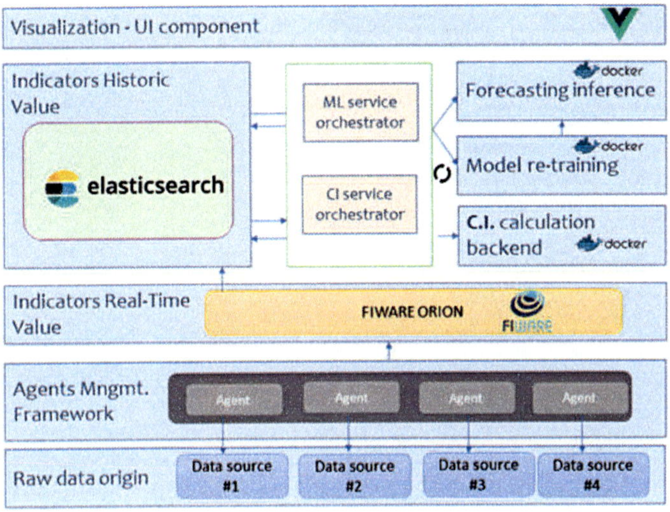

Fig. 3 Architecture deployed in the premises of a single intervening actor

view of visualization of the results, the visualisation techniques can provide precise information of the whole process in a comprehensive way.

3.2 *Data Gathering Strategy*

The novel data gathering approach aims to offer a framework for an efficient management of the data collected and to facilitate the exchange of this information among the participants. This requires the adoption of common APIs and the definition of common data models. In addition, the vocabulary providers manage and offer vocabularies which can be used to annotate and describe datasets [14].

The agents translate data from the original format into a Context Broker-compliant format following FIWARE Data Models guidelines [19]. In addition, they calculate the "raw indicator" associated to each gathered data piece. They send the information to a Semantic Broker which receives the information and interacts with the Agents Management Framework. This interoperability broker offers a common API to access to functionalities, metadata description and data in common formats.

The Agents Management Framework provides a web-based tool to manage and run agents that integrates data into the platform and a SDK wizard to create a template to reduce the effort to develop and deploy this agents. It deals with the requirements about how to get the data (from a Database, API, a sensor, a file...), the way to access to the data ("go for it or wait for it"), how often the data must be collected and shared, volume, format and whether there is already a data model structure or a current semantic translation being applied.

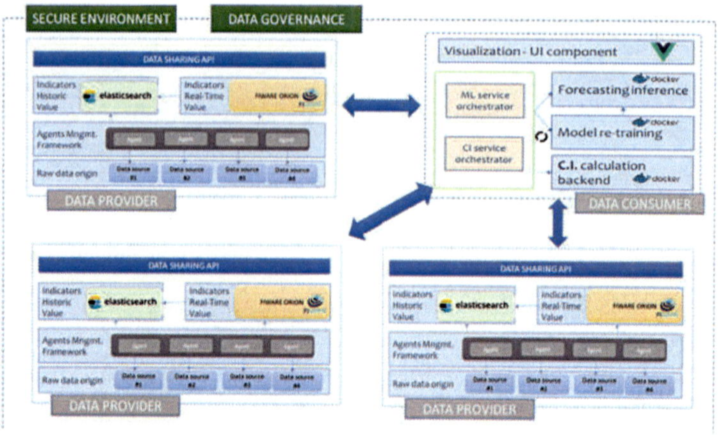

Fig. 4 Conceptual rationale behind the proposed solution

3.3 Data Sharing Strategy

The target users of this approach can be divided into data providers and data consumers. Data providers offer their data sources, translating their information into "raw indicators" values, and aim to take advantage from the composite index calculation and its visualization dashboards. Data consumers are interested in analyzing the composite index values and making decisions based on the results obtained from this analysis. Figure 4 presents an example of how the proposed solution could make use of indicators from different organizations to calculate and predict the values of the composite indices. The described schema is known as "Data Spaces" approach.

Data Spaces bring technical means for guaranteeing that participants in the exchange of data can trust each other and exercise sovereignty over the data they share. This requires the adoption of common standards for managing the identity of participants, the verification of their truthfulness and the enforcement of policies agreed upon data access and usage control. IDS Connectors are responsible for such connectivity and usage control. They allow the execution of trusted apps in an isolated Identity Provider environment. IDS Connectors register the description of their data endpoints at an IDS Broker. The Data Consumers can look up available data sources and data in terms of their content, structure quality, actuality and other attributes. The Identity Providers offer a range of services to create, maintain, manage and validate identity information of and for IDS participants [14].

4 Candidate Open Source Technologies for Implementation

From the data provider side, the Data Processing Services would be implemented using several tools. Orion Context Broker is the reference implementation of the Context Broker GE of FIWARE, which constitutes the main component of a Powered by FIWARE platform. This component provides a NGSI interface to manage context information, enabling updates, queries, registrations and subscriptions [22]. In the proposed solution, Orion manages the data obtained from the data sources by the agents. Pyngsi [23] is a Python framework for the development of FIWARE NGSI agents. This framework provides an easy way to connect the different data sources to the Context Broker. Cygnus manages the persistence of the context information, providing a historical view of the data. In the proposed solution, Cygnus receives the data from Orion and stores it dynamically in Elasticsearch [21], which is a distributed database and full-text search engine over JSON documents. Elasticsearch enables storage and queries over the historical indicators values in the proposed solution.

Regarding the Data Governance services, the FIWARE GEs Keyrock, Wilma and AuthZForce would enable security based on OAuth 2.0 authentication and authorization policies defined following the XACML standard [20]. Keyrock provides identity management, while the access control is enabled by Wilma and AuthZForce. In addition to the aforementioned components, an IDS connector is needed to enable trusted data sharing between organizations [25, 26]. There are several open-source implementations of the IDS connector [27, 28]. One of them, the FIWARE TRUE connector, is an incubated FIWARE GE [24].

In addition to these Fiware related tools, the novel solution should integrate the open-source components developed in several projects. Firstly, the tool from a previous research work [18] could be used to calculate, create and visualize the information of a specific Composite Index Calculation scenario. Secondly, the Data Access and Semantic Interoperability framework from DataPorts Project [15] could create and manage the corresponding agents. In addition, the Data Abstraction and Virtualization capabilities and Analytic Services offered by DataPorts Platform could be integrated. Finally, the Information Hub and the Dashboard capabilities from PIXEL [12] could be used to visualize the data and results.

5 Future Work and Conclusions

The proposed design, based on widely adopted open source provisions, has devised a (mature-enough) baseline to build IDS-compatible, CEF-Building Block-compatible solutions on. It may accelerate the delivery of data spaces to the market while collaborating with other initiatives like GAIA-X in Europe. The next step is the materialization of this design to deploy a trusted, multi-agent Composite Indicator-service solution in real environment. Some potential scenarios have been identified inspired on currently-on-going initiatives such as PIXEL's Port Environmental Index (PEI) or H2020 DataPorts' use-cases. Whereas PEI is actually a composite index service rooting on the usage of IoT and pyngsi-based agents, it lacks the secured, trusted provisions. The latter is, in contrast, what DataPorts' use-cases do entail (which provide solutions for IoT, SDK for Agents implementation, AI and an IDS approach). Therefore, a combination of the two (and other potential additions) might become a complete experimentation of the proposed solution.

Finally, some prospective research lines related to Machine Learning and Distributed Ledger Technologies will be analyzed as natural continuation of the work initiated in this manuscript. This might offer a framework to forecast future index calculation results or explore new technical ways to share the information.

Acknowledge of funding This research was funded by the European Commission via the H2020-project DataPorts—grant number 871493—and via the H2020-project PIXEL—grant number 769355.

References

1. Erna Ruijer, Designing and implementing data collaboratives: A governance perspective, Government Information Quarterly, 2021, 101612, ISSN 0740-624X
2. Mazziotta, M., Pareto, A.: Synthesis of indicators: the composite indicators approach (2017). https://doi.org/10.1007/978-3-319-60595-1
3. Smidova, Z.: Building the evidence for OECD integrity and anti-corruption agenda: the current situation and avenues for future analysis. OECD Economics Department Working Papers, No. 1614, OECD Publishing, Paris (2020). https://doi.org/10.1787/80ebe6e6-en
4. https://europa.eu/youreurope/business/product-requirements/standards/standards-in-europe/index_en.htm
5. https://internationaldataspaces.org/why/data-sovereignty/
6. https://techjury.net/blog/how-many-iot-devices-are-there/#gref
7. https://inter-iot.eu/
8. https://european-iot-pilots.eu/project/create-iot/
9. https://www.marketwatch.com/press-release/iot-integration-market-report-2021-to-2030-globally-exploring-sales-volume-revenue-and-price-by-types-2021-07-09?siteid=bigcharts&dist=bigcharts&tesla=y
10. Saisana, M., Tarantola, S.: State-of-the-art report on current methodologies and practices for composite indicator development, EUR 20408 EN. European Commission-JRC, Italy (2002)
11. Hudrliková, L.: Composite indicators as a useful tool for international comparison: the Europe 2020 example. Prague Econ. Pap. **22**, 459–473 (2013). https://doi.org/10.18267/j.pep.462
12. https://pixel-ports.eu/
13. http://www.citykeys-project.eu/
14. Otto, B., Lohmann, S., Steinbuß, S., Teuscher, A.: IDS Reference Architecture Model Industrial Data Space Version 2.0
15. https://dataports-project.eu/
16. https://www.gaia-x.eu/
17. https://i4trust.org/
18. Lacalle, I., Belsa, A., Vaño, R., Palau, C.E.: Framework and methodology for establishing port-city policies based on real-time composite indicators and IoT: a practical use-case. Sensors **20**(15), 4131 (2020). https://doi.org/10.3390/s20154131
19. https://smartdatamodels.org/

20. Sarabia-Jácome, D., Palau, C.E., Esteve, M., Boronat, F.: Seaport data space for improving logistic maritime operations. IEEE Access **8**, 4372–4382 (2019)
21. https://github.com/FIWARE/catalogue/tree/master/data-spaces
22. https://github.com/FIWARE/catalogue
23. https://pypi.org/project/pyngsi/
24. https://github.com/Engineering-Research-and-Development/fiware-true-connector
25. Alonso, Á., Pozo, A., Cantera, J.M., De la Vega, F., Hierro, J.J.: Industrial data space architecture implementation using FIWARE. Sensors **18**(7), 2226 (2018). https://doi.org/10.3390/s18072226
26. Munoz-Arcentales, A., López-Pernas, S., Pozo, A., Alonso, Á., Salvachúa, J., Huecas, G.: Data usage and access control in industrial data spaces: implementation using FIWARE. Sustainability **12**(9), 3885 (2020). https://doi.org/10.3390/su12093885
27. https://github.com/International-Data-Spaces-Association/DataspaceConnector
28. https://industrial-data-space.github.io/trusted-connector-documentation/

MACHINE LEARNING (Main Track)—Chair: Antonio Guerrieri

Using Evolving Ensembles to Deal with Concept Drift in Streaming Scenarios

Diogo Ramos, Davide Carneiro, and Paulo Novais

Abstract In a time in which streaming data becomes the new normal in Machine Learning problems, to the detriment of batch data, new challenges arise. In the past, a data source would be static in the sense that all data were known at the moment of the training of the model. A model would be trained and it would be in use for relatively long periods of time. Nowadays, data arrive in real-time and their statistical properties may also change over time, rendering trained models outdated. In this paper we propose an approach to deal with the concept drift problem with minimal computational effort. Specifically, we continuously update an ensemble with new weak learners and adjust their weights according to their performance. This approach is suitable to be used in real-time in the form of an ever-evolving model that adapts to change in the data.

1 Introduction

Real-world Machine Learning problems often attempt to model complex realities and patterns. These patterns are sometimes also not static. That is, the relationship between the variables, their statistic properties, or the variables themselves, may change over time. This is known as *concept drift*, which should not be confused with noise [12]. The challenge of learning with concept drift is thus often twofold [12]: it needs to learn new patterns, and to forget old ones. Not doing the latter would result in over-complex models.

D. Ramos · D. Carneiro (✉)
CIICESI, ESTG, Politécnico do Porto, Felgueiras, Portugal
e-mail: dcarneiro@estg.ipp.pt

D. Ramos
e-mail: 8150101@estg.ipp.pt

P. Novais
Algoritmi Centre/Department of Informatics, Universidade do Minho, Braga, Portugal
e-mail: pjon@estg.ipp.pt

D. Camacho et al. (eds.), *Intelligent Distributed Computing XIV*, Studies in Computational Intelligence 1026, https://doi.org/10.1007/978-3-030-96627-0_6

One of the most classic examples is that of concept drift in market basket analysis [11]: buyers' preferences and buying habits change over time. Part of this change is periodic and can be modeled through the inclusion of additional variables such as *month* or *season*. However, this change can also be partly caused by the natural and normal evolution of buyers' preferences in light of the also evolving market offer.

There are different types of concept drift (i.e. abrupt, incremental, gradual or recurring) [3] and two major ways to deal with it using ML: incremental learning [6] and batch learning [5].

This paper describes an incremental approach to learning, combined with a meta-learning scheme in which not only the model is adapted over time but also its hyper-parameters, to continuously adapt to concept drift. More specifically, the work described in this paper falls into the category of ensemble learning. We propose a bagging method [2] in which the meta-learning task is to find the best way to combined the weak-learners. The meta-learning task is implemented with a Genetic Algorithm, that continuously optimizes the weights of each weak learner to adapt to changes in concepts. Not only that, but the ensemble also changes over time through the exclusion of outdated models and the inclusion of new ones [10]. This way, the model can continuously adapt to concept drift with minimal computational effort.

2 An Evolving Ensemble to Deal with Concept Drift

The methodology to deal with concept drift proposed in this paper is based on a key principle: to replace the weak learners that constitute a Machine Learning ensemble over time, and to optimize their weights in the voting scheme of the ensemble, so that the ensemble's predictions continuously adapt to changes in the statistical properties of the stream of data. It can thus be seen as a real-time data pipeline for training and predicting on an ever-evolving complex Machine Learning model.

The main components of the proposed system are thus defined:

- Data sources—these represent multiple sources of data, such as hard or soft sensors, social networks, etc.
- Streaming platform—a platform able to receive streaming data, temporarily store it in a structured way, and make it available for the other components the system in real-time upon request.
- Model pool—a memory space which holds the weak learners that constitute an ensemble at a given moment in time.
- Model metadata—a database containing the metadata of all models trained as well as the current configuration of the ensemble.

The following sections describe the main phases of the proposed approach.

2.1 Initialization

The whole process starts with the initialization phase, which is triggered when a new learning problem is configured. The learning problem is configured by choosing one or more topics from the streaming platform, and identifying each one as a dependent or target variable. These topics dot not need to be necessarily new. That is, one topic can be shared by multiple learning problems.

In the configuration of a new problem, the user also sets the frequency f at which the model should be potentially updated. This frequency can be defined either in terms of time or in terms of number of messages (data instances) received. For instance, the user can determine that a new weak learner should be trained at each 1000 messages. The inclusion of the weak learner on the dataset however depends on whether it would or not improve the performance of the ensemble, as described further below.

The window of data selected for training the new weak learner is however not restricted to the data that arrived since the training of the previous one. Indeed, the window of data used can be configured in several ways:

- f—frequency at which new models are trained (in number of instances or seconds)
- ws—the standard size of the window (in number of instances)
- wv—the random positive variation allowed for the window (%). A larger value will allow higher overlap of data with previously trained models

Figure 1 visually depicts a representation of what several windows of data used for training different models of the ensemble could look like, considering a scenario in which $f = 1000$ (instanaces), $ws = 1000$ and $wv = 100\%$. That is, this configuration would generate windows containing between 1000 and 2000 instances of the most recent data.

The superposition that exists between data frames is purposeful and makes the ensemble more robust to noise. The balance between the ability to adapt to concept drift and the robustness to noise, already addressed in Sect. 1, can be controlled by adjusting the hyper-parameters f, ws and wv. Namely, sensitivity to noise increases with smaller values of f, ws and wv.

When a new window is selected, the corresponding data are read from the streaming platform and split into test and training datasets. The training dataset is used to train the new weak learner that will potentially be added to the model pool. In order to make each model in the ensemble a weak learner, each is trained with a subset of

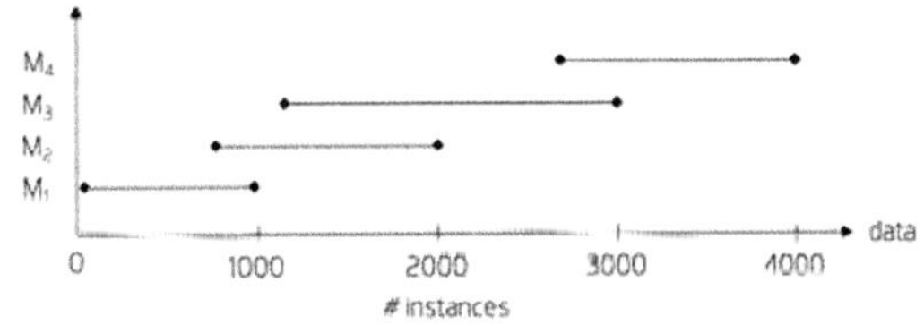

Fig. 1 Visual depiction of a possible distribution of data for the different weak learners as they are trained over time, for $f = 1000$, $ws = 1000$ and $wv = 100\%$

the feature vector. The learner also becomes *weak* as a result of being trained with a relatively small amount of data. The main hyperparameters to control the complexity of each learner are thus the sample rate sr (which results from the size of the window generated when selecting data) and the column sample rate csr.

The test dataset is used to evaluate the performance of this new weak leaner. Once the model is trained its data is added to the database that stores the models' metadata. This includes model properties (i.e. the model configuration) and model performance metrics (e.g. RMSE, Accuracy, Precision). The data in the test dataset is also added to a test data pool, that is used for evaluating the fitness of the whole ensemble, so that each of its weak learners is evaluated on data that was not known during training.

The following measures are calculated for each solution, depending on the type of problem: accuracy, precision, recall, F1-score and RMSE. These indicators are stored in the metadata database, together with the unique identifier of the model, the date of training, and the range of instances used in the training.

During this initialization process there is still no evolutionary process. Models are added to the pool as they are trained. If a prediction is requested from the ensemble during this phase, a bagging method is used. That is, during the initialization phase, the predictions of each weak learner all have the same weight to compute the final prediction.

The initialization phase continues until a minimum number of models mm is in the pool of models, after which the process advances to the evolutionary phase. The whole process of initialization can thus be summarized as follows. Data received from the streaming platform is used to train new weak learners at certain intervals. New models are added to the pool of models, and each new version of the ensemble is evaluated and its performance metrics and other metadata saved in the database. The process of initialization ends when enough models exist in the pool (Fig. 2).

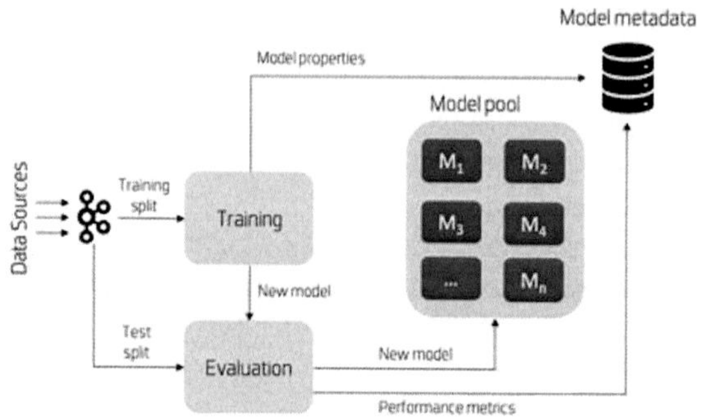

Fig. 2 Data flow during the initialization phase in which new models are progressively added to the model pool

2.2 Evolutionary Phase

Once the model pool reaches the established number of weak learners mm, the evolutionary phase of the process starts. During this phase, the number of weak learners in the ensemble is kept constant. Thus, whenever a new weak learner that would improve the performance of the ensemble is trained, the learner with the weakest performance currently in the ensemble is removed, and replaced by the new one. Hence, during the evolutionary phase, the training of new weak learners continues at the same rate as during the initialization phase.

However, during this phase, asides from existing weak learners being replaced by better ones, their weights are also continuously optimized so that better learners have a higher weight in the voting scheme of the ensemble. This makes it possible that the ensemble can make better predictions with the existing weak learners, without requiring the training of a new model. This is especially useful if one aims to minimize the training of models, namely due to resources restrictions, or if data distribution changes at a very fast pace.

The evolutionary phase thus has as main goal to find the optimum weight of each weak learner in a given ensemble. It follows the standard process of a Genetic Algorithm [4]. The evolutionary process starts with the creation of n random solutions for the problem. A solution for the problem, in this case, e a group of weights, one for each weak learner in the ensemble. Thus, solutions are represented as weights vectors w of size mm. The initial population is thus constituted by n weights vectors.

This first group of solutions constitutes the first generation g_1 of a group of G generations that will be created throughout the process. Each weight $w_{g,s,i}$, $g \in [1..G]$, $s \in [1..n]$, $i \in [1..t]$ represents the weight of the weak learner i in the solution s of generation g ($\sum_{i=1}^{t} w_{g,s,i} = 1, \forall s \in [1..n], \forall g \in [1..G]$).

The fitness of each solution in the current generation is then calculated. This depends on the type of the ensemble. In a classification problem, the fitness is given by the accuracy of the predictions, while in a regression problem the fitness is given by the RMSE. Both are calculated using the *test* dataset, which acts as holdout. This results in a fitness vector f of size t, in which each fitness value $f_{g,s}$, $g \in [1..G]$, $s \in [1..n]$ represents the fitness of a given solution s in generation g.

After the fitness is calculated, the b solutions with the highest fitness are selected to be reproduced, thus originating the first solutions of a new generation. A group of m new solutions are added to this new generation, based on the mutation operator. This operator takes as input a solution (selected randomly from the b best solutions), generates a random number r, $r \in [0..mr]$ (in which mr denotes the mutation rate), and randomly selects two positions in the weights vector i and j. In this scheme, the mutation rate mr is one of the hyper-parameters and can be set by the user in the interval [0.01..0.5]. The solution is thus mutated according to $w_{g,s,i} \leftarrow w_{g,s,i} + mr$ and $w_{g,s,j} \leftarrow w_{g,s,j} - mr$. Since there is the restriction that $0 \leq w_{g,s,i} \leq 1$, this operation is repeated for the same chromosome until all its weights are valid.

In each iteration, another group of r new solutions are also added to this new generation. These solutions are created completely random, following the same process as in the initialization. The aim of introducing these new solutions in each generation is to maintain a certain level of diversity in each generation and to try to cover the search space as much as possible, while still focusing on the best solutions.

Once the new generation is complete, including not only the b best solutions but also the m solutions generated by mutation and the r solutions generated randomly ($b + m + r = w$), the process returns to the fitness evaluation stage and repeats.

The process stops when a limit of generations l is reached, or when the value of the best fitness found so far has not improved by δ over the last x generations. The output of the process is the weights vector $w_{g,s}$ with the highest fitness value in all generations ($g \in [1..G]$, $s \in [1..n]$).

The hyper-parameters that can be tuned when using this algorithm are thus:

- sr—sample rate
- csr—column sample rate
- n—number of solutions in each generation
- t—number of trees in the ensemble
- b—number of best solutions to consider in each generation
- mr—mutation rate, defining how much each solution should mutate
- m—number of new solutions to generate through mutation, in each generation
- r—number of new random solutions to generate, in each generation
- l—maximum number of iterations (stopping criterion)
- δ and n—define the reaching of convergence (stopping criterion).

3 Results

In order to assess the validity of the proposed methodology, a synthetic stream of data was created that encodes different types of concept drift. The stream is composed of 100.000 instances of 3 variables: 2 independent variables a and b and one target variable $f(a, b)$. Concept drift in this stream occurs in several different ways. The average value of a increases linearly with time, from 3 in the beginning of the stream to 12.5 in the end. The standard deviation of b increases linearly from 0.2 to 1.2. Finally, the relationship between the independent variables a and b and the target variable $f(a, b)$ also changes over time. At the beginning $f(a, b) = 0.8a + 0.2b$, while at the end, $f(a, b) = 0.2a + 0.8b$.

Figure 3 shows the plot of each of the three variables over time. Table 1 provides basic descriptive statistics of the stream in three moments in time t_1, t_2 and t_3, which represent respectively the beginning, the middle, and the end of the stream.

In order to assess the validity of the proposed approach, two different experiments where compared: one in which the approach descried in this paper is used, an another in which there is no evolution of the ensemble after it is first trained. In both scenarios, the dataset presented previously is streamed an used at every 1000 instances to train a new model and/or evaluate the ensemble. Decision trees were used as weak learners in the ensemble, in a pool of size 25.

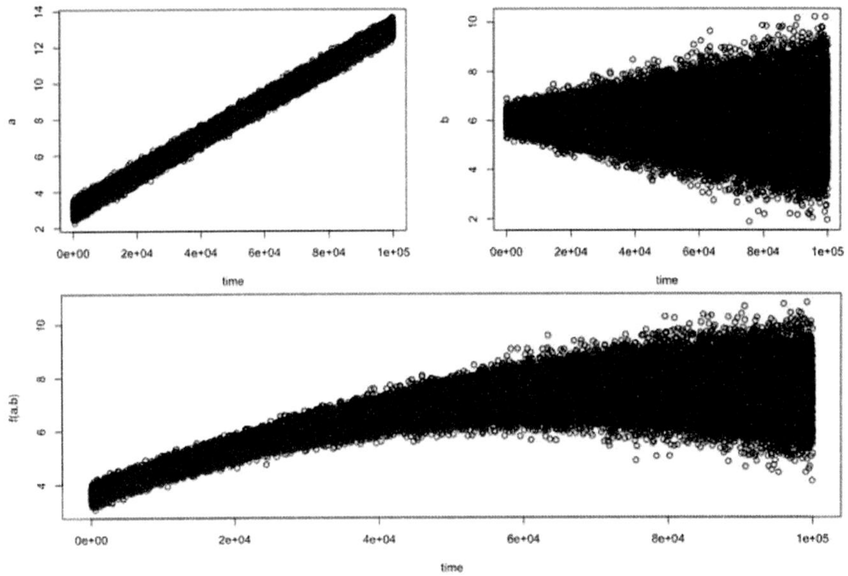

Fig. 3 Plot of the two independent variables, *a* and *b*, and of the target variable, over time

Table 1 Statistics of the 3 windows of size 1000 of the data stream, at the beginning, middle and end

	t_1		t_2		t_2	
	$\overline{x}$	σ	$\overline{x}$	σ	$\overline{x}$	σ
a	3.04	0.20	8.00	0.20	12.50	0.20
b	6.00	0.21	5.98	0.69	5.95	1.19
f(a, b)	3.64	0.16	6.99	0.36	7.37	0.95

In the first experiment, the dataset is randomly split into two sets (train and test) at each 1000 instances, with a proportion of 75/25%. A new model is trained with the train test. This model may or may not be added to the pool of models, depending on its performance metrics and those of the models currently in the pool. Nest, one instance of the Genetic Algorithm is run to optimize its weights. The test set is used to compute the rmse and assess the quality of the model at the time. This process goes on until data finishes arriving.

Figure 4 shows the evolution of the weights for some of the models in the pool as the ensemble evolves. It shows how stronger predictors, with higher weights, tend to last longer until they finally disappear by becoming outdated. The image also shows when models were trained and added to the pool and when they where removed from the pool. Only some of the models are shown for the sake of clarity, and the figure

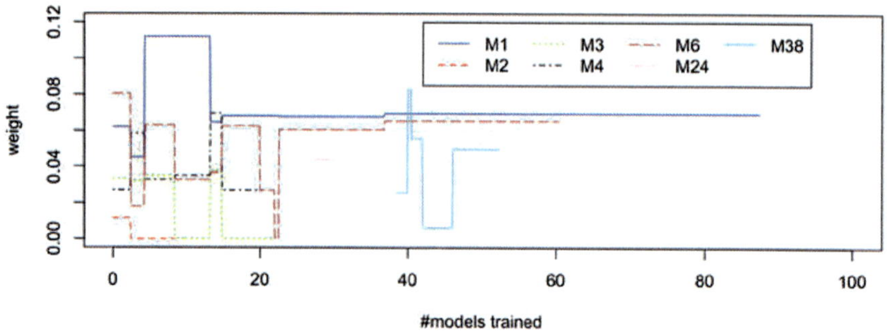

Fig. 4 Evolution of the weights of some of the models during the evolutionary phase

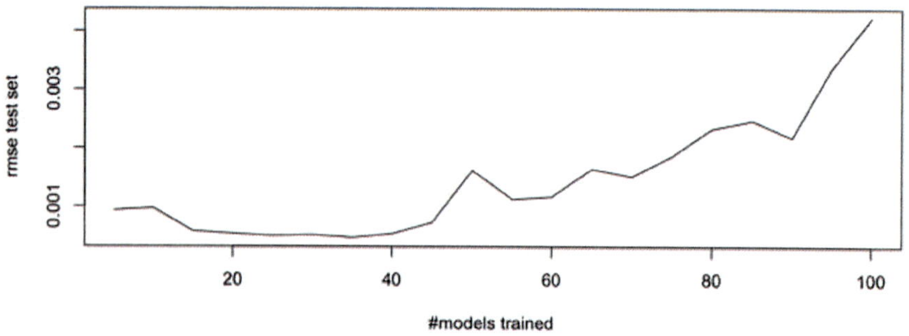

Fig. 5 Evolution of the rmse measured on the test set throughout time as the ensemble changes (experiment 1)

only shows the data from the evolutionary phase since the weights are not adjusted during the initialization phase, as described before.

If an equal distribution of weights was used, as in a regular ensemble, each model would have a weight of 0.04. The figure shows of some models have lower weights and tend to be removed quickly from the ensemble (e.g. M2, M3), while others have fairly higher weights and tend to last longer (e.g. M1, M2). In this figure, each position and vertical section can be seen as the best solution found for this learning problem, at that time, i.e., a group of models and corresponding weights.

Figure 5 shows the evolution of the rmse through time, as the ensemble evolves. Although it tends to increase over time, the increase is very small and negligible (≈ 0.003). This shows that the proposed approach is able to maintain a constant performance throughout time, and adapt to concept drift.

In the second experiment, the training of new models and the evolutionary process were not included. Thus, an ensemble is trained with data from the beginning of the stream during the initialization and it is used throughout the whole experiment. This means that each model in the ensemble has the same weight (0.04). It also means that the training data is discarded, and the initial ensemble is evaluated on the streaming test data throughout the experiment.

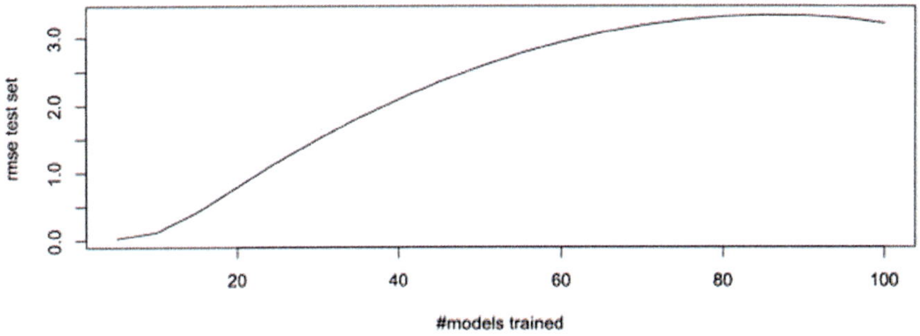

Fig. 6 Evolution of the rmse measured on the test set without the evolutionary phase (experiment 2)

Figure 6 shows how the rmse of the ensemble starts out with values that are similar to those found in the first experiment. However, the rmse quickly increases to more than 3.0, as the changes in the streaming of data occur. This shows, in comparison with the first experiment, how the ensemble quickly becomes outdated once the data starts to change.

4 Conclusions and Future Work

Machine Learning problems are becoming increasingly complex as the data in which it is based also becomes more complex. Namely, we nowadays need to deal not only with very large datasets, but also with a streaming scenario in which data arrives continuously rather than in a single batch.

The complex of streaming is not only the technological one of receiving and storing the data. It also implies that data and its statistical properties may change. The main consequence of this, from a Machine Learning perspective, is that models may quickly become outdated as they were trained on data that is potentially different from the current one.

In this paper we presented an approach for dealing with these challenges. Specifically, we use a ensemble composed of weak learners which are adapted throughout the time, as the data arrives. Not only that, but the weights of each weak learner in the ensemble are also optimized so that better models have a larger weight, and worse models eventually are replaced. Specifically, we show how the error measure (rmse) evolves during the stream when the proposed approach is used or not, with significant differences.

We are now working on only training a new model when it is necessary. Indeed, while in this dataset concept drift is linear and constant, that is not always the case, and there may be long periods in which there is no concept drift on the data. In such periods, the computational effort of training new models and evolving their weights would hold few improvements. Thus, the goal now is to monitor performance metrics and detect significant changes, and only update the ensemble in those moments, in order to minimize the necessary computational effort.

Acknowledgements This work was supported by the Northern Regional Operational Program, Portugal 2020 and European Union, trough European Regional Development Fund (ERDF) in the scope of project number 39900—31/SI/2017, and by FCT—Fundação para a Ciência e Tecnologia within projects UIDB/04728/2020 and UID/CEC/00319/2019.

References

1. Brazdil, P., Carrier, C.G., Soares, C., Vilalta, R.: Metalearning: applications to data mining. Springer Science & Business Media (2008)
2. Breiman, L.: Bagging predictors. Mach. Learn. **24**(2), 123–140 (1996)
3. Gama, J., Zliobaite, I., Bifet, A., Pechenizkiy, M., Bouchachia, A.: A survey on concept drift adaptation. ACM Comput. Surv. (CSUR) **46**(4), 1–37 (2014)
4. Haupt, R.L., Ellen Haupt, S.: Practical genetic algorithms (2004)
5. Klinkenberg, R.: Learning drifting concepts: example selection versus example weighting. Intell. Data Anal. **8**(3), 281–300 (2004)
6. Minku, L.L., Yao, X.: DDD: a new ensemble approach for dealing with concept drift. IEEE Trans. Knowl. Data Eng. **24**(4), 619–633 (2011)
7. Pfahringer, B., Bensusan, H., Giraud-Carrier, C.G.: Meta-learning by landmarking various learning algorithms. In: ICML, pp. 743–750 (2000)
8. Romera-Paredes, B., Aung, H., Bianchi-Berthouze, N., Pontil, M.: Multilinear multitask learning. In: International Conference on Machine Learning, pp. 1444–1452 (2013)
9. Schapire, R.E., Freund, Y.: Boosting: foundations and algorithms. Kybernetes (2013)
10. Sivanandam, S., Deepa, S.: Genetic algorithms. In: Introduction to Genetic Algorithms, pp. 15–37. Springer, Berlin (2008)
11. Waghmare, V., Mukhopadhyay, D.: Mobile agent based market basket analysis on cloud. In: 2014 International Conference on Information Technology, pp. 305–310. IEEE (2014)
12. Widmer, G., Kubat, M.: Learning in the presence of concept drift and hidden contexts. Mach. Learn. **23**(1), 69–101 (1996)

Applying of Machine Learning for Analyzing Network Traffic in the Conditions of an Unbalanced Data Sample

Babyr Rzayev and Ilya Lebedev

Abstract The article provides a solution to the problem of identifying anomalous situations in information and telecommunication systems, based on artificial intelligence methods. The presented method for identifying an anomalous situation is based on processing the received tuples of network traffic packet values using various classification models. The proposed solution improves the identification accuracy and makes it possible to use classification algorithms optimized for different types of events and anomalies, trained on various training samples, presented in the form of tuples of network packet parameters. The difference between the algorithms is achieved by introducing an imbalance in the training samples. The paper describes the experiment using Naïve Bayes, Hoeffding Tree, J48, Random Forest, Random Tree, REP Tree machine learning classification algorithms, and the Multilayer Perceptron neural network. The method can be applied in information security monitoring systems when analyzing network traffic. A feature of the proposed solution is the possibility of its scaling and combination by adding new algorithms for classification of machine learning.

Keywords Anomaly detection · Network traffic · Information security

1 Introduction

The functioning of information and telecommunication systems (ITS) requires continuous monitoring of the emergence of various failures and collisions related to a network traffic processing. The development of the Industrial Internet and the

B. Rzayev (✉)
S. Seifullin Kazakh Agrotechnical University, 010000 Nur-Sultan, Republic of Kazakhstan

I. Lebedev
St. Petersburg Federal Research Center of the Russian Academy of Sciences, 199178 St. Petersburg, Russian Federation
e-mail: isl_box@mail.ru

D. Camacho et al. (eds.), *Intelligent Distributed Computing XIV*, Studies in Computational Intelligence 1026, https://doi.org/10.1007/978-3-030-96627-0_7

Industrial Internet of Things (IoT) concepts causes the need to assess the performance, functional safety of individual network devices and network segments formed by them. The status analysis is carried out using various monitors processing internal and external information containing statistical data of network packets and their processing indicators. As a result, multidimensional time series and tuples of values appear that contain a time-variable parameters reflecting the functioning of the system. The growth and the constant accumulation of network traffic information in various segments of Industrial networks and IoT necessitates the use of artificial intelligence methods for processing time series and tuples of values in order to detect anomalies.

Depending on the tasks of analyzing the status of information security of communication channels and the ITS itself, a number of directions based on classification, clustering and forecasting can be distinguished [1–4].

The implementation of such approaches causes certain difficulties. During the operation of the IoT devices, conflicts may occur both at the level of the information system and a separate network segment or device that influence the processes of receiving and transmitting information, increase the loading of communication channels, reduce the performance and processing speed of commands and messages [5–7]. Detection and preventing such situations requires improving models, methods for monitoring the state, allowing to analyze causal relationships and transitions. Adaptation of methods of statistical analysis, the formation of precedent, event models makes it possible to forecast to prevent negative consequences [5–10].

The detection of an anomalous situation in a network segment is based on statistical information of traffic in various modes and states and is carried out using neural networks, Markov models, machine learning methods, and others. The formed tuples of features have the form of patterns intended for analyzing the behavior of the device in various modes of operation [11–14].

The main disadvantage of existing approaches is that the systems that implement them may not always be effective in the face of constant changes in configuration and architecture. In this regard, it becomes necessary to develop and adapt methods for analyzing information that are resistant to changing operating conditions, providing a given completeness and accuracy of detecting anomalous situations. The purpose of this work is to improve the quality indicators of event identification in network traffic by using an ensemble of algorithms trained on various unbalanced training samples. The proposed solution is aimed at achieving a variety of classifiers. The resulting effect of the spread of the algorithms' answers is smoothed out by applying the voting procedure.

2 The Proposed Approach

Modern network traffic analysis in most cases is based on machine learning methods. Automatic, without the participation of an expert, feature extraction and versatility are undoubtedly positive qualities of such approaches. However, due to the specifics of the field of information security, attacks on devices, networks and telecommunications used by an attacker each time have a certain "novelty" and uniqueness. On the other hand, the share of secure connections (P2P services, HTTPS), where traffic is compressed and encrypted, is increasing. Consequently, problematic situations may arise associated with the functioning of algorithms, input parameters, the extraction of analyzed features and their interpretation for various kinds of information influences and attacks. In this regard, a method for identifying an anomalous situation in an information and telecommunication system is proposed, based on bagging, which ideally makes it possible to analyze the destructive effects on the ITS by comparing the results of several classifiers.

A formalized description of the construction of an ensemble of classifying algorithms is presented in [11, 15–17].

In general, the problem looks like this. There is a sample where the values of the studied parameters are obtained for various states, which allows us to associate a set of tuples from the set X with each state.

$X_i = (x_1, \ldots, x_n)$ contains a tuple of $n \geq 2$ length values.

The set of states is defined by tuples $\{X_1, X_2, \ldots, X_m\} \in X$ where m is the number of records in the sample, reflecting the behavior of the process in various states. Each state is assigned a label of a binary set of a subset of dangerous C_1 and safe C_2 states.

Thus, there is a labeled final training set. It is necessary for the input tuple of values X_i to construct a classification algorithm $a = \{b_1, b_2, \ldots, b_k\}$, mapping $X \rightarrow C$, where k is the number of basic classifying algorithms in the ensemble.

During the operation of the classification algorithm, an error occurs, which can be smoothed out by a sequence of independently trained basic classifiers $b_q, q = 1, \ldots, k$ [17, 18].

The response of the algorithm $b_q(X_i)$, obtained as a result of the work, allows us to determine the class of subsets C_j, $j = 1, 2$, belonging to the binary set of classes of dangerous and safe states C.

The result of the operation of the algorithms $b_q, q = 1, ..., k$ separately is the probability that the input tuple X_i belongs to the class of states:

$$b_q(X_i) = max_{j=1,2} P_q(C_j|X_i). \quad (1)$$

The resulting class predicted by the ensemble of classifiers for a tuple X_i can be determined based on the values of the functions F_1 and F_2 for the binary subset C_1 and C_2:

$$F_1(b_1(X_i), \ldots, b_k(X_i)) = 1/k \sum_{q=1}^{k} P_q(C_1|X_i), \quad (2)$$

$$F_2(b_1(X_i), \ldots, b_k(X_i)) = 1/k \sum_{q=1}^{k} P_q(C_2|X_i). \quad (3)$$

The implementation of the ensemble of basic algorithms is described by the expression:

$$b(X_i) = f(F_1(b_1(X_i), \ldots, b_k(X_i)), F_2(b_1(X_i), \ldots, b_k(X_i))), \quad (4)$$

where f is the decision rule that allows you to determine the probabilistic estimate and establish the class number.

Thus, the given solution uses a binary classification. The formalization is summarized by a decision rule that translates the grade into a class number. To calculate the results of the ensemble under consideration, an auxiliary set (space of estimates) is used. In order to achieve the difference between the algorithms included in the model, their training occurs independently of each other, both on randomly selected and unbalanced subsets of the training sample. The work examines the quality indicators of the ensemble when an unbalanced sample is used.

3 Experimental Study of an Ensemble of Classifiers

Experimental evaluation of the considered approach was carried out on the NSL-KDD dataset which having 41 attributes and one more attribute for class. The dataset contains more than 30 types of attacks. These attacks can be divided in four different types with some common properties. These categories attacks include Denial of Service (DoS), Probe, Remote to Local (R2L), User to Root (U2R) attacks. Detailed description of the data set is presented in the works [18, 19]. As part of the experiment, a binary classification of the states of a telecommunication system (identification of parasitic and normal traffic) was carried out.

One of the problematic issues was the composition of the ensemble of classifiers. There are only some recommended rules related to the formation of training samples. Most of the research is devoted to the use of "weak" classifying algorithms in the ensemble. In a number of works, in order to ensure conditions aimed at achieving the "difference" of classifiers, it is recommended not to use training subsamples that are resistant to change [20, 21]. The use of exact classifiers in conjunction with relatively weak classifiers [22] is considered, combinations of ensembles, in which there are algorithms unequal in terms of quality indicators, are investigated [22–25]. Approaches using stacking, blending, multilevel stacking are presented [18, 21, 22].

To date, there are no regulated rules for forming the composition of an ensemble, which requires additional research and experiments. The choice of the model is based on the problem, where the composition of the algorithms is determined in advance, and its change will be difficult. In the case of functioning in an autonomous mode, depending on the conditions, arising anomalies, training opportunities and the formation of a training sample, classifiers can be both strong and weak. In this regard, it is necessary to investigate the characteristics in different situations.

The formation of a training sample is an important component for the development of a successful model [18, 26, 27]. There are certain recommendations for the formation of training sets, but in practice their implementation is difficult. It is not always possible to obtain the probability of the appearance of objects and signs equal to the probability of their appearance in the general population. Problematic issues in the formation of training samples are associated with the detection, separation and correct interpretation of background and significant patterns, the absence of training objects of a certain type and elements of the feature system, inaccurate ranges of values of variables, imbalance, the appearance of external patterns associated with the conditions for the formation of a training set [26].

In the experiment, the case of voting was considered, when both strong and weak classifiers participated in the ensemble, the diversity of which was formed by the training sample. The assessment was carried out for classifiers: Naïve Bayes (NB), Hoeffding Tree (HT), J48, Random Forest (RF), Random Tree (RT), REP Tree (REP). The analysis results were obtained using the free Weka software.

The sample was marked up and divided into two parts, one of which was used as training, and the other was used for testing. The data structure vector included more than 40 attribute values, which are described in [19]. The resulting dataset of the network traffic is considered as a sequence of tuples of values (1) of heterogeneous packet, about 50% of which reflects the normal traffic and 50% is the parasitic. The set was divided in proportions 20/80, 40/60, 60/40 and 80/20, where the upper part of the proportion shows the value of the sample for training, and the lower part for testing.

The classification algorithms were trained using the standard Weka settings. The correspondence of the tuples of the tested sample to one of the classes of dangerous or safe states was determined by expressions (2), (3) and (4). The training samples were formed artificially. Initially, it was assumed that the structure is unbalanced; for each classifier, individual events are presented with different frequencies.

Figure 1 shows an example of sample modification, when each classifier was given its own "competence area".

A typical situation was simulated when the classifier can show good results on the training set, but the quality indicators fall when working with test (in the simulated case) or real data, i.e., unaccounted features appear and the training set does not fully reflect the general population.

In the first part of the experiment, classification was carried out using machine learning algorithms - NB, HT, J48, RF, RT, REP.

Assessment of classifiers was carried out on the basis of the area under the ROC-curve (AUC) [26] for the test set (Fig. 2).

The figure shows the sample size of experimental data from the dataset and the quality of algorithms for binary classification. The histogram (figure, b) makes it possible to determine the "weak" and "strong" classifiers that arose after the use of unbalanced training samples as a result of incomplete features, incorrect determination of dependencies, etc.

The classifiers were assessed based on the values of the Accuracy, Precision, Recall, F-Measure parameters.

The obtained experimental values are presented in Fig. 3.

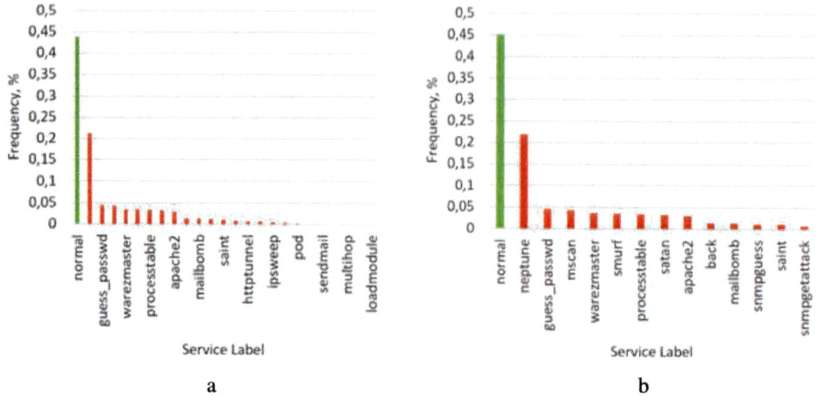

Fig. 1 Sample modification. **a** is a whole sample, **b** is an example of an unbalanced sample for a single classifier

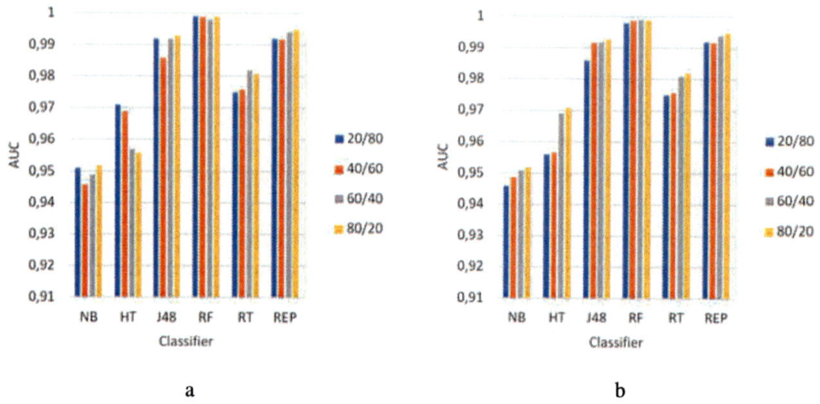

Fig. 2 Assessment of the quality of classifiers by the area under the ROC-curve with balanced (**a**) and unbalanced (**b**) training sample in ratios 20/80, 40/60, 60/40, 80/20

In the second part of the experiment, an ensemble of classifiers is implemented (3). The incoming data were simultaneously processed by all algorithms, and on the basis of expressions (2) and (3), the class of the subset was determined, which was compared with the pre-labeled test set.

For the analysis of the ensemble of classifiers, AUC and an indicator of the overall accuracy of the sequence of classifiers were selected. The results obtained after using the ensemble are shown in Fig. 4a. The third part of the experiment uses the Multilayer Perceptron (MLP) neural network. The results are shown in Fig. 4b.

A comparative table for the AUC parameter for all classifiers is shown in Fig. 5.

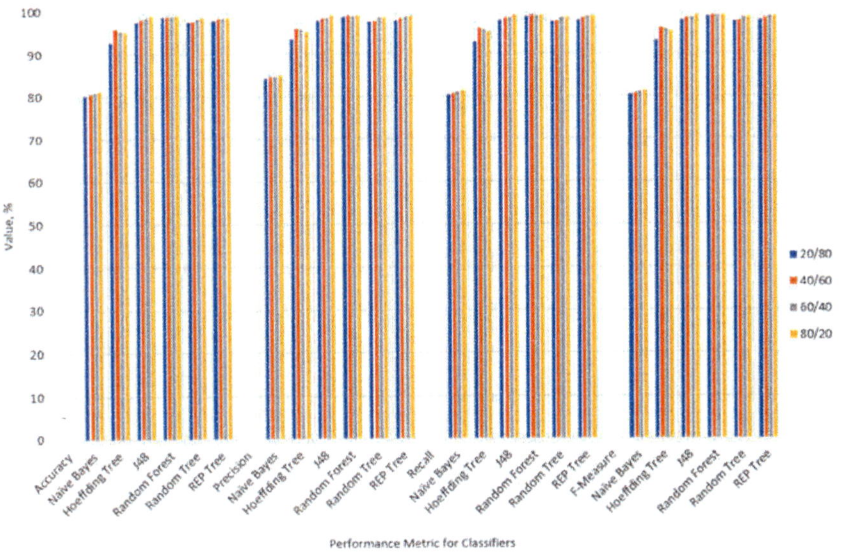

Fig. 3 Accuracy, precision, recall and F-Measure for individual classifiers

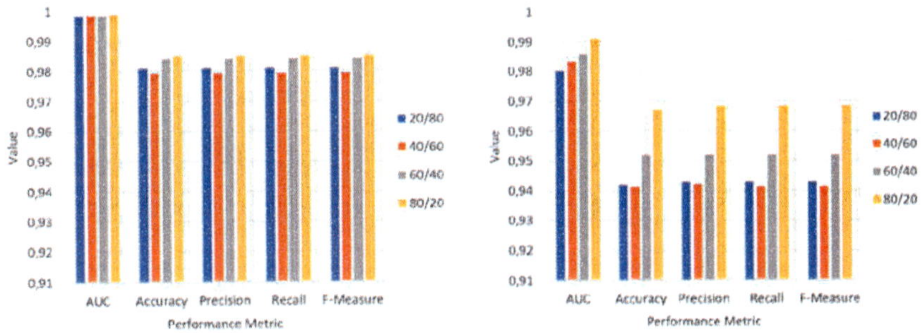

a b

Fig. 4 Results of applying Bagging of Classifiers (**a**) and MLP (**b**) in ratios 20/80, 40/60, 60/40, 80/20

The results of testing the open NSL-KDD dataset with machine learning classifiers implemented in the Weka application, using bagging, showed an accuracy of more than 99%, even on a relatively small training set.

The main advantage of the proposed solution is that a relatively small sample is sufficient to achieve results comparable with other methods [19, 20, 23]. In contrast to traditional approaches to the formation of a training set for algorithms, imbalanced training samples are created, which makes it possible to achieve a "variety" of classifiers [18, 21, 25, 26]. The classification algorithms are adjusted for different types of events and anomalies, and the resulting effect of the scatter of responses is smoothed

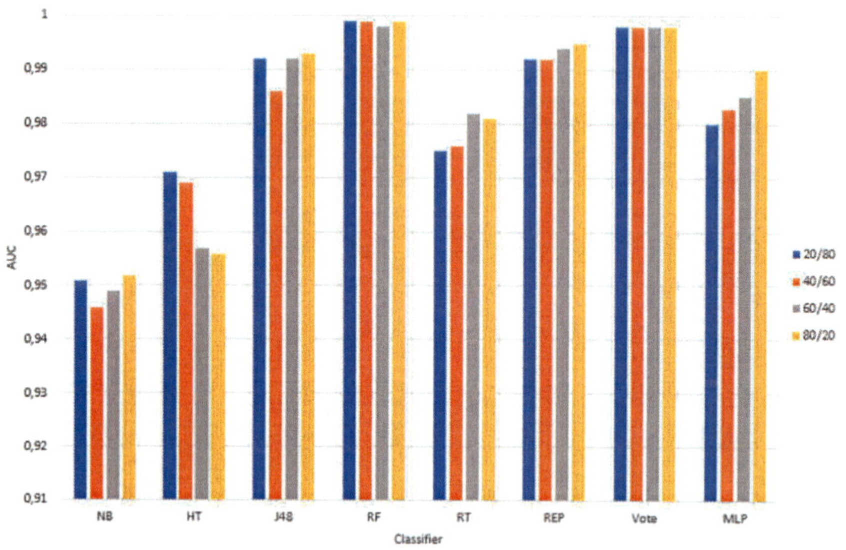

Fig. 5 Assessment of the quality of classifiers by the area under the ROC-curve in ratios 20/80, 40/60, 60/40, 80/20

by an ensemble. Each classifier "specializes" on a certain part of events, which allows it to be adapted to different conditions of network segments functioning.

Among the disadvantages of the proposed approach, it is necessary to note the sensitivity of the classification algorithms to the bias of responses. It is necessary to analyze the data, feature space and classifiers in advance for the possibility of this effect. In the case of its strong influence, the results of the ensemble can be significantly reduced.

4 Conclusion

The growing volume of network traffic necessitates the development of models, methods of its analysis in order to identify destructive influences and anomalous situations. The number of new types of attacks, malicious sites, and methods of introducing unauthorized software is constantly increasing. In this regard, it is necessary to analyze a large number of parameters of network information exchange.

One of the main indicators of monitoring systems is the accuracy of state identification. The paper proposes a method based on bagging classifiers to identify anomalous situations in network traffic. Taking into account the large number of processed indicators, the proposed approach allows obtaining results acceptable in terms of accuracy, smoothing out potential errors of heterogeneous classifiers. The variety of classifiers is achieved by using unbalanced training samples. When using the ensemble, as the volume of the training sample increases, an increase in accuracy is observed.

The method has a limitation associated with the possible manifestation of the effect of displacement of responses by classifying algorithms. In this regard, before use, there is a need for additional research of data and classifiers.

The advantage is the ability to scale and combine it by adding new classifying algorithms, taking into account the parameters of network traffic in various network segments.

References

1. Ahlgren, B., Hidell, M., Ngai, E.: Internet of things for smart cities: interoperability and open data. IEEE Internet Comput. **20**(6), 52–56 (2016). https://doi.org/10.1109/MIC.2016.124
2. Salehi, H., Burgueño, R.: Emerging artificial intelligence methods in structural engineering. Eng. Struct. **171**, 170–189 (2018). https://doi.org/10.1016/j.engstruct.2018.05.084
3. Oikarinen, E., Tiittanen, H., Henelius, A.: Detecting virtual concept drift of regressors without ground truth values. Data Min. Knowl. Disc. **35**(3), 821–859 (2021). https://doi.org/10.1007/s10618-021-00739-7
4. Gokhale, A., McDonals, M.P., Drager, S., McKeever, W.: A cyber physical system perspective on the real time and reliable dissemination of information in intelligent transportation systems. Netw. Protoc. Algorithms **2**(3), 116–136 (2010). https://doi.org/10.5296/npa.v2i3.480
5. Maletzke, A., dos Reis, D., Batista, G.: Combining instance selection and self-training to improve data stream quantification. J. Braz. Comput. Soc. **24**(12), 123–141 (2018). https://doi.org/10.1186/s13173-018-0076-0
6. Kwon, D.W., Ko, K., Vannucci, M., Reddy, A.L.N., Kim, S.: Wavelet methods for the detection of anomalies and their application to network traffic analysis. Qual. Reliab. Eng. Int. **22**(8), 953–969 (2006). https://doi.org/10.1002/qre.781
7. Semenov, V.V., Lebedev, I.S., Sukhoparov, M.E.: Approach to classification of the information security state of elements for cyberphysical systems by applying side electromagnetic radiation. Sci. Tech. J. Inf. Technol. Mech. Opt. **18**(1), 98–105 (2018). (in Russian). https://doi.org/10.17586/2226-1494-2018-18-1-98-105
8. López, V., Fernandez, A., Garcia, S., Palade, V., Herrera, F.: An insight into classification with imbalanced data: empirical results and current trends on using data intrinsic characteristics. Inf. Sci. **250**(7), 113–141 (2013). https://doi.org/10.1016/j.ins.2013.07.007
9. Genkin, D., Shamir, A., Tromer, E.: Acoustic cryptanalysis. J. Cryptol. **30**(2), 392–443 (2017). https://doi.org/10.1007/s00145-015-9224-2
10. Semenov, V.V., Lebedev, I.S., Sukhoparov, M.E., Salakhutdinova, K.I.: Application of an autonomous object behavior model to classify the cybersecurity state. In: Internet of Things, Smart Spaces, and Next Generation Networks and Systems, pp. 104–112 (2019). https://doi.org/10.1007/978-3-030-30859-9_9
11. Palacios, A., Sanchez, L., Couso, I.: Combining Adaboost with preprocessing algorithms for extracting fuzzy rules from low quality data in possibly imbalanced datasets. Int. J. Uncertain. Fuzzines Knowl.-Based Syst. **20**(2), 51–71 (2012). https://doi.org/10.1142/S0218488512400156
12. Sethi, T., Kantardzic, M.: Handling adversarial concept drift in streaming data. Expert. Syst. Appl. **97**, 18–40 (2018). https://doi.org/10.1016/j.eswa.2017.12.022
13. Jin, J., Gubbi, J., Marusic, S., Palaniswami, M.: An information framework for creating a smart city through internet of things. IEEE Internet Things J. **1**(2), 112–121 (2014). https://doi.org/10.1109/JIOT.2013.2296516

14. Sukhoparov, M.E., Semenov, V.V., Salakhutdinova, K.I., Lebedev, I.S.: Identification of anomalous functioning of Industry 4.0 devices based on behavioral patterns: information security problems. Comput. Syst. (1), 96–102 (2020). (in Russian)
15. Semenov, V., Lebedev, I., Sukhoparov, M.: Identification of the state of individual elements of cyber-physical systems based on external behavioral characteristics. J. Appl. Inform. **13**(5, 77), 72–83 (2018). (in Russian)
16. Sukhoparov, M.E., Lebedev, I.S.: Identification the information security status for the internet of things devices in information and telecommunication systems. Syst. Control, Commun. Secur. (3), 252–268 (2020). (in Russian). https://doi.org/10.24411/2410-9916-2020-10310
17. Rzayev, B.T., Lebedev, I.S.: Applying bagging in finding network traffic anomalies. Sci. Tech. J. Inf. Technol., Mech. Opt. **21**(2), 234–240 (2021). (in Russian). 10.17586/2226–1494-2021-21-2-234-240
18. Ingre, B., Yadav, A.: Performance Analysis of NSL-KDD dataset using ANN. In: Proceedings of the 4th International Conference on Signal Processing and Communication Engineering Systems (SPACES), pp. 92–96 (2015). 10.1109/SPACES.2015.7058223
19. Dhanabal, L., Shantharajah, Dr. S.P.: A study on NSL-KDD dataset for intrusion detection system based on classification algorithms. Int. J. Adv. Res. Comput. Commun. Eng. **4**(6), 446–452 (2015). https://doi.org/10.17148/IJARCCE.2015.4696
20. Vorontcov, K.V.: Lectures on algorithmic compositions. http://www.machinelearning.ru/wiki/images/0/0d/Voron-MLCompositions.pdf. Last accessed 11 May 2021
21. D'yakonov, A.G.: Solution methods for classification problems with categorical attributes. Comput. Math. Model. **26**(3), 408–428 (2015). https://doi.org/10.1007/s10598-015-9281-2
22. Zhou, Z.-H.: Ensemble Methods: Foundations and Algorithms, p. 222. CRC Press, New York (2012)
23. Zhou, Z.-H., Feng J.: Deep forest. Natl. Sci. Rev. **6**(1), 74–86 (2019). arXiv:170208835v4. Last accessed 20 May 2021
24. Khan, S., Yairi, T.: A review on the application of deep learning in system health management. Mech. Syst. Signal Process. **107**(1), 241–265 (2018). https://doi.org/10.1016/j.ymssp.2017.11.024
25. Pedersen, T.: A simple approach to building ensembles of Naive Bayesian classifiers for word sense disambiguation. In: NAACL 2000: Proceedings of the 1st North American chapter of the Association for Computational Linguistics Conference, pp. 63–69 (2000)
26. Kaftannikov, I.L., Parasich, A.V.: Problems of training set's formation in machine learning tasks. Bulletin of the South Ural State University. Series Computer Technology, Aotimatic Control, Radio Electronics **16**(3), 15–24 (2016). (in Russian). https://doi.org/10.14529/ctcr160302
27. Fawcett, T.: An introduction to ROC analysis. Pattern Recognit. Lett. **27**(8), 861–874 (2006). https://doi.org/10.1016/j.patrec.2005.10.010

A WNN-Based Approach for Network Intrusion Detection

Mario Di Mauro, Giovanni Galatro, and Antonio Liotta

Abstract Nowadays, Intrusion Detection Systems (IDSs) are becoming more and more effective since they can benefit from the flexibility offered by Machine Learning (ML) techniques. In this work we investigate the potentiality of the Weightless Neural Networks (WNNs) as a classification method of network attacks. Traditionally, WNNs have been exploited in the image classification field and are implemented through the *WiSARD* algorithm. Interestingly, our analysis reveals that, applied to the IDS realm, WNNs offer surprising results in terms of performance/time complexity trade-off with respect to other ML-based techniques. The experimental assessment is carried on by considering one of the most updated datasets (CIC-IDS) in the field of the intrusion detection, where two exemplary attacks to be detected are considered: Distributed Denial of Service (DDoS) and PortScan.

1 Introduction

The usage of Intrusion Detection Systems (IDSs) is crucial for guaranteeing the security of all classes of networked systems, from big network providers to LANs available in small offices. Today, such tools can also interact with ML-based frameworks to learn statistical features characterizing the traffic flows (e.g. distribution of packet length, mean flow duration, average round-trip time, etc.) with the aim of classifying them according to specific criteria. This appears to be essential in revealing malicious traffic since the attacker has the possibility of hiding the *deterministic* features (e.g.

M. Di Mauro (✉)
University of Salerno, Fisciano, Italy
e-mail: mdimauro@unisa.it

G. Galatro
Amazon AWS, Seattle, Ireland

A. Liotta
Free University of Bozen-Bolzano, Bolzano, Italy
e-mail: Antonio.Liotta@unibz.it

D. Camacho et al. (eds.), *Intelligent Distributed Computing XIV*, Studies in Computational Intelligence 1026, https://doi.org/10.1007/978-3-030-96627-0_8

IP address, port number, protocol type, etc.) but not the statistical ones [1–7]. One of the most critical problems in applying ML into IDS domain is the huge variety of techniques which exhibit different performance both in terms of classification accuracy and in terms of time complexity. In this work we propose an intrusion detection approach based on the Weightless Neural Networks (WNNs) [8], a particular class of artificial neural networks (ANNs) whose functioning is inspired to RAM memories. In particular we will focus on WiSARD, the most representative algorithm belonging to the WNNs family, which has been traditionally applied to the image classification domain, but that offers very interesting outcomes in the realm of network intrusion detection. The work is organized as follows: Sect. 2 offers an overview of the affine literature where main differences with our work are highlighted; Sect. 3 presents the classic architecture of a WNN and some peculiarities of the WiSARD algorithm; Sect. 4 is aimed at describing the CIC-IDS dataset that we have exploited to perform the experiments; Sect. 5 is entirely devoted at an experimental assessment where a comparative analysis among many neural-based techniques (including Deep-based approaches) and the WNNs is proposed; finally, Sect. 6 draws some conclusions.

2 Related Research

The scientific literature exploring the application of ML techniques to the detect anomalies in various domains (including the IDS realm) is quite rich [9–11]. This notwithstanding, there are some differences with the approach that we adopt here.

The first difference pertains to the performance analysis. Many works are limited to the classic evaluation of ML techniques in terms of accuracy classification but neglect a time complexity analysis which is crucial when revealing fast attacks such as DDoS which typically make damages in the order of few tens of minutes. This is the case of the following works: [12] (compared methods: J48, ANN, SVM, and others), [13] (compared methods: Naïve Bayes, Bagging, J48, and others), [14] (compared methods: ANN, trees, and others). In contrast, we present a detailed analysis of both performance and time complexity of the investigated techniques so as to highlight the trade-off between such two metrics.

A second difference concerns the usage of specific classification datasets. For instance, one of the most used datasets is the so-called KDD99 [15] or its updated version NSL-KDD [16]. Unluckily, such two datasets contain attacks signatures no longer available today (e.g. satan probing, smurf attacks, warezmaster, etc.) being outdated. Among recent works exploiting such datasets we find: [17–24] where is demonstrated the effectiveness of exploiting IDSs in conjunction with ANNs to improve the quality of traffic detection. Conversely, we adopt the CIC-IDS dataset [25] which contains signatures of modern attacks such as: DDoS, PortScan, TOR, Android-based malwares, and many others.

Finally, it is worth noticing that many comparative studies (including big surveys) on the ML and IDS just provide pointers to other works but do not offer any experimental analysis [26–30]. On the other hand, all the experiments (both for performance and time analysis) are conducted in this proposal on a real computer architecture (Intel CoreTM i5-7200U 2.50 GHz, 16 GB of RAM).

3 WiSARD: A WNN-Based Implementation

The WiSARD (**Wi**lkes, **S**tonham and **A**leksander **R**ecognition **D**evice) algorithm [31, 32] has been originally designed to solve image classification problems, but it has recently been proven to be effective in many other fields.

WiSARD is the most known implementations of WNNs, where the functions evaluated by single neurons are not saved in the weights (as occurs for classic ANNs) but in structures called *lookup tables* inspired to the RAM functioning. The basic idea, in fact, is that the input data is transformed into a binary form which results in reduction of time complexity. Figure 1 shows a simplified structure of a WNN *discriminator*, a structure able to learn the binary patterns associated to a specific class. The *RAM neurons* behave as a classic RAM circuit with 2^n memory locations addressed by n address lines (inputs) representative of the neuron connectivity. As occurs for most ML-based techniques also WNNs are characterized by two phases:

Training phase: all the RAM neurons learn the occurrences of n-tuple vectors extracted from a training pattern, and store them in a memory cell. The training phase

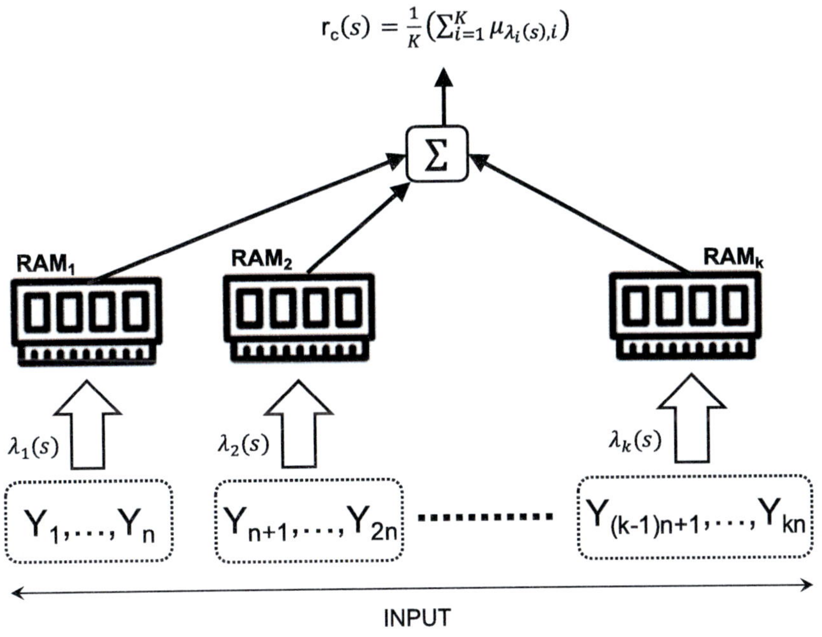

Fig. 1 Basic architecture of a WNN discriminator

is equivalent to the *RAM writing* operation. Specifically, given $\mu_{\lambda,i}$ the memory cell with address λ in the i-th RAM (initially empty), we have the following update rule:

$$\mu_{\lambda,i} = \theta \left(\sum_{j \in \mathcal{J}} \delta_{\lambda,\lambda_i(j)} \right), \tag{1}$$

where: $\theta(x)$ is a function amounting to x if $0 \leq x \leq 1$, and to 1 if $x > 1$; j is a pattern defined in the training set $\mathcal{J}$; $\lambda_i(j)$ is the address generated starting from pattern j; δ is the Kronecker delta function amounting to 1 if $\lambda = \lambda_i(j)$, and to 0 elsewhere.

Classification phase: it means to classify an unseen pattern s by assigning s to a class c, provided that the corresponding discriminator exhibits the highest output, namely

$$\arg\max_{c} \left(\sum_{i=1}^{K} \mu_{\lambda_i(s),i} \right), \tag{2}$$

whereas, the response of the c-th class discriminator on pattern s is:

$$r_c(s) = \frac{1}{K} \sum_{i=1}^{K} \mu_{\lambda_i(s),i}. \tag{3}$$

The classification phase is equivalent to the *RAM reading* operation. Interestingly, since WiSARD is made of a set of discriminators, the overall response of a set of discriminators trained on N classes produces a response vector $\mathbf{r}(s) = [r_{c_1}(s), \ldots, r_{c_N}(s)]$.

4 The CIC-IDS Dataset

In this section we provide an overview on the dataset adopted to carry on the experimental assessment. The CIC-IDS dataset (publicly available at [25]) is a collection of features characterizing many modern network attacks. Such features (up to 80, but they can be further reduced by applying feature selection techniques [33]) can be categorized in:

- **Coarse-grained features**: Protocol Type, Source and Destination Ports, Source and Destination IP Addresses;
- **Flow-based features**: Length/Size of a flow (min, max, mean, etc.);
- **Time-based features**: Forward/Backward inter-arrival times between flows, duration of an idle flow (min, max, mean, std), duration of an active flow (min, max, mean, std), etc.;
- **Byte-based features**: Forward/Backward number of bytes within a flow, Forward/Backward bytes used for headers, etc.;

- **Packet-based features**: Forward/Backward number of packets in a flow, Forward/Backward length/size of packets in a flow (min, max, mean, std, etc.);
- **Flag-based features**: Number of packets with active TCP/IP flags (SYN, FIN, PUSH, RST, URG, etc.).

As regards the attack types, we choose two exemplary families: DDoS and PortScan. A DDoS attack is aimed at impairing the resources of a network target through a huge amount of legitimate (thus, not suspicious) requests [34–36]. In contrast, PortScan is aimed at monitoring potential open ports at the attacked side. In this way, DDoS and PortScan appear to be complementary: the former is characterized by a large number of connections towards a single port of a network target (e.g. the port 80 in case of an HTTP-based DDoS attack); the latter considers a single node (or a few nodes in case of simultaneous Portscans) which opens a huge number of connections towards multiple ports of a certain network target.

5 Experimental Results

In this section we show the results of an experimental analysis aimed at comparing a series of neural based techniques against WNNs represented by the WiSARD algorithm. Each technique is evaluated in terms of performance (Accuracy/F-Measure) and in terms of classification time [37]. We choose two families of neural based techniques to compare against WNN:

- *Deep-based techniques*: two standard deep networks with 2 and 3 hidden layers respectively (Deep-2 and Deep-3), a Convolutional Neural Network (CNN), a Recurrent Neural Network (RNN), a Long-Short Term Memory (LSTM) network;
- *Codebook-based techniques*: 3 variants of Learning Vector Quantization techniques (LVQ1, LVQ2, LVQ3) [38] and Self-Organizing Maps (SOM).

The choice of optimized hyperparameters (see Table 1) for the various aforementioned techniques has been made through a grid search.

The results of performance analysis are reported in Table 2. The first thing that catches the eye is the great performance of all the deep-based techniques in classifying both types of attacks (DDoS and PortScan). Interestingly, there isn't any value (both for Accuracy and F-Measure) lying under 0.99, and this is in line with the power of deep methods which typically perform excellently when many samples are available.

In contrast, the codebook-based techniques seem to suffer in terms of performance as appears from the values in Table 2, where we reported in italic all the values under 0.98. Such a behavior is basically due to the use of the Euclidean distance (a metric frequently used in codebook-based approaches) which does not perfectly capture the properties of the two different attacks.

Table 1 Adopted hyperparameters

Algorithm	Optimized hyperparameters and models info
Deep-2, 3	· LR (learn. rate) = 0.001
	· Number of weights = 2000
	· Exp. Decay (first moment estimate) = 0.9
	· Neurons per hidden layer: Deep-2(23,10); Deep-3(20,16,11)
CNN	· 7 filters (4X1), 8 neurons fully connected
	· 1 Pooling layer with Pooling size = 2
	· 1 Dropout layer dropout rate = 0.3
RNN	· 18 Recurrent units (RNN), 10 neurons fully connected
	· 6 LSTM units, 8 neurons fully connected
	· 8 GRU units, 10 neurons fully connected
WiSARD	· Batch size = 100
	· Resolution (in bit) per neuron: 8
LVQ(1,2,3)	· Batch size = 100
	· LR = 0.3
	· Codebook vectors = 20
	· Window size (for LVQ2 and LVQ3) = 0.3
SOM	· Batch Size = 100
	· LR = 0.3
	· Neighborhood function: Gaussian

Table 2 Performance analysis (Accuracy/F-Measure)

Method	Accuracy	F-Measure	Method	Accuracy	F-Measure
Deep-2	0.9986	0.9987	Deep-2	0.9996	0.9995
Deep-3	0.9988	0.9989	Deep-3	0.9996	0.9995
CNN	0.9921	0.9931	CNN	0.9967	0.9970
RNN	0.9990	0.9991	RNN	0.9915	0.9923
LSTM	0.9980	0.9982	LSTM	0.9983	0.9985
WiSARD	**0.9953**	**0.9959**	WiSARD	***0.9770***	***0.9796***
LVQ1	*0.9248*	*0.9348*	LVQ1	*0.7956*	*0.7069*
LVQ2	*0.8963*	*0.9073*	LVQ2	*0.8996*	*0.8763*
LVQ3	*0.8917*	*0.9025*	LVQ3	*0.8104*	*0.7322*
SOM	*0.9300*	*0.9403*	SOM	*0.8924*	*0.8707*
	DDoS Dataset			PortScan Dataset	

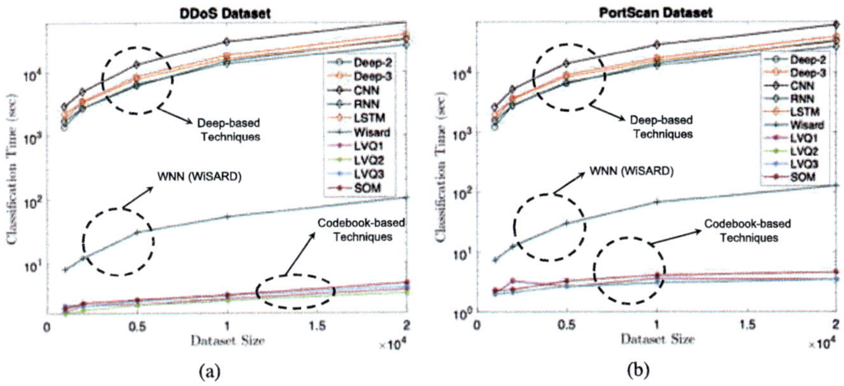

Fig. 2 Classification time of the considered techniques for: **a** DDoS dataset; **b** PortScan dataset

Finally, WiSARD seems to be quite closer to the performance of deep techniques for the DDoS dataset. A bit decrease in performance is observed for the PortScan dataset (Accuracy $= 0.9770$, F-Measure $= 0.9796$) but nothing of particularly dramatic.

Let us discover how such techniques behave in terms of classification times. Figure 2 shows on a semi-log scale the classification time (in seconds) versus the dataset size for all the considered techniques and for the two datasets (Fig. 2a-DDoS case, 2b-PortScan case.). The first thing to notice is that, as expected, as the dataset size grows the classification time increases for all the considered techniques. It is immediate to notice that the deep-based methods are the slower since they rely on a complex structure of fully connected neurons. Conversely, codebook-based techniques are very fast since only a limited subset of neurons (the so-called winner neurons) are enabled to update their weights.

Finally, we see that the WNNs are reasonably fast (more close to the classification times of codebook-based techniques than the deep-based methods). Such a behavior is basically due to the RAM-like structure of a discriminator which can operate directly on a binarized version of input data, so as to boost the whole process.

In conclusion, we can safely state that the WNNs exhibit the best performance/time complexity trade-off over the other competing techniques. Probably, one of the most interesting result is that WNNs have hardly ever been considered in the domain of data traffic classification, whilst they seem to be a very interesting candidate in the online intrusion detection, where the real-time requirement is crucial.

6 Conclusion

This work proposes to exploit the Weightless Neural Networks (WNNs) as a method to classify and reveal anomalous (in particular malign) traffic flows. Surprisingly, such a technique has been rarely considered in the intrusion detection realm, being almost always adopted in the field of the image classification. Our experimental evaluation relies on modern datasets (CIC-IDS) which, differently from the outdated KDD99, contains signatures of modern attacks (e.g. DDoS, PortScan, etc.). Moreover, our analysis takes into account both performance (in terms of Accuracy and F-Measure) and time complexity which is often neglected in most literature. Some interesting outcomes emerge from our assessment. For instance, deep-based techniques show outstanding performance in each experiment (Accuracy and F-Measure never lies under 0.99) but exhibit high classification times basically due to the slow convergence of the underlying back-propagation algorithm. In contrast, the prospect is reversed for the codebook-based techniques where the scarce performance is counterbalanced by very challenging classification times. Finally, the WNNs seem to embodying benefits of both deep-based and codebook-based algorithms, by exhibiting a good performance (Accuracy and F-Measure never lying under 0.97) and interesting classification times (between 10 and 100 s). For this reason, future works may be focused on a more fine tuning of WNNs so as to make them particularly apt to support the online learning classification.

References

1. Di Mauro, M., Di Sarno, C.: Improving SIEM capabilities through an enhanced probe for encrypted Skype traffic detection. J. Inf. Secur. Appl. **38**, 85–95 (2018)
2. Di Mauro, M., Di Sarno, C.: A framework for internet data real-time processing: a machine-learning approach. In: 2014 International Carnahan Conference on Security Technology (ICCST), pp. 1–6 (2014)
3. Di Mauro, M., Longo, M.: Revealing encrypted WebRTC traffic via machine learning tools. In: 2015 12th International Joint Conference on e-Business and Telecommunications (ICETE), vol. 04, pp. 259–266 (2015)
4. Di Mauro, M., Longo, M.: Skype traffic detection: a decision theory based tool. In: 2014 International Carnahan Conference on Security Technology (ICCST), pp. 1–6 (2014)
5. Addesso, P., Cirillo, M., Di Mauro, M., Matta, V.: ADVoIP: adversarial detection of encrypted and concealed VoIP. IEEE Trans. Inf. Forensics Secur. **15**, 943–958 (2020)
6. Matta, V., Di Mauro, M., Longo, M., Farina, A.: Cyber-threat mitigation exploiting the birthâĂŞdeathâĂŞimmigration model. IEEE Trans. Inf. Forensics Secur. **13**(12), 3137–3152 (2018)
7. Addesso, P., Barni, M., Di Mauro, M., Matta, V.: Adversarial kendallâĂŹs model towards containment of distributed cyber-threats. IEEE Trans. Inf. Forensics Secur. **16**, 3604–3619 (2021)
8. Aleksander, I., Morton, H.: Introduction to Neural Computing. Chapman and Hall, London (1990)
9. Cauteruccio, F., Fortino, G., Guerrieri, A., Liotta, A., Mocanu, D.C., Perra, C., Terracina, G., Vega, M.T.: Short-long term anomaly detection in wireless sensor networks based on machine learning and multi-parameterized edit distance. Inf. Fusion **52**, 13–30 (2019)

10. Erhan, L., Ndubuaku, M., Di Mauro, M., Song, W., Chen, M., Fortino, G., Bagdasar, O., Liotta, A.: Smart anomaly detection in sensor systems: a multi-perspective review. Inf. Fusion **67**, 64–79 (2021)
11. Cauteruccio, F., Cinelli, L., Corradini, E., Terracina, G., Ursino, D., Virgili, L., Savaglio, C., Liotta, A., Fortino, G.: A framework for anomaly detection and classification in multiple IoT scenarios. Futur. Gener. Comput. Syst. **114**, 322–335 (2021)
12. Yin, C., Zhu, Y., Fei, J., He, X.: A deep learning approach for intrusion detection using recurrent neural networks. IEEE Access **5**, 21954–21961 (2017)
13. Ravipati, R., Munther, A.: A survey on different machine learning algorithms and weak classifiers based on KDD and NSL-KDD datasets. Int. J. Artif. Intell. Appl. **10**, 01–11 (2019)
14. Azwar, H., Murtaz, M., Siddique, M., Rehman, S.: Intrusion detection in secure network for cybersecurity systems using machine learning and data mining. In: 2018 IEEE 5th International Conference on Engineering Technologies and Applied Sciences (ICETAS), pp. 1–9 (2018)
15. Kdd cup 1999 data.: http://kdd.ics.uci.edu/databases/kddcup99/kddcup99.html. Last accessed: 10 Sept 2021
16. Tavallaee, M., Bagheri, E., Lu, W., Ghorbani, A.A.: A detailed analysis of the KDD CUP 99 data set. In: 2009 IEEE Symposium on Computational Intelligence for Security and Defense Applications, pp. 1–6 (2009)
17. Khan, R.U., Zhang, X., Alazab, M., Kumar, R.: An improved convolutional neural network model for intrusion detection in networks. In: 2019 Cybersecurity and Cyberforensics Conference, pp. 74–77 (2019)
18. S. T. F. Al-Janabi and H. A. Saeed. A neural network based anomaly intrusion detection system. In *Developments in E-systems Engineering*, pages 221–226, 2011
19. Taher, K.A., Jisan, B.M.Y., Rahman, M.M.: Network intrusion detection using supervised machine learning technique with feature selection. In: 2019 International Conference on Robotics,Electrical and Signal Processing Techniques (ICREST), pp. 643–646 (2019)
20. Papamartzivanos, D., Marmol, F.G., Kambourakis, G.: Introducing deep learning self-adaptive misuse network intrusion detection systems. IEEE Access **7**, 13546–13560 (2019)
21. Fernando, Z.T., Thaseen, I.S., Kumar, C.A.: Network attacks identification using consistency based feature selection and self organizing maps. In: First International Conference on Networks Soft Computing, pp. 162–166 (2014)
22. McElwee, S., Cannady, J.: Improving the performance of self-organizing maps for intrusion detection. In: SoutheastCon 2016, pp. 1–6 (2016)
23. Li-ying, C., Xiao-xian, Z., He, L., Gui-fen, C.: A network intrusion detection method based on combined model. In: International Conference on Mechatronic Science, Electric Engineering and Computer, pp. 254–257 (2011)
24. Al-Sultani, Z.N., Naoum, R.S.: Learning vector quantization (LVQ) and k-nearest neighbor for intrusion classification. World Comput. Sci. Inf. Technol. J. **2**(3), 105–109 (2012)
25. The CSE-CIC-IDS2018 Dataset.: http://netflowmeter.cal. Last accessed: 10 Sept 2021
26. Nguyen, T.T.T., Armitage, G.: A survey of techniques for internet traffic classification using machine learning. IEEE Commun. Surv. Tutor. **10**(4), 56–76 (2008)
27. Boutaba, R., Salahuddin, M.A., Limam, N., Ayoubi, S., Shahriar, N., Estrada-Solano, F., Caicedo, O.M.: A comprehensive survey on machine learning for networking: evolution, applications and research opportunities. J. Internet Serv. Appl. **9**(1), 16 (2018)
28. Hindy, H., Brosset, D., Bayne, E., Seeam, A., Tachtatzis, C., Atkinson, R., Bellekens, X.: A taxonomy and survey of intrusion detection system design techniques, network threats and datasets. CoRR (2018). arXiv:abs/1806.03517
29. Khraisat, A., Gondal, I., Vamplew, P., Kamruzzaman, J.: Survey of intrusion detection systems: techniques, datasets and challenges. Cybersecurity **2**(1), 20 (2019)
30. Aldweesh, A., Derhab, A., Emam, A.: Deep learning approaches for anomaly-based intrusion detection systems: a survey, taxonomy, and open issues. Knowl.-Based Syst. **189**, 105–124 (2020)
31. Bowden, P., Alexander, I., Thomas, W.: Learning deep architectures for AI. Sens. Rev. **4**(3), 120–124 (1984)

32. De Gregorio, M., Giordano, M.: An experimental evaluation of weightless neural networks for multi-class classification. Appl. Soft Comput. **72**, 338–354 (2018)
33. Di Mauro, M., Galatro, G., Fortino, G., Liotta, A.: Supervised feature selection techniques in network intrusion detection: a critical review. Eng. Appl. Artif. Intell. **101** (2021)
34. Matta, V., Di Mauro, M., Longo, M.: Botnet identification in randomized DDoS attacks. In: Proceedings of the 24th European Signal Processing Conference, pp. 2260–2264 (2016)
35. Cirillo, M., Di Mauro, M., Matta, V., Tambasco, M.: Botnet identification in DDoS attacks with multiple emulation dictionaries. IEEE Trans. Inf. Forensics Secur. **16**, 3554–3569 (2021)
36. Cirillo, M., Di Mauro, M., Matta, V., Tambasco, M.: Application-layer DDoS attacks with multiple emulation dictionaries. In: IEEE ICASSP, pp. 2610–2614 (2021)
37. Di Mauro, M., Galatro, G., Liotta, A.: Experimental review of neural-based approaches for network intrusion management. IEEE Trans. Netw. Serv. Manage. **17**(4), 2480–2495 (2020)
38. Kohonen, T.: Self-Organizing Maps, 3rd edn. Springer, Berlin, Heidelberg (2001)

A Force Field Reinforcement Learning Approach for the Observation Problem

Jamy Chahal, Amal El Fallah Seghrouchni, and Assia Belbachir

Abstract This paper studies the framework of a multi-agent system for the surveillance of zones. In such systems, patrolling agents are deployed to observe mobile targets. The observation problem consists in maximizing the number of viewed mobile targets by at least one agent of the mas. However, formal methods relying on *potential field* to solve this problem, such as the *A-CMOMMT*, that implement observation strategies, do not adapt them to the target's behavior. In this article, we propose a trained method using a reinforcement learning (RL) approach to cope with naive and evasive targets in order to improve the observation of mobile targets, while protecting the patrolling agents from collisions. This paper compares our force field reinforcement learning (FFRL) method with some significant formal observation methods on various scenarios. It shows that FFRL has a better target's observation than the studied methods for both naive and evasive targets.

1 Introduction

Patrolling problem consists in visiting an area as frequently as possible in order to seek targets. A team of agents is used to optimize a performance criterion generally based on the notion of idleness. The latter represents the time difference between two visits of a same location by at least one agent. The concept of idleness was formalized by the authors in [9] in the context of a graph representation, where each location is represented by a node.

In this paper we are interested in observation problems, where it consists in positioning agents to maximize the number of viewed targets by at least one agent. In that

J. Chahal (✉) · A. E. F. Seghrouchni · A. Belbachir
Sorbonne Université, LIP6—UMR 7606 CNRS; Institut Polytechnique des Sciences Avancées—IPSA, Department of Aerospace System, Ivry-Sur-Seine, Paris, France
e-mail: jamychahal@hotmail.com

A. E. F. Seghrouchni
e-mail: amal.elfallah@lip6.fr

A. Belbachir
e-mail: assia.belbachir@ipsa.fr

D. Camacho et al. (eds.), *Intelligent Distributed Computing XIV*, Studies in Computational Intelligence 1026, https://doi.org/10.1007/978-3-030-96627-0_9

sense, an approach called Cooperative Multi-Robot Observation of Multiple Moving Targets (CMOMMT) was defined in [10]. Teams of robots and targets are positioned in an enclosed region, where a solution called A-CMOMMT was proposed to maximize the number of viewed targets. Each patrolling agent is attracted by targets on the basis of local force vector and repulsed by other agents within a protection range. Along with the CMOMMT formalism, the authors [10] proposed to define the random target's behavior and the evasive target's behavior, where targets move in the opposite direction of the patrolling agent. The authors in [5] spotted a problematic situation of the A-CMOMMT, where agents can lose the observation of multiple targets if these targets are separating in a symmetrical way. To overcome this situation, which may appear more often with evasive targets, the authors have designed a method called Behavior CMOMMT (B-CMOMMT). The agent can either follow a target, explore, but also ask for help in case of a detected potential loss of target tracking. However, the B-CMOMMT assumes the possibility to identify each target by a unique tag. Besides the observation problem, we have considered also the patrolling problem for new targets search. This method, called Idleness CMOMMT (I-CMOMMT [3]), shows a compromise between patrolling and observation, considering only random targets. In this paper, we propose an improvement of I-CMOMMT [3] that introduces reinforcement learning technique based on force fields called Force Field Reinforcement Learning (FFRL). In our approach, each agent develops its own strategy to improve the observation of the targets when facing random or evasive target's behavior. In this paper, we show that FFRL improves the results of the observation problem thanks to the combination of force field and reinforcement learning. This paper is organized as follows. Section 2 states the related work on observation and reinforcement learning methods. Section 3 describes our FFRL approach for observation problem. Section 4 presents the experiments. Finally, Sect. 5 concludes our work and presents some perspectives.

2 Overview of Reinforcement Learning Method Applied to the Observation Problem

Formal observation methods propose explicit strategies such as A-CMOMMT. However, these methods do not adapt their strategies according to the target's behavior, which may lead to a loss of effectiveness. Reinforcement learning methods allows agents to develop their own strategy. Learning methods are guided by rewards, where the agent performs several experiments to assess its behavior and attempts to improve it. As highlighted in [2], the collaboration between agents in the context of multi-agent reinforcement learning (MARL) is still a challenge which also applies to the observation problem.

A distributed lazy Q-learning has been proposed [14] to consider the local and incomplete environment's observability inherent to the observation problem. The cooperation between agents is based on a shared reward to the whole agent's group according to the number of observed targets. Although the effectiveness of the proposed method is better than the random policy, it does not exceed the A-CMOMMT.

In [4] the authors implement the Evolutionary Nearest Neighbor Classifier -Q-learning (ENNC-QL) algorithm to improve the agents coordination. The authors have shown that the collaborative approach performs better than the individual approach and overcomes the distributed lazy Q-learning.

Deep reinforcement learning has been evaluated by Yan et al. [15] to process input images from UAV's sensors. This method is based on the Cooperative Multi-UAV Observation of Multiple Moving Targets (CMUOMMT) formalism [6], where agents and targets move on a discrete map. Thus, each agent observes the targets within the surrounding cells. The agents interact with each other by showing their position on one of the input images. The results show a similar efficiency than the A-CMOMMT.

In contrast with the continuous action space used by Touzet [14], defined by a rotation and forward motion, the methods of Fernandez et al. [4] and Yan et al. [15] use a discrete action space which are the 8 cardinal directions. This discrete actions can undermine a real life application.

However, to our knowledge, none of the previous RL methods deals with evasive target behavior. In this paper, we focus our work on a continuous action within a continuous environment representation, with both random and evasive targets. In the next sections, we explain the Force Field Reinforcement learning approach (FFRL) that is implemented to improve our developed approach I-CMOMMT.

3 A Force Field Reinforcement Learning Approach (FFRL)

This section develops the FFRL method, inspired by the force field of the A-CMOMMT [10], to solve the observation problem. We explain the studied force field framework and the environment definition with the learning approach.

3.1 Force Field Framework

In this FFRL method, an agent a_i, at a time t, undergoes a desired force from its policy F^{π} and a protection force F^{P}:

$$F(a_i, t) = F^{\pi}(a_i, t) + F^{P}(a_i, t) \tag{1}$$

The protection force prevents the agent to collide with another agent and from leaving the environment limit. The agent undergoes a repulsion force f^{r} from all other agents k within its safety range and a repulsion force f^{b} when it gets closer to the environment boundaries:

$$F^P(a_i, t) = f^b + \sum_{k=1}^{m} f^r_{ik} \tag{2}$$

With m the number of agents. The repulsion forces act on the agent when the distance with another agent, or with the map's boundaries, is lower than a define safety range (SR). The magnitude of f^b and f^r_{ik} increases linearly until reaching the maximum value of 1 when the distance is equal or below a defined dangerous range (DR). Thus, in addition to the learning, the protection forces compel the agent to not take critical action that could lead to a dangerous behaviour. This FFRL enables safer experimentation with real multi-robot, and especially multi-UAV system.

3.2 Environment Definition

Solving the observation problem by reinforcement learning requires a dedicated environment, which includes defining the *observation space*, the *action space* and the *reward* of the experimented method. This new environment, called *obs_env*, relies on the multi-agent library PettingZoo [12], which integrates the RL framework of gym [1].

obs_env is based on the *simple_adversary* environment, created by Lowe et al. [8] within the Multi-agent Particle-World Environment (MPE).

obs_env considers a set of m agents and n targets, observing and acting in the environment in sequential manner. Besides, the environment is created according to a set of parameters, such as its size, or its maximum time duration. Agents are considered as homogeneous, sharing the same setting such as their capabilities to see other targets through an observation range, to communicate with each other through a communication range, and to avoid collision through a safety range.

In the reinforcement learning context, at each time step, the agent interacts with the environment as follows (see Fig. 1):

- *An observation or state*, represents the perceived information from the environment by an agent. The observation space structures and describes the type of observation returned by the environment. For instance, an observation can include an inner-state (e.g. a position, a speed) or a feedback from a sensor (e.g. a video feed, or an obstacle detector).
- *An action* describes how an agent can act on the environment. The action space lists explicitly all the actions that an agent can perform. A discrete representation is described by all the possibilities (e.g. go up, down, right, left), while a continuous representation is defined as a set of operational range (e.g. the speed is between 0 and 1).
- *A reward* is the utility that an agent gets after performing an action. Its value can be positive or negative.

The following sections describe how these elements are defined in the *obs_env* environment.

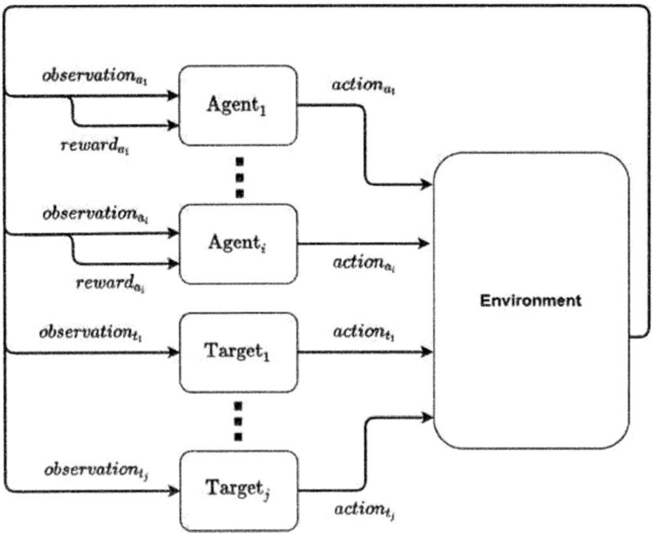

Fig. 1 Environment interaction diagram for the *obs_env* environment

3.2.1 The Observation Space

In *obs_env* environment, we propose to structure for each agent the observation as a vector of eighteen elements, described as follow:

- The euclidean position (x_{self}, y_{self}) of the agent within the environment frame.
- The euclidean relative position $(\delta x_{a_i}, \delta y_{a_i})$ of the four closest agents within the communication range. Extra spaces are replaced by zeros.
- The euclidean relative position $(\delta x_{t_j}, \delta y_{t_j})$ of the four closest targets within the observation range. Extra spaces are replaced by zeros.

For the reinforcement learning method, we propose to normalize the observation space for a better scalability. In this way, the agent's learning would be less disturbed by a change in the environment's parameters. This process is done by dividing each element of the observation vector by its boundaries:

- The position (x_{self}, y_{self}) is divided by the size of the map, where $(x_{self}, y_{self})_n \in [-1, 1]^2$.
- The relative agent's position $(\delta x_{a_i}, \delta y_{a_i})$ is divided by the communication range, where $(\delta x_{a_i}, \delta y_{a_i})_n \in [-1; 1]^2$.
- The relative target's position $(\delta x_{t_j}, \delta y_{t_j})$ is divided by the observation range, where $(\delta x_{t_j}, \delta y_{t_j})_n \in [-1; 1]^2$.

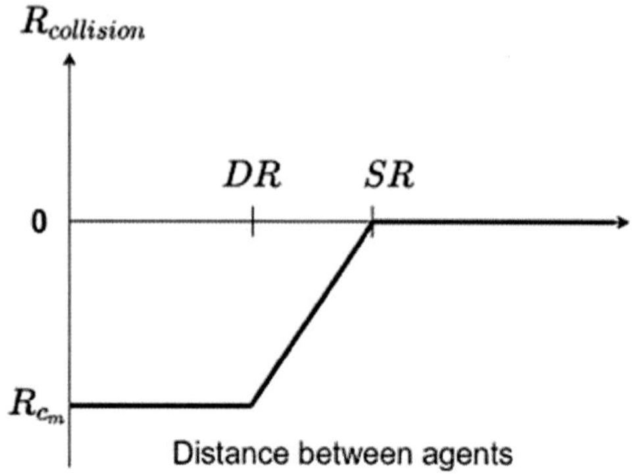

Fig. 2 Illustration of the collision reward

3.2.2 The Action Space

At each step, each agent is able to move in the environment. We propose to define this movement as the two components of the F^{π} force: (1) the desired force along the x axis and (2) the desired force along the y axis. These actions are set as a continuous value between -1 and 1, representing the normalization of the maximum agent speed.

3.2.3 The Rewards

In order to get a high reward, the agent a_i has to avoid collision with agents and maximize the observation of targets. The reward $R(a_i)$ is computed as follow:

$$R(a_i) = R_{obs}(a_i) + R_{collision}(a_i)$$

The collision reward $R_{collision}(agent_i)$ is described in Fig. 2, with SR the safety range, DR the dangerous range and R_{c_m} the maximum negative reward. The reward function is continuous to consider the credit assignment problem and potentially improve the learning.

We defined the observation reward $R_{obs}(agent_i)$ as follow:

Individual reward:

Each agent obtains a reward equal to the number of target within its observation range:

$$R_{obs_{ij}} = \begin{cases} 1 & \text{if target j is observed by agent i} \\ 0 & \text{otherwise} \end{cases}$$

3.2.4 The Targets

We defined in the environment two kinds of targets' behavior:

- Naive behavior: The target randomly selects a desired position and goes straight to reach it.
- Reactive or evasive behavior: The target uses repulsive force field to go in the opposite direction of the surrounding agents. If there is no agent around, the target follows the naive behavior.

4 Training and Simulation Results

In this section, we will explain the training parameters and the defined scenario in order to evaluate our observation method.

4.1 Training Model

The training is based on parameter sharing, where a single agent model is trained to learn one policy for all the homogeneous agents. This provides a policy that can be applied on several scales. As explained in [13], the parameter sharing technique is in contrast with a fully independent learning, where each single agent learns its own policy. During the training session, the initial position of all the agents and targets are randomly set to avoid over-fitting learning. However, the environment's configuration is fixed. Thus the environment's size is $30u \times 30u$, the observation's range is $3u$ and the communication's range is $4u$, with u standing for an abstract distance unit.

Table 1 PPO's parameters. SGD stands for Stochastic gradient descent

Parameters	Value
Number of training iterations	4000
Time horizon per training iteration	1000
Mini-batch size	256
Clipping ratio	0.3
Gamma	0.99
Entropy coefficient	0
Hidden layers	$128 \times 128 \times 128$
Lambda	1.0
Number of SGD iterations	30
Stepsize of SGD	5.10^{-5}

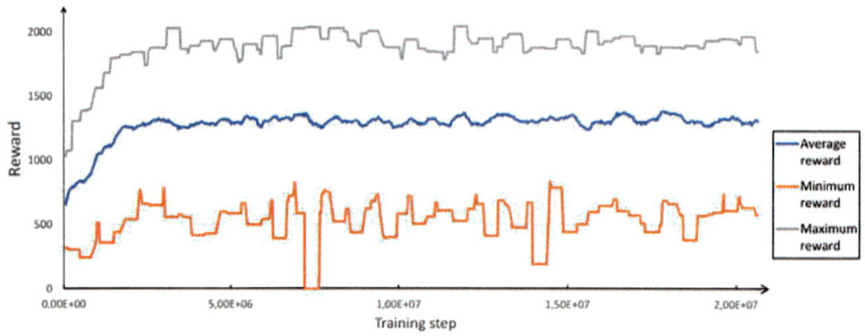

Fig. 3 Evolution of Rewards During the Training

The chosen training algorithm is the Proximal Policy Optimization (PPO) [11], available within the library Rllib [7] with the Actor-Critic model (A3C) implementation. PPO has the advantage to be compatible with either discrete or continuous action spaces and to bounds the policy updates to avoid radical policy change. The PPO's parameters are discribed in Table 1. We can see a policy convergence around 3.10^6 steps in Fig. 3.

Table 2 Simulation's parameters. u stands for unit and ts for time step

Category	Variable name	Value
Environment	Environment's size	$40 \times 40u$
	Experience duration T	3 600 ts
	Observation's range	$5u$
	Communication's range	$6u$
	Target's maximum speed $v_{o_{max}}$	2 u/ts
	Agent's maximum speed $v_{a_{max}}$	2 u/ts
I-CMOMMT	σ	$0.8 \times T$
I-CMOMMT and A-CMOMMT	(do_1, do_2, do_3,predictive tracking range)	$(1u, 2u, 4u, 5u)$
	(dr_1, dr_2)	$(1u, 2u)$
RL method	DR	$1u$
	SR	$2u$
	R_{c_m}	-1

4.2 Results

The simulation's parameters to evaluate the methods are presented in Table 2. Each methods is run 30 times per configuration, defined by the number of agents and the number of targets, to get an average. To evaluate the RL method, the environment size, the observation's range and communication's range have been changed compared to the trained parameter.

The methods are compared with the A metric, defined in [10] as the average of seen target by at least one agent during the whole mission duration. As we can see on the Figs. 4 and 5, for several configurations, the FFRL method shows a better observation average than the other formal methods, which are the A-CMOMMT, the I-CMOMMT and the random policy.

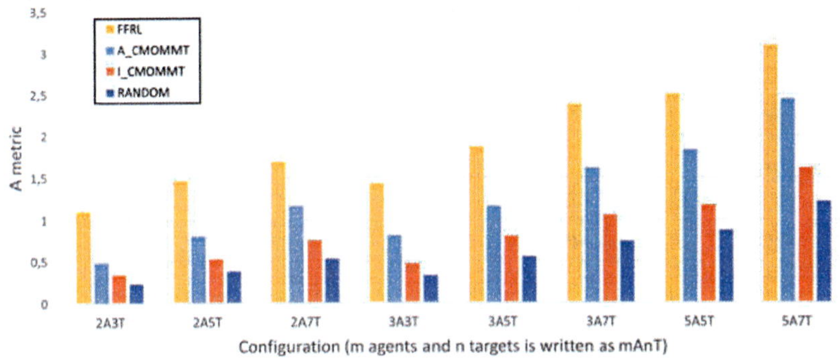

Fig. 4 Result of the A metric for several configurations with naive target's behavior

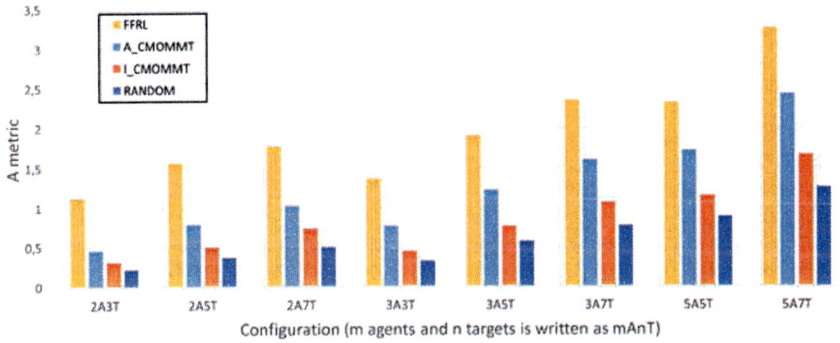

Fig. 5 Result of the A metric for several configurations with evasive target's behavior

5 Conclusion and Future Work

In this paper, we proposed a novel multi-agent reinforcement learning approach for tackling observation problems where the objective is to maximize the number of observed mobile targets by at least one agent of the multi-agent system. We proposed a method (FFRL) relying on the potential field and we trained our observation method using reinforcement learning. Then the paper compared our FFPL method with three existing formal methods namely A-CMOMMT, I-CMOMMT and Random method using an A-metric. To evaluate the observation approaches, we defined two types of targets, (1) naive targets that do not have any strategy (2) evasive targets that have a simple strategy. Using the A-metric, the paper showed that FFRL has a better target's observation than the studied methods for both naive and evasive targets.

As a future work, we are interested to implement a collective reward in the context of multi-agent system, and evaluate the result with comparison to the individual reward. The behavior of the targets will also be refined to introduce more realistic constraints.

References

1. Brockman, G., Cheung, V., Pettersson, L., Schneider, J., Schulman, J., Tang, J., Zaremba, W.: Openai gym (2016)
2. Canese, L., Cardarilli, G.C., Di Nunzio, L., Fazzolari, R., Giardino, D., Re, M., Spanò, S.: Multi-agent reinforcement learning: a review of challenges and applications. Appl. Sci. **11**(11) (2021). https://doi.org/10.3390/app11114948, https://www.mdpi.com/2076-3417/11/11/4948
3. Chahal, J., Belbachir, A., Seghrouchni, A.E.F.: I-CMOMMT: a multiagent approach for patrolling and observation of mobile targets with a continuous environment representation. In: Proceedings of the International Conference on Software Engineering and Knowledge Engineering, SEKE (2021). https://doi.org/10.18293/SEKE2021-135
4. Fernández, F., Borrajo, D., Parker, L.: A reinforcement learning algorithm in cooperative multi-robot domains. J. Intell. Rob. Syst. **43**, 161–174 (2005)
5. Kolling, A., Carpin, S.: Cooperative observation of multiple moving targets: an algorithm and its formalization. Int. J. Robot. Res. **26**(9), 935–953 (2007). https://doi.org/10.1177/0278364907080424
6. Li, X., Chen, J., Deng, F., Li, H.: Profit-driven adaptive moving targets search with uav swarms. Sensors **19**(7) (2019). https://doi.org/10.3390/s19071545, https://www.mdpi.com/1424-8220/19/7/1545
7. Liang, E., Liaw, R., Moritz, P., Nishihara, R., Fox, R., Goldberg, K., Gonzalez, J.E., Jordan, M.I., Stoica, I.: Rllib: abstractions for distributed reinforcement learning (2018)
8. Lowe, R., Wu, Y., Tamar, A., Harb, J., Abbeel, P., Mordatch, I.: Multi-agent actor-critic for mixed cooperative-competitive environments. Neural Inf. Process. Syst. (NIPS) (2017)
9. Machado, A., Alessandro, A.: Multi-agent movement coordination in patrolling. In: First Workshop on Agents in Computer Games, at The 3rd International Conference on Computers and Games (CG'02) (2002)
10. Parker, L.E.: Distributed algorithms for multi-robot observation of multiple moving targets. Auton. Robot. **12**, 231–255 (2002). https://doi.org/10.1023/A:1015256330750
11. Schulman, J., Wolski, F., Dhariwal, P., Radford, A., Klimov, O.: Proximal policy optimization algorithms (2017)

12. Terry, J.K., Black, B., Grammel, N., Jayakumar, M., Hari, A., Sulivan, R., Santos, L., Perez, R., Horsch, C., Dieffendahl, C., Williams, N.L., Lokesh, Y., Sullivan, R., Ravi, P.: Pettingzoo: gym for multi-agent reinforcement learning (2020). arXiv:2009.14471
13. Terry, J.K., Grammel, N., Hari, A., Santos, L., Black, B.: Revisiting parameter sharing in multi-agent deep reinforcement learning (2021)
14. Touzet, C.F.: Distributed lazy q-learning for cooperative mobile robots. International Int. J. Adv. Robot. Syst. **1**(1), 1 (2004). https://doi.org/10.5772/5614
15. Yan, P., Jia, T., Bai, C.: Searching and tracking an unknown number of targets: a learning-based method enhanced with maps merging. Sensors **21**(4) (2021). https://doi.org/10.3390/s21041076, https://www.mdpi.com/1424-8220/21/4/1076

Macro Architecture for Microservices: Improving Internal Quality of Microservice Systems

Lars Braubach, Kai Jander, and Alexander Pokahr

Abstract Microservices have established themselves as a common software engineering pattern for large-scale systems. However, while the focus of the approach on maximum flexibility for development of individual services has increased software development efficiency, the inter-service architecture of microservice-based systems has received little attention, resulting in systems with a multitude of confounding and poorly planned interactions between services, impeding initial development, maintenance and continued development of applications. This paper identifies three areas that currently require improvement in order to address the global architectural challenges of microservice systems. Furthermore, a solution is proposed for each of the areas with an implementation demonstrating how the proposed solution can improve development efficiency for and internal quality of microservice systems.

1 Introduction

Microservices are currently one of the most favored approaches for the development of large-scale service-oriented architecture (SOA)-based applications. By vertically slicing an application into microservices, each responsible for its own data, logic and sometimes also user interface layer, allows for rather independent development and enhancements of the services. This is further organizationally supported by having small teams that develop the services in a DevOps manner and in this way can release frequently and react to changing requirements very quickly. Even

L. Braubach (✉)
Bremen City University of Applied Sciences, Bremen, Germany
e-mail: lars.braubach@hs-bremen.de

K. Jander
Brandenburg University of Applied Sciences, Brandenburg, Germany
e-mail: jander@th-brandenburg.de

A. Pokahr
Actoron GmbH, Hamburg, Germany
e-mail: ap@actoron.com

D. Camacho et al. (eds.), *Intelligent Distributed Computing XIV*, Studies in Computational Intelligence 1026, https://doi.org/10.1007/978-3-030-96627-0_10

though these advantages led to widespread adoption, the global architecture level of microservice systems usually overlooked. This leads to complex configurations, unclear responsibilities and dependencies among services. Architectural deficiencies are considered to result in reduced internal quality for software. Internal quality in software engineering refers to the ability to continue development on a project while external quality deals meeting customer requirements [5]. As a result, internal quality is key from a development perspective as it directly influences development efficiency. In this paper, we will identify three important architectural challenges of current microservice approaches and address them with the goal of improving internal software quality.

In the next Sect. 2 the challenges are put forward each also discussing relevant related work explaining current solutions. Thereafter, in Sect. 3, an implementation for the approach is presented and evaluated using practical examples. Finally, in Sect. 4 concluding remarks and a short outlook on future work is given.

2 SOA Challenges and State of the Art

Microservices are an architectural pattern aiming at supporting scalability on multiple levels. E.g., on a technical level functional decomposition allows for separate scaling and load-balancing of each functionality. On an organizational level independent DevOps teams that can use their preferred programming languages and tools and new technology can quickly be adopted for new modules without having to migrate existing ones. The independence of microservices compared to, e.g., modules of a monolithic application, thus provides many benefits but also comes with a number of drawbacks (cf. e.g. [15]) that pose challenges when developing microservice applications. In the following three important developer-oriented challenges will be discussed.

2.1 Inter-service Architecture Challenges

In microservice architectures generally no restrictions are placed on visibility of services, e.g. every microservice is technically capable of invoking any other microservice. While this approach grants a large degree of flexibility and freedom for implementing each microservice, it often leads to a confusing global architecture that is difficult to analyze and debug. This issue is often demonstrated using Deathstar visualizations (see Fig. 1a), which shows interactions between individual microservices with services in a spherical shape. The resulting overview shows almost no structure at the global application level. While the Deathstar architecture style sometimes argued as the price for flexibility, special management approaches are required to deal with the high complexity of the system [14].

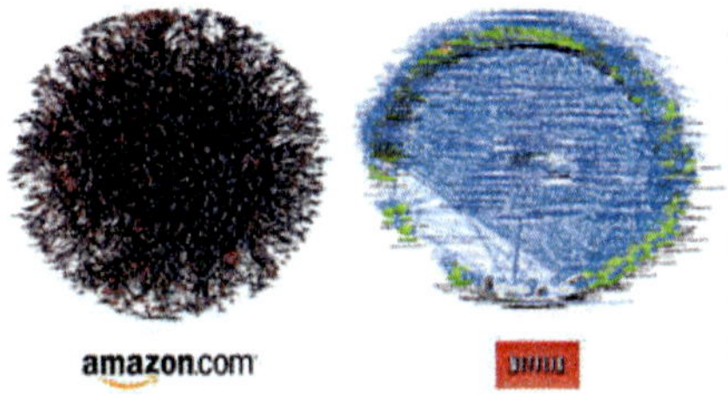

Maintainability (ISO 25010)	Intra-Service Level (Local)	Inter-Service Level (Global)
Modularity	+ (package / module mechanisms)	- (coherence and coupling of services, decomposition problems)
Reusability	+ (libraries, helper classes)	= (granularity and domain orientation of services, decomposition problems)
Analysability	+ (low complexity)	- (no explicit system-level architecture)
Modifiability	+ (low complexity)	- (unlcear service dependencies and versions)
Testability	+ (limited functionality and scope)	- (integration and end-to-end tests)

Fig. 1 **a** "Deathstar" visualization of the Netflix microservice architecture (from [2]) and **b** Maintainability challenges

Figure 1b further illustrates challenges to important aspects of maintainability according to the ISO 25010 software quality standard [6] when comparing intra- and inter-service levels. Regarding intra-service levels, good solutions for all aspects are available, while at the inter-service level remain challenging. To achieve modularity and reusability for a single service, all well-known software engineering mechanisms like libraries are applicable. Analysability and modifiability can be attained easily due to the low complexity of a single microservice (it is simple by definition). In addition, for testability the scope is narrow (unit tests and service interface tests).

The inter-service level is far more demanding (cf. e.g. [15]). Modularity depends on the resulting coherence and coupling of services. Even if developers identify perfect services, they are missing a hierarchical decomposition concept, so service use and reasons for it remain unclear. Reusability primarily depends on granularity and domain orientation. If done right, services are flexible to orchestrate. However, with great number of services, it becomes difficult to identify services to be combined due to non-existing grouping and/or composition methods. Analysability remains ok for small systems but partially falls apart as the system grows. Such evolution also harms the modifiability of the system. Typically, microservice systems are said to be easily extendable by adding services implementing desired customer requirements. However, when functionality of services become intertwined and services are not removed, systems tend to be prone to the lava flow antipattern [1] leaving unused code and thus blurring the architecture. Finally, testability of microservices is hard at the integration and end-to-end layer [11]. Besides non-SOA-typical problems like increased setup complexity and time consumption, responsibility problem arise - who is responsible for tests and when are they executed? If, e.g. the tests are done as part of a build pipeline, how can buggy services avoid disturbing building others service? If mocks reduce the problems somewhat, a new problem arises that original and mocked service must be in sync. This led to partially drastic solutions in practice such as omitting tests altogether in favor of testing with synthetic transactions [11].

Few approaches for developing inter-service level architectures exist so far. In [9] the "monolith first" idea is mentioned, which suggests a monolith prototype before deriving the architecture. This requires additional development effort and it remains unclear if results are transferable (e.g. monolithic layered architectures slice systems vertically instead of horizontally), though suggestions exist on migration,

	REST	RPC	Publish Subscribe	Service Mesh
Implementations	Jersey, Express	gRPC, Thrift	Kafka, ActiveMQ, Mule	Istio, linkerd, Consul
Interface Expressiveness	CRUD	full, defineable via IDL	events, no interface	CRUD
Typing	weak	strong	weak	weak
Infrastructure	decentralized	decentralized	centralized	centralized
CallDistribution/Load Bal.	no	no	yes	yes
Service Discovery	necessary	necessary	not necessary	not necessary
Request-Reply	yes	yes	indirectly yes	yes
Subscription	yes (via long polling)	typically no	yes	typically no
Complex Interactions	no	no	yes	no

Fig. 2 Comparison of interaction approaches

e.g. [12]. Domain Driven Design (DDD) [3] is available for migration and greenfield projects, allowing for decomposing a system into so-called bounded contexts, which often allow straight-forward mapping to services. While DDD is useful and often an important success factor for microservice systems [11], its approach does not consider which services should exist or interact. In a whitepaper from MuleSoft,[1] six important micro service design patterns have been identified. One is called "layered APIs over fine-grained SOA" and proposes to layer services in order to cope with increasing complexity. Although service composition may lead to slight performance penalties, a layered API can lead to simpler designs due to clear scopes and responsibilities of services.

2.2 *Interaction Challenges*

Interactions between services are challenging because there are different communication means available each with different characteristics. The common approaches are shown and compared with respect to several dimensions in Fig. 2. Considering interface expressiveness, RPC allows for using the most natural style well-known and established in object-oriented systems. Typically, a programming language independent interface definition language (IDL) is employed that allows for specifying methods with parameter types and return values in the spirit of CORBA [13] thus ensuring strong typing of interfaces. IDL descriptions are typically used to automatically generate code stubs for the specific implementation languages. In contrast to RPC, REST has a so called uniform interface [4], which uses CRUD (create, read, update, delete) semantics with reinterpreted HTTP methods on resources. Although many use cases can directly be implemented based on that scheme it is a restriction of expressiveness and for some use cases workarounds such as virtual resources or extra resource properties have to be employed. In addition, as REST uses HTTP, parameter and response types remain rather implementation than specification aspects, even though approaches like OpenAPI improve the situation. As today, service mesh middleware relies on REST the same restrictions apply for them as well. Using publish-

[1] https://www.mulesoft.com/lp/whitepaper/api/top-microservices-patterns.

subscribe approaches like message brokers or ESBs, there are no interfaces at all. Instead, event types need to be specified and service consumption and production of event types must be documented.

REST and RPC are decentralized solutions that require no middleware being present. Instead, communication is performed directly between a service provider and a consumer. This is in contrast to middleware-oriented solutions like message brokers and service meshes. Though some implementations of brokers allow for redundant functionalities [10], system communication is nevertheless transmitted using a dedicated middleware infrastructure instead of point-to-point communication. While it sounds like the decentralized solutions are advantageous due to avoiding a single point of failure and allowing for a simpler system setup, this only partially true. Middleware solutions simplify service discovery and may offer features like automatic load balancing or recovery of services by implementing robustness patterns like circuit breaker.

Considering the possible interaction patterns, publish subscribe is the most generic solution allowing all sorts of protocols being realized as sequence of events (including request-reply and subscriptions). As only events are distributed, services become less coupled and do not need to know which other services depend on them. Although direct coupling is lower, the services still depend on specific types of events being produced, which makes coupling more indirect and less visible on a system level. Thus, changing services with respect to event generation will not lead to errors on the compile level but defer problems to the runtime layer and possibly harm to process execution. Using REST or RPC emphasizes the request-reply interactions and typically needs extensions or workarounds to handle more complex patterns like subscriptions. Especially problematic are protocols including multiple steps like e.g. auctions. Those require interfaces with methods for the different protocol steps and additionally imply that those methods have to be called in the right order for the protocol being properly executed.

2.3 *Service Discovery*

Getting microservices to know each other is a surprisingly hard challenge. Figure 3 compares common discovery approaches with respect to important developer characteristics, more specifically, the following approaches are taken into consideration: Hard coding addresses means that no discovery mechanism is used at all and services must be known and are selected at build time. Registry approaches come in different two variants: Simple data stores typically are key value based and are primarily general purpose but are sometimes used for service registration. Service registries represent software specifically developed to store and query service data.

Also event based approaches come in two different flavors: Message brokers realize a publish subscribe infrastructure while ESBs add further integration functionalities on top of communication layer. These approaches strongly emphasize decoupling services, allowing communication without explicitly knowing the partic-

	No Discovery	Simple Data Store / Registry	Message Broker / ESB	Mesh
Implementations	-	Etcd, ZooKeeper / Eureka, Consul	Kafka, ActiveMQ / Mule, BizTalk	Istio, linkerd, Consul
Programming model	Hardcoding service addresses in code	Send queries and directly interact with services	Send events to indirectly interact with services	Invoke side car services via hardcoded names
Simple and concise prog. experience	-	-	-	=
High expressiveness for service selection	-	- / +	+	-
High invocation performance	+	+	=	=
Further features available (load bal.,..)	-	-	- / +	+

Fig. 3 Comparison of discovery challenges

ipants. Instead, services require knowledge which events are produced and consumed by other services. Finally, a service mesh uses so called side-cars for each service which represent proxies attached to a middleware routing invocations to suitable services.

Each of these approaches appear to exhibit different strength and weaknesses: For example, only dedicated registries and ESB offer the means for selecting services according to complex criteria. Meanwhile, service mesh approaches instead tend to resolve a service name to a service address, resulting in a rather implicit approach for service selection. Service invocation performance is good for all approaches directly invoking selected services, except for ESB and mesh where each call is routed indirectly via middleware to the target, leading to some performance penalty compared to direct calls. However, this minor disadvantage is usually accepted because middleware approaches like ESB and mesh allow for integration of further functionalities such as load-balancing, monitoring, tracing or fault detection and recovery via circuit breakers without burdening developers with separate infrastructure for each of these tasks.

Each of the approaches result in a particular SOA programming model used by programmers to develop systems based on these approaches: Hardcoding service addresses in code requires little to no infrastructure or setup, however, the code quickly becomes unmaintainable. For example, a single service address change could instantly break the whole system. On the other hand, registry-based approaches require developers to first issue queries to the registry whenever a service is needed (a.k.a. "SOA Triangle"). While this may sound simple, the resulting code quickly becomes complex due to edge cases such as a query returning no results. In general, temporarily absent services are an unavoidable fact for distributed system due internal errors or network partitions, but also in situations like a system startup, where services depend on each other but have varying initialization times. As a result, developers are forced to implement repeated but temporarily spaced service searches within an appropriately chosen grace period in order to guarantee robust behavior in which the service can either be acquired in a timely manner or it aborts occurs after a defined number of retries.

For message-broker-based systems, the SOA programming model becomes event-oriented instead of service-oriented. Instead of APIs, service invocation occurs as reaction to consumed events. Proponents of this approach often claim that using an event broker leads to loosely coupled systems and fosters independent service development. However, while services do not require direct knowledge of each other, this direct relationship is instead substituted by requiring knowledge of event types that are emitted and consumed. As a result, the API of a service is represented by a set of consumed and produced events, often hidden in code (sometimes as literals). Furthermore, if events are used by multiple services, changes to events could easily break a system without clear indication regarding the cause. As a result, clear responsibilities for event definitions have to be established among teams.

In contrast to the other approaches, a mesh simplifies the programmer experience by trying to strike a balance: Services are invoked by name but dynamically resolved by the infrastructure. The model assumes that services under the same name are always replicas and developers do not need to care which one is invoked. When this is not the case, workarounds have to be found, e.g. by using different service names for the same type of service.

2.4 Implications

First, many microservice approaches have a weak or incoherent inter-service architecture. This is e.g. illustrated by the strong focus on single service development, while there are very few proposals for modeling or designing the top-level SOA architecture of a microservice system, leading to a number of undesirable consequences such as unclear dependencies and unintended and unnecessary retention of obsolete services and code. As a result, an approach providing *concepts for modeling the global architecture* while retaining the flexibility and low-threshold implementation style of current microservice systems is desirable.

Second, while multiple approaches for service interactions are available, however, each is limited in its own way. For example, approaches based on Remote Procedure Call (RPC) and Representational State Transfer (REST) rarely provide more than simple request-response type interactions and, if provided, the more complex interaction types are limited. While approaches such as Enterprise Service Bus (ESB) provide more flexibility in terms of interaction complexity and loose coupling, the interactions tend to be implicit based on event types leading to unclear interaction patterns. Additionally, using an ESB the API of services cannot easily be seen anymore and dependencies between services becomes buried in the implementations. Thus, an approach is lacking that supports *type-safe and comprehensible complex interaction patterns*.

Finally, service discovery approaches either focus on transparency or control over the service selection. E.g. in ESB and service meshes, service invocations are largely transparent to the client developer and support sophisticated features such as elasticity and circuit breakers. On the other hand, when the client developer wants more control over the service selection, she has to resort to, e.g., hard-coded addresses or manual lookups using DNS or registries resulting in tedious and error-prone code that explicitly has to deal with e.g. partial failures and mostly lacks the advanced features of ESB and service meshes. Therefore, an approach is needed to *combine the expressive power of manual service selection with automatically handling issues of distributed systems, such as race conditions, partial failures, etc.*

3 Concepts and Implementation

Figure 4 provides an example how the novel approach improves microservice architectures by addressing the outlined shortcomings of current approaches in a combined manner. In general, our contribution is based on three major changes:

- Services can be organized hierarchically with encapsulation support using publication scopes
- Services are automatically retrieved using persistent queries with minimal code
- Communication between services is based on strongly typed asynchronous RPC calls also supporting more complex interaction forms

The example demonstrates an application providing logistics services to external customers via a *Logistics* service type. Meanwhile, the logistics application relies on internal services to provide its functionality. These internal services represent three different types of services: The *Dispatch* type for organizing parcel pickup, the *Delivery* type for transporting the parcel to its destination and *Monitoring* for logging events and monitoring system performance. The service types are backed by service instances, which provide the actual service functionality for each type of service. Multiple instances of each service type can be available: For example, there are two instances of the delivery service, *log.del1* and *log.del2*, both being encapsulated by the instance of the logistics application *log*. Multiple service instances can be either

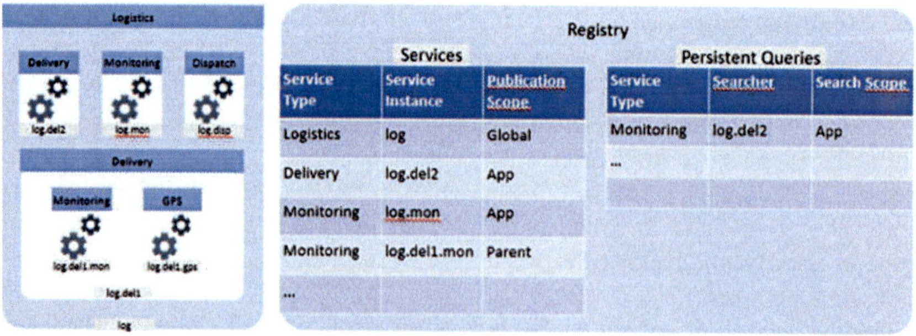

Fig. 4 Approach

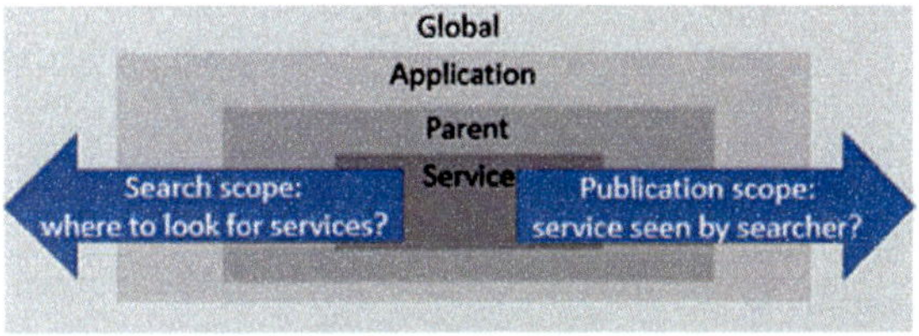

Fig. 5 Search and publication scopes

used to provide redundancy for resilience or horizontal scaling, or, alternatively, to provide service reuse in different contexts: The service instance *log.del1.mon* provides a separate monitoring service as internal service for the delivery service instance *log.del1*. Each service instance can limit visibility of its functionality through the use of publication scopes: For example, *log.del2* sets its publication scope to *App*, only search queries issued by service instances that are part of the *log* application will match. Furthermore, *log.del2* needs a monitoring service and has thus issued a persistent query to the registry for that type. It can be seen in the Figure that the search scope of that query is App, meaning that all services visible in the log application are potential candidates. In this case only *service log.mon* will be found, because the publication scope of *log.del1.mon* is *Parent*, which makes the service invisible outside of the delivery context.

In Fig. 5 the concept of scopes is visualized. While *Service* scope represents the local view of a service, *Parent* scope also allows sibling service visibility. *Application* scope represents all services belonging to a complete application while global scope allows services of different applications to find a service. The figure also illustrates the difference between search and publication scopes. The first is used to limit the search for services, whereas the latter acts as a visibility restriction and determines if a service is visible to a searching service.

Next, the programming model of persistent queries will be explained in more detail. Issuing a persistent query to the registry has the effect that the registry saves the query and automatically tests newly added services whether they match the query. When a match is found, the searcher will be notified, which can be realized very conveniently in a programming model using callback methods that initiate processing for matching new services. Taking this further, our model also allows implicitly defining queries by labeling (callback) methods with an @OnService annotation containing query content such as the search scope. The service type is also implicitly derived from the method parameter. In Fig. 6, this is illustrated with a code snippet used to set up a persistent query for monitoring services in our example scenario. Aside from low overhead, the programming model also exhibits further advantages such as more fine-grained control of the search, e.g. by specifying tags attached to a service or by using quality of service attributes that have to be met. In addition, recovery from service errors is simplified: If multiple services meet a query, the method will be invoked for each discovered instance which can be stored and used as backup services. Even if no service is currently available, the method is called immediately once one becomes available (without further programming effort).

```
@OnService(scope="App")
public void foundMonitoring(Monitoring mon) {
  // handle the newly found service, e.g. store in variable
}
```

Fig. 6 Persistent query example

Due to the layered architecture, service interfaces are a vital factor as they can describe not just micro element functionality (i.e. services) but subsystems and whole applications (otherwise the facade pattern via a web gateway is often used). As a result, a strongly-typed approach is chosen akin to gRPC. All methods declared as asynchronous using future return values. This would normally restrict calls to request-reply interactions, however, an extended set of futures types supporting more complex interaction patterns within the context of one call is made available. For example, a subscription future can be used, which is able to retrieve a potentially unbounded stream of values. In addition, iterator futures allow transmission of multiple values but invert responsibilities by allowing receivers to fetch on request.

Communication between services is facilitated without additional explicit programming effort and little configuration using an overlay network with multiple communication approaches (TCP, WebSockets, ...) and optionally using third-parties as intermediaries to facilitate communication between two isolated services. Routes between services are discovered dynamically and automatically. Data is serialized using a compact binary form (alternatively JSON) and is end-to-end encrypted by default, authenticated using a variety of schemes or opportunistically, allowing for secure communication between services even when interacting through third parties [7].

A running example of the approach has been implemented as part of the Jadex Active Components framework.[2] A number of examples demonstrating the approach are included in the framework, including an example and tutorial showing the approach for a time server system.[3] A global service registry is available and configured by default, however, users can deploy and configure their own registry system, if desired. Details about the registry architecture and realization can be found in [8].

4 Conclusion and Outlook

In this paper three important challenges with respect to microservice systems have been identified: First, the software architecture is typically based on a flat set of a potentially huge number of services, leading to Deathstar antipattern structures. Second, interaction is facilitated using different schemes, each exhibiting their own drawbacks and third, service discovery is either explicit with error prone trial and

[2] https://www.activecomponents.org/.

[3] https://github.com/actoron/jadex/blob/master/docs/tutorials/quickstart/01%20Introduction.md.

error programming models or implicit with a simple programming model but no control over service selection.

To solve these issues an integrated approach has been presented based on **layered services** and a simple, yet fully controllable programming model using **persistent queries**. Decomposition in service layers allow for hierarchical system structures, in which top-level services comprise the system level API, while lower-level services represent subsystems and provide partial functionalities for higher-level ones. The service discovery supports this model by offering publication and search scopes discovery of contextually meaningful services. In addition, persistent queries allow processing and discovering services the moment they become available. This simplifies the programming model for developers by retaining the expressive power of manual queries and combining it with easy handling of typical distributed system issues. Finally, interaction of services is based on **asynchronous RPC-calls** extended with capabilities for directly handling complex interaction types such as subscriptions and iterators. The proposed concepts have been implemented within the Jadex framework.

Currently, the implementation is limited by requiring that services to be implemented in Java, therefore limiting the use of heterogeneous technology stacks for different services. As future work we intend to change this by introducing a polyglot API similar to the approach used by vert.x [4] and REST-based access for the registry.

References

1. Brown, W.J., Malveau, R.C., "skip" McCormick, H.W., Mowbray, T.J.: AntiPatterns: Refactoring Software, Architectures, and Projects in Crisis: Refactoring Software, Architecture and Projects in Crisis. Wiley (1998)
2. Cockcroft, A.: Microservices workshop: Why, what, and how to get there (2016)
3. Eric Evans: Domain-Driven Design: Tackling Complexity in the Heart of Software. Addison-Wesley (2004)
4. Fielding, R.T.: Architectural styles and the design of network-based software architectures. Ph.D. thesis (2000). AAI9980887
5. Freeman, S., Pryce, N.: Growing Object-Oriented Software, Guided by Tests, 1st edn. Addison-Wesley Professional (2009)
6. ISO/IEC 25010: ISO/IEC 25010 2011, systems and software engineering - systems and software quality requirements and evaluation (square) - system and software quality models (2011)
7. Jander, K., Braubach, L., Pokahr, A.: Defense-in-depth and role authentication for microservice systems. In: Proceedings 9th International Conference on Ambient Systems, Networks and Technologies. Procedia Computer Science (open-access), pp. 456–463. Elsevier Science (2018)
8. Jander, K., Pokahr, A., Braubach, L., Kalinowski, J.: Service discovery in megascale distributed systems. In: Ivanovic, M., Badica, C., Dix, J., Jovanovic, Z., Malgeri, M., Savic, M. (eds.) Intelligent Distributed Computing XI, pp. 273–284. Springer (2017)
9. Nadareishvili, I., Mitra, R., McLarty, M., Amundsen, M.: Microservice Architecture: Aligning Principles, Practices, and Culture, 1st edn. O'Reilly Media, Inc. (2016)
10. Narkhede, N., Shapira, G., Palino, T.: Kafka: The Definitive Guide: Real-Time Data and Stream Processing at Scale. O'Reilly UK Ltd. (2017)

[4] https://vertx.io/.

11. Newman, S.: Building Microservices - Designing Fine-Grained Systems. O'Reilly Media (2015)
12. Newman, S.: Monolith to Microservices: Evolutionary Patterns to Transform Your Monolith. O'Reilly Media, Incorporated (2019)
13. Object Management Group (OMG): The Common Object Request Broker: Architecture and Specification, revision 2.0 edn. (1995)
14. Ragan, T.: Navigating the microservice deathstar with deployhub (2019)
15. Richardson, C.: Introduction to microservices - the drawbacks of microservices (2015)

An Artificial Neural Network to Simulate Surface Runoff and Soil Erosion in Burned Forests

Lidia Fotia, Manuel Esteban Lucas-Borja, Domenico Rosaci, Giuseppe M. L. Sarné, and Demetrio Antonio Zema

Abstract Few experiences of Artificial Neural Networks (ANNs) for hydro-logical predictions in forest soils after wildfire and post-fire treatments are available in literature. To fill this gap, an ANN model has been adapted to predict surface runoff and soil erosion in Mediterranean burned pine forests (Central Spain), and tested against hydro-logical observations at plot scale throughout 2 years. The model gave very accurate runoff and erosion predictions in burned and non-burned soils as well as for all soil treatments (mulching and/or logging or not). Although further experimental tests are needed to validate the ANN applicability to soils in burned and treated forests in other ecosystems, the use of ANN may be useful for landscape planners as decision support system for the integrated assessment and management of forests.

L. Fotia
DIMES, University of Calabria, Rende (CS), Italy
e-mail: lidia.fotia@dimes.unical.it

M. E. Lucas-Borja
Departamento de Ciencia y Tecnología Agroforestal y Genética, Universidad de Castilla La Mancha, Campus Universitario s/n, C.P. 02071, Albacete, Spain
e-mail: manuelesteban.lucas@uclm.es

D. Rosaci
DIIES, University "Mediterranea", Reggio Calabria, Italy
e-mail: domenico.rosaci@unirc.it

G. M. L. Sarné
Department of Psychology, University of Milano Bicocca, Milano (MI), Italy
e-mail: giuseppe.sarne@unimib.it

D. A. Zema (✉)
Department AGRARIA, University "Mediterranea" of Reggio Calabria, Località Feo di Vito, 89122 Reggio Calabria, Italy
e-mail: dzema@unirc.it

D. Camacho et al. (eds.), *Intelligent Distributed Computing XIV*, Studies in Computational Intelligence 1026, https://doi.org/10.1007/978-3-030-96627-0_11

1 Introduction

Prediction models are used worldwide and for several applications in the environmental field. The simulations of modelled processes are generally satisfactory, provided that suitable datasets of input parameters are available, and their performances are validated in environmental conditions that are similar as those in which the models must be used. The main advantage of the ANNs over traditional methods is the lower requirements of information about the complex processes that are instead described in a mathematical closed form. Furthermore, ANNs can generalise relationships also from a small dataset, but remain more or less robust when noisy or missing inputs are present. ANNs learn from the analysis of the available input data and do not require reprogramming, but they must be trained, optimized and tested.

In the last two decades, the ANNs had an increasing popularity for estimating and forecasting water resources. ANNs have been extensively used also for rainfall-runoff modelling, flood predictions, reservoir operations, routing of polluting compounds. In spite of several applications of ANNs for modelling surface runoff worldwide, only a few studies have analysed the ANN performance in soil erosion modelling [1] and, even, ANN has not been used for hydrological predictions in burned soils. Modelling soil erosion and runoff after wildfires using ANNs is important, since accurate predictions of water and sediment flows after fire help land managers in the adoption of the most suitable actions to mitigate post-fire land degradation and rehabilitation planning.

To provide a useful prediction tool in the forest hydrology field, this study adapted an ANN for applications in forests subject to wildfire and soil treatments. Its hydrological prediction capability is evaluated in a pine forest of the Mediterranean semi-arid environment, using a hydrological database collected in instrumented plots for model calibration.

2 Materials and Methods

2.1 *Theoretical Approach About the Artificial Neural Networks*

In this work a standard feedforward neural network has been used to simulate the hydrological response of forest soil. A standard feedforward neural network [2] is composed by a set of N nodes N and a set of M arcs A (Fig. 1). The nodes are partitioned into L groups, called layers, with $L > 2$. The first layer is a set of I input nodes N_I called input layer; then, there are $L - 2$ hidden layers, of which each hidden layer h_t, with $t = 1, \dots, L - 2$ is a set of H nodes NH_t. Finally, there is a set of O nodes NO, called output layer. Each node (denoted by o) of the output layer is connected with each node (denoted by h) of the NH_{L-2} hidden layer by an edge

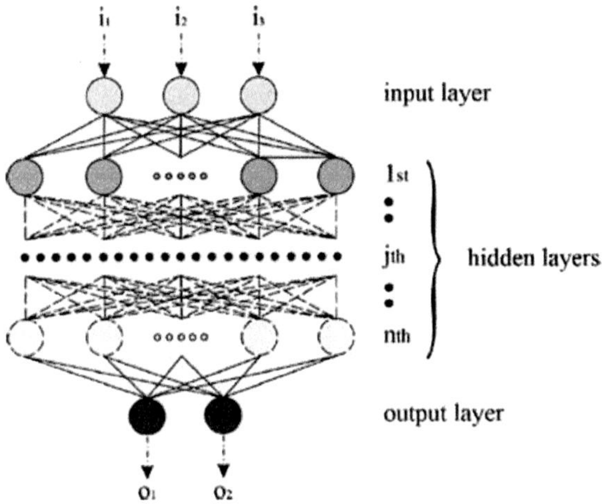

Fig. 1 The architecture of the proposed ANN

directed from o to h, and each node y of the NH_1 hidden layer is connected by an edge with each node x of the input layer by an edge directed from y to x.

For each edge of the network, we denote by i (resp. j) the source (resp. destination) node and we associate a real value W_{ij}, called weight, with the edge. The neural network is used for representing a real function. Each input layer node is associated with an input (real) value and each output layer node is associated with an output (real) value of the function. The output values are computed by the neural network using the input values. Hidden layer nodes are associated with intermediate results of the computation. The neural network computes the output values as follows. Both of each hidden and output layer node n are provided with the same function a, which is called activation function, and with a parameter Θ, which is called bias. The node j of the first hidden node NH_1 computes its associated hidden value $h_1 = a \cdot (\sum_{i=1}^{I} W_{ij} \cdot I_i - \Theta)$, where i is an input layer node, i.e., by computing the weighted sum of the values I_i of the input layer using the weights W_{ij} associated with all the connections between each input layer node i and the hidden layer node j.

The node j of each hidden layer NH_l computes its associated hidden value $h_j^l = a \cdot (\sum_{i=1}^{H} W_{ij} \cdot h_i^{l-1} - \Theta)$, where i is the $l-1$ layer node, i.e., by computing the weighted sum of the values h_i^{l-1} of the nodes of the previous layer. Analogously, each output layer node j computes its associated output value $o_j = a \cdot (\sum_{i=1}^{H} W_{ij} \cdot h_i^{L-2} - \Theta)$, where h_i^{L-2} is a hidden $L-2$ layer node. The weight W_{ij} associated with the edges of the set A and the activation function parameters are suitably set by a training algorithm that tries to learn how correctly approximating the desired output. Training algorithms can be unsupervised or supervised. In the first case, the ANN autonomously learns the functional dependence between an input and its correct output. Differently, a supervised training algorithm takes advantage from the availability of a training dataset where for each input its correct output is provided; by measuring the difference between the correct and the computed ANN outputs then it is possible to tune the ANN parameters to minimize this error.

When the ANN reaches the desired precision in reproducing the outputs of the training dataset, then the learnt ends and the ANN can be considered ready to work with unknown input data. Multilayer feedforward networks are commonly used to approximate real functions, i.e. for determining weights and parameters of a given neural networks such that a set of given output data matches with a corresponding set of input data, with an approximation error. Some theoretical results have been provided in the related literature [3] to assure the possibility of approximating any real function satisfying some determined constraints. Many types of activation function a can be used with the above neural network model. In this work, we will use the well-known sigmoid function with the following formula:

$$a_x = \frac{l}{l + e^{-\beta x}} \tag{1}$$

where β is a parameter that should be appositely chosen when designing the neural network architecture.

2.2 Neural Network Architecture

We adopted the Neuroph, which is an ANN tool, and the Multi Layer Perceptron architecture, which is a feedforward ANN. This ANN model maps sets of input data into a set of appropriate output. It consists of multiple layers of nodes in a directed graph, with each layer fully connected to the next one. Except for the input nodes, each node is a neuron with nonlinear activation function. Multilayer Perceptron uses a supervised learning technique called backpropagation for the training stage. It is a modification of the standard linear Perceptron, which is not able to distinguish data that not linearly separable, as in our case. We set multi-layer Perceptron's parameters. The number of input and output neurons was the same as in the training set. Then, we had to choose number of hidden layers, and number of neurons in each layer. The topology of our ANN was chosen as the result of a preliminary study, where several alternatives in terms of number of hidden layers and number of neurons for layer were tested. At the end of this study, the best performance architecture resulted in two hidden layers with 20 neurons in each layer (Fig. 2).

Then we adopted a 'Sigmoid' for transfer function, while, for learning rule, we chose a 'Backpropagation with Momentum'. The momentum is a real value added to speed up the process of learning and to improve the efficiency of the algorithm.

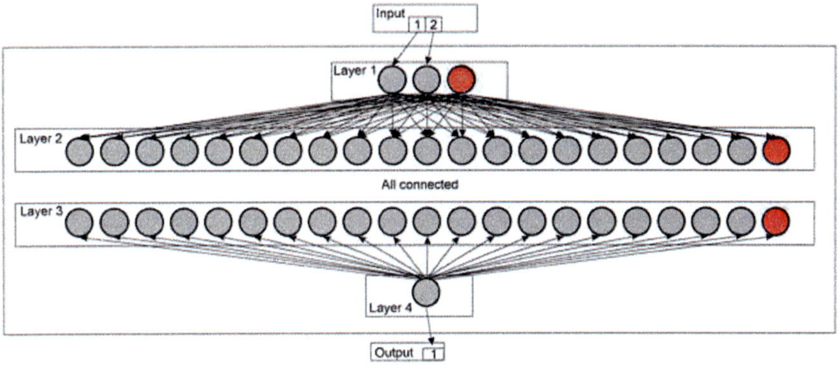

Fig. 2 The ANN with two hidden layers with 20 following neurons

2.3 Neural Network Training

After we have created training set and set the parameters of the neural network, we started to train it. In our case the maximum error was set to 0.0001, learning rate was set to 0.2 and momentum was set to 0.7. In the first phase, we calculated the total Mean Square Error (MSE). For that purpose, the following formula was used:

$$MSE = \frac{l}{n} \cdot \sum_{i=1}^{n} (Y_i - \widehat{Y_i})^2 \tag{2}$$

where MSE is the arithmetic mean of the squares of the errors $(Y_i - \widehat{Y_i})^2$. In the following, we will refer to the MSE as the Total Network Error. When this Total Network Error value dropped below the max error, the training was complete. The smaller the error is, the better the obtained approximation is.

2.4 The Hydrological Database

The hydrological observations used to prepare the ANN were carried out in the Sierra de las Quebradas forest (Liétor, Castilla-La Mancha region, Central Spain). The climate is hot dry Mediterranean (mean annual rainfall = 282 mm and mean annual temperature = 16 °C). The forest land mainly consists of Pinus halepensis M. stands, at a land elevation is 520 and 770 m. The soil texture is sandy loam. Immediately after the wildfire (July 2016), twelve experimental plots each one of 27 m^2 were installed immediately after the fire in the burned area, and three other plots were located in a close unburned area (control). As post-fire treatments, in September 2016, straw mulching was manually spread to six plots in the burned area. Moreover, salvage logging was conducted in December 2016 in six plots, of which three non-mulched and three mulched. Therefore, the experimental design consisted of the following soil conditions in relation to the wildfire: (1) "Non-Burned, NB" (three plots); (2) Burned+Mulching+No-Logging (B+M+NL, six plots); (3) Burned+No-Mulching+No-Logging (B+NM+NL, six plots); (4) Burned+Mulching+Logging (B+M+L, three plots); (5) Burned-No-Mulching+Logging (B+NM+L, three plots). The plots were equipped with metallic fences, pipelines and 25-L tanks, to water (for surface runoff) and sediments (for soil erosion) after precipitations (the latter measured by a weather station with a tipping bucket rain gauge. More details about the experimental site and design can be drawn from these studies [4–6].

2.5 ANN Implementation

In these experiments, we used the Neuroph framework for training the ANN on a data set of real hydrological information. The data set contains 243 tuples of four attributes, namely (i) treatment, (ii) precipitation (mm), (iii) runoff (mm) and (iv) soil loss (kg/ha). The treatment assumes the following discrete values: Burned+Mulching+No-Logging, Burned+No-Mulching+No-Logging, Non-Burned, Burned+Mulching+Logging and Burned+No-Mulching+Logging. The attributes (i) and (ii) are considered as the neural network inputs, while (iii) and (iv) are used as neural network outputs.

2.6 Data Pre-processing

First, we have processed the data to obtain a suitable dataset to train the neural network. The value of treatment has been transformed into an integer number that takes values between 1 and 5. In particular, Burned+Mulching+No-logging = 1, Burned+No-mulching+No-logging = 2, Non-burned = 3, Burned+Mulching+Logging = 4 and Burned+No-mulching+Logging = 5. Since some pair of inputs <treatment, precipitation> were associated with different outputs (due to the fact that the same precipitation can produce different runoff volumes, because of many factors, such as the variability of precipitation intensity, soil characteristics in time and space), we averaged in those cases the values of the surface runoff and soil loss. Then, the data set had to be normalized. Normalization implies that all values from the dataset should take values in the range from 0 to 1. For this purpose, we used the following formula:

$$X_n = \frac{X - X_{min}}{Xmax - X_{min}} \tag{3}$$

where X is the value that should be normalized, X_n is the normalized value, X_{min} is the minimum value of X and X_{max} is the maximum value of X.

3 Results and Discussions

First, we train the neural network for the first output. After 250000 iterations we obtained a Total Mean Square Error drop down to a specified level of 0.0001, which means that training process was successful.

3.1 *Neural Network Approximation*

A Total Mean Square Error of 1.965 e^{-4} in simulating the runoff volume was achieved, which certainly is a very good result, because our goal is to get the total error to be as small as possible. Looking at the individual errors, we can observe that most of them are at the low level, below 0.1. MAE was equal to 0.025 mm. So we can conclude that this type of neural network architecture is the best choice. We used the same neural network shown in Fig. 2. Also in this case, we set the maximum error to 0.0001, the learning rate to 0.2 and the momentum to 0.7. After 175000 iterations we obtained a total MSE drop down to a specified level of 0.0001, which means that training process was successful and that now we can exploit this trained neural network (Fig. 3). The Total Mean Square Error for this second neural network was 1.78 e^{-4}. The relative error on the individual soil loss between the observations and the simulations was lower than 17.1 kg/ha while MAE was equal to 3.57 kg/ha.

3.2 *Evaluation of the ANN Prediction Capability*

The scatter plots of Fig. 4a, b show a very close agreement between the predictions provided by ANN and the corresponding observations collected at the plots for both surface runoff volumes and soil loss for all the experimental conditions (control, burned and treated/not treated soils).

In general, when the ANN performance is evaluated by aggregating all the soil conditions, the differences between the mean values of simulated and observed surface runoff and soil loss were very low (<3.0%). A more detailed analysis of the ANN performance, carried out separately for the individual soil conditions (burned/unburned) and treatments (mulching/logging) highlighted that: (i) the observed and predicted mean values of both runoff and soil loss are practically the same and the maximum difference (16.2%, however acceptable) is detected for soil loss prediction in B+M+L plots; (ii) the lower agreement between observations and predictions was found in the maximum runoff (with differences lower than 32%) and in the minimum soil loss (below 112%). ANN showed a very slight tendency to overestimate or underestimate the hydrological observations (for instance, overestimation of runoff in B+NM+L and B+NM+L plots as well as underestimation of soil loss in B+NM+NL and B+NM+L). For the B+M+L plots, the worst performance of the ANN was found

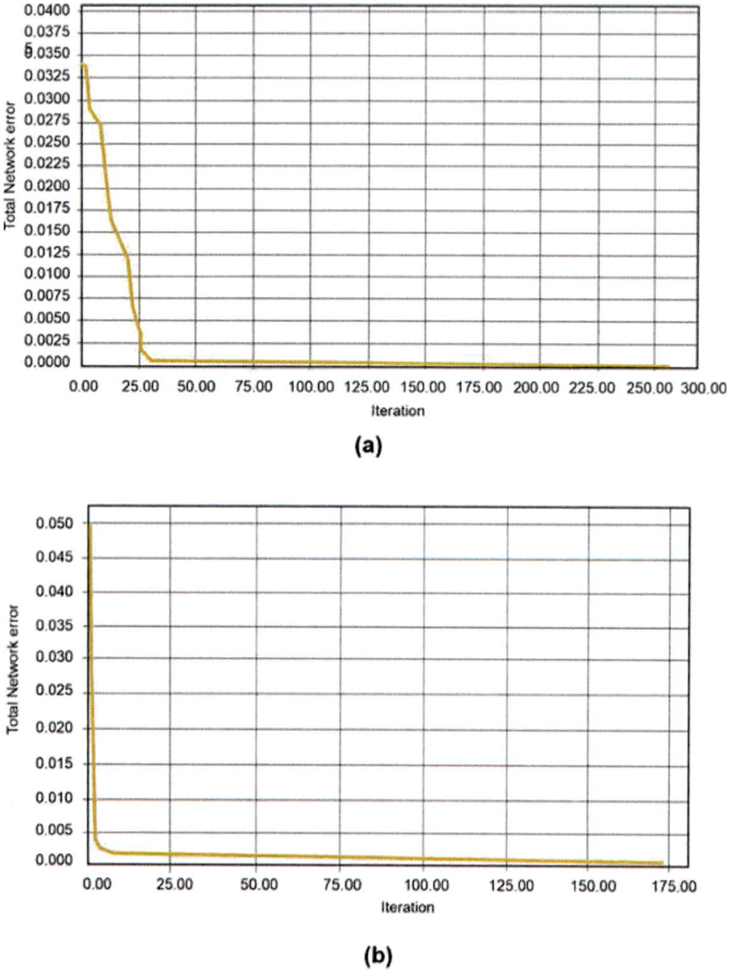

Fig. 3 Total Network Error (equal to the Total MSE, Mean Square Error) for runoff volume (**a**) and soil loss (**b**) simulated by the ANN

for both runoff and erosion predictions. Presumably, in soil subjected to logging, the impacts of machinery wheels on soil determine the formation of small rills, in which small volumes of water and sediments are stored and do not feed runoff. Since, in general, many models find difficulties in modelling rill erosions, this behaviour could be common with ANN. Overall, the prediction capability of the ANN can be considered as good for both runoff and soil loss. This indicates that a soil disturbance due to more than two factors (in our case wildfire, mulching and logging) founds some difficulties in being simulated by ANN, which however does not compromise the generally good model performances. The ANNs, which require only precipita-

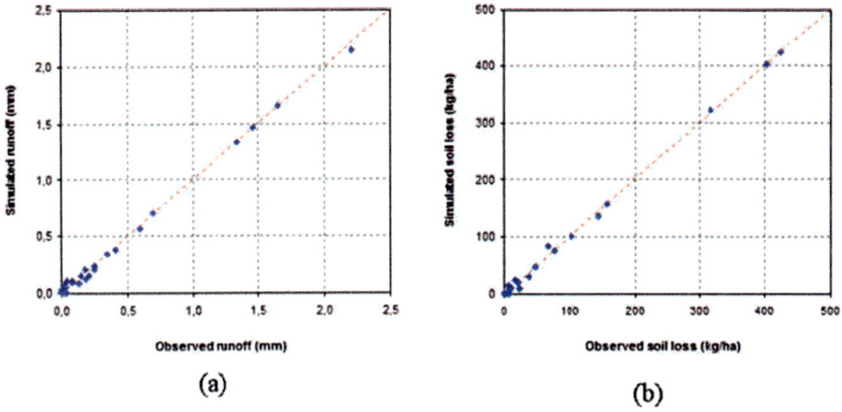

Fig. 4 Scatter plot of the observed versus simulated (by ANN) runoff volumes (**a**) and soil loss (**b**)

tion as input, but use a complex mathematical structure, were successful in capturing the output hydro-logical variables from the observational input data, as shown by the very good prediction capacity detected for the ANNs in the experimental conditions of this study. Therefore, the main advantages of the ANN use are in such environmental contexts are the low input requirement in comparison to the more complex physically-based models and, at the same time, the prediction accuracy in comparison to the simpler empirical models. This is appreciated by land planners and forest managers, who have a powerful prediction tool easy to be used in data-poor environment, as often the Mediterranean forests are.

4 Conclusions

The evaluation of the ANN for hydrological modelling in the forest plots subject to wildfire showed that the runoff and erosion prediction capability is in general very good. The ANN performance was exceptionally high for all the experimental conditions, although a slight underestimation of the observations was detected. The ANN proposed is also very robust, in the sense that its performance is exceptionally high for all the experimental conditions (burned or non-burned soils) and treatments (mulching and/or logging or not), with only one exception (that is, in the condition where the soil disturbance is higher). Overall, this modelling approach only needs precipitation data (whose measuring equipment are available also in forestlands) as well as a reasonable set of rainfallrunoff observations to train the ANN. Therefore, the use of ANNs for hydrological predictions in burned areas of Mediterranean forests seems to be a useful decision support system for the integrated assessment and management of forested watersheds.

References

1. Gholami, V., Booij, M.J., Nikzad Tehrani, E., Hadian, M.A.: Spatial soil erosion estimation using an artificial neural network (ann) and field plot data. Catena **163**, 210–218 (2018)
2. Haykin, S., Network, N.: A comprehensive foundation. Neural Netw. **2**(2004), 41 (2004)
3. Hecht-Nielsen, R.: Kolmogorov's mapping neural network existence theorem. In: Proceedings of the International Conference on Neural Networks, vol. 3, pp. 11–14. IEEE Press, New York (1987)
4. Lucas-Borja, M.E., González-Romero, J., Plaza-Álvarez, P.A., Sagra, J., Gómez, M.E., Moya, D., Cerdà, A., de Las Heras, J.: The impact of straw mulching and salvage logging on post-fire runoff and soil erosion generation under Mediterranean climate conditions. Sci. Total Environ. **654**, 441–451 (2019)
5. Zema, D.A., Lucas-Borja, M.E., Fotia, L., Rosaci, D., Sarnè, G.M.L., Zimbone, S.M.: Predicting the hydrological response of a forest after wildfire and soil treatments using an Artificial Neural Network. Comput. Electron. Agric. **170**, 105280 (2020)
6. Zema, D.A., Nunes, J.P., Lucas-Borja, M.E.: Improvement of seasonal runoff and soil loss predictions by the MMF (Morgan-Morgan-Finney) model after wildfire and soil treatment in Mediterranean forest ecosystems. Catena **188**, 104415 (2020)

FITS (Special Session)—Chair: Maria Nadia Postorino

Cooperative Negotiation in Connected Vehicles for Mitigating Traffic Congestion

Tri-Hai Nguyen, Gen Li, Hyoenseong Jo, Jason J. Jung, and David Camacho

Abstract Traffic congestion has an impact on traffic efficiency and the quality of life. To address this issue, this paper proposes a distributed, cooperative negotiation method for connected vehicles in traffic flow optimization. In particular, when the connected vehicles obtain the traffic congestion alerts from the roadside units, they exchange their routing information and distribute the traffic flows across the roads by using a collective learning algorithm that does not rely on a centralized controller. Results exported from Simulation of Urban Mobility show that the proposed method outperforms traditional routing methods. In a high traffic demand scenario, the average travel time of the proposed method decreases by 35% and 12% compared with the shortest path routing and the dynamic traffic routing methods, respectively.

1 Introduction

With the growth of urbanization, traffic congestion has gotten increasingly severe in most metropolises. It not only has an impact on traffic efficiency, but it also creates major environmental pollution and traffic incidents, posing a significant risk to people [12, 13]. With present urban space capacity growth constraints, how to fully utilize existing road traffic resources to reduce urban congestion has arisen as the main subject in the development of intelligent transportation systems. With the introduction of new means of transportation and technological advancements, urban mobility

T.-H. Nguyen · G. Li · H. Jo · J. J. Jung (✉)
Department of Computer Engineering, Chung-Ang University, Seoul, Korea
e-mail: j3ung@cau.ac.kr

T.-H. Nguyen
e-mail: haint93@cau.ac.kr

H. Jo
e-mail: joo9245@naver.com

D. Camacho
Departamento de Sistemas Informáticos, Universidad Politécnica de Madrid, Madrid, Spain
e-mail: david.camacho@upm.es

D. Camacho et al. (eds.), *Intelligent Distributed Computing XIV*, Studies in Computational Intelligence 1026, https://doi.org/10.1007/978-3-030-96627-0_12

has altered drastically in recent years. The development of modern communication technologies, in particular, gives rise to the idea of the Internet of Things (IoT), which refers to a collection of physical things that connect to the Internet to gather and share data [7, 14]. Recently, as an application of IoT in urban transportation, connected vehicles have been currently considered as promising solutions for the realization of intelligent traffic management systems in the future [11]. Benefited from vehicle-to-everything (V2X) communications including vehicle-to-vehicle (V2V) and vehicle-to-infrastructure (V2I) communications, connected vehicles can share information with each other and with the infrastructure within their communication ranges. To relieve traffic congestion, two major issues that are intersection traffic control and traffic routing can be addressed. Traffic control at isolated intersections has been well discussed in recent years [2, 3]. However, traffic routing in road networks remains to be further investigated since it is complex to plan for traffic demand when there is high uncertainty on the road such as accidents, road maintenance, hazardous weather, special events, and obstacle vehicles [9]. The existing traffic routing methods rely on centralized traffic control systems, which collect all traffic information and then can assign the optimal routes for vehicles. However, the centralized methods entail immense computing capital because a central controller is deployed to coordinate all vehicles' trajectories in the road network. Besides, the centralized controller can fail to provide the optimal routes in real-time due to the complexity and uncertainty of the traffic networks.

As V2X communication advances, there is a potential to use cooperation among vehicles to address the congestion issue. If vehicles could collaborate opportunistically when they are near, they might effectively spread themselves out among different routes, preventing traffic jams on the same paths. In this work, we propose a traffic negotiation model, in which vehicles acting as smart agents undertake cooperative routing using V2X communications. The two main contributions of the paper are as follows: (i) proposing the distributed negotiation method, wherein each vehicle plans its route preferences toward the destination and resolves conflict situations with other vehicles as they arise along the way; and (ii) demonstrating the effectiveness of the proposed method in the simulations based on Simulation of Urban Mobility (SUMO).

2 Related Work

Traffic routing is the process of assigning traffic demand to routes on a road network. It is a challenge due to the complexity and uncertainty of the traffic networks. Urban transportation has changed dramatically in recent years with the emergence of new transport modes and technological advances [4, 9, 12, 13]. However, some existing works still rely on a centralized controller and predetermined paths. The centralized controller can obtain a full overview of traffic information. However, it results in a high computational burden, scalability issues, and potentially a single point of failure. In dynamic traffic routing, the rerouting of vehicles can be rerouted to avoid the congested roads [4, 12]. However, if many vehicles reroute to the same path, it can create traffic congestion in others places. With the connected vehicle technology, decentralized approaches can distribute the workload among connected vehicles in exchange for a higher effort for synchronizing communication. Swarm intelligence can be utilized to identify and mitigate traffic congestion in order to achieve decentralized traffic routing systems. Techniques based on swarm intelligence, such as ant pheromone, bee foraging, and fish schooling, are influenced by the unique characteristics of species (ants, honey-bees, fish) [14]. These approaches help with traffic flow predictions and traffic organization through indirect communication, resulting in well-coordinated vehicle movement in a decentralized way. For example, ant colony optimization algorithms are investigated in the dynamic traffic routing problem in the connected environment [12, 13]. While the vehicles are moving on the roads, they send their digital pheromones, which usually represent the travel speed, travel time, and traffic density in the current roads, to the corresponding roadside units. The incoming vehicles obtain the pheromone information and perform route choice decision-making to disperse traffic flows over the road network and minimize trip time.

In multi-agent systems, automated negotiation between intelligent agents is a typical technique [8]. When agents negotiate, they may alter their local plans or relax their constraints to reach an agreement with other agents in the system. As V2X communication evolves, connected vehicles can be considered as agents that can work together to solve the traffic congestion problem. Automated negotiation is used in different problems, e.g., supply chain management [6], intersection traffic control [1], vehicle route allocation [5], and smart home energy management [1]. It can be shown that distributed agents can collaborate locally, either directly or indirectly, while still regulating the overall outcomes in the form of global optimization or an acceptable solution.

3 Cooperative Negotiation in Traffic Routing

3.1 Problem Formulation

As shown in Fig. 1, the road network is abstracted as a directed graph $G = \langle N, E \rangle$, where N is a set of nodes or intersections and E is the set of edges or road segments between two intersections. The number of intersections is n and $i, j \in N$ are the intersections. The road segment is defined as (i, j), which has the length l_{ij}, the speed limit s_{ij}^{max}, and its roadside unit (RSU). Two separate directions of a real road are regarded as two distinct road segments. Assuming that the vehicles are connected and they can communicate with one another and with the RSU. Letting V be the set of all connected vehicles entered the road network, each vehicle $v \in V$ is represented by a tuple $v = \langle ID_v, src_v, dst_v, p_v \rangle$, where ID_v is the vehicle identifier, src_v, dst_v,

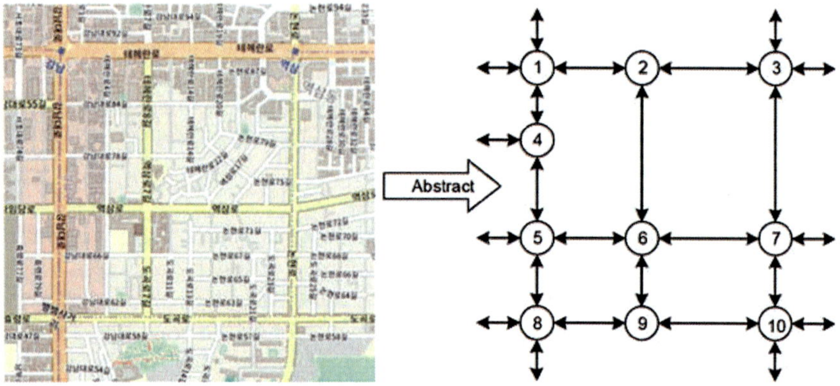

Fig. 1 A directed graph for the road network

and p_v are the source, destination, and current travel path, respectively. The objective is to identify the travel paths for all vehicles with a minimum of travel times as

$$\begin{aligned} &\text{minimize} \quad \sum_{i,j,v} x_{ij}^{v} \cdot t_{ij}^{v} \\ &\text{subject to} \quad x_{ij}^{v} \in \{0, 1\}, \forall i, j, i \neq j \in N, \forall v \in V \end{aligned} \tag{1}$$

where $x_{ij}^{v} \in \{0, 1\}$ is a binary variable that is 1 if and only if v passes (i, j) and t_{ij}^{v} is travel time of the vehicle v on (i, j).

3.2 *Cooperative Negotiation Model*

We focus on the method of cooperative traffic routing among connected vehicles. We assume that the network communication is secure and reliable. When a vehicle chooses a route, it affects the traffic density of that route. If all vehicles choose the same routes without negotiating, they may cause new traffic congestion on the same route. To find the optimum route while avoiding traffic congestion, each vehicle needs to evaluate the routing decisions of the other vehicles based on preference and the real-time traffic conditions on alternate routes.

3.2.1 Travel Time Estimation

Initially, each vehicle selects a preferred route based on estimated travel time. The Greenshield model is used to estimate the travel time [16]. It considers the linear relationship between speed and traffic density as

$$\bar{s}_{ij} = s_{ij}^{max} \cdot \left(1 - \frac{V_{ij}}{C_{ij}}\right) \tag{2}$$

where $\bar{s}_{ij}$ denotes the estimated average speed of (i, j), V_{ij} is the current number of vehicles moving on (i, j), and C_{ij} is the capacity of (i, j). C_{ij} is calculated as

$$C_{ij} = \frac{l_{ij} \cdot lanes_{ij}}{l_v + l_{gap}} \tag{3}$$

where l_{ij} is the length of (i, j), $lanes_{ij}$ is the lanes number of (i, j), l_v and l_{gap} are the average length of vehicles and safety gap between vehicles, respectively. According to the average speed, the estimated travel time is calculated as

$$t_{ij} = \frac{l_{ij}}{\bar{s}_{ij}} \tag{4}$$

Hence, the total travel time of the vehicle v from origin to destination may be represented as

$$tt_v = \sum_{i,j} x_{ij}^v \cdot t_{ij}^v \tag{5}$$

where $(i, j) \in v_r$ is the road segment that belongs to the current travel route v_r.

3.2.2 Traffic Congestion Detection

Every vehicle periodically transmits a message including its ID, position, speed, current route, and timestamp, to the nearest RSU. Based on the received information, the RSU calculates the average speed $\bar{s}_{ij}$ and the total number of vehicles V_{ij} on (i, j) [13]. We use the road density in conjunction with the average road speed to determine the different types of traffic situations, as shown in Table 1. The proportion of average speed to the speed limit of the road is used to classify speed. The density is determined by the percentage of the number of vehicles on the road in relation to the maximum capacity of that road.

Table 1 Traffic congestion classification

Speed	Traffic density			
	Low	Medium	High	Very high
Fast	Free-flow	Free-flow	Free-flow	Slight
Medium	Free-flow	Slight	Slight	Moderate
Slow	Free-flow	Slight	Moderate	Moderate
Very slow	Slight	Moderate	Moderate	Severe

3.2.3 Cooperative Negotiation

Each RSU updates the traffic condition on a regular basis. When there is congestion, RSU sends congestion alert messages to approaching vehicles to facilitate cooperative negotiation in rerouting decisions. To solve this issue, we used a collective learning algorithm, the Iterative Economic Planning and Optimized Selections (I-EPOS) [15]. It utilizes a structured network topology (tree-topology) with synchronous learning iterations and communication between agents.

Each vehicle is considered as an agent and has a set of local plans (alternative paths) when it receives the congestion alert message from RSU. The vehicles self-organize in a tree topology as a means of organizing the interactions and communications. I-EPOS performs consecutive learning iterations, which are divided into two phases: the bottom-up (leaves to root) phase and top-down (root to leaves) phase. At each iteration t, vehicle v chooses the path $p^t_{v,s}$ to achieve the objective

$$p^t_{v,s} = \arg \min_{k=1}^{|P_v|} \left((1-(\alpha+\beta))\left(f_G\left(p^t_{v,k}\right)\right) + \beta\left(f_L\left(p^t_{v,k}\right)\right) + \alpha\left(f_U\left(p^t_{u,k}\right)\right)\right) \quad (6)$$

where $|P_v|$ represents the number of local plans or alternative paths for vehicle v, $f_G(p^t_{v,k})$ denotes the global cost of selecting $p^t_{v,k}$, which is the variance of traffic load across different paths. Each path has a travel time or local cost $f_L(p^t_{v,k})$. $f_U(p^t_{u,k})$ represents the unfairness, measured by the dispersion of the local cost of the selected paths over all vehicles, with lower values indicating more equal distribution of the local cost across all vehicles. The α, β, and $1-(\alpha+\beta)$ parameters represent the preferences of vehicles for unfairness, local cost, and global cost, in that order. Vehicle v selects the path $p^t_{v,s}$ for travel after the maximum of number of iterations F is met. In addition, Contract Net Protocol is adopted to support exchanging the solutions of alternative paths between vehicles [1].

4 Evaluation

4.1 *Simulation Settings*

We conducted the simulations on a real urban road network in SUMO simulator with the Traffic Control Interface (TraCI) via Python[1] [10]. Figure 2 illustrates the SUMO interface with a real traffic network obtained from OpenStreetMap.[2] We compared the proposed method with two non-cooperative algorithms.

- Cooperative Negotiation Traffic Routing (NegoTR) method: This is the proposed method. We assume an ideal scenario in which all vehicles follow the route guid-

[1] https://www.eclipse.org/sumo/.

[2] https://www.openstreetmap.org/.

Fig. 2 SUMO interface with a real traffic network obtained from OpenStreetMap at the coordinates of (37.4991, 127.0473), Gangnam District, Seoul, South Korea

ance. Hence, the $\alpha = \beta = 0$, in which the vehicles are altruistic and the objective is to minimize the global cost.

- Shortest path traffic routing (STR) method: All vehicles take the route with the shortest travel time, which is the free-flow travel time estimated by the length and speed limit of the route, ignoring road capacity and traffic density.
- Dynamic traffic routing (DynTR) method: When an RSU detects traffic congestion in a road segment, the incoming vehicles find new routes with respect to the new estimated travel times [4, 13].

We considered three performance metrics: average travel time (s), average travel distance (m), and average carbon dioxide (CO_2) emissions (g). The CO_2 emissions are calculated using the EMIssions from Traffic (EMIT) model implemented in SUMO. Traffic demands (vehicles per hour, veh/h) are generated by the Random Trips tool following a Poisson process with an arrival rate. In 3600 s of each simulation, the number of vehicles was generated up to 5500 veh/h, which is the maximum capacity of the road network. The other parameters are default settings in SUMO 1.9.2.

Fig. 3 Average travel time

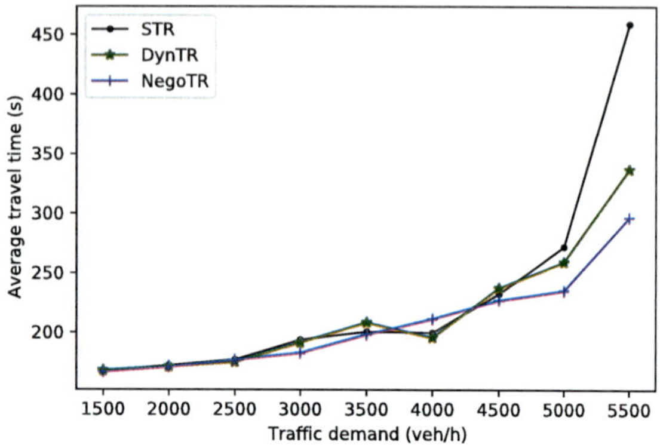

4.2 Results

Figure 3 shows the performance of the proposed method NegoTR in terms of average travel time compared to other traffic routing methods. The average traveling time of NegoTR is mostly the shortest in scenarios with different traffic demands. Compared to STR, DynTR can reduce the average travel time by periodically updating traffic information and rerouting the vehicles; however, it can still lead to new congestion, because many vehicles can be rerouted through the same route. In NegoTR, the vehicles try to distribute throughout the road network; hence, the average travel time is reduced greatly. In the high traffic demand scenario, the average travel time of NegoTR decreases by 35% and 12% compared with STR and DynTR, respectively.

Since the DynTR and NegoTR reroute the vehicles to alternative paths, the new paths may be longer than the original paths. Their average travel distance values are slightly higher than that of the STR, as shown in Fig. 4. The results of the emissions are directly related to the travel time and travel distance. NegoTR outperforms the other two methods as shown in Fig. 5. In the highest traffic demand scenario, NegoTR

Fig. 4 Average travel distance

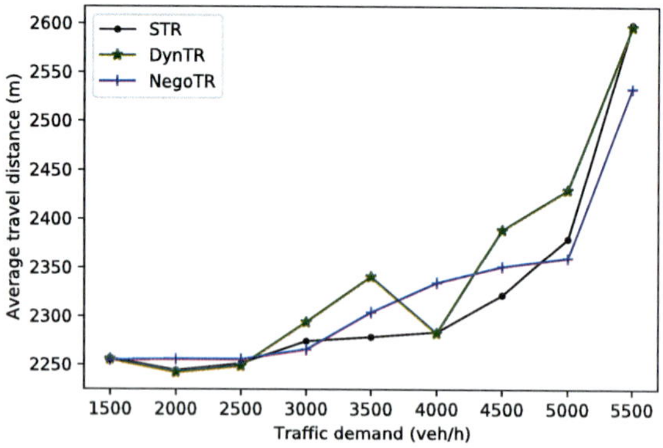

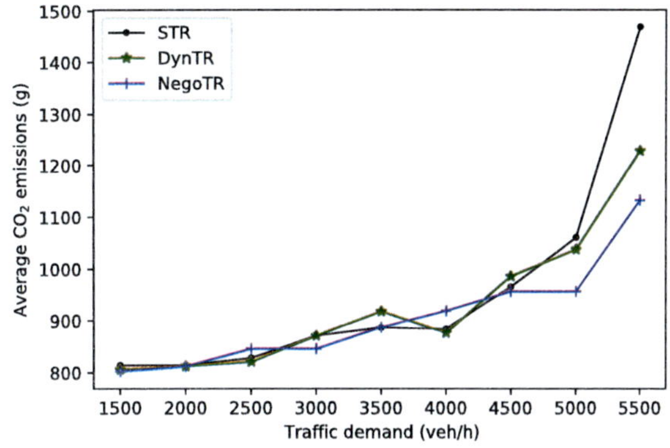

Fig. 5 Average CO_2 emissions

decreases the average CO_2 emissions by 22.85% and 7.80% in comparison with STR and DynTR, respectively.

5 Conclusions

In this paper, we proposed the distributed, cooperative negotiation method in connected vehicles to prevent traffic congestion in the urban road scenario. By exchanging and negotiating traffic routes, the vehicles can travel on the congestion-free paths so that the travel time, emissions, and fuel consumption can be reduced. Simulation results show the effectiveness of the proposed method by decreasing the average travel time by approximately 35% and 12% compared to the shortest path traffic routing and dynamic traffic routing methods in the high traffic demand, respectively.

Acknowledgements This research was supported by the MSIT (Ministry of Science and ICT), Korea, under the National Program for Excellence in SW (2017000100005100l) supervised by the IITP (Institute of Information and communications Technology Planning and Evaluation) in 2021. Also, this work was supported by the National Research Foundation of Korea (NRF) grant funded by the Korea government (MSIP) (NRF-2019K1A3A1A80113259).

References

1. Bui, K.H.N., Jung, J.J.: Computational negotiation-based edge analytics for smart objects. Inf. Sci. **480**, 222–236 (2019). https://doi.org/10.1016/j.ins.2018.12.046
2. Bui, K.H.N., Camacho, D., Jung, J.E.: Real-time traffic flow management based on inter-object communication: a case study at intersection. Mob. Netw. Appl. **22**(4), 613–624 (2017). https://doi.org/10.1007/s11036-016-0800-y
3. Bui, K.H.N., Jung, J.E., Camacho, D.: Game theoretic approach on real-time decision making for IoT-based traffic light control. Concurr. Comput. Pract. Exp. **29**(11), e4077 (2017). https://doi.org/10.1002/cpe.4077

4. Cao, Z., Jiang, S., Zhang, J., Guo, H.: A unified framework for vehicle rerouting and traffic light control to reduce traffic congestion. IEEE Trans. Intell. Transp. Syst. **18**(7), 1958–1973 (2017). https://doi.org/10.1109/tits.2016.2613997
5. Desai, P., Loke, S.W., Desai, A., Singh, J.: CARAVAN: congestion avoidance and route allocation using virtual agent negotiation. IEEE Trans. Intell. Transp. Syst. **14**(3), 1197–1207 (2013). https://doi.org/10.1109/tits.2013.2256420
6. Fang, F., Wong, T.: Applying hybrid case-based reasoning in agent-based negotiations for supply chain management. Expert Syst. Appl. **37**(12), 8322–8332 (2010). https://doi.org/10.1016/j.eswa.2010.05.052
7. Fortino, G., Savaglio, C., Spezzano, G., Zhou, M.: Internet of things as system of systems: a review of methodologies, frameworks, platforms, and tools. IEEE Trans. Syst. Man Cybern. Syst. **51**(1), 223–236 (2021). https://doi.org/10.1109/tsmc.2020.3042898
8. Kiruthika, U., Somasundaram, T.S., Raja, S.K.S.: Lifecycle model of a negotiation agent: a survey of automated negotiation techniques. Group Decis. Negot. **29**(6), 1239–1262 (2020)
9. Li, G., Nguyen, T.H., Jung, J.J.: Traffic incident detection based on dynamic graph embedding in vehicular edge computing. Appl. Sci. **11**(13), 5861 (2021). https://doi.org/10.3390/app11135861
10. Lopez, P.A., Wiessner, E., Behrisch, M., Bieker-Walz, L., Erdmann, J., Flotterod, Y.P., Hilbrich, R., Lucken, L., Rummel, J., Wagner, P.: Microscopic traffic simulation using SUMO. In: 2018 21st International Conference on Intelligent Transportation Systems (ITSC), IEEE, IEEE, pp. 2575–2582 (2018). https://doi.org/10.1109/itsc.2018.8569938
11. Lu, N., Cheng, N., Zhang, N., Shen, X., Mark, J.W.: Connected vehicles: solutions and challenges. IEEE Internet Things J. **1**(4), 289–299 (2014). https://doi.org/10.1109/jiot.2014.2327587
12. Nguyen, T.H., Jung, J.J.: Multiple ACO-based method for solving dynamic MSMD traffic routing problem in connected vehicles. Neural Comput. Appl. **33**(12), 6405–6414 (2020). https://doi.org/10.1007/s00521-020-05402-8
13. Nguyen, T.H., Jung, J.J.: Swarm intelligence-based green optimization framework for sustainable transportation. Sustain. Cities Soc. **71**, 102947 (2021). https://doi.org/10.1016/j.scs.2021.102947
14. Nguyen, T.H., Nguyen, L.V., Jung, J.J., Agbehadji, I.E., Frimpong, S.O., Millham, R.C.: Bio-inspired approaches for smart energy management: state of the art and challenges. Sustainability **12**(20), 8495 (2020). https://doi.org/10.3390/su12208495
15. Pournaras, E., Pilgerstorfer, P., Asikis, T.: Decentralized collective learning for self-managed sharing economies. ACM Trans. Auton. Adapt. Syst. **13**(2), 1–33 (2018). https://doi.org/10.1145/3277668
16. Roess, R.P., Prassas, E.S., McShane, W.R.: Traffic Engineering. Pearson/Prentice Hall (2004)

New Challenges for Urban Air Mobility Systems: Aerial Cooperative Vehicles

Chiara Caterina Ditta and Maria Nadia Postorino

Abstract In recent years, attention has been devoted to Urban Air Mobility (UAM) systems, based on autonomous and cooperative aircraft development as well as on some basic airspace architectures – included air corridors enabling safe procedures and operation, particularly separations among aerial vehicles. In this context, each aerial vehicle would represent an "intelligent" entity that must be provided with suitable computational, sensing and communication capabilities to cooperate with other similar entities. To this purpose, data coming from both the infrastructure and the other aerial vehicles would be autonomously processed by each entity and compared with the current location and the planned destination in order to complete the trip efficiently and safely. Moreover, communication network such as Multi-Group Ad-Hoc Network, would ensure connection and cooperation among Aerial Cooperative Vehicles, also supported by algorithms and software, which will compute and manage flight path in real time to avoid collisions.

1 Introduction

Over the recent years, research in physics, computer science, drone operations and transportation systems have paved the way to a new air transportation system, the so called Urban Air Mobility (UAM), which could improve and redefine urban and interurban mobility. UAM is an emergent mobility concept, which would create a faster, cleaner, safer, and more integrated transportation systems using both the third space dimension and "intelligent" technologies. UAM systems are thought to provide trips for both passengers and goods by advanced and autonomous aircraft flying in lower airspace. In fact, improvement in power electronics, computer science, telecommunications, data analytics, as well as large cost reduction in producing some of the needed components, have opened opportunities for new aircraft genera-

C. C. Ditta (✉) · M. N. Postorino
DICAM, Alma Mater Studiorum University of Bologna, Bologna, Italy
e-mail: chiaracaterina.ditt2@unibo.it

M. N. Postorino
e-mail: marianadia.postorino@unibo.it

D. Camacho et al. (eds.), *Intelligent Distributed Computing XIV*, Studies in Computational Intelligence 1026, https://doi.org/10.1007/978-3-030-96627-0_13

tion [19] known in the literature as electric Vertical Take-Off and Landing (eVTOL), small Unmanned Air System (sUAS), Unmanned Air Vehicles (UAVs), depending on applications (e.g., passenger transport, freight delivery, data collection) and prototypes (e.g., manned and unmanned). Hereinafter, they will be generically named Aerial Cooperative Vehicles (ACV). UAM is expected to produce societal and environmental advantages thanks to the joint use of ACV and advanced telecommunication devices, which will provide direct paths, quick connections and point-to-point links without the need, or with limited use, of roads or ground infrastructures [10]. Focusing on UAM passenger transport, the expected flying parameters of ACV are: minimum cruise speed 240 km/h; range 95 km + reserve range of almost 10 km; 300 m Above Ground Level, AGL) [4]. Their features – mainly high precision on-board sensors, such as radars, Detection and Avoidance (DAA), controlling battery systems and vehicle stabilization – will decrease dependency on pilot skills, while keeping high levels of inflight safety. Moreover, high-performance wireless communication links will transmit and receive data through different frequency band (WLAN, LTE, 5G), and will ensure reliable Vehicle-to-Vehicle (V2V) and Vehicle-to-Environment (V2X) communications [13]. To summarize, UAM would be a new air transport system using ACV to increase and optimize the urban and extra-urban connection capacity. Some preliminary studies (e.g. [12, 14]) have simulated agent-based UAM scenarios, which combine ground and urban air transport. The results have showed that UAM trips below 50 km range would have travel time savings compared to car trips and that UAM advantages would depend on trip origin/destination points, as well as on location and distribution of the so-called vertiports where ACV will take-off and land. For a UAM system to be operational, some key issues must be discussed and solved. Among them, airspace control and flight management rely on the capabilities of ACV to communicate and possibly cooperate in order to keep safe conditions for both on-board users and citizens. However, the development of such a system requires setting (1) the features of the airspace in which ACV are moving, together with the main management and control criteria, (2) the structure of the virtual aerial network, where ACV will link origin/destination points and then (3) the cooperative, distributed and automated ACV properties coherent with the previous two aspects. To summarize (Fig. 1), each vehicle represents an "intelligent" entity provided of suitable computational, sensing and communication capabilities that has to cooperate with other similar entities in order to achieve safe trips. The features of the airspace where such vehicles are allowed and the structure of the air network within such airspace define the environment in which the aerial vehicle is moving, or, in other words, the rules that have to be followed. Data coming from both the infrastructure and the other aerial vehicles are autonomously processed by each entity and compared with the current location and the planned destination in order to complete the trip efficiently and safely. Feedback among Airspace, Air Transport Network and Vehicles in the UAM system will be used to improve both trip safety and efficiency.

This paper wants to introduce the above aspects in the perspective of providing a suitable reference framework where cooperative, distributed and automated ACV can be developed. In this light, the paper is organized as follows. Section 2 presents the

Fig. 1 Relationships between UAM airspace, UAM air transport networks and UAM vehicles

latest infrastructure and systems in the Very Low-Level (VLL) airspace, which are being developing for hosting ACV satisfying UAM requirements. Section 3 describes some services and architectures proposed to integrate flight vehicles and UAM concept in current transport systems. Section 4 describes how automated and connected ACVs should be thought to be coordinated and cooperative. Finally, in Sect. 5 some conclusions are drawn.

2 Management of UAM Infrastructure

For an efficient and effective UAM system to be developed, it is necessary to set a suitable airspace infrastructure in order to allow safe flight management and operation. Currently, it is assumed that early UAM systems will be manned and operated under Visual Flight Rules (VFR) at Very Low Level (VLL) airspace, i.e. 0–500 AGL. Therefore, by considering ICAO airspace classes (ICAO, Annex 11), UAM flights might have origins and destinations in class E (controlled, under VFR) or class G (uncontrolled) airspace. However, UAM is expected to be developed in metropolitan areas to realize fast connections between city areas and airports or among close airports in the region; then, it is be expected that UAM flights will operate also in classes B, C and D [20]. The development of drones and automated aerial vehicles for civil applications drive researchers to identify an air traffic management framework to control daily flight vehicle activities. NASA has presented a Concept of Operations (ConOps) to set safe procedures and activities of Unmanned Aerial Systems (UAS) in the airspace infrastructure and then UAS Traffic Management (UTM) system have been identified. The aim of UTM is to allow safe and efficient low-altitude airspace operations by providing services such as airspace dynamic configuration, dynamic geo-fencing, weather information, route planning, flight separation and congestion management [6]. The UTM architecture is supported by UAS remote pilots, control centers, vehicles and airspace infrastructures, which are interconnected by three different communication methods: wide-area wireless, wireline, and vehicle-to-vehicle (Fig. 2). The UTM architecture was initially designed for drones, but it is evolving to support aerial transport networks with fully autonomous vehicles that will be able to manage their journey independently. Moreover, the UTM system concept would be described by several subsystems: UTM Client, UTM Services, UTM System Automated Services and UTM Manager. The UTM Clients – achieving relevant flight

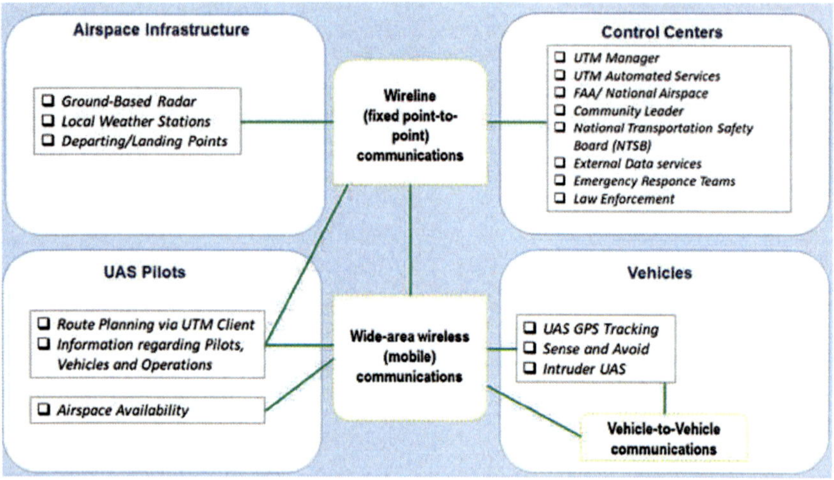

Fig. 2 UTM architecture (Source: 5)

information by Ground-based radars, GPS and local weather stations – are web-based software applications that will permit remote pilots and controllers to access critical airspace data and to plan and review flight paths. UTM Services will allow pilots to submit flight plans, to view real-time vehicle location, and access relevant data to ensure safe operations. In addition, UTM Automated Services will prepare routes for autonomous vehicles and review pilot flight plans. UTM Managers oversee user and vehicle authorisation, off-nominal event management and quality assurance of UTM automated service [5].

NASA defined an UTM platform characterized by four UTM Technical Capability Levels (TCL-1, TCL-2, TCL-3 and TCL-4) [7], which are expected to represent and regulate the complex airspace environment from rural to urban areas. Each increasing level includes and improves the applications and methods of the previous one. Mostly, TCL-4 will aim to manage operations involving simultaneously all autonomous and connected ACV in urban and high-density environment. Operations will include data collection, delivery and on-demand taxi use. Some studies verified that TCL-4 could be employed for UAM operations and the systems could be improved through the implementation of air services [22]. In Europe, a similar approach has been adopted by introducing the U- space, developed in 2017 by SESAR Joint Undertaking (SJU). The U-space involves a series of specific procedures and services, identified to guarantee safe and secure drone activities in airspace through high-level digitalized and automated functions. Moreover, the U-space will support Air Traffic Management (ATM), service providers and authorities to manage operations in all types of airspace (particularly very low level airspace). The U-space service system consists of 4 levels with increasing automation and connection, which require improvement of data exchanges and interaction with the environment [17].

Currently, resources are committed for developing and designing the most advanced U-space services (U3/U4), which will empower UAM activities in complex metropolitan areas. U-space systems should be responsible for providing a suitable airspace separation among manned and unmanned aircraft, controlling functionalities during weather events, and relevant hazards, avoiding unsafe conditions through appropriate operations [18]. UTM and U-space services will aim to bear and control ACV flight plan procedures, also to guarantee that flights will be within an allowed geo-fence area. The service activities could be worsened by the latency and loss of packets due to LTE mobile network performances. To overcome this problem, the framework proposed in [2] is a high-level informatics architecture for a Multi-access Edge Computing (MEC), which would reduce the latency and increase the reliability of the communications between connected ACVs and UTM services [1]. The use of MEC in ACV traffic management and networks would reduce issues due to ineffective communication. The model consists of: (1) the cloud domain, which organizes the management services; (2) the core and transport network, which ensure the communication traffic between MEC-hosted services and the cloud domain, and (3) the MEC nodes, which will host the ACV Flight Controller service. This service will monitor and control the routes of autonomous ACV, will collect information and instruction from cloud-hosted services, to calculate and adapt ACV flights [2].

3 UAM Integrated Network and Flight Planning

Services designed for low level airspace and small ACV operations, through UTM and U-space, have to be implemented with specific architectures in order to provide safe trips. In addition, it may be difficult to adapt traditional Air Traffic Control (ATC) rules and current separation standards to UAM aerial vehicles. Therefore, it is necessary to define a suitable "air network" framework, which will ensure separations among users in controlled low airspace. Traditional ground transportation networks are modelled by using graph theories. The modelled transport network represents infrastructures (e.g., roads) and/or services (e.g., bus lines) where users will move to get their destination from a given origin. Ultimately, users will load paths between origin/destination pairs and congestion effects might be detected along them. Similarly, in UAM networks ACV will move between origin/destination pairs along some "paths", which occur at several levels in the low airspace. Particularly, multimodal and connected transport networks are expected to be developed to allow both ground and urban air mobility. At a first attempt, some approaches are using procedures to generate a flight route (the "path") linking origin/destination pairs that will be followed by ACV in order to get quickly the planned destination. The designed path will avoid potential conflicts by ensuring separations of ACV trajectories. As an example, Bertram et al. (2019) proposed a computational algorithm with collision avoidance capabilities, by using Markov Decision Processes (MDP). Through on-board sensors, ACV will collect information on surrounding obstacles – included the positions of other ACV – and the algorithm will operate online sequential decisions to select actions in real-time. The algorithm, called Memoryless, solves efficiently the problem and has no dependence on the size of the state space. This algorithm would compute the routes in a free flight environment for autonomous on-demand flight in UAM [21]. Other approaches are focusing on models that will provide a structured framework to integrate different operations and systems, such as using airways and corridors similarly to VFR flyways concept [3, 8]. The framework, proposed by Lascara et al. (2019) and extended by Nguyen (2020), combines four operational components to allow successful integration of automated UAM systems into the National Airspace System (NAS) in USA, by placing minimal changes to current procedures: Augmented VFR; Dynamic Delegated Corridors (DDCs); Automated Decision Support Services; Performance-Based Operations. The Dynamic Delegated Corridors (DDCs) are airspace volumes designated for flights using procedures and rules, which could facilitate UAM procedural separation from traditional aircrafts by assuming that ACV will be equipped with flight automation technologies. Contrary to VFR flyways, DDCs use a dynamic approach, therefore size and operating time may change according to the needs of air traffic and the received information. These corridors, combined with ACV features (i.e. hardware and software to support automated cooperation), could be directed by a certified, automated traffic management that would supervise the entry and exit clearance also within the boundaries of different airspace classes. These air volumes are expected to be supported by a high-level software system, to ensure efficient data transmission and operation control among ACV in real time, thus reducing risks. In this perspective, the Automated Decision Support Services would allow information exchange and safe autonomous operations in DDCs so that ACV would be able to exchange information with each other in order to manage the routes autonomously. In addition, it would provide information regarding DDC status, obstacles locations and traffic coordination. The architecture would be similar to UTM architectures and will support flight path control. As opposite to UTM, which currently focuses on non-passenger journeys and Beyond Visual Line Of Sight (BVLOS) operations below 400 ft, the proposed services would be applied for automated aircraft operations with longer flight distances and higher altitudes. In addition, the Performance-based operation concept is addressed to facilitate the growth of UAM operations provided that ACV would be supplied with better performing technologies such that they could fly the most efficient route. Indeed, some DDCs could be open to ACV equipped with a specified level of navigation precision, which allows to provide more direct routing options to these vehicles. This concept enables ACV equipped with high level technologies (i.e. navigation precision, DAA, noise reduction, and V2V communications) to operate in some flight area by using more precise routing options and lower separation requirements [11]. Particularly, to integrate UAM technologies into existing transport systems, some studies have hypothesized to use a multilayers model in uncontrolled airspace, including corridors/airways. In class G space, the separation among layers is guided by the Containment Limit (CL), a volume where the aircraft is within at 95%, at any time of its stated position, horizontally and vertically. The multilayers airspace model is defined by using an elevation map achieved through a topological dataset of static obstacle-free points in the different layers. These points will build an obstacle-free airspace volume, such as a 3D network, which comprises Airways and Nodes – Airways are corridors connecting Nodes within a layer or between layers. When flying in these corridors, ACV will not need to communicate with the UTM, but mainly they will flight by using information exchanged by V2V communications, also thanks to software which allow cooperative operations. The Flight Mode (horizontal flight, vertical flight and hovering) will be changed within nodes. Finally, in this model the separate layers will allow different speed ranges, which increase with altitude [15].

4 Cooperative and Automated Aerial Vehicles

The aforementioned systems will require an Information and Communication Technology (ICT) infrastructure to guarantee effective coordination and control of operations, ACV-to-ACV and ACV-to-infrastructure communications. Automation and connection characteristics will increase the interoperability in order to manage the flow of information related to passengers' demand, freight distribution and emergency situations. The transmission will allow ACV identification, lateral flight speed, location and times. Indeed, by using control algorithms, actuators and integrated software, the aerial vehicles could detect and perceive the environment, analyse detected information, communicate, plan and take decisions independently. Researchers have assumed that the communication scheme would have two architecture network approaches, according to the aircraft level of automation and the support needed by operators: the cellular networks and ad-hoc networks [20]. The cellular networks (e.g. LTE) depend on access points, as base station, by which ACV will be connected through dedicated communication links. Ad-hoc networks do not rely on connections to base stations, already used for V2V communications. It is also suitable for communication application in ACV, e.g. Flying Ad-Hoc Networks (FANET). In ad-hoc networks, aerial vehicles can communicate each other through direct links. If a receiving ACV is out of the communication range of the sending one, other ACV within the range can act as relays and forward the transmission over multiple "hops" [16]. This architecture will support UTM systems and the development of cooperative multi-ACV systems, by supporting ACV independency. Indeed, ACV-to-ACV connections allow (i) coordinating actions, (ii) identifying a framework where the aerial vehicles could handle data information about traffic and environment conditions and make routing decisions, by using airways in VLL. In fact, the active information exchange and the anti-collision sensors (DAA) placed in each aerial vehicle will allow ACV to control autonomously their operations and use the available VLL airspace capacity. Recent studies proposed a multi-group and multi-layer ACV network, based on ad-hoc architecture network to ACV communication.

The multi-group ACV Network (Fig. 3) would consider a backbone ACV connected to the ground station, which also will create connections with other ACV. Intra-group communications will occur within the ACV ad hoc network, while inter-group communications will be achieved through respective backbone ACV and the

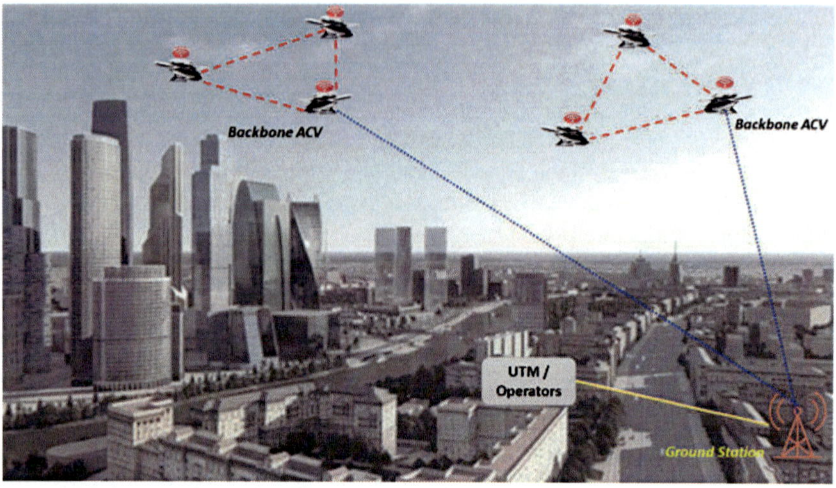

Fig. 3 Multi-group ACV network

ground station. The ground station will allow aerial vehicles to receive information and guideline by UTM/U-space. This ACV-to-ACV architecture is suitable for a scenario that involves many ACV with different flight or communication features. Thus, it could be an efficient solution in UAM networks, which will be a heterogeneous and complex environment where different flights will be involved. The Multi-Layer ACV Network would be described through a lower layer where an individual group realizes ACV ad hoc network, and the upper-layer ACV ad hoc network is composed of the backbone ACV of all groups connected to the ground station. In this network, information exchanged between any two ACV groups does not need to be routed through the ground station. The ground station computation load would be reduced as only data from backbone ACV will be processed. The multi-layer ACV ad hoc network architecture is important for implementing the one-to-many ACV operation mode [9]. In order to reach acceptable cooperation among aerial vehicles in these connected systems, the software features will be crucial for achieving a suitable management of the data flow and for elaborating them in order to ensure safety during flight and approach phases. Due to their automated features, in an integrated and safe UAM network multiple cooperative ACV might guarantee flight paths and vehicle separations. Yang and Wei (2020) presented a message-passing decentralized computational guidance algorithm, which realizes separation capability and runs on each cooperative aerial vehicle. The model is formulated as a multi-agent MDP (MMDP), solved by the Monte Carlo Tree Search (MCTS) algorithm that can run on-board the vehicle. By using MMDP structures and by incorporating information (such as position, speed, destination) by wireless connections, ACV will decide immediately actions to reach the optimal state and separations from other ACVs. The model would allow to (i) avoid potential Loss of Separation (LOS) events among aerial vehicles in the same airspace level, and (ii) guide ACV quickly toward their destinations. The problem would be solved in two dimensions because the vehicles are supposed to fly at the same altitude. In addition, airspace sectorization is proposed with the purpose of reducing the computation time. This framework would be suitable in free-flight conditions and would provide a potential solution for ensuring distributed separation by supporting safe, efficient, and scalable operations in high-density urban air scenarios [23].

5 Discussion and Conclusion

Urban transport systems are undergoing a major revolution. Large, autonomous, connected aircraft/drones for passengers and goods would realize the concept of an urban air transport system and overcome the limitations of ground infrastructures. The realization of ACV and UAM networks require facing many challenges to define safe operations and traffic management. In this paper, we have analysed some recent developments in low airspace infrastructures, both in USA and Europe, and the fundamental requirements of this new transport system, particularly in terms of vehicle cooperation and data processing. We presented the current state of the art of UTM and U-space systems, which in expected to provide a mix of distributed and centralized procedures and services to ensure safe and efficient activities in VLL airspace, based on high level of digitization and computerized functions of ACV. UTM architectures will include several subsystems, particularly communication links and connection technologies (such as wide-area wireless, wireline, and vehicle-to-vehicle) to receive and send data to ACV, in order to support autonomous operations in low airspace. Services supplied by UTM/U-space and supported by MEC architecture will allow ACV to exploit suitably computerized procedures in order to improve and control their route in real time. We focused on MDP computing methods to generate flight paths that would be used by ACV to move safely in uncontrolled airspace. Particularly, ACV would move in specific multilayer networks and volumes, like Dynamic Delegated Corridors, by adopting UAM procedural separations from other vehicles and by autonomously adjusting flight features and paths according to the information collected in the surrounding environment. Suitable data transmission among aerial vehicles will enable autonomous and independent operations, in order to modify speed and horizontal/vertical positions in real time for avoiding collisions and guarantee safe trips for both passengers and freight. Some network architectures for connected and cooperating aerial vehicles have been presented, such as the Multi-Group Ad-Hoc Network that could be appropriate and successful for UAM network applications involving many ACV with different functions and communication characteristics. As a result of a multi-agent MDP, aerial vehicles could cooperate and compute their position to achieve optimal separations among them, by ensuring efficient operations in low airspace. Current approaches are based on free flight environment and consider 2D structures. In order to define a real application, it will be necessary to adapt these algorithms in structured airspace and to consider also vertical flight separations. Finally, UAM and ACV discussed in this paper involve many aspects in the field of information technology, telecommunications, and smart transport systems, which have to be further investigated for proposing suitable UAM scenarios to be implemented in the next future.

References

1. Ali, B., Gregory, M., Li, S.: Multi-access edge computing architecture, data security and privacy: a review. IEEE Access (2021)
2. Bekkouche, O., Taleb, T., Bagaa, M.: Uavs traffic control based on multi-access edge computing. In: 2018 IEEE Global Communications Conference, pp. 1–6. IEEE (2018)
3. Bertram, J., Yang, X., Brittain, M., Wei, P.: Online flight planner with dynamic obstacles for urban air mobility. In: AIAA Aviation 2019 Forum, pp. 3625 (2019)
4. Holden, J., Goel, N.: Uber elevate: fast-forwarding to a future of on-demand urban air transportation. uber technologies. Inc., San Francisco, CA (2016)
5. Jiang, T., Geller, J., Ni, D., Collura, J.: Unmanned aircraft system traffic management: concept of operation and system architecture. Int. J. Transp. Sci. Technol. **5**(3), 123–135 (2016)
6. Kopardekar, P.: Unmanned aircraft system (uas) traffic management (utm): Enabling civilian low-altitude airspace and unmanned aerial system operations (2016)
7. Kopardekar, P., Rios, J., Prevot, T., Johnson, M., Jung, J., Robinson, J.: Unmanned aircraft system traffic management (utm) concept of operations. In: 16th AIAA Aviation Tech., Integration, and Operations Conf., pp. 1–16. AIAA (2016)
8. Lascara, B., Lacher, A., DeGarmo, M., Maroney, D., Niles, R., Vempati, L.: Urban air mobility airspace integration concepts: Operational concepts and exploration approachs. Technical report, MITRE CORP MCLEAN VA MCLEAN (2019)
9. Li, J., Zhou, Y., Lamont, L.: Communication architectures and protocols for networking unmanned aerial vehicles. In: 2013 IEEE Globecom Work, pp. 1415–1420. IEEE (2013)
10. Moore, M.: Concept of operations for highly autonomous electric zip aviation. In: 12th AIAA Aviation Tech., Integration, and Operations Conf., pp. 1–15 (2012)
11. Nguyen, T.: Dynamic delegated corridors and 4d required navigation performance for urban air mobility (uam) airspace integration. J. Aviat./Aerosp. Educ. Res. **29**(2), 57–72 (2020)
12. Postorino, M.N., Sarné, G.M.L.: Reinventing mobility paradigms: flying car scenarios and challenges for urban mobility. Sustainability **12**(9), 3581 (2020)
13. Rohde, Schwarz: An overview of current research and developments in urban air mobility–setting the scene for uam introduction. In: Report (2020)
14. Rothfeld, R., Fu, M., Balać, M., Antoniou, C.: Potential urban air mobility travel time savings: an exploratory analysis of munich, paris, and san francisco. Sustainability **13**(4), 2217 (2021)
15. Samir Labib, N., Danoy, G., Musial, J., Brust, M., Bouvry, P.: Internet of unmanned aerial vehicles—a multilayer low-altitude airspace model for distributed uav traffic management. Sensors **19**(21), 4779 (2019)
16. Schalk, L.: Communication links for unmanned aircraft systems in very low level airspace. In: 2017 Integrated Communications, Navigation and Surveillance Conference (ICNS), pp. 6B2–1. IEEE (2017)
17. SESAR. U-space blueprint (2017). https://www.sesarju.eu/sites/default/files/documents/reports/U-space%20Blueprint.pdf
18. SESAR. Supporting safe and secure drone operations in europe (2020). https://www.sesarju.eu/sites/default/files/documents/u-space/U-space%20Drone%20Operations%20Europe.pdf
19. Straubinger, A., Rothfeld, R., Shamiyeh, M., Büchter, K., Kaiser, J., Plötner, K.: An overview of current research and developments in urban air mobility-setting the scene for uam introduction. J. Air Transp. Manag. **87**, 101852 (2020)

20. Thipphavong, D., Apaza, R., Barmore, B., Battiste, V., Burian, B., Dao, Q., et al.: Urban air mobility airspace integration concepts and considerations. In: 2018 Aviation Tech., Integration, and Operations Conf., p. 3676 (2018)
21. U.S. Department of Transportation. Order jo 7210.3w section 5. charted vfr flyway planning chart program. FAA (2010)
22. Verma, S., Monheim, S., Moolchandani, K., Pradeep, P., Cheng, A., et al.: Lessons learned: using utm paradigm for urban air mobility operations. In: 39th Digital Avionics Systems Conf., pp. 1–10. IEEE (2020)
23. Yang, X., Wei, P.: Scalable multi-agent computational guidance with separation assurance for autonomous urban air mobility. J. Guid. Control Dyn. **43**(8), 1473–1486 (2020)

Combining Coordination Strategies for Autonomous Vehicles in Intersections Networks

Marco Gambelli, Stefano Mariani, Giacomo Cabri, and Franco Zambonelli

Abstract Different strategies have been proposed to date to coordinate autonomous, connected vehicles in crossing an isolated intersection, and their performances analysed. However, few works assess what happens in a network of intersections each managed with a different strategy, a situation likely to occur during the transitory phase when not all intersections in a network can (or are worth to) be equipped with a dedicated computational infrastructure to apply a specific strategy. Accordingly, we analyse the impact of combining different strategies in an intersections network. The key result we achieve is that, even in the presence of a few intersections managed by strategies different from currently mainstream ones (such as traffic lights or precedence), significant improvements to global traffic are measured.

1 Introduction

Autonomous (i.e., self-driving) vehicles [8] will soon relieve us from the duty of driving and paying attention, thus making it possible to exploit travel time in other activities. Furthermore, they will notably reduce crashes, now mostly due to bad human behaviours and errors, they will reduce the number of circulating vehicles,

Work supported by the Italian MIUR PRIN 2017 Project "Fluidware".

M. Gambelli (✉)
University of Modena and Reggio Emilia, Reggio Emilia, Italy

S. Mariani · F. Zambonelli
Department of Sciences and Methods of Engineering, University of Modena and Reggio Emilia, Reggio Emilia, Italy
e-mail: stefano.mariani@unimore.it

F. Zambonelli
e-mail: franco.zambonelli@unimore.it

G. Cabri
Department of Physical, Computer, and Mathematical Sciences, University of Modena and Reggio Emilia, Modena, Italy
e-mail: giacomo.cabri@unimore.it

D. Camacho et al. (eds.), *Intelligent Distributed Computing XIV*, Studies in Computational Intelligence 1026, https://doi.org/10.1007/978-3-030-96627-0_14

and, consequently, traffic and pollution [10]. Finally, they will pave the way to innovative solutions in the provisioning of mobility services [13] with greater quality and efficiency than today.

Most of the applied research in the area to date is about the methods and tools to enable *individual* vehicles to hit the road safely. However, solutions based on coordinating *connected* autonomous vehicles are required in a number of situations [1, 14], such as crossing intersections, entering a motorway, platooning, organising urban deployment and rides for fleets of ride/car-sharing vehicles, improving parking occupancy and reduce parking times. In this paper, we focus on the different coordination strategies that can be adopted for intersection crossing, and in particular on analysing the global impact on a network of intersections that the combined adoption of different strategies can have. Indeed, autonomous vehicles make it possible to exploit innovative intersection crossing strategies, beyond traditional traffic lights and right of way (precedence), such as *reserving* trajectories via a centralised intersection manager, and engaging in competitive *auctions* to gain access to the intersection.

The greater effectiveness of such strategies over traditional ones has been already assessed in several studies [6, 17, 19]. However, an analysis of what happens when, in an urban network of intersections, such strategies are *progressively deployed* only to a fraction of the intersections, is still missing. Indeed, this is the most likely situation that will occur in the near future: first, dedicated infrastructure cannot be deployed to existing intersections all at once for practical reasons; second, some intersections may not be even worth to upgrade to advanced strategies, e.g. due to low traffic; third, legacy approaches may be still reserved to human-driven vehicles.

Accordingly, our contribution is to assess the global impact, at the level of intersections network, of progressively deploying different strategies to different portions of the intersections network. Key results achieved include:

- reservation-based strategies, beside performing better than other strategies at the level of individual intersections (as already known [6]), have the best global impact on the intersections network;
- even in the case where only a few intersections are managed through reservation, significant global improvements to the traffic situation are observed;
- not only the *number*, but also the *locations* where reservation-based strategies are deployed has an impact.

The paper is organised as follows. Section 2 sketches some background about the coordination of autonomous vehicles, presenting the problem and some related approaches. Section 3 presents the simulation environment, the implemented experiments, and the results achieved. Section 4 concludes the paper.

2 Background and Related Work

The problem of coordinating a collective of self-driving, connected vehicles can be modelled as a decision making process, involving vehicles themselves and possibly some infrastructural entities, aimed at orchestrating vehicles' actions so as to achieve a goal which cannot be achieved, or not optimally, by each vehicle in isolation. Depending on the specific nature of the coordination problem, the goal may be: *(i)* shared among a collective of vehicles, such as allocating parking slots to the vehicles of a company fleet, making the coordination problem *collaborative*, or *(ii)* individual, and possibly contrasting with the ones of the other vehicles of the collective, such as finding a parking slot in the presence of multiple private cars looking for a slot in the same zone, making the coordination problem *competitive*.

Intersection crossing is the most representative coordination problem in urban areas, definitely the most studied in the literature, and the one we adopt as case study in this article. Intersection crossing concerns the need of coordinating vehicles while concurrently crossing intersections [12]. As such, it is a *competitive* problem in which vehicles are self-interested agents willing to obtain the right of way as soon as possible across the *shared resource* represented by the intersection. A proper solution to intersection crossing should enable vehicles to *safely* cross an intersection by avoiding collisions, eventually giving each vehicle the right of way, while possibly minimising the average *delay* experienced by vehicles in waiting for the right of way.

2.1 *Traditional Strategies*

Currently, intersection crossing is managed either by a central controller, the traffic light, or by imposing to vehicles (i.e., to their drivers) pre-defined coordination rules (precedences) to be obeyed. Traffic lights are relatively safe (beside hazardous red-crossing drives) and ensure avoidance of starvation. However, they do not promote efficiency, as they leave the intersection area under-exploited, and may make vehicles wait unnecessarily. Precedence is somewhat more efficient in exploiting the intersection resource. Yet, it can induce starvation (e.g., for left-turning vehicles), and delegates safety to individual vehicles (i.e. their drivers). Autonomous, connected vehicles open up new possibilities.

2.2 *Innovative Strategies*

In the presence of connected autonomous vehicles: *(i)* the traffic light can be substituted by a central software controller in charge of communicating to vehicles when and how to safely enter an intersection; *(ii)* traditional precedence-based strategies can be made safer and more efficient by enabling vehicles to communicate, negotiating and agreeing upon which vehicles will pass first.

Reservation-based strategies consider the presence of a central computational authority, the intersection manager, bearing alone the burden of decision making, thus: *(i)* receiving information from vehicles approaching the intersection (i.e., origin, destination, speed); *(ii)* elaborating a set of collision free trajectories, which may require some vehicles to slow down or change lane; and *(iii)* instructing, or directly commanding, the vehicles about what to do, or informing them about what constraints they must abide to while crossing.

The seminal proposal by Dresner and Stone [7], that we implemented in our simulations (see Sect. 3.1), in particular, lets incoming vehicles request assignment of a time slot for crossing to the intersection manager, who computes decisions based on a local control policy. Simulations show that similar approaches dramatically reduce the waiting time for vehicles with respect to traditional approaches based on stop signs or traffic lights [7], because: *(i)* the occupancy of the intersection is maximised and *(ii)* non-colliding vehicles can cross the intersection simultaneously.

A problem of centralised approaches is requiring the presence of a dedicated infrastructural element (the intersection manager) and the capability of vehicles to interact with it at all times. Thus, they can hardly be applicable in the wild, and adopting them in a road network requires a progressive implementation covering increasing portions of the network at different times.

Negotiation-based strategies let vehicles participate in a protocol aimed at establishing the intersection crossing order. Such protocol, given the competitive nature of the problem, can take the form of an *auction*, where vehicles place a "bid" to "buy" the portions of the intersection they require for crossing, for the time required to cross. A broker collects the bids, gives the right of way to the set of vehicles that are in a collision free trajectory and, amongst those that are in collision, to the ones having placed the highest bid.

Examples of auction-based protocols for intersection crossing are described in [2, 4, 18]. There, different policies to resolve the auction are analysed, based on the different strategies of the bidding vehicles, as well as the different strategies adopted by the broker in establishing the winners. In general, auction-based strategies tend to offer performances similar to precedence-based approaches, but let vehicles try to avoid long waiting times by tuning their own bidding strategies.

2.3 *Intersection Networks*

The aforementioned literature aims to optimise traffic at the level of individual intersections. In our work, we are interested in understanding the impact of different coordination strategies at the level of a network of intersections.

Besides centralised, optimisation-oriented approaches [11], decentralised solutions involve the distributed and dynamic identification of the vehicles' routes, so as to balance the traffic load in the network and avoid congestions. Among the many proposals in this area, there are some based on route negotiation among vehicles [5], or on nature-inspired self-organising vehicles movements [3, 15]. Our study, although still aimed at traffic optimisation, has a specific focus on intersection management rather than on route management. The study presented in [18] analyses the impact of competitive auctions on an intersections network, showing that the auction mechanisms, and the preference for vehicles to reduce the overall price paid in their trip, can lead to effective distribution of traffic. Yet, this study does not address the impact of intersections managed by different strategies. In [9], instead, it is shown that the presence of reservation-based intersections can induce improvement of global traffic in complex intersections networks where also traditionally managed intersections exist. Although analysing such peculiar situations is definitely of interest, our study is more focused at analysing the general impact of different crossing strategies.

3 Simulations and Results

After a careful analysis of the existing simulators in the field, we decided to use *SUMO (Simulation of Urban Mobility)*. It is a highly portable, open-source, microscopic simulator that allows, among other things, to: *(i)* carry out simulations with a large number of vehicles; *(ii)* customise the behaviour of each individual vehicle in various ways; *(iii)* create large networks with multiple lanes in each direction. For more details on SUMO and all its features, we refer the reader to the official documentation [16]. We also used Python as the scripting language to control the simulation through *TraCI (Traffic Control Interface)*, the SUMO library allowing programmers to interact with the simulations from code. All the code developed is publicly available on Github: https://github.com/zer0meister/crossing.

On SUMO we built our multi-intersection network depicted in Fig. 1. Specifically, we use a squared network of 5 intersections per side, where each road has 3 reserved lanes (right-turn, straight, and left-turn). For each simulation, the following

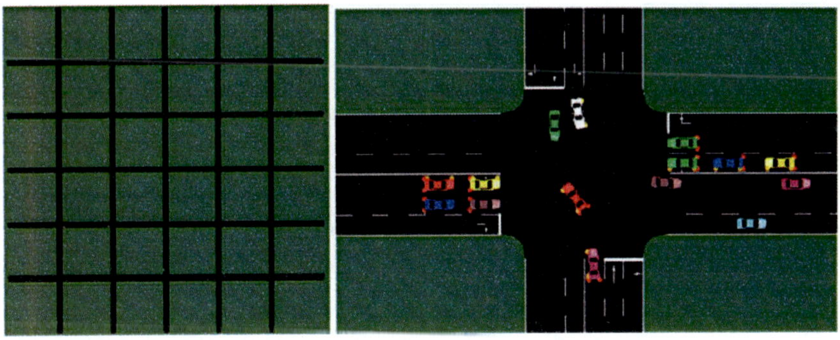

Fig. 1 Network of intersections (left) and detail of an intersection (right)

parameters have been used, common to all the analysed strategies (as described in next section): *(i) vehicles/s*, as the number of vehicles put into the road network per second; *(ii) runs*, as the number of executions that a given simulation, with the same parameters, has been launched to average out random effects such as vehicles' locations and followed routes (see next item); *(iii) simulation seed*, as the seed for the random numbers generator responsible for randomising vehicles' initial locations, routes within the network, etc. across simulations. As 10 runs for each simulated scenario have been performed, 10 seeds have been used.

3.1 Implemented Strategies

We implemented different intersection crossing strategies, and assigned them to different intersections amongst those in the road network, so as to simulate different combination of strategies and see how the global performance of the network varies, and if there is an optimal displacement of strategies. Precedence, that is used as baseline, is not explained as it is common nowadays hence widely known.

Auction-based. Auctions are carried out for groups of vehicles whose trajectories are colliding at the intersection, in two variants: the first, competitive, has the vehicle making the highest bid get its right of way, while the losing vehicles participate to the next auction; the second, cooperative, has permissions given to all participating vehicles according to the value of bids ranked in descending order. In both cases, for each lane, a group of vehicles in that lane can participate to an auction at anytime: the one at the head of a queue will make a bid, and those behind will support such bid by acting as "sponsors" (hence lending part of their money or road credits to the leading vehicle). Two main components have been implemented: the *Auction Manager*, who has the task of managing auctions from creation to resolution, and the *Crossing Manager*, who takes care of keeping track of the vehicles passing through the intersection, and to coordinate them when they get their right of way.

Reservation-based. Each intersection has been divided into a matrix of occupancy cells (see Fig. 2), each meant to track the occupancy times of the different vehicles whose trajectory crosses that cell. The algorithm estimates occupancy times for each vehicle, based on current speed and distance to the intersection, and gives the right of way to all non-colliding vehicles, and to either one of those that will occupy the same cells at the same time according to the *FCFS (First Come First Served)* policy.

3.2 Results

Figure 3 shows the impact of displacing a combination of crossing strategies in different intersections of the road network (as per description in legenda) on the time spent by vehicles as *leaders of a queue* of vehicles waiting to cross (hence, the

Fig. 2 Different subdivisions of an intersection in a matrix of cells

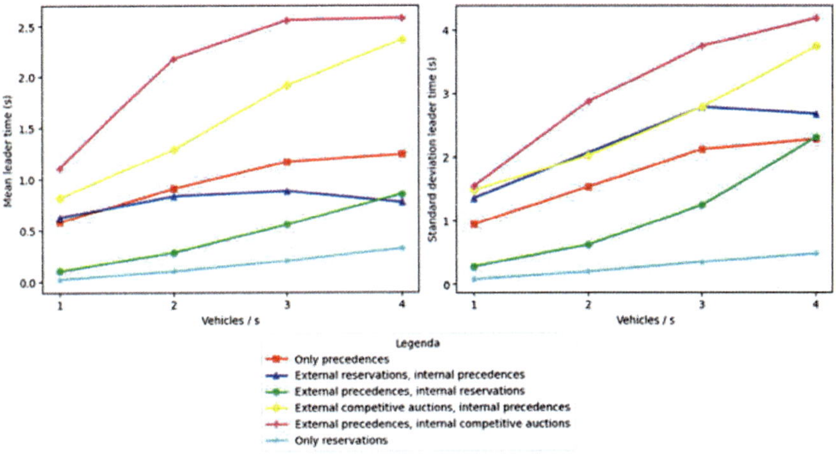

Fig. 3 Best combination of approaches for metric "time as leader of queue" is reservation-based internally and right of way externally

lower the better). The first information such a graph delivers, is the confirmation of literature results about reservation-based approaches being the best performing ones, even when considering a network of intersections (compare red, square-dotted line with cerulean, diamond-dotted one).

We then simulated more heterogeneous scenarios. In fact, it may be not possible to have *all* intersections equipped with an intersection manager at once: on the one hand, the installation of an intersection manager will require money and work, so their introduction will likely be carried out gradually; on the other hand, there could be little benefit in equipping intersections usually experiencing low traffic or that do not influence the overall traffic in the network much.

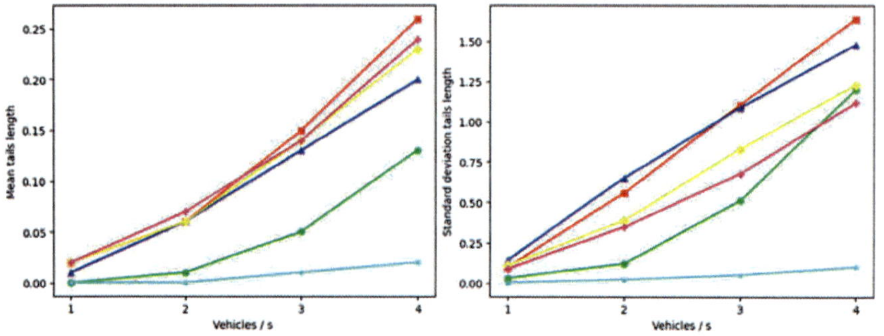

Fig. 4 Best combination of approaches for metric "queue length" is reservation-based internally and right of way externally (legend as per Fig. 3)

For these reasons, we simulated mixed scenarios where different approaches are combined (other lines in Fig. 3). A novel result is that by replacing some of the reservation-based intersection managers with other approaches, placed strategically across the road network (as per description in legenda), the overall performance of the intersections network degrades gracefully, depending on two factors: *(i)* the location of the intersections replaced, and *(ii)* the specific crossing strategy adopted.

In Fig. 3, for instance, the second best approach, following the purely reservation-based one, is the one exploiting reservation in internal intersections and precedence in external ones (with reference to the grid shown in Fig. 1). For comparison, the combination of precedence with competitive auctions is the worst. It is worth emphasising that an approach using only precedence is still worse than one mixing precedence with reservations, indicating that progressively deploying intersection managers in nowadays intersections would still provide benefits gradually until reaching the optimum (all reservations).

Figure 4 confirms the results, as for the performance metric averaging the *length of queues* across the intersections network, the best approach still is precedence externally and reservation-based internally. Also, notice the standard deviation, which is the smallest one when the reservation approach is adopted. Even if this does not affect the general performance, it can lead to fair distribution of the resources (in our case the intersections), avoiding very fast vehicles and very delayed ones. This can be taken into consideration by a city administration, for instance.

Finally, it is interesting to study whether there is a "critical mass" of intersections that should be managed by reservation to gain benefits, or whether improvement of performance is more linear. In other words, whether a third factor determines the aforementioned graceful degradation of performance: the number of intersections managed through reservation, irrespective of their locations. Figure 5 shows the impact of progressively deploying more and more reservation-based strategies across the intersections network (from 1 to 10, randomly displaced): the more the intersections managed through reservation, the more efficient the overall network.

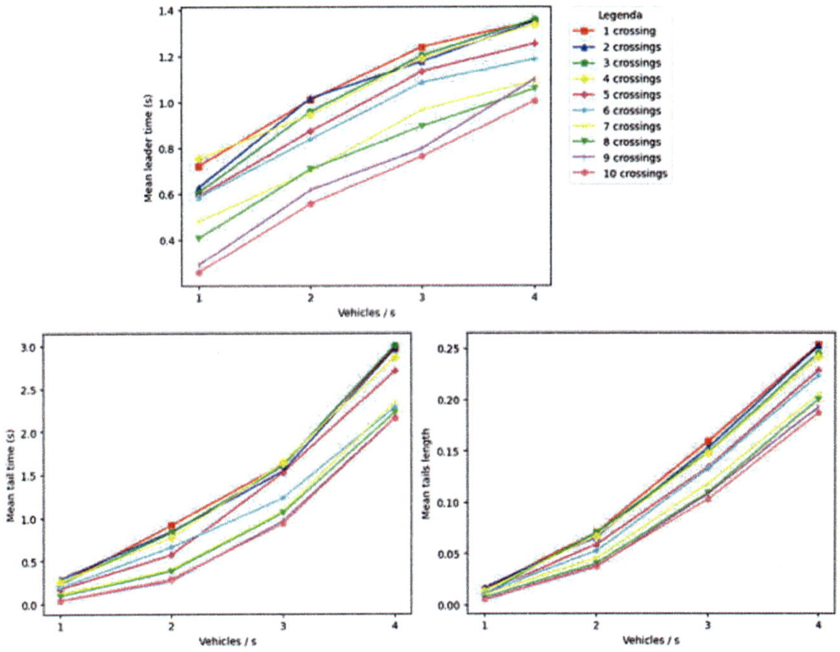

Fig. 5 Effects of progressively deploying reservation-based strategies

Results show that the number of reservation-based intersections linearly improve all the performance metrics already considered in the previous experiments.

Summing up, the experiments show that, in presence of different approaches, both the number of and the choice where to place the intersection management approaches influence the performance of the network. As a city administration can have a limited number of intersection managers to deploy, having criteria for choosing where to put them is important. Our results show that placing the intersection managers in the most "crowded" intersections (those towards the centre of the road network) leads to the best results. Moreover, the results confirm that the approach based on reservation, even in combination with precedence, is the best one with regard to performance, and that increasing the number of intersections managed through reservation has a linear (positive) impact on performance.

4 Conclusions

In this paper, we have analysed the impact of deploying different coordination strategies to different intersections within a network.

From the simulation results we can state that: *(i)* the reservation-based strategy presents the best performances, even mixed in with different strategies; *(ii)* the choice of the strategy for each intersection, the number of intersections with that strategy, as well as their location, influence the overall performance; *(iii)* the reservation approach presents the smallest standard deviation leading to fairer distribution of the delays.

In our work we have considered all autonomous vehicles, which is a long-term scenario; in a mid-term scenario we must consider also human-driven vehicles, and an appropriate model is due. Second, we have considered a fixed approach for each intersection, while the approach can change depending on different factors; it would be very interesting to have self-adapting intersections, i.e. intersections that choose the best approach in a dynamic way.

References

1. Abeywickrama, D.B., Mamei, M., Zambonelli, F.: Engineering collectives of self-driving vehicles: the SOTA approach. In: Leveraging Applications of Formal Methods, Verification and Validation. Lecture Notes in Computer Science, vol. 11246, pp. 79–93. Springer (2018)
2. Cabri, G., Gherardini, L., Montangero, M.: Auction-based crossings management. In: Proceedings of the 5th EAI International Conference on Smart Objects and Technologies for Social Good, GoodTechs '19, pp. 183–188. ACM, New York (2019). https://doi.org/10.1145/3342428.3342689
3. Camurri, M., Mamei, M., Zambonelli, F.: Urban traffic control with co-fields. In: Environments for Multi-Agent Systems III, Third International Workshop, E4MAS 2006, Hakodate, Japan, May 8, 2006, Selected Revised and Invited Papers. Lecture Notes in Computer Science, vol. 4389, pp. 239–253. Springer (2006)
4. Carlino, D., Boyles, S.D., Stone, P.: Auction-based autonomous intersection management. In: 2013 16th International IEEE Conference on Intelligent Transportation Systems-(ITSC), pp. 529–534. IEEE (2013)
5. Desai, P., Loke, S.W., Desai, A., Singh, J.: Caravan: Congestion avoidance and route allocation using virtual agent negotiation. IEEE Trans. Intell. Transp. Syst. **14**(3), 1197–1207 (2013)
6. Dresner, K., Stone, P.: Sharing the road: autonomous vehicles meet human drivers. In: Proceedings of the 20th International Joint Conference on Artifical Intelligence, pp. 1263–1268. Morgan Kaufmann Publishers Inc (2007)
7. Dresner, K., Stone, P.: A multiagent approach to autonomous intersection management. J. Artif. Intell. Res. **31**, 591–656 (2008)
8. Fagnant, D.J., Kockelman, K.: Preparing a nation for autonomous vehicles: opportunities, barriers and policy recommendations. Transp. Res. Part A: Policy Practic. **77**, 167–181 (2015)
9. Levin, M.W., Boyles, S.D., Patel, R.: Paradoxes of reservation-based intersection controls in traffic networks. Transp. Res. Part A: Policy Practic. **90**, 14–25 (2016)
10. Menon, N., Barbour, N., Zhang, Y., Pinjari, A.R., Mannering, F.: Shared autonomous vehicles and their potential impacts on household vehicle ownership: an exploratory empirical assessment. Int. J. Sustain. Transp. 1–12 (2018)
11. Qian, G., Guo, M., Zhang, L., Wang, Y., Hu, S., Wang, D.: Traffic scheduling and control in fully connected and automated networks. Transp. Res. Part C: Emerg. Technol. **126**, 103011 (2021)
12. Rios-Torres, J., Malikopoulos, A.: A survey on the coordination of connected and automated vehicles at intersections and merging at highway on-ramps. IEEE Trans. Intell. Transp. Syst. **18**(5), 1066–1077 (2017). https://doi.org/10.1109/TITS.2016.2600504
13. Röth, T., Pielen, M., Wolff, K., Lüdiger, T.: Urban vehicle concepts for the shared mobility. ATZ worldwide **120**(1), 18–23 (2018). https://doi.org/10.1007/s38311-017-0163-4
14. Sassi, A., Zambonelli, F.: Coordination infrastructures for future smart social mobility services. IEEE Intell. Syst. **29**(5), 78–82 (2014). https://doi.org/10.1109/MIS.2014.81

15. Slager, G., Milano, M.: Urban traffic control system using self-organization. In: 13th International IEEE Conference on Intelligent Transportation Systems, pp. 255–260 (2010). https://doi.org/10.1109/ITSC.2010.5625236
16. SUMO Development Team: Sumo Documentation. https://sumo.dlr.de/docs/SUMO_at_a_Glance.html
17. Tonguz, O.K.: Red light, green light–no light: tomorrow's communicative cars could take turns at intersections. IEEE Spectr. **55**(10), 24–29 (2018). https://doi.org/10.1109/MSPEC.2018.8482420
18. Vasirani, M., Ossowski, S.: A market-inspired approach for intersection management in urban road traffic networks. J. Artif. Int. Res. **43**(1), 621–659 (2012)
19. Zhang, Y.J., Malikopoulos, A.A., Cassandras, C.G.: Optimal control and coordination of connected and automated vehicles at urban traffic intersections. In: 2016 American Control Conference (ACC), pp. 6227–6232 (2016). https://doi.org/10.1109/ACC.2016.7526648

Using Boundary Objects and ANT Concepts to Implement Multi-agent Systems as Drivers of Innovation in Logistics

Check for updates

Yury Iskanderov and Mikhail Pautov

Abstract This paper introduces application of the boundary objects (BO) method combined with the actor-network theory (ANT) to the problems of overcoming knowledge boundaries between organizations when designing processes for inter-organizational knowledge integration aiming to achieve common knowledge and understanding. The case study is based on the process of propagation of technological and organizational innovations based on knowledge sharing and integration between the MAS R&D community and various logistics systems.

1 Introduction

This paper is using the paradigm of boundary objects first introduced in [2] to reflect on the innovative role played by the multi-agent systems (MAS) in nowadays logistics. The case study is inspired by the field research conducted by a group of researchers from the Technical University of Berlin [1] still pending publication, where one of us has participated as an expert to inform investigation of the potential of use of multi-agent systems in modern logistics. The goal of the research group was to identify different application scenarios of multi-agent systems for logistics, their potential, and estimate the point in time of widespread implementation [1]. It is important to emphasize that the innovation projects on the way of their translation into the final practical innovations (adopted new technologies and/or organizational and managerial policies and practices) normally encounter significant resistance of the actors. This resistance to innovation emerges at the interfaces between the R&D communities and logistics systems. The reason behind this resistance is that the logistics theorists and the developers of the novel technologies and organizational methods destined to real logistics systems do not always avail of enough knowledge

Y. Iskanderov
The St. Petersburg Institute for Informatics and Automation of RAS, 39, 14-th Line, St. Petersburg, Russia

M. Pautov (✉)
Foscote Group, 23A Spetson St. 102A, 4000 Mesa Geitonia, Limassol, Cyprus
e-mail: mikhail.pautov@foscote.net

D. Camacho et al. (eds.), *Intelligent Distributed Computing XIV*, Studies in Computational Intelligence 1026, https://doi.org/10.1007/978-3-030-96627-0_15

in practical logistics, and the logistics experts, in turn, often treat the developments and suggestions provided by the researchers and developers of the novel technologies (IT being the champion in innovations) with skepticism and mistrust. It is worth mentioning that the innovative logistics solutions (e.g. route optimization, resource planning and management etc.) face similar resistance of enterprises using traditional practices. As the 'social worlds' (in terms of the boundary objects paradigm [2]) the MAS community and the logistics experts representing various branches of the logistics industry manifest their specific worldviews with their characteristic systems of values, beliefs, interests and goals that support and constrain thinking and action in a particular direction [3]. These worldviews introduced as institutional logics (a concept with roots in institutional theory) may cooperate or compete with each other, resulting in incoherencies across the regime [3]. When an innovation attempts to transcend the boundaries of a target system, a natural tension emerges between the new and the old. This tension manifests itself in a competition materialized in terms of the actor-network theory through mobilization of allies by the traditional technologies on the one hand and innovations on the other hand in their struggle for survival and favorable decisions of the decision-makers of the target system. In case of a successful adoption of the full package of innovations on a small segment of the system, or in case of their selective adoption on the entire system with satisfactory return on investment (ROI) and other user characteristics (e.g. convenience in use, increased efficiency of the processes, improvement of the economic parameters, elimination of system breaches and failures, improvements in general security etc.) the decision-makers can gradually extend the segmental innovations to the entire system, or test the additional functions provided by the new technologies. In general terms, a transition in a sociotechnical system can be conceptualized as a transformation of the relative influence of its constituent institutional logics [3]. A compromise solution could be a dialog, a search for a common language of knowledge presentation to form common discourse to create consortia for consolidated development, implementation and use of the innovations (resulting in adequate and commonly accepted domain logic used in assembling and automated management of logistics systems). The basic question posed here is how the innovation is conducted into the logistics systems through cross-organizational knowledge integration and implementation of multi-agent technologies overcoming resistance of the logistics systems to innovation. The method of boundary objects presented in the next section in combination with the actor-network concepts provides a theoretical framework for this problem.

2 Boundary Objects and ANT

The boundary objects were introduced in [2] as the objects which simultaneously inhabit several intersecting social worlds and satisfy the informational requirements of each of them. "Boundary objects are objects which are both plastic enough to adapt to local needs and the constraints of the several parties employing them, yet robust enough to maintain a common identity across sites" [2]. The boundary objects may be concrete or abstract (metaphorical and intangible [4]). They have different semantics in the different social worlds but their structure is common to these social worlds to make them a means of translation [2]. "The creation and management of boundary objects is a key process in developing and maintaining coherence across intersecting social worlds" [2]. Boundary objects can be understood as the "stuff of action" [6], i.e. a sort of arrangement allowing different groups to work together without consensus. In other words, boundary objects support the collaboration of heterogeneous groups by letting them communicate and work together on a target that is not yet mutually perceived by these groups [4]. Boundary objects represent artefacts (things, concepts, processes, narratives etc.) which can be differently interpreted in different social worlds, however constitute a solid basis for communication and interaction between different sociotechnical systems [3]. The boundary objects concept appears fruitful for identification of informative artefacts (books, documents, records, citations or other informative representations) as socio-material forms [5]. As representational devices boundary objects function as central actors in the "structuring of practices and technologies as alignments of both material and discursive practice" [5]. A formal presentation of the idea of a boundary object is offered in [3] based on insights from semiotics. A boundary object is presented as a signifier (a discursive or material artefact or a process) characterized by its large interpretive flexibility [2]. As a sign a boundary object can be formally defined as a tuple of the three components [7]: $<n, p, M>$, where n is the name of the sign; p is the portrait (image) of the sign; $M = \{m_i\}$ is the set of meanings or 'sign significances' [8] of the sign. The set of meanings M reflects the ability of a boundary object to adopt different (concrete) signifieds m_i inside different institutional logics (this adoption of different signifieds approximates various translation scenarios in terms of the actor-network theory), while keeping its identity as a unique signifier to support collaboration across institutional logics without equalizing its perception by the various actors [3]. Some authors perceive boundary objects as interfaces between multiple social worlds facilitating the flow of resources (information, concepts, skills, materials) among multiple social actors [5]. As a translational device, boundary objects perform critical work at organizational and intellectual boundaries [5]. The word 'boundary' denotes the shared space between institutional logics, and the word 'object' refers to the shared aim of action rather than a material entity [3]. There is an obvious connection between the boundary objects paradigm and the actor-network theory. Text [2] recognized as canonic for the boundary objects method was created in close discussions and polemics with Latour, Callon and Law—the 'fathers' of ANT. The two methods recognize the central role played by the process of translation. Also, the concept of obligatory passage point (OPP) defined as the subject and driver of the translation (an entity intending to speak on behalf of other actors of the actor-network translating in a way that "fits together their interests and behaviours" [9]) is fundamental for both: ANT and the BO approach (although the authors of [2] suggest the n-way nature of the interessement phase of the translation process with multiple passage points, which will be discussed in the next section). Boundary objects are needed and should be taken into account to overcome the possible knowledge boundaries between organizations when designing processes for inter-organizational knowledge integration aiming to achieve common knowledge and understanding [4]. The authors of [3] provide some examples of boundary objects given in the literature: computer models, IPCC reports on climate changes, ecosystem services, innovation narratives, sustainability transitions. In the next section we will define and discuss the boundary objects used for integration of knowledge on MAS among multiple actors and the potential of its use in logistics.

3 Implementing Multi-agent Systems as Drivers of Innovation in Logistics

The work [1] done by the research group from the Technical University of Berlin (the results of which, we believe, will be published by its authors in due course) gives a vivid example of how the MAS R&D community and the diverse logistics systems (representing manufacturing, local distribution, online taxi services, airports, supply chains, logistics networks etc.) interact with each other overcoming the communicational and knowledge barriers with the aim of implementation of MAS as the efficient driver of technological and organizational changes in logistics. It is the process of propagation of innovation in logistics through the widespread implementation of MAS that we focus on and interpret using the language of boundary objects and ANT. Boundary objects are considered the mediating actors—carriers of innovation and knowledge integration between the MAS R&D community and diverse logistics systems. The authors of [3] suggest a seven-phase life cycle of a boundary object in its transition. Though we appreciate the originality of this idea, we deem that it should specifically apply to the disruptive regimes (like the case of sustainability transitions studied in [3]). In a normal (non-disruptive) regime like the case of inter-organizational knowledge integration and propagation of innovations we suggest to consider a 'life cycle' of boundary objects as the four-phase macro-process of translation as it is conceptualized in ANT [10] thus highlighting the organic nexus between the two methods (the boundary objects paradigm and the actor-network theory) having common theoretical roots. An overview of the translation process in actor-networks was given by us in recent publications (e.g. [10]). Here we just briefly recapitulate the basic definitions. Process of translation in actor-networks can be defined as a delegation of powers of representation from a set of actors (actor-networks) to any particular (black-boxed) actor or actor-network in a particular program of actions: $A = T(A_1, \ldots, A_n)$, where T is the translation of actors $A_1, \ldots, A_n$ to A [10]. "A translates B" means A defines B, regardless of whether B is human or nonhuman, a collectivity or an individual [11]. Translation equalizes actor-network actions in various space–time areas and various meta-levels of presentation (e.g. when behaviour of an actor-network is translated through texts: graphs, diagrams, algorithms, formulae etc.). Any association is possible if this association is encoded as heterogeneous connections established through the operation of translation [12]. Any translation is in principle reversible. Irreversibility of translation means its ability to resist reversed translation (retranslation) and competing translations [10]. The more numerous and heterogeneous the interrelationships in an actor-network the greater the degree of network coordination and the greater the probability of successful resistance to alternative translations [11]. The process of translation in actor-networks can be presented as a tuple of consequent operations: $T = <P, I, E, M>$, where P is problematisation, I is interessement, E is enrolement and M is mobilization [10]. These four operations described one by one below reflect the four phases of the life cycle of the boundary objects considered in this paper.

Problematisation. Problematisation is the first phase of the translation process, because according to Latour every action wants to solve a problem [9]. Problematisation can be defined as something that is indispensable and where one or more key actors try to define the exact nature of the problem as well as the roles of other actors that could fit with the proposed solution [13]. At this first stage the problem and the key actors are defined and the boundary objects are constructed and described as follows.

Problem: Assessment of the potential of multi-agent systems in logistics and development of a technology roadmap for relevant scenarios [1].

Key actors: RND—the MAS R&D group targeting the logistics sector; $LOG_1 \ldots LOG_n$—logistics systems representing multiple branches of the industry (manufacturing, distribution, supply chains, airports, airlines, shipping lines, online taxi services, logistics networks etc.); MAS_TUTORIAL, MAS_SURVEY—boundary objects defined as follows:

MAS_TUTORIAL: the brief MAS tutorial introducing the MAS basics to the logistics experts, prefaced by an invitation to future cooperation in the form of a survey. It provides key definitions in a language more or less adapted to the worldview of the logistics experts (i.e. in the most 'popular' style destined to the general public), e.g. definition of an agent as "a computer system that is situated in some environment, and that is capable of autonomous action in this environment in order to meet its design objectives" [14], highlighting the four basic characteristics of an agent: autonomy, reactivity, pro-activeness, social ability. MAS_TUTORIAL further provides a brief history of MAS and their basic attributes (decentralization, modularity, flexibility and robustness). Then the tutorial defines the characteristics of the agents' functional environment in a way the researchers find most relevant for the world of logistics as follows: non-deterministic, inaccessible, dynamic, continuous. Finally, the tutorial gives a SWOT (strengths, weaknesses, opportunities, threats) matrix for MAS applications in logistics and concrete examples of the successful practical applications of the agent-based adaptive scheduling and other multi-agent technologies in various branches of the logistics industry with indexes reflecting efficiency of these applications (reduction of lost sales, increase in service level and profitability), e.g. real-time transport optimization for transport logistics and 3PL; real-time taxi scheduling for online taxi services; airport ground handling management (combination of MAS and RFID technologies) for airports; manufacturing control and better adaptation of production line to changing conditions for manufacturing; real-time adaptive event-driven scheduling for replenishing products for manufacturers and distributors; production process planning and monitoring, process and quality control with self-adaption at local and global level for various industries; ground floor and route optimization for each cargo item and vehicle for airlines. MAS_TUTORIAL is an 'ideal type' boundary object as it is defined in [2]: it serves as a means of communication and cooperation between actors symbolically, i.e. a 'good enough' roadmap for all parties [2] unambiguously perceived by RND and $LOG_1 \ldots LOG_n$: RND shares MAS-related knowledge integrated in logistics context, $LOG_1 \ldots LOG_n$ absorb this knowledge.

MAS_SURVEY: the interactive tool for knowledge exchange and integration between RND and $LOG_1 \ldots LOG_n$ activated after the basic knowledge on MAS and their applications in logistics is integrated into $LOG_1 \ldots LOG_n$ through MAS_TUTORIAL. The standard procedure is based on a multi-round interactive questionnaire sent by RND to $LOG_1 \ldots LOG_n$. The method used in [1] is Delphi survey method for small groups of experts based on the three-round anonymous questionnaire with the option for experts to share feedback without any discretions and the group summary handed back to the participants to re-evaluate their answers. This method is regarded as an exploratory and detailed procedure with long-term forecasting accuracy higher versus the conventional methods and the focus of study determined and adjusted through the answers of the participants [1]. Unlike MAS_TUTORIAL considered a one-way knowledge integrator (designed to integrate the MAS knowledge into $LOG_1 \ldots LOG_n$), MAS_SURVEY is a two-way knowledge integration tool where $LOG_1 \ldots LOG_n$ reciprocate with expert knowledge integrated into RND. Formally speaking, the three rounds of the survey separated in time can be considered boundary sub-objects of MAS_SURVEY: MAS_SURVEY_1, MAS_SURVEY_2, MAS_SURVEY_3. In the first round of the survey the experts ($LOG_1 \ldots LOG_n$) were asked to name application scenarios, potential and key barriers for implementation of MAS in their branch of logistics. In the second round the experts were asked to assess the main MAS application scenarios selected by the group in the first round based on their potentials (productivity increase, logistic network resilience), complexity, key barriers (standardization, maturity, change management, cost, measure and show benefits, applications) and estimated time of implementation determined by the group. In the third round the experts were asked to re-evaluate their assessments from the second round by letting the experts see the group average (arithmetic mean or median) on each question [1]. The results were combined and analyzed by RND. MAS_SURVEY is a 'standardized forms' type boundary object defined in [2] as methods of common communication across dispersed work groups.

Another important actor is knowledge shared between RND and $LOG_1 \ldots LOG_n$ through the boundary objects. The problematisation phase ends when MAS_TUTORIAL penetrates into $LOG_1 \ldots LOG_n$.

Interessement. Synonymous with interposition this stage of translation process represents a group of actions where an actor tends to impose and stabilize identities of other actors determined at the problematisation stage [10]. To make a group of actors "interested" means to create a "virtual device" [15], which can be placed between this group and all other entities who strive to re-identify this group. In other

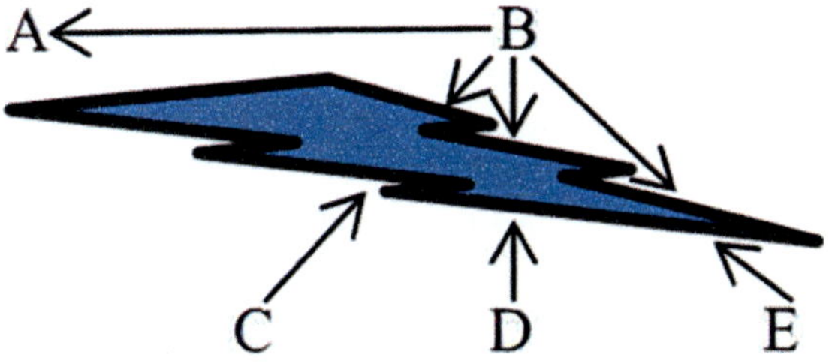

Fig. 1 Interessement of actor B by actor A in actor-network AN = {A, B, C, D, E} [15]

words, A_1 makes A_2 "interested" when it breaks or weakens all links between A_2 and group A_3, A_4, A_5, …, A_n of n-2 actors who may tend to liaise with A_2. Figure 1 schematically demonstrates the process of interessement of actor B by actor A in actor-network AN = {A, B, C, D, E} [15].

Successful outcome of the interessement confirms (more or less completely) the efficiency of the problematisation and supposed actor alliances (such heterogeneous alliances are often viewed as human-nonhuman quasi-objects or hybrids in ANT texts). Besides, this stage tries to break all competitive liaisons and build a system of alliances within actor-network. At this stage the socio-technological communities are formed and fixed. Both defined boundary objects (MAS_TUTORIAL and MAS_SURVEY) play the role of a virtual device of interessement in our study. The initial number of the logistics systems ($LOG_1 \ldots LOG_n$) pre-selected for the survey according to [1] was n = 461. The effect of MAS_TUTORIAL had brought this number down to n = 32 before MAS_SURVEY_1 was launched. Eventually, the number of those who accepted MAS_SURVEY_3 was n = 18 [1]. The effect of the interessement phase is measured not only by the number of the interested experts. It is also the diversity of the logistics specialities represented by $LOG_1 \ldots LOG_n$ in MAS_SURVEY that was affected. The interessement phase clearly demonstrates how the same boundary objects are differently translated in different institutional logics of $LOG_1 \ldots LOG_n$. 93% of the initial pool of prospective participants fell apart after MAS_TUTORIAL had been launched. Most probable reasons for actor $LOG_i \in \{LOG_1 \ldots LOG_n\}$ to reject MAS_SURVEY were: resistance of the traditional (or alternative to MAS) technologies and methods allied with LOG_i; irrelevant interpretation of MAS_TUTORIAL; communication issues. Interessement is the last phase of the life cycle of MAS_TUTORIAL as an autonomous boundary object. It may live through the remaining two phases only in alliance with MAS_SURVEY.

Enrolement. The core function of the enrolement is the determination and coordination of roles of actors aimed at creation of a steady network of alliances [13]. As we highlight it in [10] the enrolement stage of the translation process is interconnected with the MAS concepts where one of the forms of actor obligations is the role accepted by or assigned to an actor [16]. The notion is that actors have general roles to play in the collective effort, and by using knowledge of these roles the actors can make better interaction decisions. This notion can be explicitly manifested in an organizational structure, which defines roles, responsibilities and preferences for the actors within a cooperative society, and thus in turn defines control and communication patterns between them [17]. The roles taken by the interested $LOG_1 \ldots LOG_n$ included: consumers of MAS knowledge provided by RND (the effect of MAS_TUTORIAL); providers of valuable knowledge for RND (the effect of MAS_SURVEY); potential partners and clients of RND (as co-creators and users of multi-agent technologies in their respective domains of logistics); conductors of MAS-related knowledge to a broader professional community (e.g. supply chain partners); commentators and generators of ideas inspired by the process of inter-organizational knowledge transition and integration and propagation of innovations (like us, the authors of this paper).

Mobilization: The verb "mobilize" should literally mean to give mobility to immobile entities. Through step-by-step appointment of representatives and establishment of a series of equivalences heterogeneous actors are moved and then "reassembled" at a new place/time. This stage completes translation and certain actors start acting as representatives (delegates) of other actors [15]. The mobilization concept of ANT can significantly invest into understanding of the team behavior of actors in MAS [10] (where it is considered as something more than just a set of coordinated individual actions of the actors [16]). The mobilization concept suggests new semantics for translation-like effects discussed in the framework of MAS studies [10] where a collection of actors needed to accomplish a task frequently includes humans who have delegated tasks to the (nonhuman) actors and/or humans who will be performing some of the work [18]. The basic problem of the team work organization in MAS is to organize the work of actors as a team in situations where every individual actor has but limited information on his own team, environment and competition. At the same time the actor has to materialize his own intentions through individual actions performed simultaneously or consequently with the actions of other actors [16]. Knowledge sharing and integration between RND and $LOG_1 \ldots LOG_n$ paves the way to future formation of sociotechnical alliances which can be materialized in concrete innovative projects aimed at development, implementation, use and continuous improvement of multi-agent technologies in logistics systems. Selected and formalized knowledge provided by $LOG_1 \ldots LOG_n$ and translated through the boundary objects is mobilized by RND as a 'delegate' of $LOG_1 \ldots LOG_n$ in the academic world, while the MAS-related knowledge is mobilized by $LOG_1 \ldots LOG_n$ representing this knowledge and RND in the world of logistics.

4 Conclusions

The boundary objects paradigm and ANT are increasingly used today as an analytical framework in the information systems and innovation studies. These two cognate methods play a role of the theoretical integrator between the technical dimension focused on new technologies on the one hand, and the social dimension dealing with human behaviours that impact technological solutions [13] on the other hand, which is essential for the investigation of how new technologies shape organizational structures of the sociotechnical systems they are incorporated into. An intelligent

IS (e.g. MAS-based) incorporated into a sociotechnical (e.g. logistics) system can be considered an assemblage of hardware, IT, humans, data and knowledge integrated for the purpose of information sourcing, storage, processing, transmission and performance [19]. Implementation of multi-agent technologies in a logistics system is considered a process of innovation in the context of the general technological renovation of the logistics system, formation of new organizational structures and implementation of new policies and practices. Since the implementation of multi-agent solutions in a logistics system is inevitably connected with changes, and since MAS represent certain novelty (at least for some users), the R&D work must involve the methods of inquiry into technological innovations. In this paper we have discussed from the boundary objects and ANT perspective how the innovation is conducted into the logistics systems through cross-organizational knowledge integration and implementation of multi-agent technologies overcoming resistance of the logistics systems to innovation. Based on survey [1] we have defined two basic boundary objects MAS_TUTORIAL and MAS_SURVEY transcending the boundaries of systems-actors RND (the MAS R&D group) and $LOG_1 \ldots LOG_n$ (the various logistics systems involved in the survey) and translating knowledge. The phases of the life cycle of these boundary objects were defined as the four stages of the process of translation in actor-networks: problematisation, interessement, enrolement, mobilization. This paper continues our research programme aimed at the development of applied ANT-based method of modelling and simulation of sociotechnical systems and has a target to familiarize broader research community with the boundary objects and ANT concepts at work.

References

1. Nitsche, B., Straube, F., Gebhardt, J.: Assessment of the Potential of Multi-Agent Systems in Logistics. Development of a Technology Roadmap for Relevant Scenarios. Competence Center International Logistics Networks. Technische Universität Berlin (2021)—UNPUBLISHED (contact authors at: nitsche@tu-berlin.de)
2. Star, S.L., Griesemer, J.R.: Institutional Ecology, "translations" and boundary objects: amateurs and professionals in Berkeley's Museum of Vertebrate Zoology, 1907–39. Soc. Stud. Sci. **19**, 387–420 (1989). https://doi.org/10.1177/030631289019003001
3. Franco-Torres, M., Rogers, B.C., Ugarelli, R.M.: A framework to explain the role of boundary objects in sustainability transitions. Environ. Innov. Soc. Trans. **36**, 34–48 (2020). ISSN 2210–4224. https://doi.org/10.1016/j.eist.2020.04.010
4. Mäenpää, S., Suominen, A.H., Breite, R.: Boundary objects as part of knowledge integration for networked innovation. Technol. Innov. Manag. Rev. **6**(10), 25–36. (2016). https://doi.org/10.22215/timreview/10254
5. Huvila, I., Anderson, T.D., Jansen, E., McKenzie, P., Worrall, A.: Boundary Objects in Information Science, vol. 156. FIMS Publications (2016). https://ir.lib.uwo.ca/fimspub/156
6. Star, S.L.: This is not a boundary object: reflections on the origin of a concept. Sci. Technol. Hum. Values **35**, 601–617 (2010). https://doi.org/10.1177/0162243910377624
7. Kiselev, G.A., Panov, A.I.: Sign-based approach to the task of role distribution in the coalition of cognitive agents. In: SPIIRAS Proceedings, Issue 2(57), 161–187 (2018). ISSN 2078–9181 (print), ISSN 2078–9599 (online). https://doi.org/10.15622/sp.57.7

8. Kiselev, G.A., Panov, A.I.: Synthesis of the behavior plan for group of robots with sign based world model. In: Ronzhin, A., Rigoll, G., Meshcheryakov, R. (eds.), Interactive Collaborative Robotics. ICR 2017. Lecture Notes in Computer Science, vol 10459. Springer, Cham (2017)
9. Fischer, E.: Socio-Technical innovations in urban logistics: new attempts for a diffusion strategy. In: 16th Conference on Reliability and Statistics in Transportation and Communication, RelStat'2016, 19–22 Oct. 2016, Riga, Latvia. In Procedia Engineering vol. 178 (2017), pp. 534–542 (2017)
10. Iskanderov, Y., Pautov, M. Agents and multi-agent systems as actor-networks. In: Rocha, A., Steels, L., van den Herik. J.P. (eds.), Proceedings of the 12th International Conference on Agents and Artificial Intelligence ICAART 2020, 22–24 Feb. 2020, vol. 1, pp. 179–184 (2020)
11. Callon, M. Techno-economic Networks and Irreversibility. Sociol. Rev **38**(1_suppl), 132–161 (1990)
12. Iskanderov, Y., Pautov, M.: Heterogeneous engineering in intelligent logistics// Advances in intelligent systems and computing/ In: Abraham, A., Shandilya, S., Garcia-Hernandez, L., Varela, M. (eds.), Hybrid Intelligent Systems. HIS 2019, vol. 1179, pp. 83–91. Springer, Cham (2020). https://doi.org/10.1007/978-3-030-49336-3_9
13. Silic, M.: Using ANT to understand dark side of computing—computer underground impact on eSecurity in the dual use context. University of St Gallen. Institute of Information Management, Jan. 2015
14. Wooldridge, M.: An Introduction to Multiagent Systems. Wiley (2009)
15. Callon, M. Some Elements of a Sociology of Translation; Domestication of the Scallops and the Fishermen of St.Brieuc Bay // Power, Action and Belief/ J. Law (ed.). L.: Routledge, pp 196–223 (1986)
16. Gorodetskii, V.I., Karsayev O.V., Samoylov V.V., Serebryakov S.V.: Applied multiagent systems of group control. Sci. Tech. Inf. Process.T **37**(5). C, 301–317 (2010)
17. Durfee, E.: Distributed Continual Planning for Unmanned Ground Vehicle Teams, vol. 20(4), pp. 55–61. AI Magazine, Winter (1999)
18. Cohen, P., Levesque, H.: Communicative actions for artificial agents. In: Proceedings of ICMAS'95 C, pp. 65–72 (1995)
19. Tatnall, A.: Actor-network theory as a socio-technical approach to information systems research. Socio-Technical and Human Cognition Elements of Information Systems, pp. 266–283. IGI Global, USA (2003)

Traffic Flow Modelling When Autonomous Vehicles Coexist with Human Driven Vehicles: Perspectives and Challenges

Giacomo Cabri, Serena Crisci, and Manuela Montangero

Abstract The problem of mathematically modeling traffic flows has been tackled, with different approaches, since the first decades of the last century. The classic framework that accounts only for human-driven cars is bound to change due to the rise of autonomous driving technologies, which will deeply transform mobility, as well as general driving rules and infrastructures. It is therefore urgent to design novel approaches for modeling the transition phase, where autonomous vehicles and human-driven vehicles will have to coexist, and to devise appropriate solutions for their coordination. In this paper, we provide a brief overview of the most recent studies on the transition scenario, and outline the main challenges and perspectives that are likely to appear in the near future for this case.

1 Introduction

Mobility will pass through deep transformations in the next future in response to ever-changing innovative technologies and to a growing society's demand for a more inclusive and less polluted environment. We expect a *transition phase* for the years

---

Serena Crisci—Member of the Indam research group GNCS.

G. Cabri (✉) · M. Montangero
Dipartimento di Scienze Fisiche, Informatiche e Matematiche, University of Modena and Reggio Emilia, Modena, Italy
e-mail: giacomo.cabri@unimore.it

M. Montangero
e-mail: manuela.montangero@unimore.it

S. Crisci
Dipartimento di Matematica e Fisica, University of Campania Luigi Vanvitelli, Caserta, Italy
e-mail: serena.crisci@unicampania.it

D. Camacho et al. (eds.), *Intelligent Distributed Computing XIV*, Studies in Computational Intelligence 1026, https://doi.org/10.1007/978-3-030-96627-0_16

ahead where both *automated* and *human-driven* vehicles will coexist: this scenario presents many challenges, in particular regarding the coordination of vehicles at intersections and their mutual interaction.

In such a context, *traffic flow models* could provide a mathematical framework for analysing traffic dynamics, with the aim to understand driving behaviours, make traffic predictions and develop novel strategies to optimize resources (parking, fuel consumption, etc). Classic models for the conventional scenario, i.e. human-driven cars, may represent the basis for studying the impact of autonomous vehicles (AVs) within the mixed scenario that is going to arise in the near future; one of the key issue required to this framework is the problem of modeling the interactions between *different types* of vehicles. Indeed, if on the one hand human-driven cars introduce randomness, on the other autonomous vehicles may be equipped with different kind of technologies, presenting different levels of automation.

The first attempts to provide a mathematical framework for analysing traffic dynamics date back to the 30s of the last century; at that time, Bruce D. Greenshields, a traffic engineering scientist, started his empirical studies based on collecting information on the observed vehicular traffic by means of a photographic measurement method [12]. He also introduced a key concept in traffic flow theory, the *fundamental diagram*, which expresses the relationship between traffic flow and vehicle density and is still a relevant tool to empirically explain some features of real traffic phenomena. In the subsequent years, with the precipitous growth of traffic volume and all the related issues, many different mathematical models were developed. In particular, since traffic can be viewed as a system of particles interacting each other, two main paradigms were identified and most of the existing models generally fall into one of two main classes, i.e. microscopic description and macroscopic description. Despite the vast literature, theoretical models for the conventional scenario are still far from being highly consistent with the complex reality of traffic dynamics, since they are not able to capture the intrinsic unpredictable nature of these systems, especially the high level of randomness introduced by the presence of human drivers. In similar way, the advent of AVs has pushed the research towards the development of ad hoc microscopic models that, however, are still at an early stage. Hence, it would be desirable to combine principled models with empirical studies based on observed traffic and measured data, in order to validate the models and make further progress towards the achievement of more reliable approaches.

The development of appropriate models to describe mixed traffic flow relies on the correct understanding of mutual interactions between autonomous vehicles and human-driven vehicles. In past and recent years, several work have tried to address, from different perspectives, various aspects related to this issue, by investigating psychological and behavioral factors that influence driving process in human vs autonomous vehicles interactions, as well the problems of reciprocal coordination and communication [7, 8, 32, 33]. However, these dynamics are still not well-studied and the first attempts for their mathematical formalization are based on conventional approaches. Starting by these premises, the paper presents a concise overview of the most recent studies and modeling approaches for the transition scenario, i.e. when mixed traffic conditions are considered, with a discussion on future challenges.

The paper is organized as follows. General notions on traffic flow models are introduced in Sect. 2. In Sect. 3, we provide a brief summary of the most recent developments on mixed traffic flow and discuss some challenging issues arising within this framework. Conclusions are drawn in Sect. 4.

2 Classical Models for Traffic Flows

In this section, we briefly recall the main features of mathematical modeling for traffic flow in the conventional scenario, i.e. the case where only human-driver vehicles are considered, in order to provide a general idea of classic mathematical results on traffic flow. Indeed, despite the wide literature on this subject, the key concepts underlying state-of-the-art models are based on a few fundamental mathematical frameworks that were developed in past decades, starting from the former studies devised in the mid 1950s. For a more extensive discussion and review we refer to [3, 13, 16].

Traffic flow models can be grouped into two main categories

Microscopic Models describe the dynamics of individual vehicles interacting with other vehicles, generally stated in terms of large systems of ordinary differential equations;

Macroscopic Models describe aggregate behaviour of traffic flows, regarded as a continuum similar to a fluid or a gas; they are stated via partial differential equations.

For each category the dynamics can be described by means of first-order models or second-order models, corresponding to the cases where the system state is vehicle positions or positions and speed, respectively. The choice of the proper model depends on the specific features of the case that has to be represented; for example, first-order dynamics are generally used to reproduce shock waves that occurs in traffic flows when the state of the front line changes (e.g. accident, red/green lights). Furthermore, microscopic and macroscopic framework are related in the sense that macroscopic models can be viewed, under suitable conditions, as the limit of the time discretization of microscopic models, as the the number of vehicles increases [1].

2.1 Microscopic Models

In microscopic models each vehicle is determined by its position and its speed at each time step A simple framework for this type of models is *car-following*, which describes the behaviour of a vehicle only depending on the vehicle in front. The first car-following models were developed in the 1950s and 1960s and, since then, they have continued to receive attention. Indeed, many enhanced versions of the

first prototype models have been proposed for more than half a century, in order to overcome their limitation, improving their ability to reproduce realistic traffic conditions. For a historical review of car-following models, we refer to [4]. In this framework, the most well-known approaches are represented by the so-called Follow-the-leader-models, which date back to the early sixties of the last century, starting from the work of Gazis, Herman and Rothery [9]. These models are based on a stimulus-response equation that describes the acceleration (*response*) of a vehicle after a reaction time as a function of the relative speed (*stimulus*) with respect to the vehicle ahead multiplied by a (*sensitivity*) parameter. Different functionals can be considered to express the sensitivity, the choice of a special type with respect to another one aims at reflecting actual traffic phenomena.
Other popular microscopic models are

- Optimal Velocity Model [20], where the acceleration equation of a vehicle is expressed by a function depending on the difference between its speed and the *legal velocity*, defined as a monotonically increasing function of the relative position between the vehicle and some neighbouring vehicles. Despite its ability in describing the formation of stop-and-go waves and emergent traffic jams, this model might exhibit unrealistic results related to acceleration and deceleration behaviours that are not able to prevent accidents.
- Generalized Force Model [14], whose acceleration equation is described by the sum of the so-called *generalized forces* expressing a trade-off between the motivation to reach a desired speed and the motivation to keep safe distance to the vehicle in front. The aim, indeed, is to express vehicle dynamics taking into account different factors that might influence it, like perceptions, psychological motivations and reactions, or social behaviours, as well. The limiting cases of this model are the optimal velocity model and the follow-the-leader model.

Inspired by these models, subsequent improvements have been made in car-following theory, with the aim to embed more realistic dynamics in the mathematical description, taking into account different effects that can emerge in real observed traffic (see, e.g., [15, 34–36]). A comprehensive review of the most significant car-following models can be found in [23].

Many well-known software for computer simulation are based on car-following microscopic models [26], for example: Aimsun (Gipps' model) [10], Vissim (Wiedemann model), and SUMO (Krauss model [17], Intelligent Drive model).

Finally, another paradigm for microscopic models is *lane-changing*, which deals with the way in which vehicles change lanes based on the vehicles around them, for example to avoid obstacles, or when they are following a slower vehicle. For a review we refer to [31].

2.2 Macroscopic Models

In the macroscopic framework, the fundamental variables can be recovered by local averages of the microscopic state of the vehicles. These models are used to describe aggregate traffic flow conditions, on the basis of a conservation equation. In this context, a classical model is the Lighthill-Whitham-Richard (LWR) Model [19, 30], a first-order model developed independently by Lighthill and Witham (1955) and Richards (1956). The key assumption in this model is that the mean speed at each instant of time is given by the equilibrium value of the vehicle density. As a consequence this model is not useful to describe traffic jam and stop-and-go waves.

A widely used model for describing traffic dynamics on highways is the Payne-Whitham (PW) Model [29] that is a second-order model incorporating a sort of car-following rule. The general differential equation for the PW model is the basis for many macroscopic models, and, unlike LWR model, is able to take into account also non-equilibrium situations. The need to overcome some unrealistic effects deriving from PW models leads to the development of the Inhomogeneous Aw-Rascle-Zhang (AZR) Model [2], which introduced a correction by means of a convenient choice for the pressure law, which takes into account drivers' reactions to the state of traffic in front of them.

To conclude this section, we mention another class of traffic flow models, i.e. *cellular automata*. These models originated from physics and were developed starting from the the work by Nagel and Schreckenberg [27]. These are discrete in space, time and state variables and are designed in a way that allows convenient computer simulations.

2.3 Machine Learning Approaches for Traffic Simulation

In the last decades, alternative approaches to the classic traffic models were developed exploiting machine learning and deep learning methodologies, especially in the context of traffic simulation, as well as for real-time prediction of traffic conditions. The main purpose of these novel strategies is to improve the driver agent realistic simulation of the vehicle and optimize traffic management in future smart urban environments. Indeed, such research directions are tailored for being integrated in Intelligent Transportation Systems (ITS) and in the broader context of IoT technologies. Machine learning models are based on real-time traffic data, collected by different types of sensors and detectors. Hence, a crucial requirement is the availability of open and public data sets. Some reviews of state-of-the-art machine learning approaches for traffic prediction and broader ITS application can be found in [28, 37].

3 Modeling Traffic Flow in a Mixed Scenario: Perspectives and Challenges

The still growing and advanced scientific research on autonomous vehicles made it necessary to develop novel theoretical models including AVs. The presence of automated vehicles that are capable of communicating with other vehicles and/or with city infrastructures will have a significant impact on the transportation system and will contribute to a deep transformation of driving behaviours and traffic conditions, inspiring new modeling approaches for their description and analysis.

In 2020, Gora et al. [11] published an extensive analysis on new microscopic traffic simulation models for Connected and Automated Vehicles (CAVs). The paper provides an overview of the state-of-the-art literature related to microscopic simulation approaches for CAVs and leads to the conclusion that there is no universal model and a fair comparison of different models and their outcomes is difficult to achieve, due to the lack of empirical data that could be used to validate the models as well as the variety of different tools and methodologies used by AVs manufacturers.

We remark, however, that most of the traffic models and simulation tools reviewed in [11] refer to a scenario that involves exclusively autonomous vehicles that move within a smart environment. As we know, this ideal situation is not going to appear soon. To ensure a successful and optimized coexistence during the long transition phase where both self-driving vehicles and human-driven vehicles will populate our streets, it is extremely important to analyse theoretical and practical aspects of modelling the transition scenario; in particular, new ideas are needed to plan a rearrangement of urban environments in a hybrid perspective, that means, for example, equipping the existing infrastructures with supplementary technologies that are able, at the same time, to manage human-driven vehicles and to exploit the potential of autonomous vehicles.

The presence of small numbers of AVs in a mixed scenario has been recently investigated in a microscopic framework with the purpose of regulating human driven traffic to dissipate stop-and-go waves, with positive effect on fuel consumption [24]. At this regard, there are theoretical findings showing that a mixed traffic system can be stabilized by AVs. In [38] analytical results on the AVs capability to suppress unstable traffic waves and to induce higher speed of traffic flow are established, in both the cases of a mixed traffic with one single AV and multiple AVs. In particular, the dynamics of human-driven cars is modeled by means of an optimal velocity model, while additional equations describing a car-following dynamics for AVs are considered, assuming that AVs have access to global traffic information related to all other human-driven vehicles. In this scenario, authors proved that AVs can be used as mobile actuators to control traffic flow internally and introduced an optimal control strategy to smooth mixed traffic flow.

Besides this results, as pointed out in [21], a complete macroscopic theory for control of bulk traffic through AVs is still missing, although, the need for such a theory is motivated by dimensionality issues related to microscopic models, which might prevent control when the number of vehicles increases.

Future challenges
Despite these first attempts to model mixed traffic flows, the transition period is a research field mostly unexplored, which needs to be more deeply and carefully analysed, since it poses many open problems and challenges, mainly due to the unpredictability of the human driver's behavior and the heterogeneity of the inner technology of AVs systems. In the following, we discuss some relevant issues that should be addressed.

- *Vehicles coordination at road intersections* is a crucial task. In this context, the main open questions to be answered are: how to model this problem and how to design appropriate strategies in presence of traffic lights or traffic management devices created ad hoc for the mixed scenario. Auction-based approaches for coordinating AVs and human-driven vehicles are proposed in [5, 25], inspired by the seminal paper [6]. The idea is to adopt an auction system to decide the priority order when crossing an intersection, based on different strategies for auction resolution and bid payment. These approaches borrow from market theory the idea to value drivers' interest according to their urgency to pass the crossing and their budget.
- Many other *coordination problems* (e.g. parking, platooning, ramp merging, etc.) will require a special analysis and appropriate solutions for the mixed traffic case. A taxonomy for the different classes of coordination problems of autonomous vehicles can be found in [22], where the authors also proposed a unifying conceptual framework for possible solutions. To our knowledge, a similar analysis is still missing for the case involving human-driven cars among self-driving cars equipped with different levels of automation.
- *Driver reliability.* Autonomous vehicles have the advantage of being programmed to follow precise instructions and perform specific tasks, which makes their behaviour much more reliable than human-driven vehicles. On the other side, human drivers are sometimes able to resolve critical circumstances using their driving experience and instinct. From this point of view, it would be interesting to study the impact that standardised actions have on AVs' reaction times with respect to the reaction times and actions typical of human drivers. Such analysis would help to answer the question of whether it is desirable to introduce a little amount of randomness to autonomous driving systems in order to simulate some positive results of human driving behaviour (see, e.g., [18]).

4 Conclusions

In this paper we presented an outline of recent approaches for modeling traffic flows in the transition phase, in which both human-driven cars and autonomous vehicles are considered. We also highlighted some open issues that must be properly addressed from both theoretical and practical viewpoints. Some of these challenging goals started to be addressed in recent times, but we are still far from reaching a solid theory capable to explain such a complex system. Indeed, the dynamic of mutual interactions between autonomous vehicles and human-driven vehicles is still unclear and deserves a more accurate investigation. At this regard, standard approaches may reveal too simplified or inappropriate to describe mixed traffic, hence, there is a need to change the perspective, in order to improve existing models and devise a novel paradigm.

Acknowledgements The work of Serena Crisci has been funded by the project "Coordination of autonomous vehicles" under the Regional Operational Program ROP-ESF 2014/2020.

References

1. Aw, A., Klar, A., Rascle, M., Materne, T.: Derivation of continuum traffic flow models from microscopic follow-the-leader models. SIAM J. Appl. Math. **63**(1), 259–278 (2002)
2. Aw, A., Rascle, M.: Resurrection of second order models of traffic flow. SIAM J. Appl. Math. **60** (2000)
3. Bellomo, N., Dogbe, C.: On the modeling of traffic and crowds: a survey of models, speculations, and perspectives. SIAM Rev. **53**(3), 409–463 (2011)
4. Brackstone, M., McDonald, M.: Car-following: a historical review. Transport. Res. F: Traff. Psychol. Behav. **2**(4), 181–196 (1999)
5. Cabri, G., Gherardini, L., Montangero, M.: Auction-based crossings management. In: Proceedings of the 5th EAI International Conference on Smart Objects and Technologies for Social Good, GoodTechs '19, pp. 183–188. Association for Computing Machinery, New York (2019)
6. Carlino, D., Boyles, S.D., Stone, P.: Auction-based autonomous intersection management. In: Proceedings of the 16th IEEE Intelligent Transportation Systems Conference (ITSC) (2013)
7. Dresner, K.M., Stone, P.: Autonomous vehicles meet human drivers: Kurt M Dresner and Peter Stone. Sharing the road. Ijcai **7**, 1263–1268 (2007)
8. Färber, B.: Communication and communication problems between autonomous vehicles and human drivers. In: Autonomous Driving, pp. 125–144. Springer (2016)
9. Gazis, D.C., Herman, R., Rothery, R.W.: Nonlinear follow the leader models of traffic flow. Oper. Res. **9**(4), 545–567 (1961)
10. Gipps, P.G.: A behavioural car-following model for computer simulation. Transp. Res. Part B: Methodol. **15**(2), 105–111 (1981)
11. Gora, P., Katrakazas, C., Drabicki, A., Islam, F., Ostaszewski, P.: Microscopic traffic simulation models for connected and automated vehicles (cavs) – state-of-the-art. Proc. Comput. Sci. **170**, 474–481 (2020). The 11th International Conference on Ambient Systems, Networks and Technologies (ANT)/The 3rd International Conference on Emerging Data and Industry 4.0 (EDI40)/Affiliated Workshops
12. Greenshields, B.D.: A study of traffic capacity. HRB Proc. **14**, 448–477 (1934)
13. Helbing, D.: Traffic and related self-driven many-particle systems. Rev. Modern Phys. (2001)
14. Helbing, D., Tilch, B.: Generalized force model of traffic dynamics. Phys. Rev. E **58**, 133–138 (1998)
15. Jiang, R., Qingsong, W., Zhu, Z.: Full velocity difference model for a car-following theory. Phys. Rev. E **64**, 017101 (2001)
16. Kessels, F.: Traffic Flow Modelling: Introduction to Traffic Flow Theory Through a Genealogy of Models. EURO Advanced Tutorials on Operational Research. Springer (2019)
17. Krauss, S.: Microscopic modeling of traffic flow: investigation of collision free vehicle dynamics (1998)

18. Li, L., Ota, K., Dong, M.: Humanlike driving: empirical decision-making system for autonomous vehicles. IEEE Trans. Veh. Technol. **67**(8), 6814–6823 (2018)
19. Lighthill, M.J., Whitham, G.B.: On kinematic waves: a theory of traffic flow on long crowded roads **229**, 317–345 (1955)
20. Bando, M., Hasebe, K., Nakayama, A., Shibata, A., Sugiyama, Y.: Dynamical model of traffic congestion and numerical simulation. Phys. Rev. E **51**, 1035–1042 (1995)
21. Liard, T., Garavello, M., Goatin, P., Piccoli, B.: A multiscale model for traffic regulation via autonomous vehicles. J. Differ. Equ. **269**(7), 6088–6124 (2020)
22. Mariani, S., Cabri, G., Zambonelli, F.: Coordination of autonomous vehicles: taxonomy and survey. ACM Comput. Surv. **54**(1) (2021)
23. Matcha, B.N., Namasivayam, S.N., Fouladi, M.H., Ng, K.C., Sivanesan, S., Eh Noum, S.Y.: Simulation strategies for mixed traffic conditions: a review of car-following models and simulation frameworks. J. Eng. **2020** (2020)
24. Delle Monache, M.L., Liard, T., Sternan, R., Rat, A., Bhadani, R., Seibold, B., Sprinkle, J., Work, D.B., Piccoli, B.: Feedback Control Algorithms for the Dissipation of Traffic Waves with Autonomous Vehicles, pp. 275–299. Springer International Publishing (2019)
25. Muzzini, F., Capodieci, N., Montangero, M.: Exploiting traffic lights to manage auction-based crossings. In: Proceedings of the 6th EAI International Conference on Smart Objects and Technologies for Social Good, GoodTechs '20, pp. 199–204. Association for Computing Machinery, New York (2020)
26. Rohani, M.M., Nor Azlan, N.N.: Overview of application of traffic simulation model. MATEC Web Conf. **150**, 03006 (2018)
27. Nagel, K., Schreckenberg, M.: A cellular automaton model for freeway traffic. J. Phys. I **2**(12), 2221–2229 (1992)
28. Nagy, A.M., Simon, V.: Survey on traffic prediction in smart cities. Pervasive Mob. Comput. **50**, 148–163 (2018)
29. Payne, H.: Models of Freeway Traffic and Control. Mathematical Models of Public Systems, Simulation Council Proceedings, pp. 51–61 (1971)
30. Richards, P.I.: Shock waves on the highway. Oper. Res. **4**(1), 42–51 (1956)
31. Rahman, A., Chowdhury, M., Xie, Y., He, Y.: Review of microscopic lane-changing models and future research opportunities. IEEE Trans. Intell. Transp. Syst. **14**, 1942–1956 (2013)
32. Rahmati, Y., Hosseini, M.K., Talebpour, A., Swain, B., Nelson, C.: Influence of autonomous vehicles on car-following behavior of human drivers. Transp. Res. Record **2673**(12), 367–379 (2019)
33. Saifuzzaman, M., Zheng, Z.: Incorporating human-factors in car-following models: a review of recent developments and research needs. Transp. Res. Part C: Emerg. Technol. **48**, 379–403 (2014)
34. Xu, D.J., Mao, B.-H., Chen, S.-K., Bai, Y.: Improved generalized force model considering the comfortable driving behavior. Math. Probl. Eng. (2015)
35. Hao, X., Liu, H., Gong, H.: Modeling the asymmetry in traffic flow (a): microscopic approach. Appl. Math. Model. **37**(22), 9431–9440 (2013)
36. Yu, Y., Jiang, R., Qu, X.: A modified full velocity difference model with acceleration and deceleration confinement: calibrations, validations, and scenario analyses. IEEE Intell. Transp. Syst. Mag. **13**(2) (2021)
37. Yuan, T., da Rocha Neto, W.B., Rothenberg, C.E., Obraczka, K., Barakat, C., Turletti, T.: Machine learning for next-generation intelligent transportation systems: a survey. Technical report, Nov. 2020, working paper or preprint (2019). https://hal.inria.fr/hal-02284820
38. Zheng, Y., Wang, J., Li, K.: Smoothing traffic flow via control of autonomous vehicles. IEEE Internet Things J. **7**(5), 3882–3896 (2020)

Distributed Non-monotonic Reasoning for Transportation Network Congestion Monitoring: Theoretical Foundation

Azedine Boulmakoul, Lamia Karim, Ghyzlane Cherradi, and Bharat Bhushan

Abstract In this work, we formulate premises of a non-monotonic and distributed logical theory to model the traffic control actions of urban networks. The distributed intelligent agent transaction mechanisms based on distributed default logic are used. We demonstrate the validity of our approach and the contribution of the revisable reasoning formalism and the knowledge representation of multi-agent systems for the problems of traffic congestion management and supervision according to a route or arterial approach. In distributed and collaborative environments, revisable reasoning processes provide necessary facts to activate default rules, which gives the reactive agent the best use of its local and external knowledge. These mechanisms allow the network's symbolic control system to operate efficiently in a dynamic and hypothetical environment, where basic facts and observations are often problematic.

1 Introduction

The development of logical theories for system's diagnosis and control is a key issue in the field of artificial intelligence and modelling of complex systems. The logical dimensions in the representation problematic of knowledge and reasoning are multiple. Its role in the modelling of uncertainty and the dynamic management of knowledge is of great help in modelling revisable reasoning.

Inevitably, non-monotonic logics are qualified for the development of inference systems enabling modelling of revisable reasoning. Various theories and practices issues have been developed in the literature on artificial intelligence. In particular,

A. Boulmakoul · G. Cherradi (✉)
LIM/IOS FSTM- Hassan II University of Casablanca, Casablanca, Morocco
e-mail: azedine.boulmakoul@fstm.ac.ma

L. Karim
LISA Lab ENSA Berrechid - Hassan 1st University, Settat, Morocco

B. Bhushan
Department of Computer Science and Engineering, School of Engineering and Technology, Sharda University, Greater Noida, India

D. Camacho et al. (eds.), *Intelligent Distributed Computing XIV*, Studies in Computational Intelligence 1026, https://doi.org/10.1007/978-3-030-96627-0_17

we will cite the main classes of non-monotonic logic, namely: default logics [1–4], non-monotonic logics [5], auto-epistemic logics [2] and logics based on the principle of the closed world and the principle of circumscription [2, 5, 6]. In logical modeling of diagnosis with non-monotonic logics, the work of Reiter (1987) [7] and Kleer et al. (1992) [6] are mainly the most famous. Segerberg et al. (2013) [8] offers a brief self-sufficient synthesis on the logic of action in philosophy, language, computer science and artificial intelligence. Using the work of Reiter (1980, 1987) [3, 7], we model the theory of action defined as a complement to the diagnosis theory modeled by the default theory. The rest of this article is organized as follows: Sect. 2 offers brief reminders on logical default theory and diagnosis theory based on first principles, it also presents a new formalization of symbolic control. Section 3 gives the beginnings of a distributed multi-agent architecture based on the distributed default theories for the network's symbolic control. Finally, Sect. 4 concludes and guides future developments.

2 Default Logic for Diagnosis and Symbolic Control

In this section, we recall the preliminary elements relating to default logic theory. We then define a new approach to formalize the symbolic control associated with a diagnosis, based on default logic theory.

2.1 Default Logic Theory

The default logic was introduced by Reiter (1980) [3], to formalize the simply consistent reasoning. Hereinafter we give succinctly the formalism underlying this logic. Let $\mathcal{G}$ denote a first-order predicate language. A default rule F is an expression of the form $\frac{\alpha(x):M\beta_1(x), M\beta_2(x), \ldots, M\ \beta_m(x)}{\gamma(x)}$, or $\alpha\,(x)\,, \beta_1\,(x)\,, \beta_2\,(x)\,, \ \ldots, \ \beta_m\,(x)\,, \ \gamma\,(x)$ are formulas of $\mathcal{G}$.

– $\alpha\,(x)$ is called the prerequisite of default F,

– $\beta_1\,(x)\,, \beta_2\,(x)\,, \ \ldots, \ \beta_m\,(x)$ is called justification of the default F,

– $\gamma\,(x)$ is called consequent of the default F,

– M is a symbol of metalanguage.

A default theory is a pair (W, Δ), where W is a set consisting of closed formulas of $\mathcal{G}$ and Δ a set of default rules.

2.2 Default Theories Extensions

A default theory (W, Δ) implies a certain number of sets of beliefs that can be consistently inferred from the set W. These sets of beliefs are called extensions of the default theory. Let $\Theta_{\mathcal{G}}(X) = \{w|w \in \mathcal{G}, w$ is closed and $X \vdash w\}$, it is the set of closed formulas which can be inferred from X in a valid way by the classical inference laws of $\mathcal{G}$. Let E be a set of closed formulas of $\mathcal{G}$ and let (W, Δ) be a theory with closed defaults. Let the series of formulas E_i be constructed as follows:

$$\begin{cases} E_0 = W \\ E_{i+1} = Th_{\mathcal{G}}(E_i) \cup \{\gamma | \frac{\alpha : M\beta_1, M\beta_2, \ldots, M\ \beta_m}{\gamma} \in \Delta\} \end{cases}$$

E is an extension of (W, Δ) if and only if $E = \bigcup_0^\infty E_i$. Consider the default theory (W, Δ), where $W = \emptyset$ and $\Delta = \left\{\frac{:M\alpha}{\alpha}, \frac{:M\beta}{\neg\alpha}\right\}$. This theory has two extensions:

$$E_1 = Th_{\mathcal{G}}(\{\alpha\}) \text{ and } E_2 = Th_{\mathcal{G}}(\{\beta, \neg\alpha\})$$

2.3 New Formulation of Symbolic Control

In the theory proposed by Reiter [7], a system is described by the triplet (SD, COMP, OBS), denoting respectively the system description (SD) by a set of first order logical formulas, the components of the system (COMP), a set of propositional constants; the captured observations of the system (OBS), a set of first-order logical formulas.
Definition: Diagnosis for the system (SD, COMP, OBS) is the minimal set $\Delta \subseteq COMP$ such that:
SD $\cup$ OBS $\cup \{c \in \Delta | AB(c)\} \cup \{c \in COMP - \Delta | \neg AB(c)\}$ is consistent [6]. The predicate AB(.) means "ABnormal".

We add to this system a set ACT of first order logical formulas corresponding to the actions to be taken for a state of the system designated by its diagnosis. The interest of linking the diagnosis to the instrumentation strategy of actions in accordance with the state of the system is to manage abnormal situations in terms of system behavior. In our case, it is a question of managing congestions which are abnormal traffic situations. The process of activating actions for the state control of a system can be paraphrased by the following text: "If the process is in state X and if we apply the action U, then we observe the process in a state Y". The control of a system is intrinsically linked to its diagnosis. For each diagnosis there must be corresponding actions to be carried out to control the system towards the desired states and this according to a given strategy. Namely, strategy is a "set of coordinated actions, skillful operations, maneuvers to achieve a specific goal."
Definition A symbolic control associated with a Δ diagnosis for the system $(SD, COMP, OBS, ACT)$ is the set $A \subseteq ACT$ such as:

$$SD \cup OBS \cup \{c \in \Delta | AB(c)\} \cup \{c \in COMP - \Delta | \neg AB(c)\} \models A$$

$$SD \cup OBS \cup \{c \in \Delta | AB(c)\} \cup \{c \in COMP - \Delta | \neg AB(c)\} \cup \neg A \models \emptyset$$

Corollary 1 *Either the system (SD, COMP, OBS, ACT) or E is an extension of the default theory* $DT = (SD \cup OBS, \{c \in COMP | \frac{:\neg AB(c)}{\neg AB(c)}\})$, *then* $E \cap ACT$ *is a symbolic control of the system (SD, COMP, OBS, ACT).*

The set $\mho = \{A_i = E_i \cap ACT | E_i$ extension of DT $\}$ is the set of possible symbolic controls for the system (SD, COMP, OBS, ACT). The set of symbolic controls deduced can make use of metarules to prioritize actions and simulate their execution.

In the work of Boulmakoul et al. (1993–1994) [9, 10], our aim was to provide a formal framework for modeling symbolic control based on default theories. The formalism that we have proposed reuses the work of Reiter (1987) [7] on the logical modeling of the diagnosis of systems. Our initial formalism considered system control in the same way as diagnosis, i.e., "control is a minimal set of actions that are in an abnormal state", where an action is abnormal if its eventual activation does not comply with the observations made on the components of the system, as well as with its expected behaviors. In this work, symbolic control will be associated with a diagnosis of the system. Intuitively, it is the overall state of the system characterized by its diagnosis in Reiter's sense, which can infer the actions to be taken to achieve objectives.

3 Abductive Logical Actors of the Urban Network

Complex systems are characterized by nonlinear, stochastic, and dynamics phenomena. Systemics constitutes the paradigm of excellence for modeling this class of systems. One of the tools of the systems approach, and certainly one of the most widely used modeling techniques today is multi-agent system (MAS) modeling. An agent can be defined as an autonomous communicating reactive object able of acting in its environment, its actions being directed towards a goal, of perceiving and possibly of partially representing its environment and of interacting with other agents [11]. Within the framework of this work, the built multi-agent system is developed around the default theories. Each agent A_i is an abstract logical machine denoted by the system $(SD_i, COMP_i, OBS_i, ACT_i)$.

The agent society is expressed by distributed default theory:

$$DDT = \{(W_i, \Delta_i) | i = 1..n\}, \text{ where } W_i = SD_i \cup OBS_i \text{ and } \Delta_i = \{c \in COMP_i | \frac{:\neg AB(c)}{\neg AB(c)}\}.$$

Definition. A multi-agent system (MAS) is a collection of agents which operate in the same environment. Agents can interact with the environment and with each other using interfaces, according to predefined communication languages and protocols. The multi-agent system is asynchronous which means that agents can perform their actions at any time as long as they are in a valid state [12, 13]. To each agent of MAS is associated a logical default theory $(W_i, \Delta_i) | i = 1..n$. A distributed default theory associated with MAS is the set of theories:

$$DDT = \{(W_i, \Delta_i) | i = 1..n\}.$$

Fig. 1 Main arteries of the city of Casablanca

Fig. 2 Traffic continuity and cutoff relationship

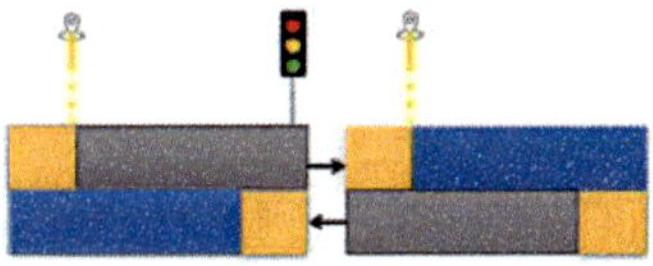

Fig. 3 A section of road is part of the knowledge of downstream and upstream crossroads

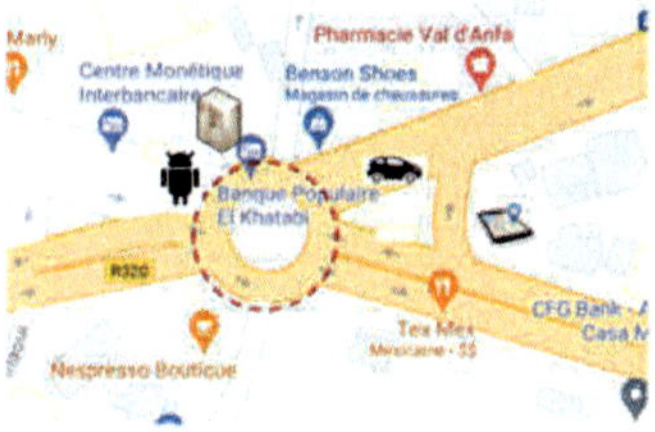

Fig. 4 Local control agent and communication architecture (connected objects)

Fusion of default theories: The multi-agent system given by its set of theories: $\mathscr{H} = \{(W_i, \Delta_i)|i = 1..n\}$. The corresponding fusion theory is given by:

$\mathfrak{I} = \{(\mathbb{C}(\sqcup W_i), \cup\Delta_i)|i = 1..n\}$. Where $\mathbb{C}(\sqcup W_i)$ corresponds to maximal consistent set obtained from $(\sqcup W_i)$.

The agent construction mechanism is developed from the global network according to the steps below:

- Generation of network arteries and association of a Crossroads-type Local Control Agent (CLCA) at each crossroads of the generated route (Figs. 1, 3, 4),
- Associate an Arterial Control Agent (ACA) with each artery generated (Fig. 5),

Fig. 5 ACA Agent: coordinator of crossroads agents of the same artery

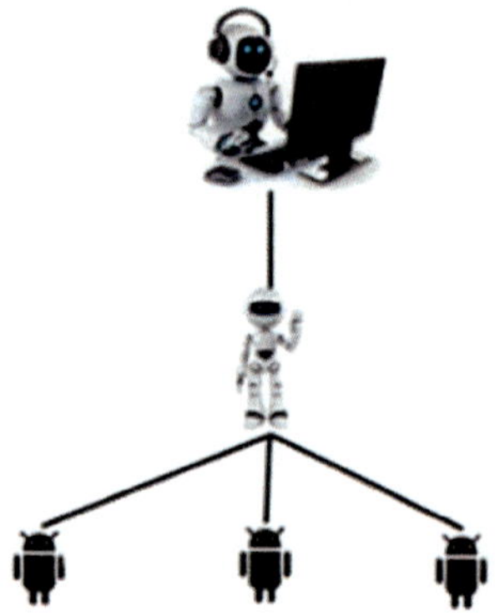

Fig. 6 Hierarchy of intelligent agents

Fig. 7 GCMA Agent: coordinator of several Arterial control agents

- Organize these arteries by the relationship of continuity of flow or intersection (Fig. 2).
- Associate a Global Control and Mediating Agent (GCMA) with all of the arteries generated (Figs. 6 and 7).

3.1 Distributed Default Theory of Intelligent Agents

An agent is defined by a holistic system appropriate to the perception of its finality. It is described by a tuple (SD, COMP, OBS, ACT), denoting respectively the description of the system by a set of first order logical formulas SD; COMP, the components of the system, a set of propositional constants; OBS, the captured observations of the system; a set of first order logical formulas; finally ACT a set of first order logical formulas corresponding to the actions to be taken by an intelligent agent.

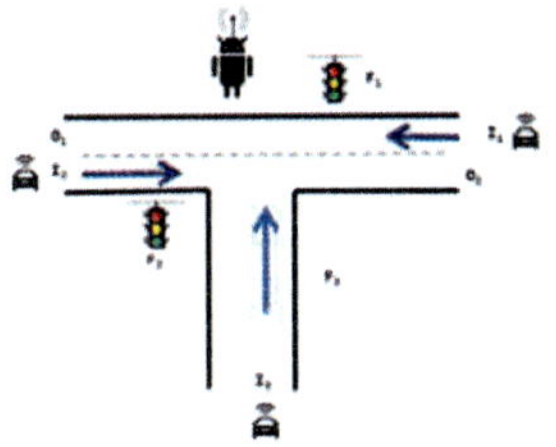

Fig. 8 Simple three-branch crossroads

Fig. 9 ACA Agent for ANFA's boulevard

Fig. 10 GCMA-Net Agent

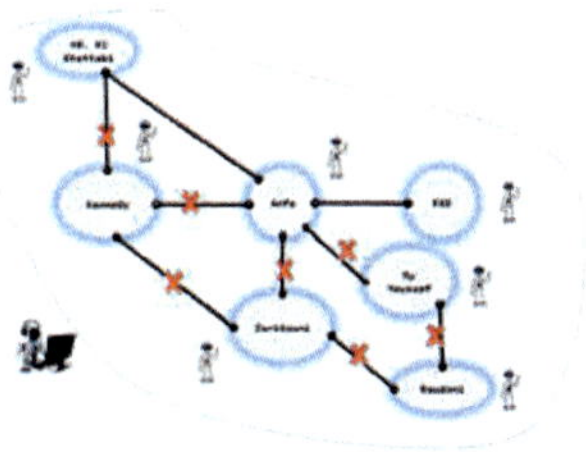

Table 1 Local theory of a crossroads type agent

Element	Content
SD	Structural relations
COMP	Entry/Exit segments, traffic lights,
OBS	State of segments, Traffic variables
ACT	Local Control Actions: micro-Control

3.1.1 Crossroads-Type Local Control Agent (CLCA)

The CLCA agent is intended to appropriate the local reality of the network through its fundamental control element, namely the crossroads. The elements of its logical theory are summarized in Table 1. Figure 8 shows a simplified three-branch junction under the care of a local junction-type agent. The associated logical system is given below. The components correspond to the input/output segments, the intersection and the traffic lights. This set is given by:

$$COMP = \left\{C, (F_i)_{i=1,2}, (I_i)_{i=1..3}, (O_i)_{i=1..2}\right\}$$

The description of this system is formulated as follows:

Input/Output segments of intersection C:

$SD = \{OUT(C, O_1), OUT(C, O_2), IN(C, I_1), IN(C, I_2), IN(C, I_3)$

Segment topology,

Successor(I_1, O_1), Successor (I_2, O_2), Successor (I_3, O_1), Successor (I_3, O_2),

Phasing of the intersection and control of segments and traffic lights,

CTRL(F_1, I_1), CTRL(F_3, I_3), CTRL(F_2, I_2),

Conflict(I_3, I_1), Conflict(I_3, I_2),

State(F_1) = State(F_2),

State(F_3) =$\neg$ State(F_1),

Phase(I_1, P_1), Phase(I_2, P_1), Phase(I_3, P_2),

Conflict(P_2, P_1)}.

The following rule can be added to SD:

$$OUT\ (C, X) \wedge AB\ (X) \wedge Successeur\ (Y, X) \rightarrow AB\ (Y).$$

To describe the normal behavior of the control attached to this junction, the following rule allows it to be specified with the predicate AB(.), which will be added to SD.

$$OUT\ (C, X) \wedge \neg AB\ (X) \wedge Successeur\ (Y, X) \wedge Phase\ (Y, P) \rightarrow \mathbb{C}\ (P, normal).$$

The predicate $\mathbb{C}\ (p, normal)$ denotes the normal control strategy for the fluid traffic regime. The above rule means *"If X is an exit from junction C and X is in a normal state (fluid) and X is the successor of segment Y and Y is controlled by a phase p is normal then the control strategy adopts the strategy of control of fluid traffic rules"*. The system's observations (OBS) concerning the signal status, programmed control strategies, traffic state of each segments and arteries etc.

$$\text{OBS} = \{\ \text{State}(I_2, Fluid), \text{State}(O_1, Saturated), \text{State}(I_3, Fluid), \text{State}(I_1, Fluid)\}$$

The ACT set of the action considered for the above example takes the following form:

$$ACT = \{\text{Activate}(P_i), \text{Cancel}(P_i), \text{Hide}(P_i), \text{Reinforce}(P_i), \text{Weaken}(P_i), \mathbb{C}\ (p_i, normal), \mathbb{C}\ (p_i, saturated)\ |i = 1, 2\}.$$

To all SD formulas, we can also add the following rule:

$$OUT\ (C, X) \wedge AB\ (X) \wedge Successor\ (Y, X) \wedge Phase\ (Y, P) \wedge Active\ (P) \rightarrow Reinforce\ (P).$$

In other words, reinforce the active phase during the rise of congestion towards intersection C.

Table 2 Theory of an agent of the arterial control

Element	Content
SD	Structural relationships between intersections
COMP	Intersections
OBS	State of intersections, intersection traffic variables
ACT	Semi-macroscopic control actions: green wave, vector method, synchronization between intersections, type operations: increase green, decrease green of an intersections axis

Table 3 Theory of an GRMA agent

Element	Content
SD	Structural relationships between arteries
COMP	The arteries
OBS	Arteries state (travel time), arteries traffic variables
ACT	Arterial Control Actions: symbolic operations such as: strengthening, reducing, balancing

3.1.2 Arterial Control Agent (ACA)

The theory of the arterial control ACA has the elements perceived at the arterial level. The nodes of the artery materialized by the crossroads and the topological and flow relationships existing between the crossroads (see Table 2). Figure 9, presented above illustrates an arterial control agent whose theory is formulated as follows. SD contains the following logical formulas:

$$\text{SD} = \{\ \text{Successor}(C1,\ C2),\ \text{Successor}(C2,\ C3),\ \text{Successor}(C3,\ C4),\ \text{etc.}\}.$$

The components correspond to the set COMP containing the intersections of the artery and the artery itself as an object; $COMP = \{A, (C_i)_{i=1,4}\}$. The observations of the ACA system concerning the state of the signaling, the control plans planned for the artery, and the traffic state of each of the intersections of the artery, etc.

$$OBS = \{State(C_2, normal), State(C_1, anormal), State(C_3, congested), State(A, free)\}$$

The ACT set of actions considered for the example below takes the following form:

$$ACT = \{\text{Fluidize}(A),\ \text{GreenWave}(A),\ \text{Holdback}(A)\}$$

3.1.3 Global Control and Mediation Agent (GCMA)

The GCMA agent is the network supervisor. This is the root of the command that integrates all ACA agents. At its level the vision only concerns the arteries of the network. Thus, the elements that feed it theory are consolidated in the Table 3.

Figure 10 illustrates an GCMA agent whose theory is formulated as follows. SD contains the following logical formulas: Continue(A_i, A_j), Cut(A_i, A_j), etc. The components correspond to the COMP set containing the arteries of the network; $COMP = \{(A_i)_{i=1,n}\}$. The observations of the GCMA system concerning the state of the arteries (travel time index, quality of service, etc.), the traffic status of each of the arteries, etc.

$$OBS = \{State(A_2, fluid), State(A_1, saturated), State(A_3, fluid)\}.$$

The ACT set of actions considered for the example below takes the following form:

$$ACT = \{ValidateFluidify(A), ValidateGreenWave(A), ValidateHold(A)\}$$

The symbolic control strategy is based in particular on meta-algorithms aligning traffic organization strategies with the objectives assigned to each road element. It also ensures a rebalancing of traffic loads according to the priorities of the axes and the functional constraints of the network. This strategy will be developed in future work. Certain technical considerations have been deliberately ignored for industrial protection considerations.

4 Conclusions

In this work, we propose a distributed default theory to model a multi-agent system for the symbolic control of a transport network. We distinguish three categories of agents. The crossroads-type local control agent (CLCA) which has local action for autonomous micro-control. The arterial control agent (ACA) which consolidates the CLCA agents attached to the intersections of the artery in question. The ACA agent at his disposal controls by route or arterial according to pre-programmed control strategies or adapted to the situation. The upper level of the control hierarchy is mandated by the global control and mediation agent (GCMA) who coordinates the ACA agents and plays the role of harmonizer of control between arteries of the network. Distributed intelligence supported by distributed non-monotonic logic is made available for modeling the problem of network control. The next step is intended to be operational for the deployment of this symbolic control pattern and to study in detail its execution profile in simulation and in a real situation.

Acknowledgements This work was partially funded by the Digital Development Agency (ADD) and the National Center for Scientific and Technical Research (CNRST) in partnership with the Ministry of Industry, Commerce and Green and Digital Economy (MICEVN) and the Ministry of National Education, Professional Training, Higher Education and Scientific Research

(MENFPESRC)# AL KHAWARIZMI Program # Intelligent & Resilient Urban Network Defender: A distributed real-time reactive intelligent transportation system for urban traffic congestion symbolic control and monitoring.

References

1. Besnard, P.: An Introduction to Default Logic. Springer, ISBN 978-3-642-08078-4 (1989)
2. Konolige, K.: On relation between default and autoepistemic logi. Elsevier, Artificial Intelligence (1988)
3. Reiter R.: A logic for default reasoning. Artif. Intell. (1980) (Special Issue)
4. Ryzko, D. Rybinski, H., Wiech, P.: Learning mechanism for distributed default logic based MAS – implementation considerations. Intell. Inf. Syst. (2008). ISBN 978-83-60434-44-4
5. Davis, M.: The mathematics of non-monotonic reasoning. Artif. Intell., vol. 13, (1980) (Special Issue)
6. De Kleer, J., Mackworth, A.K., Reiter, R.: Characterizing diagnoses and systems. Artif. Intell. (1992)
7. Reiter R.: Theory of diagnosis from first principles. Artif. Intell. (1987)
8. Segerberg, K., Meyer, J.L., Kracht, M.: The logic of action. The Stanford Encyclopedia of Philosophy (winter 2013 Edition), Zalta, E.N. (ed.). http://plato.stanford.edu/archives/win2013/entries/logic-action/
9. Boulmakoul, A., Bourbia, N.: CSN : Contrôle Symbolique par Négociation, prototypage en smalltalk. Software Engineering and its Applications, Paris, France (1994)
10. Boulmakoul, A.: Approche logique du contrôle symbolique. 9éme Congrés Reconnaissance de Formes et Intelligence Artificielle RFIA'94, vol. 2. Intelligence Artificielle, AFCET, France (1994)
11. Shoham, Y.: Agent-oriented programming. Artif. Intell. **60**(1), 51–92 (1993)
12. Gul, A.: Actors: A Model of Concurrent Computation in Distributed Systems. The MIT Press series in Artificial Intelligence (1980)
13. Bajo, J., Hernández, J.Z., Mathieu, P., Campbell, A., Fernández-Caballero, A., Moreno, M.N., ... Botti, V.: Trends in Practical Applications of Agents. Multi-Agent Systems and Sustainability, vol. 372 (2015)

IDCCPSS 1 (Special Session)—Chair: Igor Kotenko

Using Clustering Methods of Anomalies and Neural Networks to Conduct Additional Diagnostics of a Computer Network

Maxim Avilov, Yulia Shichkina, and Mikhail Kupriyanov

Abstract The article considers the problem of rapid response to emergency events. Often, based on the results of proactive monitoring, prediction an anomaly in a computer network and correct respond to a detected anomaly is not possible. In these cases, additional information and additional diagnostics are necessary. To solve this problem, the authors of the article propose an approach to organizing monitoring of a computer network using the module for additional diagnostics of anomalies and preliminary clustering of these anomalies. The additional diagnostic module is based on an artificial neural network that identifies the tools necessary to eliminate anomalies. Also, according to the results of additional diagnostics, the monitoring system makes a notification to the system engineer about the work done and about the current functioning of the computer network.

1 Introduction

The use of information and communication computer networks around the world in various fields of industry is a necessary aspect in the work support and industrial enterprises development. Along with the increase in requirements for the speed of information exchange over a computer network (CN), the number of applications working through this computer network, as well as the number of new network devices being connected, is increasing. In this regard, the problems associated with the correct operation of computer networks are also increasing.

Despite the fact the development of computer networks has a long history, studies on aspects of improving the functioning of such networks are relevant today.

A large number of studies are carried out in the field of monitoring, classification, network traffic management in various, by geographical location, computer networks [1]. Also, one of the relevant areas of research is the security of the information and communication computer network [2].

M. Avilov · Y. Shichkina (✉) · M. Kupriyanov
St. Petersburg State Electrotechnical University "LETI", St. Petersburg, Russia
e-mail: strange.y@mail.ru

D. Camacho et al. (eds.), *Intelligent Distributed Computing XIV*, Studies in Computational Intelligence 1026, https://doi.org/10.1007/978-3-030-96627-0_18

In the last decade, many studies have focused on the use of artificial neural networks in various fields of activity. In particular, artificial neural networks were investigated for the purpose of certain information processing and for the purpose of identifying network anomalies. In [3], the author gives an overview of neural networks classifiers with direct connection in static models with continuous input data.

Typically, methods for automated detection of anomalies in the operation of an information and communication computer network are based on statistical modeling, and some use machine learning to detect deviations of "abnormal" events from established "normal" events in the network. So, the article [4] describes a scheme for hybrid anomalies detecting, referred to by the authors as the Ensemble-based Classification Model for Network Anomaly Detection (EnClass). This model has three modules: clustering to determine the optimal subset of the network properties that must be used to classify network traffic; refinement of a set of functions and removal of unnecessary attributes; building a decision tree for classifying network traffic. In preparing a real data set, the authors use the KDD knowledge discovery model. This model is often used and was improved by us in 2017 [5].

The article [6] presents an approach to network anomalies detecting for systems with limited resources, based on a behavioral analysis of computer systems and/or networks. The article discusses two types of model behavior: normal behavior (expected and desired activity) and abnormal behavior (all activities that are not included in the category of normal behavior). During the diagnostic process, each case is evaluated as normal or abnormal behavior. Weight values are introduced and methods of mathematical statistics and a sliding window are applied to determine the degree of abnormality. The complexity of this approach remains in the unique classification of the event as anomalous.

In [7], the authors analyzed existing intrusion detection systems in the network and proposed an approach to creating an intrusion detection system (ANIDS) for detecting anomalies based on a backpropagation neural network (BPNN). Learning was done using a backpropagation algorithm. During the research, the authors created 48 variants of intrusion detection systems (IDS). As a result of testing on the KDD CUP '99 dataset, based on the proposed approach, the two best IDSs were identified, which showed higher accuracy and speed of anomaly detection.

In [8], the authors present a comparative analysis of the latest advances among machine learning methods regarding the detection of intrusions into distributed data processing systems. The authors consider machine learning methodologies such as J48 decision tree, Classification and Regression Trees, Mutual information algorithm and Support vector machine, Recurrent Neural Networks, Genetic algorithms, Self-organized map and k-means clustering, and others. They compare the advantages and disadvantages of methodologies based on different datasets. The authors also provide comparisons of some methods in terms of accuracy, learning rate, and false positives.

The authors in [9] present a practical solution for real-time intrusion detection using uncontrolled neural networks. They called this system RT-UNNID. This system allows to identify traffic other than "normal" by network traffic classification. They used such network packets as ICMP, TCP, UDP. A combination of anomaly and malicious behavior detection approaches can detect known and unknown attacks. One of the system components is the use of uncontrolled neural networks, which allow improving data analysis without retraining.

Recently, the fifth generation mobile technology, called 5G, has been intensively developing. This technology includes advanced communication functions. It creates new problems for cybersecurity systems. Therefore, the results of studies to identify anomalies in such networks began to appear in the literature. For example, such results are given in [10] and these results are based on the use of neural networks.

It should be noted that in comparison with the anomaly of harmful effects on the network, there are very few works devoted to other anomalies, and very few publications deal with the identification of anomalies in general.

It should also be said that today there are different software and hardware systems for general monitoring of a computer network and specialized tools for monitoring hosts. For example, monitoring tools using ICMP, SNMP, RMON, as well as tools for synchronizing and conducting polls from a monitoring center, tools for monitoring the consumption of computing resources, or tools for identifying anomalies [11] in the operation of network nodes with the subsequent output of alerts in case of unplanned critical situations.

However, when an unforeseen situation arises in a computer network, it can often be detected late or it will take a lot of time and manual additional diagnostics to establish the exact cause of the problem. The prompt determination of network failures causes directly affects the overall recovery time of the information and communication computer network in the normal mode.

In this article, we propose an approach to monitoring the information and communication computer network that allows to monitor the status of a computer network using the anomalies clustering methods and an artificial neural network to conduct additional diagnostics and prompt response to emerging events in a computer network.

2 Formulation of the Problem

The purpose of monitoring the information and communication computer network is to identify anomalies. The term "anomaly" in this study means the excess of the established threshold values of the observed parameters of the computer network nodes. As a result of the appearance of such anomalies, a computer network may partially or completely not function correctly. Computer network monitoring is divided into proactive and reactive.

Today, there are many approaches for proactive monitoring of a computer network using artificial neural network methods. However, if such monitoring has just begun work, then data for a long period of time has not yet been collected, which means that it is not possible to correctly predict events that may occur. When a critical event occurs, based on the results of computer network proactive monitoring, there can only be a notification to the system engineer about an event that has already occurred and it becomes problematic to predict the further development of the situation. Therefore, in such situations, reactive monitoring of the computer network is used.

In reactive monitoring of a computer network, a reaction to a critical situation arises after its occurrence. As a result, mechanisms are launched to inform, identify and, if possible, eliminate the critical situation that has arisen.

However, there are cases when it is necessary to "manually" conduct additional diagnostics to identify the causes of problems. For example, when the observed network node has stopped responding to the computer network monitoring system, but at the same time, the data passing through it can be successfully transmitted to neighboring network nodes. Additional diagnostics can take a long time. In order to reduce the time of additional analysis and increase the speed of response to emerging events in the operation of a computer network, it is possible to use artificial neural networks to conduct an additional analysis of the situation and launch the corresponding response "action". This is important in cases where, based on the results of proactive monitoring, it is not possible to predict an anomaly, and according to the results of reactive monitoring, it is not possible to correctly respond to an event that occurs without additional diagnostic information.

The aim of the our research was to assess the possibility of using neural networks to perform additional diagnostics of a computer network using the example of a two-layer perceptron. To solve this problem, various approaches to the formation and storage of input data were used. Methods for additional diagnostics of a computer network and clustering of computer network anomalies were developed based on the generated input data. Also, a method for monitoring a computer network using an additional diagnostics module was developed.

3 Computer Network Monitoring System with an Additional Module for Anomaly Diagnostics

Configurable, within the framework of this study, the computer network monitoring system (SM) should collect, process, store the collected data from the observed network nodes and, in case of anomalies, start additional diagnostics with direct application of scripts to the identified anomalies or with preliminary identification of tools for unknown anomalies based on neural network.

The parameters of the observed nodes can be very different, for example, such as: host names, host group names, network identification of nodes (ip-addresses, domain names), values of hardware parameters (how much CPU, RAM, ROM, etc. are loaded), time parameters (polling frequency of the network node, periods of history storage), the status of the established triggers, as well as the parameters of the monitoring system necessary to present relevant information to the system engineer, and other parameters. All values of these parameters are stored in the database (DB) of the monitoring system. For additional diagnostics, only part of these parameters is used. Therefore, there is a separate question about the use of a separate database for

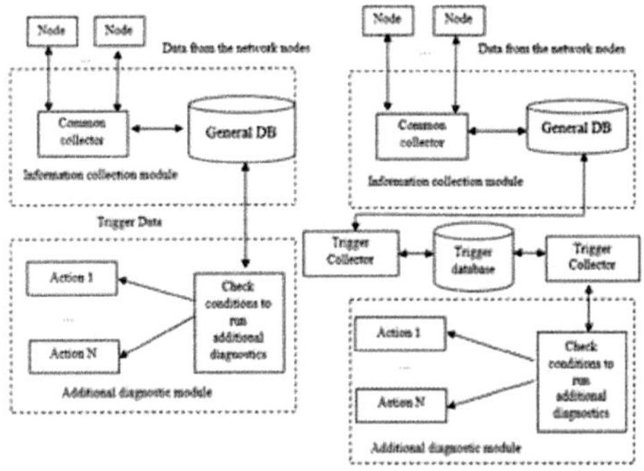

Fig. 1 Schemes of SM computer network with a common database and a separate database for triggers

storing necessary for diagnostics parameters, or for searching for these parameters in the monitoring system general database during diagnostics.

The additional diagnostics module (AD) can be activated only after training the neural network. Trigger values corresponding to parameters whose values exceeded the set threshold values are received at the input of the additional diagnostics module, thereby signaling the possible presence of an anomaly. The values of these triggers can be stored both in the database of the monitoring system together with other data for monitoring the computer network, and in a separate database. The schemes with a separate database and without it are presented in Fig. 1.

The main advantage of using a common SM database for storing all data for monitoring the CN, including data for the AD module, is saving memory due to the absence of duplicate data and the absence of overhead costs for additional computational operations to overwrite data in a separate database. But, in this case there will be a loss of time for the selection and processing of data necessary for the AD module and possible loss of data necessary for the neural network in the event of a failure in a common SM database. A system with a separate database for the AD module does not have these disadvantages. Therefore, when choosing an approach to building a computer network monitoring system with an additional diagnostic module, it is necessary to take into account a variety of aspects, such as the amount of available resources, efficiency in data processing, and information loss risks.

When anomalies appear in the operation of a computer network, a so-called trigger occurs, the values of which are processed by the additional diagnostics module.

The result of additional diagnostics may be one of the following actions:

- informing the system engineer about problem areas of the network;
- conducting additional diagnostics to identify tools for eliminating anomalies and launching the selected tools with informing the system engineer of the results of its implementation or a recommendation for the system engineer to launch the selected tools;
- execution of scripts for anomalies that do not require identification, and informing the system engineer of the scripts results.

4 Method for Monitoring a Computer Network Using the Additional Module for Anomaly Diagnostics

The neural network in the AD module solves the problem of classifying the launch of additional diagnostic tools.

Let are:

- A – the set of all observed parameters of the computer network monitoring system:

$$A = \{a_1, \ldots, a_i\}, \tag{1}$$

where i – the maximum number of observable parameters of the set A;
- B – the set of observable parameters of one network node:

$$B = \{b_1, \ldots, b_j\}, \tag{2}$$

where j – the maximum number of observable parameters of the set B.

Since the collected parameter values of each network node are part of all the monitoring system observed parameters, then:

$$A = \{B_1, \ldots, B_k\} = \{\{b_{11}, \ldots, b_{j1}\}, \ldots, \{b_{1k}, \ldots, b_{jk}\}\}, where\ b_{jk} = a_i. \tag{3}$$

Let C - the set of analyzed network nodes selected for observation:

$$A = \{B_1, \ldots, B_p\}, \tag{4}$$

where p – the number of observed parameters of the analyzed nodes.

For the analyzed nodes, not all parameters can be evaluated, but only some combination. Let d_p - interesting observable parameter of the analyzed node, then D - the set of selected observable parameters of the analyzed network nodes:

$$\begin{gathered} d_p \in B_p \\ D = \{D_1, \ldots, D_p\}, \end{gathered} \tag{5}$$

where p - the number of observed parameters of the analyzed nodes.

For each parameter d_p the system engineer sets a threshold value that is typical for the regular work of his computer network. An indicator of the output of the observed parameter value beyond the threshold value is an additional parameter - the trigger. Triggers can take the value 0 or 1, where 0 - means that the value of the observed parameter has not exceeded the set threshold value, and 1 - means that the value of the observed parameter has exceeded the set threshold value. If there is a trigger associated with each value of d_p, then the whole set D will then correspond to the set of triggers:

$$X = \{X_1, \ldots, x_p\}, \tag{6}$$

where p – the number of observed parameters of the analyzed nodes.

This set X is fed to the input of the AD module.

After the input data for the additional diagnostics module is generated, it is necessary to determine in which cases it is necessary to submit this data to the input of the neural network, and in which not. For this, it is necessary for the set X to associate a certain number (weight), which would allow it to be attributed to one of the three types of anomalies corresponding to the actions in the section on testing the AD module. Assume that the range of valid weight values is [0, 1].

As the weight V of the set X, it is possible to take the average value of its triggers:

$$V = \frac{\sum_{i=1}^{p} x_i}{p}, \tag{7}$$

where V – the average value of the triggers in set X, x_i – trigger value, p – number of triggers.

The ranges for informing, additional diagnostics and informing, running a pre-written script and informing are determined by the system engineer based on the requirements for the functioning of his computer network.

The input data for a neural network: set of triggers formed according to the rules (1)–(6).

Output data: values [0, 1], in accordance with which the action is taken on the situation in the computer network.

Method for monitoring a computer network

Step 1. Clustering of anomalies by the formula (7).

Step 2. If $V = 0$ (the entire group of network nodes is operating normally), then Step 12. Else Step 3;

Step 3. If $V = 1$ (the entire group of network nodes is unavailable), then Step 7. Else Step 4;

Step 4. If $V \in (0; a)$ (informing about problem areas of the network), then Step 7. Else Step 5;

Step 5. If $V \in [a, b)$ (launch additional diagnostic module), then Step 8. Else Step 6;

Step 6. If $V \in [b, 1)$ (running a pre-written script for critical cases), then Step 11. Else Step 12;

Step 7. Informing the system engineer about the current state of the computer network. Step 12;

Step 8. The supply of input data to the inputs of the neural network. Step 9;

Step 9. Neural network processing of input data. Step 10;

Step 10. Launch of an appropriate tool for additional diagnostics, based on the results of the operation of an artificial neural network. Step 7;

Step 11. Running pre-written script to a critical situation. Step 7;

Step 12. End of method.

5 Testing of the Module

At the moment, there are many computer network monitoring systems. As part of this study, the Zabbix monitoring system was used. The AD module was created in the Python programming language.

As part of testing the efficiency of the method, the following was done:

1. 13 network nodes were involved in monitoring the work of the CN. This number of nodes was chosen in order to clearly show the clustering of CN anomalies.
2. Data is collected via the ICMP network protocol through Zabbix on the server into a single database. The monitored parameter of a network node is the availability of a network node (by "ping" responses). Triggers are associated with observed parameters.
3. Test ranges of actions (boundaries of clustering of anomalies of CN) were set as follows:
   - 0 – do nothing;
   - (0; 0.4) – informing the system engineer about problem network areas;
   - [0.4; 0.8) – launch of an additional diagnostic module and informing the system engineer about the results of work;
   - [0.8; 1) – applying a pre-written script and informing the system engineer about the results of work;
   - 1 – informing a system engineer about a critical situation.
4. Before starting the DD module, it is necessary to train the neural network shown in Fig. 2. Training is based on the generated trigger variants with matching the correct answer. If the sum of triggers is greater than or equal to half the number of all triggers, then the second tool must be run. If the sum of the triggers is less than half the number of all triggers, then the first tool should be used. Such conditions, within the framework of demonstrating the efficiency of the method, are included in the clustering range [0.4; 0.8).
   The following tools were used as additional tools:
   - the first tool is the execution of the "ping" and "traceroute" utilities to check the transmission of data along network routes to network nodes;
   - the second tool is the execution of a previously created script to collect additional metrics.

The result of training an artificial neural network in the DD module for 700 training cycles was formed using the "matplotlib" library in Python3 and is shown in Fig. 2. The graph of the results of training the neural network shows how the accuracy of the correct answer changed. It is also possible to see that the accuracy of the answer approaches the value of absolute accuracy equal to 1. For cycles 1 to 34, learning is fast. Then deceleration begins. Only from the 670th cycle does the accuracy of the correct answer become more than 0.96. This is acceptable in this testing.

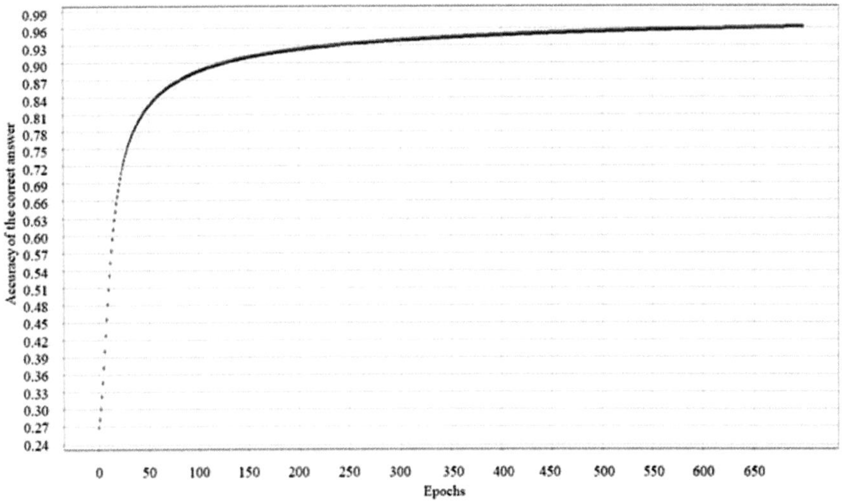

Fig. 2 The result of training a neural network

6 Conclusion

Testing the computer network monitoring method using the additional diagnostics module based on a two-layer perceptron showed positive results. This suggests that with the help of such module it is possible to carry out additional analysis and carry out a certain reaction to emerging events during the operation of a computer network without an auxiliary classification of anomalies.

Applying the approach to the operation of a computer network monitoring system with a common database of the monitoring system allows to work with data without additional data rewriting into a separate database, which reduces the amount of memory allocated for storing the collected system monitoring information. However, when applying this approach, the time for choosing the input data for the additional diagnostics module increases and additional storage of these input data is not carried out. Therefore, for safety, it is necessary to carry out automatic systematic backups of a common database.

The implementation of the proposed method can be in the form of an additional module for the integrated monitoring of the information and communication computer network. This allows to timely identify unforeseen situations in the operation of such network and make decisions on how to manage such situations. Determining the clustering boundaries allows to take into account the special properties of the computer network functioning, which the system engineer of this network knows.

Acknowledgements The research was funded by RFBR and CITMA according to the research project #18-57-34001.

References

1. Dainotti, A., Pescape, A., Claffy, K.C.: Issues and future directions in traffic classification. IEEE Netw. **4**(1), 35–40 (2012)
2. Ting, Ch., Field, R., Fisher, A., Bauer, T.: Compression analytics for classification and anomaly detection within network communication. IEEE Trans. Inf. Forensics Secur. **14**(5), 1366–1376 (2019)
3. Lippmann, R.P.: Pattern classification using neural networks. IEEE Commun. Mag. **27**(11), 47–50 (1989)
4. Garg, S., Singh, A., Batra, S., Kumar, N., Obaidat, M.S.: EnClass: ensemble-based classification model for network anomaly detection in massive datasets. In: IEEE Global Communications Conference, Singapore, pp. 1–7 (2017)
5. Degtyarev, A., Shichkina, Y., Koblov, A.: Technology of cleaning and transforming data using the Knowledge Discovery in Databases (KDD) technology for fast application of Data Mining methods. CEUR Workshop Proc. **1787**, 428–434 (2016)
6. Pelc, M., Galus, D., Zolubak, M., Ozana, S., Chlewicki, W., Cichon, K., Podpora, M., Kawala Sterniuk, A.: Behavioural approach to network anomaly detection for resource-constrained system – presentation of the novel solution – preliminary study. IFAC-PapersOnLine **52**(27), 121–126 (2019)
7. Chiba, Z., Abghour, N., Moussaid, Kh., El Omri, A., Rida, M.: A novel architecture combined with optimal parameters for back propagation neural networks applied to anomaly network intrusion detection. Comput. Secur. **75**, 36–58 (2018)
8. Eltanbouly, S., Bashendy, M., AlNaimi, N., Chkirbene, Z., Erbad, A.: Machine learning techniques for network anomaly detection: a survey. In: IEEE International Conference on Informatics, IoT, and Enabling Technologies (ICIoT), Doha, Qatar, 2020, pp. 156–162 (2020)
9. Amini, M., Jalili, R., Shahriari, H.: RT-UNNID: a practical solution to real-time network-based intrusion detection using unsupervised neural networks. Comput. Secur. **25**, 459–468 (2006)
10. Hamamoto, A.H., Carvalho, L.F., Hiera Sampaio, L.D., Abrao, T., Proenca Jr, M.Le.: Network anomaly detection system using genetic algorithm and fuzzy logic. Expert Syst. Appl. **92**, 390–402 (2018)
11. Kanaev, A.K., Kamynina, M.A., Oparin, E.V.: Methods for detecting deviations in the functioning of data transmission network elements in the interests of the control system. Bull. Results Sci. Res. (in Russian) **4**(3), 137–148 (2012)

Attack Strategies and Countermeasures in Transport-Based Time Synchronization Solutions

Diana Gratiela Berbecaru and Antonio Lioy

Abstract The security, availability, and accuracy of time information transmitted over transport networks are getting increased attention since different application domains require secure and accurate time. In this work, we classify first the security attacks affecting the transport-based time synchronization architectures. Such an architecture is currently designed in the ROOT (Rolling Out OSNMA for the Secure Synchronization of Telecom Networks) project. We indicate the attacks applying to different views of the ROOT architecture, namely the *time distribution*, *network management*, *hardware*, and *software*. We then considered the software view, and we experimented with a set of software tampering attacks on a dedicated Raspberry Pi 4 device employed for time distribution. To counteract such attacks, we exploited the Trusted Platform Module available on the device and Keylime remote attestation software to verify the integrity of time distribution software installed on the device. These tests represent a first step toward deploying the software integrity controls on the specialized nodes handling time synchronization in the ROOT project.

Keywords Cybersecurity · Time synchronization · Security attacks · Remote attestation

---

D. G. Berbecaru (✉) · A. Lioy
Dipartimento di Automatica e Informatica, Politecnico di Torino, Corso Duca degli Abruzzi 24, 10129 Torino, Italy
e-mail: diana.berbecaru@polito.it

A. Lioy
e-mail: lioy@polito.it

D. Camacho et al. (eds.), *Intelligent Distributed Computing XIV*, Studies in Computational Intelligence 1026, https://doi.org/10.1007/978-3-030-96627-0_19

1 Introduction

Smart industry, power grids, vehicular applications, critical Internet of Things, industrial automation services, as well as finance and banking applications are increasingly requiring accurate, robust, and secure time synchronization solutions. Such solutions have long been a fundamental prerequisite for the operation of telecommunications networks, and will be more important than ever in 5G networks. The main 5G synchronization requirements and solutions sketched in [1] are currently studied in the ROOT research project [2], which aims to experiment different types of cyberattacks and evaluate their impact on a time synchronization architecture.

The emerging solution for time synchronization in time-sensitive networks (TSNs) exploits the GNSS (Global Navigation Satellite System) services and the transport domain networks. GNSS services include Galileo, GPS and BDS (BeiDou Navigation Satellite System) and are offered by several space agencies. The distribution of time over transport networks has been addressed in several ITU-T recommendations [3]. These recommendations define reference synchronization networks, where the time synchronization is created by Enhanced Primary Reference Time Clocks (ePRTCs) exploiting GNSS technology and terrestrial accurate clocks, and the reference timing signal is carried across a network of clocks. The time synchronization reference is typically carried through the PTP (Precision Time Protocol) [4], or Network Time Protocol (NTP) [5] over packet switched networks. NTP is suitable for large and dynamic networks that require an accuracy of a few milliseconds over wide area networks. The PTP instead is designed for TSNs that often use PTP-aware hardware to provide an accuracy of few microseconds and even nanoseconds. PTP distinguishes between different clock types. One of these clocks is the grandmaster clock (GM), which provides the time reference for the network. The ePRTC is typically the source of time for the PTP grandmaster. The other ordinary clocks, called slaves, exchange time synchronization information with the GM.

Several papers addressed the security of PTP protocol [6–8] or of the more accurate profile White Rabbit PTP (WR-PTP) protocol [9, 10]. Nevertheless, few works addressed the classification of attacks in transport-based time synchronization networks [11]. An attacker may exploit multiple attack vectors against the network protocols exploited for the time distribution, the network elements connected to the time distribution devices, or the software (and the configuration) of time distribution devices. The work closest to ours is described in [12], in which both the network attacks as well as the *advanced internal attacks* targeting the running software or its configuration are considered. Possible attack points must be identified to select adequate countermeasures. We identified first possible security attacks applying to the time synchronization solution designed in the ROOT project [13]. Then, we tested some simple software tampering attacks on a dedicated device equipped with a Trusted Platform Module (TPM). To counteract such attacks, we exploited Keylime remote attestation software [14]. The paper is organized as follows: Sect. 2 presents the main protocols and components in time synchronization architectures, Sect. 3 details the proposed attacks classification, while Sect. 4 describes the preliminary experiments with the software attacks. Finally, Sect. 5 concludes the paper and indicates future work.

2 Transport-Based Time Synchronization Solutions

Upcoming telecom networks can respond to time synchronization requirements granting nanosecond accuracy by deploying specialized devices that support specific timing protocols, such as PTP, NTP, or WR-PTP protocols. Some possible time synchronization solutions described in [1] can be implemented either in the RAN (Radio Access Network) domain or in the transport domain. The focus of this paper is on the latter category of solutions, although, in general, a combination of techniques in both domains could be used to create a robust and reliable solution. As mentioned in [1], "distributing time synchronization over the same transport network infrastructure used for user data has the benefits of providing the same level of robustness and redundancy for timing as for the user traffic itself".

Two PTP profiles, namely G.8275.1 (PTP with full timing support from the network) and G.8275.2 (PTP with partial timing support from the network), exist for PTP use in telecom networks [3]. PTP is designed for infrastructure networks that often use PTP-aware hardware providing clock synchronization accuracy down to microsecond and even nanosecond level.

To achieve high accuracy (such as, from 65 to 130 ns for 5G front-haul applications) a high-level network architecture exploiting a centralized master clock is typically exploited [13]. A Centralized Grandmaster Clock (C-GMC) generates a time reference by combining different time sources. In particular, the C-GMC uses a GNSS receiver combined with a co-located Cesium Atomic Clock (Cs AC). Multiple reference clocks, named Distributed GMC (D-GMC), are distributed across the network to provide robustness and resilience if temporary malfunctioning or intentional attacks occur. The D-GMC combines multiple time sources to generate a time reference, typically a GNSS receiver and a (less expensive) Ove-Controlled Crystal Oscillator (OCXO), or a Rubidium (Rb) clock. Typically, at least two C-GMCs and two D-GMCs are exploited, a primary one and a secondary one, having similar configurations but operating in different physical locations. The time synchronization signal is distributed to the devices connected to the network over fiber optical channels (as shown in Fig. 1). The telecom network operators install the C-GMC and D-GMC devices at different operational levels. For example, in the architecture shown in Fig. 1 used as reference in Fig. 2, the C-GMC and D-GMC devices occur at the regional level (HL 3) and at the metro aggregation level (HL 4) of the telecom operator network. At the hierarchical level (HL 5) operates a simpler time distribution device (called BC/GM), which also exploits a GNSS receiver for time calculation, but the terrestrial clocks are not employed at this level. The HL 5 level is the most distributed one where mobile base stations connect.

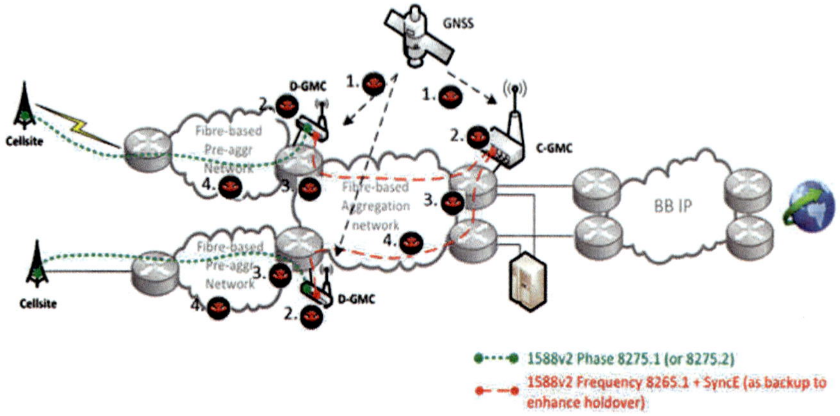

Fig. 1 High-level architecture showing possible phase synchronization distribution, by exploiting C-GMC and D-GMC devices. We indicate possible input attack points targeting both the hardware and the software of the C-GMC and D-GMC devices (shown as 2.), their communication with the GNSS satellites (shown as 1.), the network nodes, e.g., routers (shown as 3.), or the protocols (e.g., PTP or WR-PTP) employed for time distribution (shown as 4.)

3 Classification of Attacks

Attacks (unintentional or intentional) against time synchronization networks may have a devastating effect. For example, in 2013, a PTP infrastructure glitch forced Eurex (a famous international stock exchange) to postpone its market opening, because an incorrect leap second calculation caused in erroneous synchronization of their critical systems [15]. The cyberattacks have evolved in time, a recent work [16] indicate of particular concern nowadays the advanced persistent threats (APTs) that may affect the PTP networks. The APTs target a small number of power users within the target organization with malicious software, for example, malware on secondary memory devices, i.e., USB sticks. Then, they propagate themselves across the organization by exploiting software flaws. An example of a famous APT is the 2010 Stuxnet attack that compromised an industrial control system [17].

From a preliminary analysis of the architecture shown in Fig. 1, we observe that an adversary may pick one or a combination of the following security attacks:

- attacks against the communication between GNSS (satellites) to GNSS-enabled nodes (e.g., D-GCM, C-GCM, ...) such as:
  - calculate and broadcast a false GNSS signal (spoofing attacks [18]), or retransmit a real but delayed GNSS signal (meaconing attack)
  - denial of service (including jamming attacks [19])
- attacks (hardware or software) against the specific D-GCM or C-GCM devices, such as:

  - physically breach the device, then exploit the direct access to the hardware
  - modify the software running on the device, through malware, memory scraping, side channel attacks, software backdoor, or software tampering attacks
  - modify the configuration of the software running on the device
  - install malicious code/software, either through injection attacks (SQL, XSS) or through worms/trojans, ransomware, rootkits, botnets, or malicious network functions
  - exploitation of buffer overflows (in operating system or software installed)

- attacks targeting the underlying protocols (typically, PTP or WR-PTP) used for time distribution, such as denial of service (DoS) attacks, man in the middle (MITM) attacks, replay attacks, or delay attacks [9, 20].
- attacks targeting the underlying network protocols used for the management of the time distribution devices, such as DoS attacks, MITM attacks, or replay attacks. Examples of protocols employed for this purpose are SSH [21] or TLS [22] for remote access, RADIUS [23] for authentication, or SNMPv3 [24]. Network protocols are prone to manipulation of the network configuration parameters, or of the security data configuration, such as cryptography keys, digital certificates, security policies or access rules. Thus, specific protocol may be employed to check their correctness, e.g. OCSP protocol used to validate the digital certificates in real time [25]. They might even be wrongly configured or poorly configured, e.g., operate with default authentication credentials, easily guessed by the attackers.

To perform a more accurate security analysis, we have further decomposed the architecture shown in Fig. 1 into several views, namely the time distribution view, the network (management) view, the hardware view, and the software view. In classifying the attacks, we distinguished the Outsider (or External) attacks from the Insider (or Internal) ones. Outsider attacks are carried out by entities (nodes) that do not belong to the timing distribution or the network management networks. External attacks can cause serious damage, for example in case the attacker manages to inject false timing data into the network, or if the attacker can inject spurious data into the network to consume network resources and launch a DoS attack. Some attacks are less likely to occur, or even impossible for external attackers. For example, the node tampering attack assumes an attacker can have physical access to the device, so it cannot be performed by an external attacker. Insider attacks occur when an individual or a group within an organization seeks to disrupt operations or exploit organizational assets. Many attacks originate nowadays from within the network by authorized users (e.g., a disgruntled employee, an abusive administrator, or a user trying to gain access to privileged information). Internal DoS attacks appear also when the genuine nodes of the timing distribution network behave in an unintended way. An insider can cause significant damage if he can inject fake (timing) data into the timing distribution network, or if he launches other types of network attacks, such a DoS attack, or a replay attack.

Time distribution view. This view (shown in Fig. 2) deals with the logical devices and the communications links exploited for time distribution over the network. The attacks may target the individual devices at each level (HL3, HL4 or HL5), the

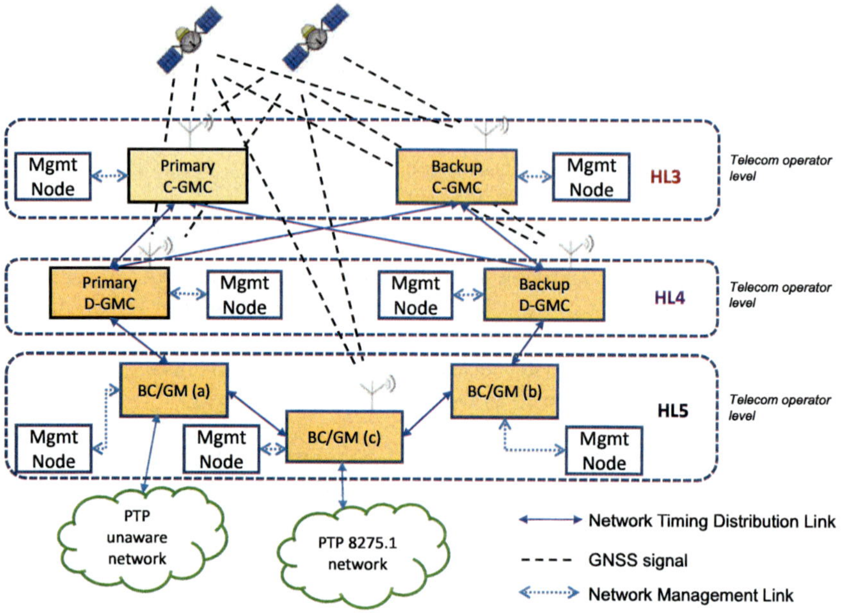

Fig. 2 Timing distribution and network management views used in the security analysis of the ROOT transport-based time synchronization architecture

timing distribution links and protocols (like PTP or WR-PTP), or the GNSS signals, e.g., spoofing, jamming [26]. The attacks against PTP are usually categorized as DoS attacks that might be carried out at various network layers, and MITM attacks, including clock masquerade, replay and filtering attacks [6, 16, 27, 28]. In addition, a novel class of insider threats has been experimentally demonstrated in [6] including two variants of DoS spamming attacks capable to incorrectly steer or permanently skew the slave's clock, as well as a master clock takeover attack.

Network management view. This view (shown in Fig. 2) is composed of the logical devices and the communications links exploited in the management of the C-GMC and D-GMC devices. The network attacks address the C-GMC and D-GMC devices reachable via the network management links from the management node(s), or the communication links themselves. The network management links are often different from the timing distribution links. They may exploit different physical cables, different input/output ports, and network interfaces of the C-GMC or D-GMC devices. An attacker could try to launch a DoS attacks against a network management link either by physically interrupting link (in case of an internal attacker), or by exploiting vulnerabilities in the network management protocol(s). Alternatively, an attacker could try to intercept the traffic exchanged between the management node and the C-GMC or D-GMC to capture sensitive data, such as usernames and passwords, or change cryptographic keys used in the configuration of the devices. Examples of attacks against network management protocols are the ARP spoofing attack, the ping flooding attack, the DNS attacks [29], or the TLS attacks [30, 31]. Typical countermeasures include the firewalls, the intrusion detection systems, and even network forensics to produce evidences with probative value in case of incidents [32]. Last but not least, advanced attacks nowadays exploit vulnerabilities in authenticating the authorized personnel in charge of management procedures. To counter such attacks, authentication with high assurance levels (such as smart-cards or one-time passwords used in combination with devices like smartphones) should be performed [33, 34].

Hardware view. Physical security is equally important because "with regards to attacking an electronic device, any successful physical breach, fundamentally compromises its security" [35]. To mitigate reverse engineering risks in which an attacker might produce circuits that impair proper functioning of electronic devices, one could identify and secure a sufficient set of individual hardware components of the device. This is also the idea behind the TPM, a computer chip broadly available in most modern computers and in several motherboards that can securely store data used to authenticate the platform.

Apart from the GNSS antenna input, other interfaces are critical and need to be adequately protected. For example, the PPS (Pulse-Per-Second) interface used by the C-GMC/D-GMC to receive external reference from a GNSS receiver, and the 10 MHz interface to receive external reference from Atomic clocks, or Rb or OCXO oscillators. In this view, the physical cables are considered as well, because they are subject to unintentional (accidental) damages, or intentional ones.

Software view. In this case, we consider the entire software stack running on the C-GMC and D-GMC devices, including the operating system, the management software (such as, for the TLS, or RADIUS protocols) and the time synchronization deamons, such as `ptpd`, `gpsd`, or `ppsi`. In general, software attacks—sometimes called low-level attacks—rely on characteristics of the hardware, compiler or operating system used to execute software programs to make these programs misbehave, or to extract sensitive information from them [36]. Possible solutions capable to detect intentional changes in the software running on the C-GMC or D-GMC devices could exploit a TPM on board for secure boot, protected storage for sensitive data (keys) and cryptographic operations, and to perform remote attestation [37].

4 Preliminary Experiments with Software Tampering Attacks

In practice, we have experimented with attacks modifying the software (or its configuration) on a device equipped with TPM employed in time synchronization. As solution, we have exploited a remote attestation software on this device.

The Trusted Computing Group establishes the TPM architecture and all its functionalities. Two versions of the TPM have been standardized and adopted, namely the version 1.2 (considered obsolete) and the version 2.0 that have been released respectively in 2003 and 2014. The TPM saves aggregates of measures that will be used in the remote attestation process. The aggregate measurements are stored in special TPM registries, called Platform Configuration Registers (PCRs). Their primary use is to provide a cryptographic record of the software state. The PCRs can be extended via one-way hash functions (e.g, SHA256), and can be read to report their state through a procedure called *quote*, which is an attestation. A TPM attestation is basically *a proof of the software state* that an attester (sometimes called Prover or Agent) can send remotely to a so-called Verifier, via a remote attestation protocol.

When the system boots up, the PCRs in the TPM are reset. The control then passes to CRTM (Core Root of Trust of Measurement), which calculates the hash of the BIOS and extends this hash to the PCRs. The control then passes to BIOS, which calculates the hash of the operating system (that is the kernel image), and extends this hash to the PCRs. Then the system boots up. Before any software daemon is started, such as the `ptpd`, `gpsd`, or `ppsi` daemons, the hash over its binary is calculated and is added to a so-called stored measurement list (SML) maintained by the kernel. This hash is also extended to the PCRs. Before any software runs, its fingerprint is recorded into the PCRs in an incremental fashion, and thus cannot be reverted back. So a system's SML has a list of hashes (one for every piece of executable), and the PCRs have the incremental hash of all these hashes. At system boot up, the kernel calculates the hash of BIOS and its own hash, and puts them into the SML. Thus, the measurement list together with PCR completely give the measurement of the runtime behavior of the system. Once the digest of the measurements is securely stored in the TPM, it is possible to create an integrity report (protected with specific keys) that will be transferred to the (remote) Verifier via a remote attestation protocol.

We measured the software stack on a dedicated Raspberry Pi 4 device equipped with an Infineon TPM v2 and pre-configured with the user space TC software (e.g., TPM2 TSS v2.4.0, TPM2 Tools v4.2, TPM2 TSS Engine v1.1.0) and the Integrity Measurement Architecture (IMA) security subsystem for Linux. For remote attestation, besides IMA, we have installed on the device the Keylime [14] software, which provides trusted computing services. In particular, we have installed the *Prover* part (or *Agent*) of the remote attestation protocol. We have used an additional PC equipped with a TPM 2.0 and Ubuntu 18.04 to install and run the Keylime software components for the software integrity verification, namely *Registrar*, used for key management, and *Verifier*. After completing the Keylime installation, we have created a whitelist and an excludelist for the software to be measured. For example, we have selected to measure the `ptpd` daemon. We modified the `ptpd` daemon executable file, by replacing it with the one of the `mv` command. As result, the *Verifier* indicated an error specifying that the hash of the executable file has not been verified. Subsequently, we modified the `ptpd` code and installed it at the same path. Also in this case, the *Verifier* has correctly indicated that the `ptpd` daemon has not been successfully verified.

5 Conclusions

This paper investigated attack strategies in transport-based time synchronization architectures. Besides the attacks dealing with time distribution in packet switched networks (described in RFC 7384), we indicate possible attacks for different architecture views, including the network management and the software ones. We describe preliminary experimental results with software tampering attacks and a possible countermeasure. The combination of different attack types, e.g., GNSS spoofing or jamming attacks, software attacks, network and timing distribution protocol attacks, targeting simultaneously various entities in transport-based time synchronization solutions has not been investigated yet, but would be analyzed in the future.

Acknowledgements This work was developed within the ROOT project (https://www.gnss-root.eu/) funded by the European GNSS Agency under the European Union's Horizon 2020—G.A. n. 101004261.

References

1. Ruffini, S., Johansson, M., Pohlman, B., Sandgren, M.: 5G syncronization requirements and solutions. Ericsson Technol. Rev. (2021). https://www.ericsson.com/en/reports-and-papers/ericsson-technology-review/articles/5g-synchronization-requirements-and-solutions
2. The ROOT (Rolling Out OSNMA for the Secure Synchronisation of Telecom Networks) Project. https://www.gnss-root.eu/
3. ITU-T Recommendations G.826x and G.827x series (G.8200-G.8299: Synchronization, quality and availability targets. https://www.itu.int/ITU-T/recommendations/index_sg.aspx?sg=15
4. IEEE 1588-2019: IEEE Standard for a Precision Clock Synchronization Protocol for Networked Measurement and Control Systems. Accessed 16 Jun 2020. https://standards.ieee.org/standard/1588-2019.html
5. Mills, D., Delaware, U., Martin, J., Burbank, J., Kasch, W.: Network Time Protocol Version 4: Protocol and Algorithms Specification. IETF RFC 5905 (2010)
6. DeCusatis, C., Lynch, R.M., Kluge, W., Houston, J., Wojciak, P.A., Guendert, S.: Impact of cyberattacks on precision time protocol. IEEE Trans. Instrum. Meas. **69**(5), 2172–2181 (2020). https://doi.org/10.1109/TIM.2019.2918597
7. Dalmas, M., Rachadel, H., Silvano, G., Dutra, C.: Improving PTP robustness to the byzantine failure. In: 2015 IEEE International Symposium on Precision Clock Synchronization for Measurement, Control, and Communication, Beijing, pp. 111–114. https://doi.org/10.1109/ISPCS.2015.7324693
8. Itkin, E., Wool, A.: A security analysis and revised security extension for the precision time protocol. IEEE Trans. Dependable Sec. Comput. **17**(1), 22–34 (2020). https://doi.org/10.1109/TDSC.2017.2748583
9. Barreto, S., Suresh, A., Le Boudec, J.: Cyber-attack on packet-based time synchronization protocols: the undetectable delay box. In: 2016 IEEE International Instrumentation and Measurement Technology Conference, pp. 1–6. https://doi.org/10.1109/I2MTC.2016.7520408
10. Girela-López, F., López-Jiménez, J., Jiménez-López, M., Rodríguez, R., Ros, E., Díaz, J.: IEEE 1588 high accuracy default profile: applications and challenges. IEEE Access **8**, 45211–45220 (2020). https://doi.org/10.1109/ACCESS.2020.2978337
11. Mizrahi, T.: Security requirements of time protocols in packet switched networks. IETF RFC 7384 (2014)

12. Alghamdi, W., Schukat, M.: Precision time protocol attack strategies and their resistance to existing security extensions. Cybersecur **4**, 12 (2021). https://doi.org/10.1186/s42400-021-00080-y
13. Pini, M., Minetto, A., Vesco, A., Berbecaru, D., Contreras Murillo, L.M., Nemry, P., De Francesca. I., Rat, B., Callewaert, K.: Satellite-derived time for enhanced telecom networks synchronization: the ROOT project. In: 2021 IEEE 8th International Workshop on Metrology for AeroSpace (MetroAeroSpace), pp. 288–293. https://doi.org/10.1109/MetroAeroSpace51421.2021.9511780
14. Keylime: Bootstrap & Maintain Trust on the Edge/Cloud and IoT. https://keylime.dev/
15. Estrela, P.V., Neusüß, S., Owczarek, W.: Using a multi-source NTP watchdog to increase the robustness of PTPv2 in financial industry networks. In: 2014 IEEE International Symposium on Precision Clock Synchronization for Measurement, Control, and Communication, pp. 87–92. https://doi.org/10.1109/ISPCS.2014.6948697
16. Alghamdi, W., Schukat, M: Cyber attacks on precision time protocol networks–a case study. Electronics **9**, 1398 (2020). https://doi.org/10.3390/electronics9091398
17. Langner, R.: Stuxnet: dissecting a cyberwarfare weapon. IEEE Secur. Priv. **9**(3), 49–51 (2011). https://doi.org/10.1109/MSP.2011.67
18. Günther, C.: A survey of spoofing and counter-measures. J. Inst. Navig. **61**(3), 159–177 (2014). https://doi.org/10.1002/navi.65
19. Falletti, E., Margaria, D., Marucco, G., Motella, B., Nicola, M., Pini, M.: Synchronization of critical infrastructures dependent upon GNSS: current vulnerabilities and protection provided by new signals. IEEE Syst. J. **13**(3), 2118–2129 (2019). https://doi.org/10.1109/JSYST.2018.2883752
20. Ullmann, M., Vögeler, M.: Delay attacks implication on NTP and PTP time synchronization. In: 2009 International Symposium on Precision Clock Synchronization for Measurement, Control and Communication, pp. 1–6. https://doi.org/10.1109/ISPCS.2009.5340224
21. Ylonen, T., Lonvick, C.: The Secure Shell (SSH) Connection Protocol. IETF RFC 4254 (2006)
22. Rescorla, E.: The Transport Layer Security (TLS) Protocol version 1.3. IETF RFC 8446 (2018)
23. Rigney, C., Willens, S., Rubens, A., Simpson, W.: Remote Authentication Dial In User Service (RADIUS). IETF RFC 2865 (2000)
24. Levi, D., Meyer, P., Stewart, B.: Simple Network Management Protocol (SNMP) Application. IETF RFC 3413 (2002)
25. Berbecaru, D., Casalino, M.M., Lioy, A.: FcgiOCSP: a scalable OCSP-based certificate validation system exploiting the FastCGI interface. Softw. Pract. Exper. **43**, 1489–1518 (2013). https://doi.org/10.1002/spe.2148
26. Borio, D., Dovis, F., Kuusniemi, H., Lo Presti, L.: Impact and detection of GNSS jammers on consumer grade satellite navigation receivers. Proc. IEEE **104**(6), 1233–1245 (2016). https://doi.org/10.1109/JPROC.2016.2543266
27. Alghamdi, W., Schukat, M.: A Detection model against precision time protocol attacks. In: 2020 3rd International Conference on Computer Applications & Information Security, Riyadh, Saudi Arabia, pp. 1–3. https://doi.org/10.1109/ICCAIS48893.2020.9096742
28. Alghamdi, W., Schukat, M.: Practical implementation of APTs on PTP time synchronisation networks. In: 31st Irish Signals and Systems Conference 2020, pp. 1–5. https://doi.org/10.1109/ISSC49989.2020.9180157
29. Kim, T.H., Reeves, D.: A survey of domain name system vulnerabilities and attacks. J. Surveill Sec. Saf. **1**, 34–60 (2020). https://doi.org/10.20517/jsss.2020.14
30. Stebila, D.: Attacks on TLS. https://www.douglas.stebila.ca/research/presentations/tls-attacks/
31. Berbecaru, D., Lioy, A.: On the robustness of applications based on the SSL and TLS security protocols. In: 4th European PKI Workshop: Theory and Practice (EuroPKI 2007), LNCS, vol. 4582, pp. 248–264. Springer, Berlin (2007). https://doi.org/10.1007/978-3-540-73408-6_18
32. Berbecaru, D.: On Creating Digital Evidence in IP Networks With NetTrack. Handbook of Research on Network Forensics and Analysis Techniques, IGI Global (2018). https://doi.org/10.4018/978-1-5225-4100-4.ch012

33. Berbecaru, D.G., Lioy, A., Cameroni, C.: Providing login and Wi-Fi access services with the eIDAS network: a practical approach. IEEE Access **8**, 126186–126200 (2020). https://doi.org/10.1109/ACCESS.2020.3007998
34. Berbecaru, D.G., Lioy, A., Cameroni, C.: On enabling additional natural person and domain-specific attributes in the eIDAS network. IEEE Access **9**, 134096–134121 (2021). https://doi.org/10.1109/ACCESS.2021.3115853
35. Kröner, U., Bergonzi, C., Fortuny-Guasch, J., Giuliani, R., Littmann, F., Shaw, D., Symeonidis, D.: Hardening of GNSS based trackers (2010). https://publications.jrc.ec.europa.eu/repository/bitstream/JRC58733/reqno_jrc58733_st_report_on_hardening_of_gnss_based_trackers_release_final.pdf
36. Piessens, F., Verbauwhede, I.: Software security: vulnerabilities and countermeasures for two attacker models. In: 2016 Design, Automation & Test in Europe Conference & Exhibition (DATE), pp. 990–999
37. Ankergård, S.F.J.J., Dushku, E., Dragoni, N.: State-of-the-art software-based remote attestation: opportunities and open issues for internet of things. Sensors **21**, 1598 (2021). https://doi.org/10.3390/s21051598

An Approach to Automated Assessment of the Image of a Territorial Entity in the Media Discourse of a Foreign States

Ksenia Namyatova, Lidia Vitkova, and Andrey Chechulin

Abstract The paper presents the results of testing the author's approach to analyzing the image of a territorial association in the media discourse of a foreign state. The proposed approach is based on a convergence approach, that is, it simultaneously uses content analysis which has already become a classic, quantitative analysis and modeling of the image of the state. A key feature is the use of automated solutions that allowed large amounts of data from eighteen foreign mass media to be collected and analyzed. The technique has demonstrated its suitability for further research.

Keywords Assessment of the image of the state in the media discourse · Media discourse · Image · Approach to analyzing the image of the state

1 Introduction

Before our eyes, in just two decades, society has been immersed in the information environment. There is a need for rapid monitoring of publications and timely response. At the same time, specialists should not view dozens or even hundreds of publications every day, but thousands of them every hour. In this case, the old methods of media discourse analysis cannot be fully effective, since it is impossible to analyze huge amounts of publications with human resources. Especially when it comes to large territorial entities.

K. Namyatova
St. Petersburg University, 13B Universitetskaya Emb, St. Petersburg, Russia

L. Vitkova (✉) · A. Chechulin
SPC RAS, 39, 14th Liniya, St. Petersburg, Russia
e-mail: vitkova@comsec.spb.ru

A. Chechulin
e-mail: chechulin@comsec.spb.ru

D. Camacho et al. (eds.), *Intelligent Distributed Computing XIV*, Studies in Computational Intelligence 1026, https://doi.org/10.1007/978-3-030-96627-0_20

At the same time, there are already many automated systems for monitoring mass media or analyzing mentions in social networks. Among them are such systems as Meltwater [27], Aylien [4] and others. In theory, these systems are quite suitable for the state image evaluation, but in reality, one faces a number of limitations. Firstly, these are commercial systems, and they are not designed to collect data of such a volume, since this requires quite serious solutions in the field of data storage and processing. Secondly, most researchers do not have the skills to use such systems and moreover are simply not familiar with them. However, the most important limitation in the use of such systems is the lack of developed approaches to analyzing the image of a territorial entity in the media discourse of a foreign state.

Summarizing, it should be noted that there is a need for researchers to develop and implement new approaches to identifying signs of state reputation deformation in global and regional areas using modern software solutions. In other words, a convergence of socio-humanitarian and technical approaches is needed. The paper describes an approach that combines content analysis, quantitative analysis, and modeling of the image of the state. A key feature of the proposed approach is the adaptation of existing methods of analyzing the media discourse of a territorial entity to automated information processing. The proposed approach makes it possible to collect and analyze large amounts of data from foreign media in a short time; to form brief conclusions about the agenda for the operator and to quickly identify the distortion of the image of a territorial unit in the media discourse of a foreign country.

The paper is structured as follows. The second section provides an overview of relevant papers, and the third section describes the approach itself and its adaptation to empirical testing. The fourth section outlines the experimental testbed, and the fifth section demonstrates the results of the experiment. The paper concludes with a section on discussion and summary.

2 State of the Art Analysis

The approach to the analysis of a territorial entity in the media discourse of a foreign state is based on methods involving the analysis of texts in the media in order to construct the image of a state or other territorial unit.

One of the first and probably the most quoted works is [13], in which the authors raise the issue of the formation of media images of the state on the example of the USA and the construction of reality through the media. In [25] the authors speculate about assessing the legitimacy of an organization by analyzing media content, but the issue of legitimacy is also important at the international level. The authors used 2 media news gathering tools—Google News and LexisNexis database [24] and in the article described a method of collecting and filtering by date range, which takes into account the number of news headlines. Further, the authors show how they used the Meaning Cloud software [8] to assess the positive and negative tone of the texts. Emotional coloring values were coded using the Janis–Fadner coefficient [12]. In the conclusions to the article, the authors point out that this approach makes it possible to assess the biases that are detrimental or beneficial to the legitimacy of institutions.

Ciechomski, W. in his work "Creating and positioning the image of a territorial unit" describes a method of evaluating the connection between marketing methods for creating the image of a city or a territorial unit and its subjective image, a unique identity [7]. In [10], the authors argue that news aggregators create algorithms for selecting news, thus shaping the information picture for the user. In their research, the authors use methods of social surveys of search engine users. In [29] the author demonstrates how the critical discourse analysis (CDA) can be used to investigate gender stereotypes in political media discourse, but in the study the authors used the InfoQuest Limited database [18] and analyzed news texts manually in expert mode. In [28], a segmentation of approaches on the technical part of text processing is proposed. In particular, the researchers specialize in one approach in their works: Natural Language Processing [6], Latent Semantic Analysis [17], Explicit Semantic Analysis [11], Sentiment Analysis [26].

Among the branding approaches, we distinguish the works of Anholt [2], Dinnie [9], Caldwell, Freire [5], Kavaratsis [19], Hankinson [16]. But we consider the approaches of Gavra D. P. and Savitskaya A. S. to be the most relevant for media research [15], the multidimensional model of Taranova and Gavra [30] and, finally, his Spherical-meaningful model of the image of the state [14]. Since they are built taking into account the analysis of information from media systems, while data collection for branding models is carried out by conducting sociological surveys. Most of the presented models consist of a number of components, such as political, economic, tourist, etc., but there are also a number of significant differences. For example, such a common model of Anholt [2] does not imply synchronization with media discourse analysis methods, since data collection is based on questions such as "the level of satisfaction with the government" and "the tourist attractiveness of the region". Such questions are too complex to translate into a plane suitable for further empirical data collection; it is difficult to imagine what key expressions in the texts should give us information about such subjective assessments. We settled on the sphere-substance model [30], since it is the most suitable for the use of convergence.

After the state-of-the-art analysis in the field of media discourse analysis, let's pay attention to the one key feature. All these methods do not imply the processing of thousands of publications, we are talking about a qualitative sample analysis of the text. At the same time, it is not taken into account that for a representative sample it is now not enough to analyze even 100 publications, since in some cases such a number of new publications may appear in 1 day. There is a contradiction: the use of purely journalistic methods in the field of media research already significantly restricts specialists, and new approaches, new methods with the use of modern IT technologies, are needed.

3 Proposed Approach

The current section describes an approach to analyzing the image of a territorial entity in the media discourse of a foreign state and adapting the proposed approach to automated information collection.

An approach to automated assessment of the image of a territorial entity in the media discourse of a foreign state.

The approach consists of five steps:

Step 1. Analysis of the media system of a foreign state. At this step, the most popular daily publications of national and regional significance are selected. To select relevant publications, it is possible to use data from monitoring centers.

Step 2. Selection of keywords and formation of the semantic core of the study, which is supposed to collect empirical material for further analysis. All key expressions are divided into spheres in accordance with the logic of the sphere-substance model of the image of a territorial entity: politics, economy, culture, sport and social components.

Modeling the image of the state is an essential part of the study. At the initial stage, when developing a data collection methodology, it is necessary to start from one or another of the selected models in order to eventually systematize the information obtained during the study and visualize it by superimposing it on the selected model. When developing the semantic core, queries are selected in such a way as to gather the most relevant publications regarding each area for a particular research period.

Step 3. Selection of time intervals, depending on the depth of the study and the purpose of analyzing the image of a territorial entity in the media discourse of a foreign state.

Step 4. Adapting key expressions for the collection system.

Depending on the number of selected media sites in step 1 and keywords in step 2, the result of step 4 will be obtained by the formula (1)

$$A_{kw}^{m} = \frac{\text{“}kw\text{"}!}{(kw - m)!} \tag{1}$$

where A is the number of placements of keywords (kw) by media sites (m).

Step 4. Identification of intonation indicators based on the key encoding matrix.

At the same time, the approach assumes that only the headlines and primary descriptions of publications in the media are analyzed. The authors of the approach use the fact that when perceiving the flow of media information, the titles and primary descriptions of papers become increasingly important. They contain a preliminary interpretation of the content by the journalist. Thus, the publication has a significant impact on the reader's perception of information even at the level of headlines. Such authors as Kulkarni et al. [22] and Leban et al. [23] analyze the influence of headlines on the perception of discourse in the mass media space.

The proposed approach allows you to create a technical task for automated information collection and develop requirements for systematization and visualization of the obtained data. Depending on the expected volume, requirements may be imposed on charts, histograms, and the depth of text analysis (only headlines, news stories, etc.).

Adapting the proposed approach to automated information collection

An example of the adaptation of the proposed approach in the framework of the study is the evaluation of an image of Russia in the media discourse of a foreign state. The objects are the publications in the Italian media devoted to the Russian Federation.

At the first step, based on the analysis of the Italian media system, the most popular daily mass media web sites of national and regional significance were selected. To select the most relevant ones, we used data from the specialized Italian state monitoring center Audipress [3]. The center publishes monthly reports and annual summaries of the audience indicators of all printed and electronic mass media, while data is collected through surveys and interviews in all regions of Italy over the years using the same methods, which allows one to most accurately track changes in behavioral preferences.

At the second step, the keywords were selected and the semantic core was formed. Key expressions/markers such as "Putin", "Lavrov", "Medvedev", and "Kremlin" were relegated to the realm of politics. The incidents included the one that occurred in Salisbury, the incident in the Kerch Strait and the Maria Butina story. The availability of information in the selected media for all key expressions was previously verified during the pilot study.

At the third step, seven annual time periods were set starting from 2012 till the end of 2018.

At the fourth stage, the initial processing of keywords and their translation for automated collection and analysis was carried out. As a result, received 4997 requests.

At the fifth step, a key encoding matrix was developed (Table 1).

An important stage of the study is the verification of the proposed approach. To do it we conducted the set of experimental studies. During the experiments we collected and systematized data, as well as determined statistical indicators and built a model

Table 1 Example of a key encoding matrix

Negative (Eng)	Negative (It)
Threat	Minaccia
Threats	Minacce
Aggression	Aggressione
Takeover, occupation	Occupazione
Takeover (power)	Conquistatore
Invader	Invasore
Conquest	Conquista
Assault	Assalto
Intervention	Intervento
Military	Militare

of Russia's image in Italian media discourse. The experimental testbed and results of analysis are presented in the next section.

4 Experimental Testbed

A system based on a microservice architecture was developed to collect information. The system's architecture is represented by the following components Fig. 1: (A) The component of forming tasks for data collection; (B) The information search component; (C) The information collection component; (D) The component of a database management.

A relational database was used for the study. The database management component starts first. At startup, the structure is checked, the necessary tables and table fields are created.

The developed database structure consists of the following tables: categories which contain names of the categories and links to the parent category, if available; statuses which contain tasks' statuses (e.g., pending, in progress, completed); requests which contain tasks for information gathering from search engines; URLs which contain tasks for information gathering from web-pages; web-pages which store the gathered web-pages.

After the database is filled with search requests, the other components can be started independently.

The component (A) for generating data collection tasks takes 2 tables in csv format as input, one of which contains query parameters (keywords, category, search time interval), and the other one contains domains and individual pages on which one need to search for the specified parameters.

The information search component (B) checks for available data collection tasks and collects URL links using open search engine interaction interfaces. During operation, the component alternately takes 1 task for the search, pre-reserving it, and then makes requests to the search engines, observing time intervals to bypass the locks on the part of search engines. The data received from search engines is converted into

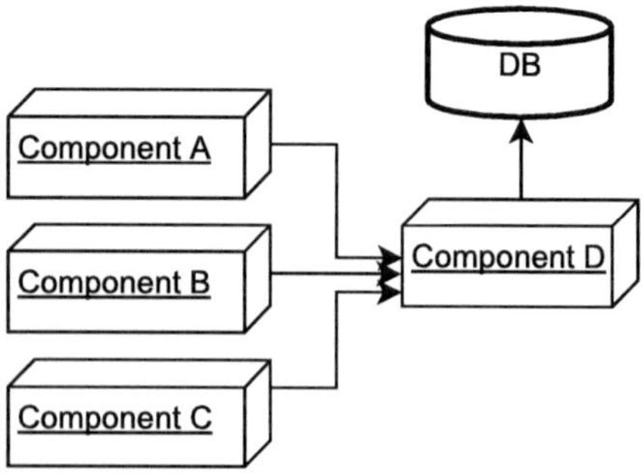

Fig. 1 The system's architecture to collect information

tasks for collecting information from URL links. In order to speed up data collection, the information retrieval component can be independently run on multiple devices using different IP addresses.

The information gathering component (C) checks for available tasks for collecting information from URL links. By reserving 1 task, information is collected from the URL link, such as the text, page description, content of html tags, etc. Similar to the information retrieval component, the collection component can be run on multiple devices to speed up collection, but the requirement for different IP-addresses is optional.

The search and gathering components go into standby mode after completing all tasks and automatically continue their work when new search and collection tasks appear, respectively.

Automated collection allows you to quickly obtain and analyze a huge array of information. Based on the obtained data, it is possible to build a model of a territorial entity in the media discourse of a foreign state.

5 Results of the Experiment

According to the experiment the following results were obtained:

1. A huge amount of information was collected and analyzed in a short time, as in total we received more than 176,000 publications. It would be difficult to analyze such volumes manually, and it would be impossible to use a single specialist.

2. A model of the image of the Russian Federation in the Italian media discourse is constructed Fig. 2. When constructing the model, we see that the main factors of the image of the Russian Federation in the Italian media discourse are political and social components. At the same time, it is worth mentioning that the social component is often no less politicized, since the government's behavior in relation to the environment is also a political agenda, not just a social one. In third place is the sports component, also influenced by the high popularity of sports publications in Italy. Economic and cultural components are not the leading agenda in shaping the image of the Russian state.

3. A more detailed analysis of the categories shows that the most popular topic for publications is life in Russia (Table 2). In this case, we are not talking directly about the image of the Russian Federation, but about the Italians living in it, since this subcategory contains publications containing the expressions "Italians in Russia" and "life in Russia". Therefore, the direct most popular component is the actions of the country's leaders and events taking place on its territory. Economic components such as exports, imports, investments and business are not key to the image of the Russian Federation.

Throughout the study period, the publications showed various interests in Russia. In 2012, the newspaper "Corriere della Sera" wrote most often about our country, but in 2018, interest in Russia became more popular among all publications. The total

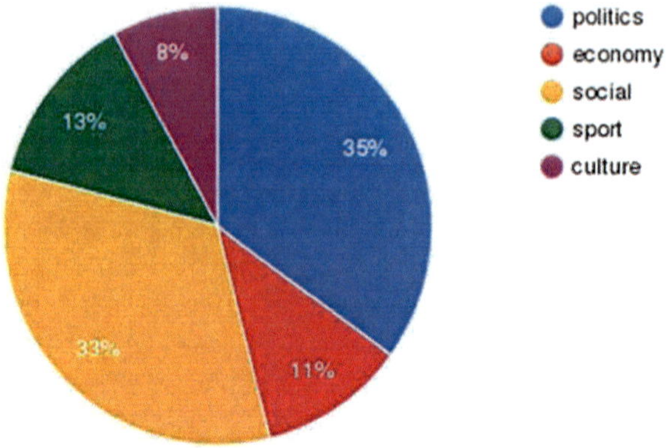

Fig. 2 The model of the image of the Russian Federation in the Italian media discourse

Table 2 Number by the sub-category

Sub-category	Sum
Life and migration to Russia	23155
Leaders	22135
Events	21577
Human capital	20575
The activities of the government	14570
Culture	13421
Sanctions	10428
Elections	10070

number of publications also confirms this from year to year. For example, in 2012, 9,638 people wrote about Russia, and in 2018 there were already 41,304 publications.

The main results of the experiment confirm the advantage of the convergent approach and the proposed approach to the analysis of the image of a territorial unit in the media discourse of a foreign state. At the same time, the use of spheres about the substantial model of the image of the state, proposed by Gavra [30], made it possible to clearly visualize the deformation of the image in the direction of the political component. Of great interest is the analysis of the evolution of the tonality of perception of the image of the state through the years. The results of the intonation analysis carried out at the last step of the author's methodology are planned to be published in subsequent papers.

6 Conclusion and Discussion

The proposed approach to the analysis of the image of a territorial entity in the media discourse of a foreign state allows us to process the maximum amount of empirical material and model the image of a territorial unit in the media discourse of any country, as well as to trace the dynamics of the tonality and activity of various media.

Summing up, we can assume that the proposed approach can help one to find and analyze signs of the state reputation deformation in the mass media discourse of another country. Thus, it can be argued that the approach is very promising and has high potential. The usage of the approach will improve the quality of media research and the level of state security by timely identifying signs of changes in the national image in the external media discourse. In turn, this can speed up the adoption of a political decision even before conducting an in-depth study of the media field and can help one to select counteraction measures. This approach was assessed and approved during a joint project with experts from the School of Journalism and Mass Communications [1].

The disadvantages of the proposed approach include the need to use dynamic IP-addressing and a distributed data collection system. Search engines usually limit the number of requests from single IP address and use blocking algorithms for a large number of similar requests in a long period of time.

In the future work we plan to test the approach for collection and analysis of publications in social networks [20] and develop our own reporting module which will use special visualization modules [21]. Another promising direction is to compare changes in the image of a territorial entity in the media discourse of different states in the same time interval. For example, we plan to conduct a comparative study of the image of Russia in the media discourse of the United States, Thailand, and Brazil.

Acknowledgements The work is performed by the grant of RSF #18-71-10094P in SPC RAS.

References

1. Higher school of journalism and mass communications of St. Petersburg State University (2021). http://eng.jf.spbu.ru/
2. Anholt, S.: Why brand? Some practical considerations for nation branding (2006)
3. Audipress: Official monitoring and statistical center of the periodical press in Italy (2021). http://audipress.it/
4. Aylien: News analytics system (2021). https://aylien.com/
5. Caldwell, N., Freire, J.R.: The differences between branding a country, a region and a city: applying the brand box model. J. Brand Manag. **12**(1), 50–61 (2004)
6. Cambria, E., White, B.: Jumping nlp curves: a review of natural language processing research. IEEE Comput. Intell. Mag. **9**(2), 48–57 (2014)
7. Ciechomski, W.: Creating and positioning the image of a territorial unit. Zeszyty Naukowe, Organizacja i Zarządzanie/Politechnika Śląska (2020)

8. Cloud M.: Text analytics - meaningcloud text mining solutions (2021). https://www.meaningcloud.com/
9. Dinnie, K.: City Branding: Theory and Cases. Springer, Berlin (2010)
10. Fletcher, R., Nielsen, R.K.: Automated serendipity: the effect of using search engines on news repertoire balance and diversity. Dig. Journal. **6**(8), 976–989 (2018)
11. Gabrilovich, E., Markovitch, S., et al.: Computing semantic relatedness using wikipedia-based explicit semantic analysis. IJcAI **7**, 1606–1611 (2007)
12. Galán-Valdivieso, F., Saraite-Sariene, L., Alonso-Cañadas, J., Caba-Pérez, M.d.C.: Do corporate carbon policies enhance legitimacy? A social media perspective. Sustainability **11**(4), 1161 (2019)
13. Gamson, W.A., Croteau, D., Hoynes, W., Sasson, T.: Media images and the social construction of reality. Ann. Rev. Sociol. **18**(1), 373–393 (1992)
14. Gavra, D., Naumenko, K.: Russian humanitarian mission in bergamo in the fight against the coronavirus pandemic: reflection in the Italian media discourse. Vestnik Moskovskogo Universiteta. Seriya 10. Zhurnalistika, pp. 77–107 (2020)
15. Gavra, D.P., Savitskaya, A.S.: Mass media in interstate conflicts: typological model "peace-conflict journalism multidimensional approach". Russ. J. Commun. **4**(3–4), 251–265 (2011)
16. Hankinson, G.: Destination brand images: a business tourism perspective. J. Serv. Market. (2005)
17. Hutchison, P.D., Daigle, R.J., George, B.: Application of latent semantic analysis in ais academic research. Int. J. Acc. Inf. Syst. **31**, 83–96 (2018)
18. Infoquest: Thailand's leading information company and news agency (2021). https://www.infoquest.co.th/company/en
19. Kavaratzis, M.: From city marketing to city branding: towards a theoretical framework for developing city brands. Place Brand. **1**(1), 58–73 (2004)
20. Kotenko, I., Saenko, I., Chechulin, A., Desnitsky, V., Vitkova, L., Pronoza, A.: Monitoring and counteraction to malicious influences in the information space of social networks. In: International Conference on Social Informatics, pp. 159–167. Springer, Berlin (2018)
21. Kotenko, I.V., Kolomeets, M., Chechulin, A., Chevalier, Y.: A visual analytics approach for the cyber forensics based on different views of the network traffic. J. Wirel. Mob. Networks Ubiquitous Comput. Dependable Appl. **9**(2), 57–73 (2018)
22. Kulkarni, V., Ye, J., Skiena, S., Wang, W.Y.: Multi-view models for political ideology detection of news articles (2018). arXiv:1809.03485
23. Leban, G., Fortuna, B., Brank, J., Grobelnik, M.: Event registry: learning about world events from news. In: Proceedings of the 23rd International Conference on World Wide Web, pp. 107–110 (2014)
24. Lexisnexis: A database with data detection, visualization and analysis capabilities (2021). https://internationalsales.lexisnexis.com/
25. López-Balboa, A., Blanco-González, A., Díez-Martín, F., Prado-Román, C.: Macro level measuring of organization legitimacy: its implication for open innovation. J. Open Innov.: Technol. Market Compl. **7**(1), 53 (2021)
26. Mäntylä, M.V., Graziotin, D., Kuutila, M.: The evolution of sentiment analysis-a review of research topics, venues, and top cited papers. Comput. Sci. Rev. **27**, 16–32 (2018)
27. Meltwater: Media monitoring system (2021). https://www.meltwater.com/
28. Omar, A., Mahmoud, T.M., Abd-El-Hafeez, T.: Comparative performance of machine learning and deep learning algorithms for Arabic hate speech detection in osns. In: The International Conference on Artificial Intelligence and Computer Vision, pp. 247–257. Springer (2020)
29. Sriwimon, L., Zilli, P.J.: Applying critical discourse analysis as a conceptual framework for investigating gender stereotypes in political media discourse. Kasetsart J. Soc. Sci. **38**(2), 136–142 (2017)
30. Taranova, Y.V., Gavra, D.P.: Imperatives of the fast and slow worlds in strategic communications of the territory. In: 2021 Communication Strategies in Digital Society Seminar (ComSDS), pp. 118–121. IEEE (2021)

FL4J—Federated Learning Framework for Java

Ivan I. Kholod, Mikhail A. Efremov, Maxim A. Kolpashikov, Andrey V. Vasilyev, Pavel L. Tabakov, and Ilya E. Aristarhov

Abstract The article suggests an approach to implementing a federated machine learning framework for the Java platform. The subject of the research is federated machine learning, a brief description is given in the introduction, and the object is frameworks for federated machine learning systems development. As an experiment, it is shown the possibility of running a system with Java FL client for training linear ANN with one hidden layer with DeepLearning4Java backend together with Python FL client with PyTorch backend. The results model's accuracies are compared.

1 Introduction

At the moment machine learning technology is one of the important topics in computer science, also because it is used in a variety of industries, from entertainment to industrial domains. But there raise several problems.

1. **The volume of data grows rapidly.** The number of personal and industrial devices, connected to the Internet and accumulating data is constantly growing. Based on this data, you can continuously perform statistical analysis and obtain more accurate models to optimize production processes. On the other hand, storing such data is an additional cost.

I. I. Kholod
Faculty of Computer Science and Technology, Saint Petersburg Electrotechnical University LETI, Saint Petersburg, Russia

M. A. Efremov (✉) · A. V. Vasilyev
Department of Computer Science and Engineering, Saint Petersburg Electrotechnical University LETI, Saint Petersburg, Russia

M. A. Kolpashikov
Department of Computer Aided Design, Saint Petersburg Electrotechnical University LETI, Saint Petersburg, Russia

P. L. Tabakov · I. E. Aristarhov
Department of Software Engineering and Computer Applications, Saint Petersburg Electrotechnical University LETI, Saint Petersburg, Russia

D. Camacho et al. (eds.), *Intelligent Distributed Computing XIV*, Studies in Computational Intelligence 1026, https://doi.org/10.1007/978-3-030-96627-0_21

2. **Data analysis requires much more computing power.** As the volume of data grows, it takes a lot of power to keep up with the processing and analysis time requirements. Taking into account the increase in the cost of semiconductors and graphics processing units (GPU) associated with the pandemic, the problem arises of optimizing and transferring computations to at least edge devices, as well as to less powerful like smartphones, but in a larger number of client devices. All this requires the organization of such calculations.
3. **Communication channel performance requirements are growing.** As the number of devices connected to the computing network of distributed machine learning grows, the amount of processed data grows, and network traffic grows accordingly. You have to either improve the network infrastructure, which is not always a possible option, or reduce the amount of data, which leads to worse analysis time, which is also may be unacceptable. The latter problem is typical for centralized systems of distributed machine learning since a large amount of data must be previously uploaded to the data center of the learning server.
4. **Requirements for privacy and law restrictions on data usage are growing.** In recent years, the governments of many countries, where the domain of data analysis is actively developing, have adopted some bills that restrict the use of personal data or increase processing costs. Examples of such laws can be the GDPR in the EU and 152th Federal Law "On Personal Data" in the Russian Federation. As a result, there is growing interest in distributed machine learning methods that don't require the transfer of any sensitive data from owners at all.

Influenced by the aforementioned problems, Google introduced the concept of federated learning (FL) [1], suggested an example of using the approach to train GBoard's predictive model—a keyboard for Android. FL is a type of distributed machine learning, during which only the parameters of models are transferred between the nodes over the network, and the data itself remains on the storage devices of their owners. Federated learning is characterized by the following features:

- transfer of computing from servers to low-performance devices;
- transferring only the parameters of the models, but never training or test data;
- taking into account the factor of device instability in the computation process.

The main goal of the work is to develop FL framework for JVM platform with Java as programming language because this choice gives several advantages:

- Allows to use well known and old in industry JVM platform.
- Allows to write federated learning systems with training clients on Android devices without any code changes.
- Allows to use modern server's frameworks and approaches such as Spring Framework for developing of federated learning applications web-backend.

This work contains a details of such framework implementation and first usage experiments with artificial neural net (ANN) as model. Several existing frameworks for federated machine learning applied systems development are discussed below.

2 Related Works

At the moment the most known FL frameworks are: TensorFlow Federated [2], FATE [3], PaddleFL [4], PySyft [5]. Frameworks are reviewed earlier [6] and several points are highlighted below.

- All of them are implemented and focused on the Python language.
- Most of them focus on deep learning models and algorithms.
- Most of them focus on horizontal data distribution.
- Most of the implementations are strictly gRPC (Google's Remote Process Calling) related.

Over the past year, at least two new frameworks have appeared: Flower [7, 8] and FEDn [9, 10]. Both are language agnostic for learning clients. Language agnostic feature consists in the use of a common intermediate serialization data format and data transfer protocol. Both frameworks, as a result, declare the possibility of client support for various libraries, frameworks and models, but in the research only artificial neural networks are used as primary experiments models currently. Due to the complexity of the implementation of vertical partitions processing, both are also focused on horizontal data partition.

The Table 1 shows a comparison of existing federated learning frameworks.

Since last year, the FL4J framework [11], focused on the JVM platform, has been developed. Like FEDn and Flower, the framework has the following goals:

- independence from the client's programming language;
- independence from certain implementations of machine learning algorithms and models;
- independence from certain transport layer stacks;
- supporting custom computational nodes topology.

The focus on the JVM is caused by the prevalence of this platform, as well as the ability not only to use already trained neural network models on Android devices, but also to train the model using DeepLearning4Java library [12]. Existing Java

Table 1 Summary overview of existed FL frameworks

	TFF	FATE	PaddleFL	PySyft	FEDn	Flower
Horizontal data	+	+	+	+	+	+
Vertical data	–	+	–	–	–	–
Cross-silo	+	+	+	+	+	+
Cross-devices	–	–	–	+	+	–
Communication	gRPC	gRPC	ZeroMQ	gRPC	gRPC	gRPC
Simulation mode	+	+	+	+	+	+
Real mode	–	+	+	+	+	+
NN support	+	+	+	+	+	+
None-NN models	–	+	+	–	–	–
Languages	Python	Python	Python	Python	Python	Python

solutions, such as TensorFlow Federated, do not provide open source tools for FL on smartphones with real mode. Also TensorFlow Lite—Android's neural nets API used by TensorFlow Federated—doesn't support model's training: running only. Android devices are important because users generate a huge amount of personal data with smartphones that is useful for further analysis.

It is also important to note the prevalence of Java for servers. This platform, taking into account the development of technologies for native compilation GraalVM, and microservice technologies for the JVM, is seen as a good choice for implementing the server side of the framework.

Further, the architecture and implementation of the FL4J framework is proposed.

3 Framework Design

3.1 Components Overview

Since the last publication, an implementation of neural networks and an algorithm for its training has been added via DeepLearning4Java library, as well as a Python FL client via PyTorch library.

FL4J allows implementing your own FLService computation nodes based on existing FLServer and FLClient. The difference between is the purpose: the server is engaged in scheduling the execution, and the client is directly executing the previously scheduled algorithm on the transmitted model with the owned data. The unit of scheduling is the Schedule and the unit of execution is the Worker.

For different topologies of nodes may be necessary to specify the mapping process for parts of the learning algorithm—blocks—with nodes for blocks execution. The Scheduler is responsible for this, and it is possible to inject your own mapping implementation into the server.

Not only execution schedule can be planned, but also the implementation of the data transferring mechanism between the nodes—FLRouter. At the moment, there are implementations of a simulation mechanism over threads with SimRouter and the second one—using gRPC.

3.2 Abstract Mining Model Design

One of the problem to be solved was the creation of an abstract data structure to describe models. The basis of the concept was described earlier [6].

Model $m \in M$ is represented as an array of model elements (1).

$$m = [e_0, e_1 \ldots e_{l-1}] = [E] \tag{1}$$

A model element e_i is defined as a hierarchy of entities, which one denotes a component of the model. For example, for artificial neural networks the elements are net's layers. Elements of the e_i model can have subelements of an arbitrary nesting depth level $e_{i_0,i_1 \ldots i_k}$, where k is the depth of the element in the hierarchy. Elements from the model are accessed over an ordered set of indices I.

The model defines a function (2) for joining local models into one global—aggregation model.

$$join(m_0, m_1 \ldots m_{n-1}) : m_0 \ldots m_n \in M \rightarrow m_{aggr} \in M \quad (2)$$

To do this, all models are sliced from with the same indices I_k and for these elements the (3) function is used to merge all elements into one.

$$merge(e_{0,I_k}, e_{1,I_k} \ldots e_{n-1,I_k}) \rightarrow e_{aggr,I_k} \in E \quad (3)$$

Where I_k denotes indices set from k slice, n is the count of models for joining, e_{aggr,I_k} is slice's element which aggregated over all models. As a result of applying (3) over all elements of all models, the global model m_{aggr} will be obtained—an aggregated model—which can be applied at the next step of the algorithm, the next round of federated learning, or could be the final result of learning.

Each element of the model must implement the $merge([M])$ function. Any of the known models aggregation methods can be selected as implementation options, for example FedAvg [13], FedMA [14], etc.

This approach allows to implement different complex models which can perform one of the main stages of FL—the merging of local models into the global.

The framework as the implementation of the approach supports the abstract classes MiningModel and MiningModelElement, inheriting from which you can describe the required models, an ANN as example.

The model parametrization is available through settings objects—inheritors of the FunctionSettings class. The settings may contain, for example, model initialization parameters: the number of layers and their characteristics, in the case of ANN implementation. The model settings object also includes the algorithm's settings—the inheritor of the AlgorithmSettings—for which the model is intended.

3.3 Data Mining Algorithm Design

Abstracted the model is defined, but it is important to define algorithms abstraction which describe the learning process. For this, the previously proposed [15] concept of Functional Mining Blocks (FMB) is used, through the composition of which an updated model $m_{updated}$ can be obtained.

$$f_i : d \in D, m \in M \rightarrow m_{updated} \in M \quad (4)$$

$$dma = f_{n-1} \circ f_{n-2} \circ \cdots \circ f_0 \tag{5}$$

In (4) f_i is a single algorithm block, d is a current dataset, m is a currently updating model, and in (5) dma is the complete data mining algorithm.

Thus, by implementing the FMB separately and combining it with each other, it is possible to create an arbitrary algorithm for model updating. Also the concept allows using well known code optimization techniques, such as interchange and fission for adaptation of given dma from vertical to horizontal partition or backward. The adaptation approach also described previously [15].

FL4J implements FLAlgorithm entities corresponding to dma and AlgorithmBlock, EnginedBlock corresponding to FMB. FLAlgorithm declares methods to be overridden that return dma as composition of AlgorithmBlock for model initialization, sequential learning, parallel centralized learning, distributed vertical learning, distributed horizontal learning, model evaluation.

These methods return the composition in the form of a single Sequence Block—FMB—which encapsulates a sequence of blocks for sequential or parallel (multithreaded or distributed) execution.

The heirs of EnginedBlock can act as a execution block. Such blocks do not directly implement the model update method $exercute(d \in D, m \in M)$, corresponding to (4), but delegate the engine loaded at runtime that implements LearningEngine interface. The engine is described by the full class name. This approach allows us to describe both individual FMBs and the entire algorithm in an abstract way and independently of the real executor.

Thus, it becomes possible to separate the server-side from large libraries with actual implementations of machine learning algorithms, which is important for deploying containers with FLServer nodes, and also allows implementing clients on other platforms, for example, Python with PyTorch. A similar Python client parses the name of a required engine and matches it with its implementation with PyTorch runtime.

3.4 ANN Implementation via DL4J

DeepLearning4Java is a library that is implemented of deep neural networks and algorithms over them for the JVM. ND4J is used as the implementation of tensors computation, which works outside the JVM memory, with the ability to compute over GPU. A distinctive feature is the ability not only to run neural nets on Android devices but learn it, in contrast to the widespread TensorFlow Lite. Using the framework abstractions above, a model and algorithm blocks were implemented for training neural networks. A simplified class diagram is shown in the Fig. 1, further describes in more detail the entities depicted and the relationships between them.

The neural network model is a set of layers. Thus, according to the model (1), each element e_i is a neural network layer, where e_0 is an input layer and e_{n-1} is an output layer. A neural network layer contains a matrix of weights, a vector of biases,

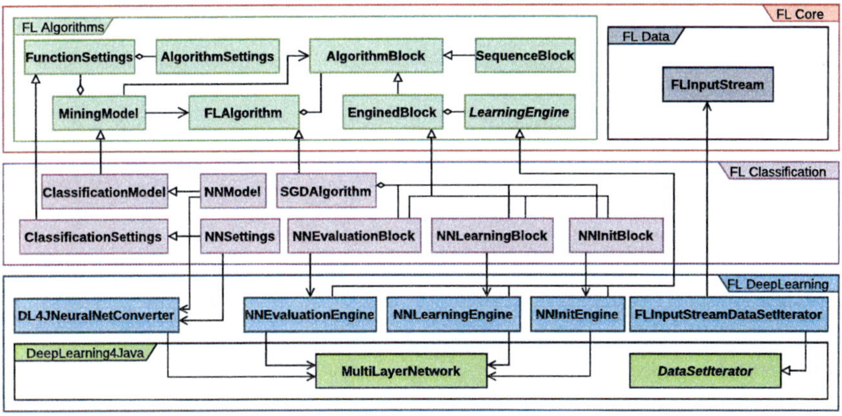

Fig. 1 Simplified class diagram for DL4J integration with FL4J framework

and an activation function. As the elements merge function (3) the weighted average approach over the weight matrix and the offset vector is used, the pseudocode of the function is shown in the Algorithm 1. The weight is a count of the processed dataset's rows and the value stores inside the model's object. To set up the model, separately implemented layer settings are used, as well as the loss function from the enumerate.

Algorithm 1 Pseudocode of weighted averaging as merge (3) function implementation for ANN dense layer

---

Require:

$e_{I_k}[]$: elements e_j for current index of elements slice I_k, equals to ANN's dense layer. Slice's index omitted below.

$w[]$: weights w_j corresponding to model $j \in [0, n-1]$.

Ensure: e_{aggr} : aggregated layer for global model m_{aggr}.

function PREPROCESSING($e[]$, $w[]$)

$weightsSum \leftarrow \sum_{j=0}^{n-1} w_j$

$wmatr \leftarrow e[].getWeights() \in List < double[][] >$

$wmatr_{aggr} \leftarrow \frac{\sum_{j=0}^{n-1} wmatr_j * w_j}{weightsSum}$

$e_{aggr}.setWeights(wmatr_{aggr})$

$bvec[] \leftarrow e[].getBias() \in List < double[] >$

$bvec_{aggr} \leftarrow \frac{\sum_{j=0}^{n-1} bvec_j * w_j}{weightsSum}$

$e_{aggr}.setBias(bvec_{aggr})$

return e_{aggr}

end function

To convert the model in terms of FL4J—NNModel—into a MultiLayerNetwork (MLN) from the DL4J library, a corresponding converter was implemented. An iterator was also implemented, which both extends DataInputStream from the framework and implements DataSetIterator interface from the library.

For the execution of the EnginedBlocks of SGDAlgorithm, the corresponding engines have implemented that call the *init*, *fit*, and *eval* methods in the engines for the converted MLN from the blocks corresponding to the initialization, training, and model evaluation.

To run the training process, you just need pass the algorithm object with a model and training settings. Training algorithm's settings contains parameters such as batch size, number of epochs, rounds of federation learning.

As a result, due to the high level of abstraction of basic entities, it was possible to quickly design a models, adapting classical algorithms for a federated environment.

4 Experiments

As an experiment, data from dumpsters vehicles sensors are used [16]: gyroscope, accelerometer and tachometer. The goal is classify an action performed by the dumpster: idle, driving, loading, dumping. Rows are glued by 40 because the data is time series. Time and speed columns are redundant. The combined set uses labels from 1 to 4 only, was normalized and balanced by classes. For the experiment combined data set was split on two balanced subsets with equal size.

For the experiment, a linear neural network with two hidden layer was used. There are 240 input and 4 output parameters, and the sizes of hidden layers are 256 and 64 respectively. The activation function for all layers except the last is—ReLU, for the last—Softmax. Loss function is Multiclass Cross Entropy. Also between layers batch normalization layers are putted. The network was initialized with Xavier method. Learning rate is 0.005, batch size is 128. The net is optimized with Stochastic Gradient Descent method.

Figure 2 shows charts of ANN's accuracies changing depends on epochs count. Presented cases:

- combined data set with only DL4J without aggregation (DL4J full data set);
- 1st and 2nd data subset with only DL4J (DL4J sub data set 1 and 2);
- 1st and 2nd data subset with FL4J with aggregation (FL4J sub set 1 and 2);

The charts are based on averaged values and display trends for the closest epochs.

When trained on a combined set using DL4J, the accuracy reaches 92% at 67 epochs. With further training, the accuracy changes slightly, but stably remains in the range from 90 to 92%. Learning on separate subsets using DL4J reaches a maximum of 90%. For the 1st subset this accuracy is achieved by the 83rd epoch, for the 2nd subset by the 45th epoch. However, in the future, the accuracy changes over a wide range. This may be due to the insufficient amount of data on which the neural network is trained and, possibly, overfitting on them.

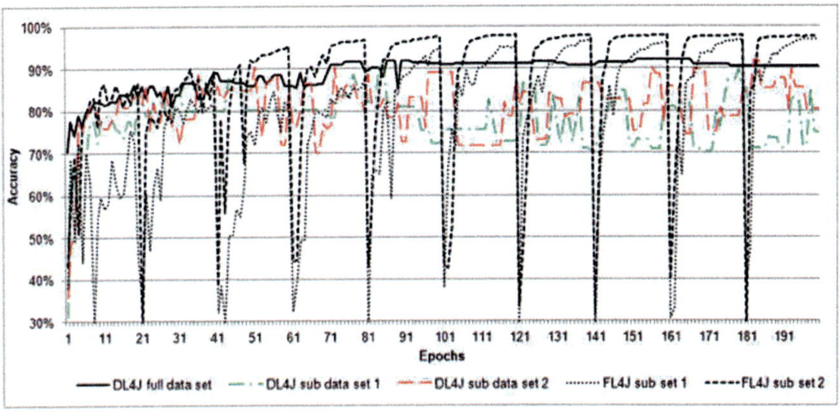

Fig. 2 Experiment results plot for cases of non-federated learning with just fitting over epochs with DL4J and federated learning with epochs, rounds and model's aggregation with FL4J

When training with FL4J, you can observe sharp drops in accuracy at the beginning of each round: on the graph this is every 20th epoch, since one round includes 20 epochs. This involves aggregating the model after each round on the server and updates the generalized model with the data. At the same time, the accuracy increases rather quickly in each round. In the first 3 rounds, the accuracy is restored during the first 5–7 epochs, and in the following rounds, the accuracy is restored in 3–5 epochs. It should be noted that for the second set (2nd client) the accuracy exceeds the training accuracy reaches 93% at the beginning of the 3rd round at 43 epochs. This is higher than training on a full set. Next, the accuracy only grows after each round and at the 5th round (95th epoch) reaches 97%. For the first set, the accuracy grows more slowly, but on the 5th round (93rd epoch) it reaches an accuracy of 93%, exceeding the maximum accuracy obtained on the combined set. At the end of the 10th round (199th epoch), the accuracy of 97% is also achieved on the 2nd client.

Thus, when using FL4J, it is possible to achieve accuracy even higher than with the combined set. At the same time, the quality of training grows steadily with each epoch, with the exception of moments after the aggregation of the model. It is aggregation that does not allow overfitting the model on the same data. Due to this, better results are achieved.

5 Conclusions

This proposal discusses the design of FL4J federated machine learning framework. The concepts of the abstract description of models and the learning algorithm were proposed, as well as the main components of the framework were described. As an example, an implementation approach of a linear neural network using the DL4J library was presented, and an experiment was also performed to compare the accuracy of federated learning with FL4J and non-federated DL4J learning. The proposal allows to use Java clients on Android systems without drastically changing the code, which would be difficult to achieve using a Python solutions. The experiment's results show netter results compare to non-federated approach.

Acknowledgements We thank Smartilizer Scandinavia AB for provided dataset and Yandex Cloud service for the platform for experiments computation purpose.

References

1. Hard, A., Rao, K., Mathews, R., Ramaswamy, S., Beaufays, F., Augenstein, S., Eichner, H., Kiddon, C., Ramage, D.: Federated learning for mobile keyboard prediction (2018). arXiv:1811.03604
2. Google LLC: TensorFlow Federated GitHub repository. https://github.com/tensorflow/federated
3. Webank's AI Department: FATE (Federated AI Technology Enabler) framework GitHub repository. https://github.com/FederatedAI/FATE
4. PaddlePaddle Baidu: PaddleFL framework GitHub repository. https://github.com/PaddlePaddle/PaddleFL
5. OpenMined: PySyft framework GitHub repository. https://github.com/OpenMined/PySyft
6. Kholod, I., Yanaki, E., Fomichev, D., Shalugin, E., Novikova, E., Filippov, E., Nordlund, M.: Open-source federated learning frameworks for iot: a comparative review and analysis. Sensors **21**(1), 167 (2021)
7. Adap GmbH: Flower federated learning framework GitHub repository. https://github.com/adap/flower
8. Beutel, D.J., Topal, T., Mathur, A., Qiu, X., Parcollet, T., Lane, N.D.: Flower: a friendly federated learning research framework (2020). arXiv:2007.14390
9. Scaleout: FEDn framework GitHub repository. https://github.com/scaleoutsystems/fedn
10. Ekmefjord, M., Ait-Mlouk, A., Alawadi, S., Åkesson, M., Stoyanova, D., Spjuth, O., Toor, S., Hellander, A.: Scalable federated machine learning with fedn (2021)
11. Efremov, M.A., Kholod, I.I., Kolpaschikov, M.A.: Java federated learning framework architecture. In: 2021 IEEE Conference of Russian Young Researchers in Electrical and Electronic Engineering (ElConRus), pp. 306–309 (2021). https://doi.org/10.1109/ElConRus51938.2021.9396508
12. Eclipse Foundation: DL4J — Open-source, distributed, deep learning library for the JVM GitHub repository. https://github.com/eclipse/deeplearning4j/
13. McMahan, B., Moore, E., Ramage, D., Hampson, S., y Arcas, B.A.: Communication-efficient learning of deep networks from decentralized data. In: Artificial Intelligence and Statistics, pp. 1273–1282. PMLR (2017)
14. Wang, H., Yurochkin, M., Sun, Y., Papailiopoulos, D., Khazaeni, Y.: Federated learning with matched averaging (2020). arXiv:2002.06440
15. Kholod, I., Shorov, A., Gorlatch, S.: Efficient distribution and processing of data for parallelizing data mining in mobile clouds. J. Wirel. Mob. Networks Ubiquitous Comput. Dependable Appl. **11**(1), 2–17 (2020)
16. Smartilizer Scandinavia AB.: Commercial vehicles sensor data set – 2 dumpers collected timestamp, speed, gyroscope, and accelerometer data. https://www.kaggle.com/smartilizer/commercial-vehicles-sensor-data-set

BLOCKCHAIN (Main Track)—Chair: Daniele Santamaria

Ontological Smart Contracts in OASIS: Ontology for Agents, Systems, and Integration of Services

Domenico Cantone, Carmelo Fabio Longo, Marianna Nicolosi Asmundo, Daniele Francesco Santamaria, and Corrado Santoro

Abstract In this contribution we extend an ontology for modelling agents and their interactions, called *Ontology for Agents, Systems, and Integration of Services* (in short, OASIS), with conditionals and *ontological smart contracts* (in short, OSCs). OSCs are ontological representations of smart contracts that permit to establish responsibilities and authorizations among agents and set agreements, whereas conditionals allow one to restrict and limit agent interactions, define activation mechanisms that trigger agent actions, and define constraints and contract terms on OSCs. Conditionals and OSCs, as defined in OASIS, are applied to extend with ontological capabilities digital public ledgers such as the *blockchain* and the smart contracts implemented on it. We will also sketch the architecture of a framework based on the OASIS definition of OSCs that exploits the *Ethereum* platform and the *Interplanetary File System*.

D. Cantone · C. F. Longo · M. Nicolosi Asmundo · D. F. Santamaria (✉) · C. Santoro
Department of Mathematics and Computer Science, University of Catania, Viale Andrea Doria, 6, 95125 Catania, Italy
e-mail: santamaria@dmi.unict.it

D. Cantone
e-mail: domenico.cantone@unict.it

C. F. Longo
e-mail: fabio.longo@unict.it

M. Nicolosi Asmundo
e-mail: nicolosi@dmi.unict.it

C. Santoro
e-mail: santoro@dmi.unict.it

D. Camacho et al. (eds.), *Intelligent Distributed Computing XIV*, Studies in Computational Intelligence 1026, https://doi.org/10.1007/978-3-030-96627-0_22

1 Introduction

One of the most important features of a decentralized and publicly shared ledger is the elimination of any third-party intermediaries, since they require clients to put total and unquestioned trust on them. This is particularly true when the ledger is also responsible for managing legal contracts in a digital platform. The *blockchain* [6] has been introduced to allow parties of a network to interact in a distributed manner without the requirement of trusted entities. The blockchain is a peer-to-peer public ledger maintained by a distributed network of computational nodes. Blockchain technologies are opening new perspectives in several key aspects such as *Internet of Things* (IoT), healthcare, insurance, energy, communications, and robotics [21], in view of the wide range of benefits they provide. For example, the blockchain guarantees the ownership, transparency, traceability, public availability, continuity, and immutability of digital assets, in an efficient and trust-less environment where censorship is not achievable. One of the most popular applications of the blockchain is a self-executable contract, also called *smart contract* (SC) [18]. The SC is a way of representing contracts into lines of immutable program codes which are allowed to be self-run on a public ledger.

Smart contracts on the blockchain are equivalent to *stored procedures* of databases, and hence they have direct access to low-level mechanisms. Abstractions of smart contracts may provide several advantages. Among these, we recall that it is easier to publish contracts that integrate and operate with several types of digital and non-digital assets, when higher-level and formal representation of their constraints and agreements are provided. In addition, high-level representations of smart contracts are easy readable by human agents, thus enabling a clear understanding of the agreements and the verification of violations outside the digital borders of the application, for example, in a lawsuit context. Finally, such contracts are platform independent, and hence they can be used and shared by several types of applications and systems, for example, in cross-chain applications.

Semantic web tools and languages aim to reach full machine interoperability, to promote common data formats, and to exchange protocols on the web, share and reuse data across applications and across enterprise and community boundaries. In the semantic vision of the web, software agents are enabled to query and manipulate information on behalf of human agents by means of machine-readable data that carry explicit meaning. Thus, data can be automatically processed and integrated by agents, and can be accessed and modified at a higher level in such a way as to increase coherence and dissemination of information. In addition, with the intervention of semantic reasoners, implicit information is processed and inferred as to gain a deeper knowledge of the domain. Moreover, automated reasoning systems allow one to also verify consistency of data and query it. The definition of a specific domain is widely called *ontology* [9, 13]. The *Word Wide Web Consortium* (W3C) recommends the *Web Ontology Language 2* (OWL 2), a knowledge representation language for web ontologies relying on the Description Logic $\mathcal{SROIQ}(\mathbf{D})$ [10].

In [4], we presented the *Ontology for Agents, Systems, and Integration of Services*[1] (in short, OASIS), a foundational OWL 2 ontology that defines a request-execute communication protocol for agents, and in particular for *Internet of Agents* (IoA), based on a mutual exchange of ontology fragments that allow a full transparent and high-level interoperability of agents. For example, such ontology fragments allow agents to send and retrieve information from other agents in a transparent way, request other agents the execution of operations without knowing their hardware and software specifications, acknowledge agents for the execution status of the requested actions, and so on. Moreover, OASIS models information about the assignment and the execution of operations, restrictions on requests, connections, and exchange of messages among agents.

In this paper, we move towards the definition of a semantic blockchain by extending OASIS so as to deal with agreements that can be established by agents and secured on the blockchain without requiring the definition of specific blockchain smart contracts. Such agreements, called *ontological smart contracts* (OSCs), are established by leveraging conditionals that are also used to abstract and formalize behavior constraints, ways of limiting, bounding, or triggering agent actions. OSCs can be secured through suitable smart contracts implemented on the blockchain and allowing storing and retrieving of RDF graphs. We will also sketch the architecture of such a system that exploits the new introduced features of OASIS and based on the *Interplanetary File System* (IPFS) [15]. The latter allows blockchain-based applications to use immutable files of large dimensions without excessively increasing the cost of transactions in terms of computation and crypto-currency. Finally, we provide an implementation on Ethereum of the designed architecture.

The paper is organized as follows. Section 2 deals with related works. Section 3 is devoted to the description of ontological smart contracts in OASIS, whereas in Sect. 4 we sketch the architecture of a framework based on OSCs exploiting the blockchain and the IPFS. Finally, Sect. 5 concludes the paper with some final considerations and hints for future work.

2 Related Work

Since 2000, several results concerning how agents enter and leave different interaction systems have been presented, by exploiting both *commitment objects* [8] and *virtual institutions* [7]. Only very recently, however, researchers have focused their interest in conjoining the blockchain and ontologies [3]. One of the areas of investigation has been the development of a characterization of blockchain concepts and technologies through ontologies. A theoretical approach looking at the blockchain with an ontological approach has been introduced in [12], whereas [17] proposes a blockchain framework for *semantic web of things* (SWoT) contexts settled as a

[1] http://www.dmi.unict.it/santamaria/projects/oasis/oasis.php.

Service-Oriented Architecture (SOA), where nodes can exploit smart contracts for registration, discovery, and selection of annotated services and resources.

Other works aim to represent ontologies within a blockchain context. In [11], ontologies are used as a common data format for blockchain-based applications, though limited to the description of implementation aspects of the blockchain. However, some of the architectural choices presented in this paper to effectively implement OSCs through blockchain are inspired from [11].

The *Blockchain Ontology with Dynamic Extensibility* (BLONDiE) project [19] provides a comprehensive vocabulary that covers the structure of different components (wallets, transactions blocks, accounts, etc.) of blockchain platforms (Bitcoin and Ethereum) and that can be easily extended to other alternative systems.

We proposed in [4] a first version of OASIS, a behavior-oriented ontology for representing agents and their interactions. As far as we know, this represents the first attempt of using semantic web technologies for defining a transparent communication protocol among agents that abstracts from their implementation details and configurations, for integrating and uniforming agents, and as a representation system for their behaviors and interactions. In [4], we also proposed a prototype version based on two modules of a domotic assistant, called PROF-ONTO, that exploits several features of OASIS in order to enable communication of IoT devices and users in a domotic environment.

3 Ontological Smart Contracts in OASIS

Before introducing the classes and properties adopted by OASIS to represent smart contracts, we first present *conditionals*. In general, conditionals are used outside the context of digital contracts to put constraints on the execution of actions and to ensure that some conditions are verified before executing a task. Conditionals are OWL sentences that have the fashion of *Semantic Web Rule Language* (SWRL) rules [20] describing operations that must be triggered when certain conditions hold and exchanged among agents. In fact, unlike SWRL rules, which check the consistency of the knowledge base and infer new sentences, OWL representation of conditionals allows OASIS-oriented applications and systems to combine constraints with behaviors, thus extending the expressiveness and providing a higher-level description and representation of agents. Moreover, the introduction of conditionals in OASIS is also justified by the fact that the satisfiability of conditionals does not coincide with the satisfiability of the knowledge base. In fact, violations of conditionals may result in a litigation among agents or may lead agents to actuate alternative plans, without affecting in any way the satisfiability of the knowledge base. In the context of smart contracts, conditionals are used to characterize contract terms, which are clearly expressed provisions that give rise to an obligation and whose breach can lead to a litigation.

Figure 1 depicts the schema of OASIS conditionals. Conditionals are constituted by a consequent (head) and an antecedent (body), both formed by a conjunctive set of atoms. Atoms in their turn comprise the subject of the conditional, the object, the operator describing the action and, possibly, an operator parameter and argument.

Conditional atoms are introduced by means of the following classes:

- *ConditionalAtom*: represents a conditional atom;
- *ConditionalHeadAtom*: represents atoms of conditional consequents and is defined as a subclass of the class *ConditionalAtom*;
- *ConditionalBodyAtom*: represents atoms of conditional antecedents and is defined as a subclass of the class *ConditionalAtom*;
- *ConditionalSubject*: represents subjects of atoms of conditional consequents or of conditional antecedents;
- *ConditionalObject*: represents objects of atoms of conditional consequents or of conditional antecedents;
- *ConditionalOperator*: represents actions of atoms of conditional consequents or of conditional antecedents;
- *ConditionalParameter*: represents parameters of actions considered by conditional consequents or conditional antecedents (the class *ConditionalParameter* includes the classes *ConditionalInputParameter* and *ConditionalOutputParameter*, representing the input and output parameter, respectively);
- *ConditionalOperatorArgument*: defines operator arguments which represent a subordinate characteristic of the conditional operator (for example, "quality check" may be represented by the operator *check* with argument *quality*);
- *ConditionalEntryTemplate*: represents templates of the features that the entities involved in the conditional, which are introduced by means of the object-property *refersAsNewTo*, must satisfy.

Conditionals are introduced in OASIS by means of the following classes:

- *Conditional*: defines conditionals, constituted by a consequent (head) and an antecedent (body), both consisting in a conjunctive set of atoms;
- *ConditionalHead*: represents consequents (heads) of conditionals;
- *ConditionalBody*: represents antecedents (bodies) of conditionals;
- *ConditionalSet*: represents conjunctive sets of conditionals.

Conditional atoms are related to a subject (instance of the class *ConditionalSubject*), an object (instance of the class *ConditionalObject*), and an operator (instance of the class *ConditionalOperator*), by means of the object-properties *hasConditionalSubject*, *hasConditionalObject*, and *hasConditionalOperator*, respectively. Possibly, conditional atoms are related to parameters (instances of the class *ConditionalParamenter*) and to operator arguments (instances of the class *ConditionalOperatorArgument*) by means of the object-properties *hasConditionalParameter* and *hasConditionalOperatorArgument*, respectively. Specifically, input and output parameters are introduced through the object-properties *hasConditionalInputParameter* and *hasConditionalOutputParameter*, respectively.

In OASIS there are two ways to connect conditional entries (i.e., instances of *ConditionalSubject*, *ConditionalObject*, *ConditionalOperator*, *ConditionalParamenter*,

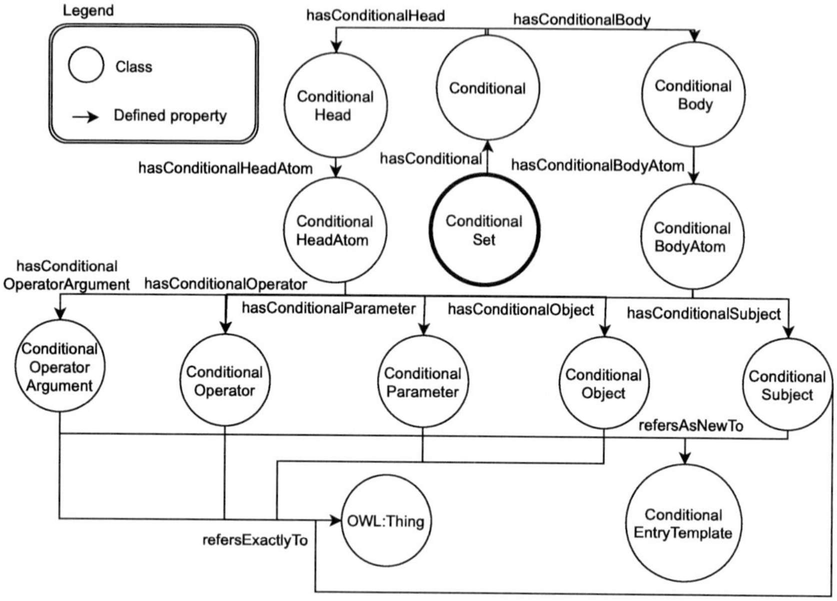

Fig. 1 Ontology schema of conditionals

and *ConditionalOperatorArgument*) to the related entities, namely by exploiting the object-property *refersExactlyTo* and the object-property *refersAsNewTo*.

The first way consists in directly indicating the individual involved in the conditional by means of the object-property *refersExactlyTo*, e.g., the address of a digital wallet of a specific person. In the second modality, the entry is unknown, but there is a set of desirable features that the entry must have in order to make the conditional satisfied when it is checked. For example, if a digital asset is sold to anyone who sends the correct amount of money to a digital wallet, the buyer (the conditional entry) is not known until he performs the transaction: completing a transaction is the feature required in order to consider an entity as a buyer. In such a case, the object-property *refersAsNewTo* is used to specify an instance of the class *ConditionalEntryTemplate*, which endows the set of features that must be satisfied by the conditional entry.

Instances of the classes *ConditionalHead* and *ConditionalBody* are related to instances of the classes *ConditionalHeadAtom* and *ConditionalBodyAtom*, representing conditional atoms through the object-properties *hasConditionalHeadAtom* and *hasConditionalBodyAtom*, respectively. The object-properties *hasConditional-HeadAtom* and *hasConditionalBodyAtom* are defined as subproperties of the object-property *hasConditionalAtom*.

Finally, conditionals are introduced in OASIS by way of instances of the class *ConditionalSet*. Such instances are linked through the object-property *hasConditional* to instances of the class *Conditional*, which in their turn are linked to instances of the classes *ConditionalHead* (representing the consequent) and *ConditionalBody* (rep-

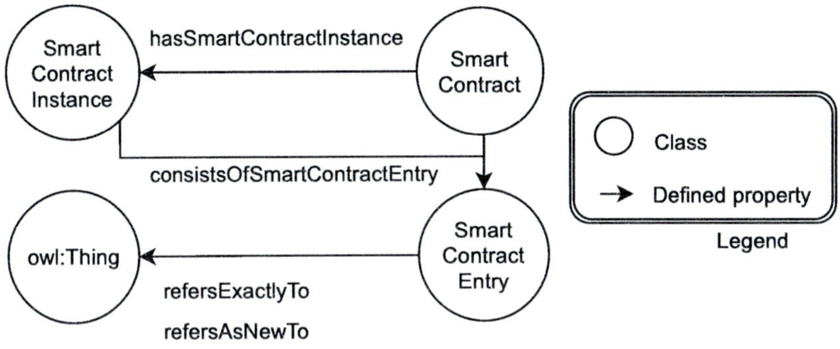

Fig. 2 Ontology schema of smart contract

resenting the antecedent) by means of the object-properties *hasConditionalHead* and *hasConditionalBody*, respectively. Example of conditionals in OASIS can be found in the extended version of this paper [5].

Smart contracts benefit from the abstraction layer provided by suitable ontological models, since they are defined at a higher level, leaving the implementation details to the underneath layer constituted by blockchain-based distributed computing platforms such as *Ethereum*.[2] With this in mind, conditionals are exploited to define ontological smart contracts in OASIS according to the schema illustrated in Fig. 2.

Anytime two parties agree to take action, in order to make an exchange or payment for something of value, a contract is created (instantiated). We call such a contract actuation a *contract instance*. For example, in the case of trading agents, the contract specifies a template of the general agreements that trader and clients shall respect, whereas the contract instance is the application of such a template, specifying how the two parties join the contract: the trader and the client instantiate a general brokerage contract, where the trader is responsible for selling stock on behalf of clients. In OASIS, smart contracts are introduced by the class *SmartContract*, whereas contract instances are modelled by means of the class *SmartContractInstance*. Smart contracts and smart contract instances are connected through the object-property *consistsOfSmartContractInstance*.

Smart contracts and smart contract instances provide a set of entries mapped by instances of the class *SmartContractEntry*, the latter containing the classes *SmartContractEntryParticipant* (including individuals referring to the participants involved in the smart contract) and *SmartContractEntryValue* (including individuals referring to values involved in the smart contract). The object-property *consistsOfSmartContractEntry* links smart contracts and smart contract instances to the corresponding entries. Entries of the contract instances are connected to the corresponding contract entries by means of the object-property *refersExactlyTo*. An example of contract in the context of trading agents can be found in [5].

[2] https://www.ethereum.org.

Conditionals are used to characterize constraints among the entries of a smart contract and to establish agreements. Agreements may be easily verified through a SPARQL query, whose result validates (or invalidates) the contract between the two parties. Ontological smart contracts model agreements among parties, whereas SPARQL query validate them: what the ontology does not guarantee is that the parties have actually agreed to the clauses of the contract, its traceability, non-repudiation, and so on, features that a blockchain framework, instead, may ensure.

4 Architectural Design of an OSC-Oriented Application

As stated above, ontological smart contracts may enjoy from a decentralized ledger such as the blockchain. However, it is prohibitively expensive to store a lot of data on it. For instance, at the time of this writing (2019), about 50 US dollars are required to store the 38 pages of the PDF version of the Ethereum yellow paper, which weights about 520Kb. In fact, according to the paper itself, approximately 20,0000 gas are required for storing 256 bit/8 bytes (1 word), namely 20 Gwei for a unit of gas (1 Gwei equals 0.000000001 Ethers). In order to reduce the cost of transactions and the time required to compute them, data-oriented applications need to rely on a decentralized server to store information. The *Interplanetary File System* (IPFS) is one of the preferred solutions. Basically, IPFS allows one to store a large amount of files, whose permanent IPFS links (CID) can be included into blockchain transactions, in such a way as to put a timestamp on the data and secure it without directly including files in the chain itself. It is sufficient to upload documents on IPFS and then to store the IPFS CID on the Ethereum blockchain. The CID is a hash obtained from the file which, hence, cannot be modified. However, IPFS provides the *InterPlanetary Naming Service* that uses the peer ID to point to a specific hash. Such a hash can change whereas the peer ID cannot. It turns out that applications can gain access to mutable content in IPFS without knowing the new hash beforehand.

Figure 3 illustrates a typical application that exploits OASIS ontological smart contracts. Ontologies are secured on the blockchain by means of transactions con-

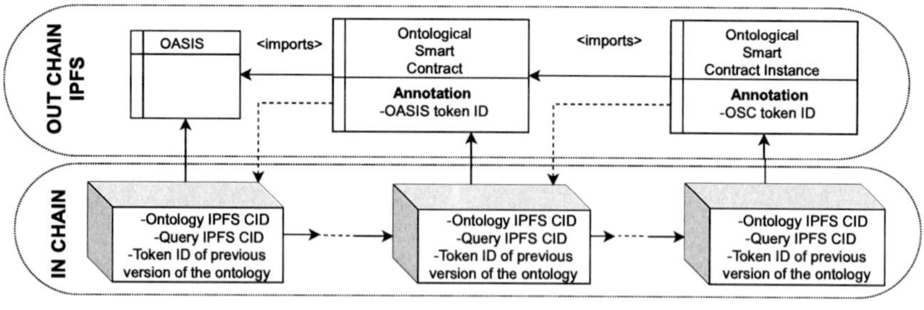

Fig. 3 Sketch of the architecture of an OASIS OSC-based application

taining only three states: the IPFS CID of the ontology, the IPFS CID of the query required to validate the ontology, and the address of the transaction that secured the previous version of both the ontology and the corresponding query. As a preliminary step, the ontology OASIS and all the ontologies required by the OSC are published on IPFS and secured through a blockchain transaction.[3] From then on, a second transaction suffices for deploying the OSC. Such a transaction secures the ontology representing the OSC, the query that validates the OSC itself, and the address of the transaction that secured, possibly, a previous version of the OSC and of the query. Analogously, any instance of the smart contract and any related ontology are secured by an additional transaction.

The IPFS CID of the ontology allows one to access the OSC and to ensure that the OSC and all the imported ontologies have not been modified after their publication, thus guaranteeing that all the ontologies exploited by the contract are exactly the ones on which an agreement has been reached. An additional source of guarantee is the SPARQL query that checks whether the ontological contract is voided by the instance under consideration. As in the case of the ontologies, the SPARQL query is published on IPFS and secured on the blockchain, in the same transaction as the OSC that the query checks. By accessing the secured OSC and the query that validates it, the participants have all the means to validate or invalidate the contract, with all the guarantee of the blockchain system and the versatility of ontologies. Moreover, in our architecture, ontologies may exploit suitable OWL annotation axioms, or alternatively the BLONDiE ontology, to refer to blockchain transactions that secured the imported ontologies. Finally, the ontological approach and the architecture model introduced in this paper may also be adopted in non-Turing-complete blockchains such as Bitcoin, since the effort required to the blockchain is limited to store at most two states pointing to resources located out of the blockchain, plus one state storing the token ID of the previous version of the ontology.

5 Conclusions and Future Work

We presented an extension of the *Ontology for Agents, Systems, and Integration of Services* (in short, OASIS), modelling ontological conditionals and smart contracts. Conditionals, classically applied to set restrictions on agent actions or to activate them when suitable conditions hold, are also used to define contract terms. Contract terms are applied, in their turn, to define ontological smart contracts, which establish responsibilities and authorizations among agents. Conditionals and smart contracts defined by OASIS are exploited to add an ontological level to the blockchain and smart contracts based on it. We also sketched the architecture of a system leveraging the blockchain and the Interplanetary File System (IPFS) to store and retrieve

[3] The smart contract is compliant with the non-fungible token standard ERC721 and is available in the Ethereum main network at the address 0x36194ab80f7649572cab9ec524950df32f638b08. A Java API to publish and retrieve OSC is available at https://github.com/dfsantamaria/CLARA.

OASIS ontological smart contracts, and we implemented it through an Ethereum smart contract.

In order to extend the integration level of OASIS with blockchains, we plan to integrate and extend BLONDiE, also by considering the ontology as a meta-model extension of OASIS. We also plan to study how OASIS can be exploited by *OntologyBeanGenerator 5.0* [2] inside the JADE framework [1] to generate code for agents and artifacts and how it can be exploited by *CArtAgO* [16], a framework for building shared computational worlds. Finally, we shall extend the set of actions and parameters provided by OASIS with the synset introduced by WordNet [14], in order to make the whole infrastructure multi-language- and meaning-oriented.

References

1. Bellifemine, F.L., Caire, G., Greenwood, D.: Developing Multi-Agent Systems with JADE. Wiley, New York (2007)
2. Briola, D., Mascardi, V., Gioseffi, M.: OntologyBeanGenerator 5.0: Extending Ontology Concepts With Methods And Exceptions. In: Proceedings of the 19th Workshop "From Objects to Agents", Palermo, Italy, June 28–29, pp. 116–123 (2018)
3. Cano-Benito, J., Cimmino, A., García-Castro, R.: Towards blockchain and semantic web. In: Abramowicz, W., Corchuelo, R. (eds.) Business Information Systems Workshops, pp. 220–231. Springer International Publishing, Cham (2019)
4. Cantone, D., Longo, C.F., Nicolosi-Asmundo, M., Santamaria, D.F., Santoro, S.: Towards an ontology-based framework for a behavior-oriented integration of the IoT. In: Proceedings of the 20th Workshop from Objects to Agents, 26–28 June, 2019, Parma, Italy, CEUR Workshop Proceeding, vol. 2404. pp. 119–126 (2019)
5. Cantone, D., Longo, C.F., Nicolosi-Asmundo, M., Santamaria, D.F., Santoro, C.: Ontological smart contracts in OASIS: ontology for agents, systems, and integration of services (extended version). In: CoRR (2021). arXiv:2012.01410
6. Christidis, K., Devetsikiotis, M.: Blockchains and smart contracts for the internet of things. IEEE Access **4**, 2292–2303 (2016)
7. Esteva, M.: Electronic institutions: from specification to development. IIIA Monograph Series. Ph.D. Thesis vol. 19 (2003)
8. Fornara, N., Colombetti, M.: A commitment-based approach to agent communication. Appl. Artif. Intell. 853–866 (2004)
9. Hofweber, T.: Logic and ontology. In: Edward N. Zalta (ed.), The Stanford Encyclopaedia of Philosophy (Summer 2018 Edition) (2018)
10. Horrocks, I., Kutz, O., Sattler, U.: The even more irresistible SROIQ. In: Doherty, P., Mylopoulos, J., Welty, C.A. (eds.) Proceedings of the 10th International Conference on Principles of Knowledge Representation and Reasoning, pp. 57–67. AAAI Press (2006)
11. Kim, H., Laskowski, M.: Toward an ontology-driven blockchain design for supply-chain provenance. Intell. Syst. Acc. Finance Manag. **25**(1), 18–27 (2018)
12. de Kruijff, J., Weigand, H.: Understanding the blockchain using enterprise ontology. In: CAiSE (2017)
13. Oberle, D., Guarino, N., Staab, S.: What is an ontology? Handbook on Ontologies. Springer, Berlin (2009)
14. Princeton University: WordNet, A Lexical Database for English (2010). https://wordnet.princeton.edu
15. Protocol Labs: The Interplanetary File Systems (IPFS). https://ipfs.io

16. Ricci, A., Piunti, M., Viroli, M., Omicini, A.: Environment Programming in CArtAgO. In: Multi-Agent Programming: Languages. Tools and Applications, pp. 259–288. Springer, Boston (2009)
17. Ruta, M., Scioscia, F., Ieva, S., Capurso, G., Pinto, A., Di Sciascio, E.: A blockchain infrastructure for the semantic web of things. In: 26th Italian Symposium on Advanced Database Systems (SEBD 2018) (2018)
18. Szabo, N.: Formalizing and securing relationships on public networks. First Monday **2**(9) (1997)
19. Ugarte Rojas, H.E.: A more pragmatic web 3.0: Linked blockchain data. In: Google Scholar (2017)
20. World Wide Web Consortium: SWRL: A Semantic Web Rule Language Combining OWL and RuleML (2004). http://www.w3.org/Submission/SWRL/
21. Xu, X., Weber, I., Staples, M., Zhu, L., Bosch, J., Bass, L., Pautasso, C., Rimba, P.: A taxonomy of blockchain-based systems for architecture design. In: 2017 IEEE International Conference on Software Architecture (ICSA), pp. 243–252. IEEE (2017). http://design.inf.usi.ch/sites/default/files/biblio/icsa2017-blockchain.pdf

Blockchains Through Ontologies: The Case Study of the Ethereum ERC721 Standard in OASIS

Giampaolo Bella, Domenico Cantone, Cristiano Longo, Marianna Nicolosi Asmundo, and Daniele Francesco Santamaria

Abstract Blockchains are gaining momentum due to the interest of industries and people in *decentralized applications* (Dapps), particularly in those for trading assets through digital certificates secured on blockchain, called tokens. As a consequence, providing a clear unambiguous description of any activities carried out on blockchains has become crucial, and we feel the urgency to achieve that description at least for trading. This paper reports on how to leverage the *Ontology for Agents, Systems, and Integration of Services* ("OASIS") as a general means for the semantic representation of smart contracts stored on blockchain as software agents. Special attention is paid to non-fungible tokens (NFTs), whose management through the ERC721 standard is presented as a case study.

This work has been supported by the ONTOCHAIN NGI European project grant agreement no. 957338. We are thankful to the ONTOCHAIN Consortium who mentored and assisted the research.

G. Bella · D. Cantone · M. Nicolosi Asmundo · D. F. Santamaria (✉)
Department of Mathematics and Computer Science, University of Catania, Viale Andrea Doria, 6, 95125 Catania, Italy
e-mail: daniele.santamaria@unict.it

G. Bella
e-mail: giampaolo.bella@unict.it

D. Cantone
e-mail: domenico.cantone@unict.it

M. Nicolosi Asmundo
e-mail: marianna.nicolosiasmundo@unict.it

C. Longo
The Sicilian Wheat Bank, Via Piazza Armerina, 30, 94100 Enna, Italy
e-mail: longo@bancadelgrano.it

D. Camacho et al. (eds.), *Intelligent Distributed Computing XIV*, Studies in Computational Intelligence 1026, https://doi.org/10.1007/978-3-030-96627-0_23

1 Introduction

The last decade reports vast interest on blockchain technology and related applications from various realms, including economic, social, business, and academic ones. Beyond the financial speculation concerning cryptocurrencies, the interest in blockchain technologies is mainly motivated by the fact that they realize decentralized and publicly shared ledgers, where third-party intermediaries demanding the client's total and unquestioned trust are no longer required. Blockchain technologies [7] were precisely introduced to allow users to interact and run programs in a distributed way without the requirement of trusted entities, yet guaranteeing ownership, transparency, traceability, availability, continuity, and immutability of digital shared assets. Applications of blockchain technologies range from the *Internet of Things* (IoT) and robotics [18], to commerce, healthcare, insurance, energy, laws, and communications.

One of the most popular applications of Turing-complete blockchains such as Ethereum [15] is the *smart contract*. Smart contracts are self-executing and immutable programs, autonomously running and verified on a distributed and decentralized public network, which implement decentralized applications on blockchain systems called *Dapps*. In 2020, Dapps have particularly grown as an exchange tool for non-fungible tokens (NFTs), namely digital certificates stored on the blockchain representing predetermined rights on certain unique assets. NFTs are mainly used as a proof of ownership of physical or digital products. Such tokens are routinely exchanged by users to witness that assets whose uniqueness is hard to demonstrate (for example, digital images) are owned in an exclusive way. At the end of 2020, the market capitalization of NFTs reached the amount of 338 millions of U.S. dollars.[1] However, one of the main limitations of blockchains is the hard-coded nature of the transactions stored on them. As a consequence, it is hard to probe a blockchain, for example, to find smart contracts trading specific NFTs that satisfy certain requirements in terms of quality or quantity. Therefore, a formal semantic knowledge representation capturing the blockchain smart contracts as well as the activities carried out on it facilitates the understanding of blockchain concepts, the interlinking with other out-of-chain information, and also formal reasoning. Moreover, a semantic representation of blockchains enables the automatic discovery of smart contracts, the interconnection of services running on different blockchains (i.e., cross-chain integration) and the integration between on-chain and off-chain services. These features turn out to be more interesting when smart contracts are implemented as mechanisms for generating and exchanging tokens. A desirable feature of token exchange systems is a precise and intelligent query mechanism capable of determining what, when, and how certain assets have been generated, exchanged or destroyed. For example, intelligent systems may be aware of the activation of smart contracts for generating tokens with specific characteristics, e.g., of the type of exchanged asset, of the exchange of particular tokens at certain conditions, or of their destruction. More in general,

[1] https://www.statista.com/statistics/1221742/nft-market-capitalization-worldwide/ (last access: 08/07/2021).

intelligent systems may be aware of the activation of smart contracts and of all the related activities over the blockchain.

Beyond a semantic representation of transactions and information stored in blocks, a real, semantically represented blockchain is effectively achievable if smart contracts are interpreted as reactive agents operating on a common environment, with a fully specified semantics of available operations, committed actions, and stored data.

The representation of blockchain actors requires ontological capabilities for fully representing agents and their interactions in a detailed way. This paper adopts the *Ontology for Agents, Systems, and Integration of Services* ("OASIS") towards the full, semantic representation of the Ethereum blockchain and the smart contracts running on it, with a special focus on the smart contracts compliant with the ERC721 standard[2] for NFTs management.

The paper is organized as follows. Section 2 presents related work. Section 3 outlines OASIS, whereas Sects. 4 and 5 depict the ontology implementing the OASIS representation of, respectively, the Ethereum blockchain and the ERC721 standard. Finally, Sect. 6 draws some conclusions and delineates future research directions.

2 Related Works

Interest in symbiotically combining semantic web technologies and blockchains is quite recent [4, 8]. One of the areas of investigation concentrates in developing a characterization of blockchain concepts and technologies through ontologies and of blocks and transactions meta-data. An ontological albeit theoretical approach at blockchain representation exists [11]. Ruta et al. propose a blockchain framework for *semantic web of things* (SWoT) contexts settled as a *Service-Oriented Architecture* (SOA), where nodes can exploit smart contracts for registering, discovering, and selecting annotated services and resources [14].

Blockchain technologies are also exploited as a secure and public storage system for small data, including linked data, and to realize a more resilient architecture for the Semantic Web [8].

Other works aim at representing ontologies within a blockchain context. In [10], ontologies are used as common data format for blockchain-based applications such as the proposed provenance traceability ontology, but are limited to implementation aspects of the blockchain.

Fill discusses blockchains applied for tracking the provenance of knowledge, for establishing delegation schemes, and for ensuring the existence of patterns in formal conceptualizations using zero-knowledge proofs [9].

A semantic interpretation of smart contracts as services bases on the *Ethon* ontology [12] exist [1]. The main limitation of that approach is the poor semantic description of smart contracts, thus hindering the discovery of unknown smart contracts and of the related operations fulfilled during their life-span.

[2] https://ethereum.org/it/developers/docs/standards/tokens/erc-721/.

The *Blockchain Ontology with Dynamic Extensibility* (BLONDiE) project [16] provides a comprehensive vocabulary that covers the structure of different components (wallets, transactions blocks, accounts, etc.) of blockchain platforms (Bitcoin and Ethereum) and that can be easily extended to other alternative systems.

Finally, in [6] the authors illustrated how the ontology OASIS is applied for describing digital contracts (called *ontological smart contracts*), intended as agreement among agents, and how they can be secured on Ethereum.

In this contribution, the definition of digital contracts in [6] is extended to include both smart contracts over blockchains, intended as programs running on the blockchain and interpreted as digital agents in the OASIS fashion, including their operational semantics, and tokens exchanged through them.

3 Preliminaries

The *Ontology for Agents, Systems, and Integration of Services* ("OASIS")[3] [5] is an OWL 2 ontology for representing agents and their activities. On one hand, the ontology models (web) agents and, in particular, the way they interact and operate in a collaborative environment, regardless of the framework and language adopted for their implementation. Agents are mainly represented by means of the mentalistic notion of agent behaviour inspired by [3], encompassing goals and tasks that are achievable (either publicly available or exposed on request) by the agent, together with actions, sensors, and actuators used to perform operations. On the other hand, OASIS is used to define actions that may be requested to other agents and their related information such as operation inputs and outputs. Such requests are submitted by exchanging suitable fragments of OASIS, whereas agents whose capabilities are compatible with the requested actions are discovered by means of SPARQL queries performed over their behaviours. OASIS was applied to build a TRL3 prototype of a home assistant that activates and manages applications, devices, and users interacting with each other within the environment [5]. OASIS was also used to define agent agreements and store these in the IPFS file system [13] in order to reduce the transaction data stored directly on blockchain [6].

OASIS models agents by publicly representing their behaviours. By exposing behaviours, agents report to the communication peers the set of operations that they are able to perform and, eventually, the type of data required to execute these and the expected output. The representation of agents and their interactions in OASIS is carried out along three main steps. The first step consists in *defining the template* for agent behaviour: templates are high-level descriptions of behaviours of abstract agents that can be implemented to define more specific and concrete behaviours of real agents. For example, a template may be designed for agents to sell and ship products to buyers, and it may be implemented by any phone seller that ships its products using the Fedex courier: templates are useful to guide developers to define

[3] The ontology is reachable at http://www.dmi.unict.it/oasis.owl.

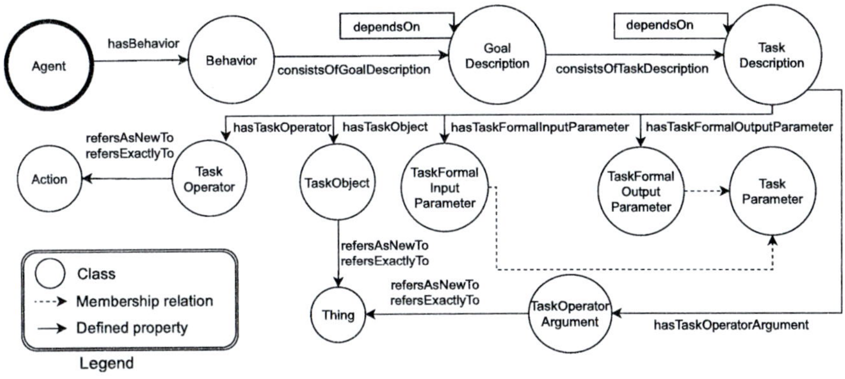

Fig. 1 Agent representation in OASIS

the most suitable representation of their agents. The second step consists in *defining a behaviour* by implementing a template, which requires a specification of the full operational details about the sought behaviour. In the final step, agent actions are represented through ontology patterns analogous to the ones of the corresponding agent behaviours and linked to them.

As depicted in Fig. 1, agent behaviours are represented by the goals to achieve, which in turn are related with their constitutional elements, namely tasks. Tasks represent atomic operations that agents execute, and are described by the actions to be performed. Actions are drawn from a shared and common vocabulary, and can be simple or composed, eventually associated with requested input parameters and expected outputs. Finally, agent behaviours in OASIS may be associated with conditionals [6], adding constraints on the execution of actions and ensuring that certain conditions are verified before or after a task is executed. OASIS conditionals are OWL sentences that have the fashion of *Semantic Web Rule Language* (SWRL) rules [17], describing operations that must be triggered when certain conditions hold.

4 Representing Ethereum Through OASIS

In this section, we describe how the Ethereum blockchain is modelled in OASIS.[4] OASIS provides a different representation of blockchains with respect to Ethon [12] and BLONDiE [16], since the description of blockchain has to be aligned with the definitions of agent and agent behaviour.

Ethereum is represented in OASIS by following the schema illustrated in Fig. 2. Ethereum blocks embedding transactions are represented by instances of the class *EthereumBlock* (subclass of *BlockchainBlock*) and connected to the transactions

[4] The ontology is reachable at https://www.dmi.unict.it/santamaria/projects/oasis/sources/ether-oasis.owl.

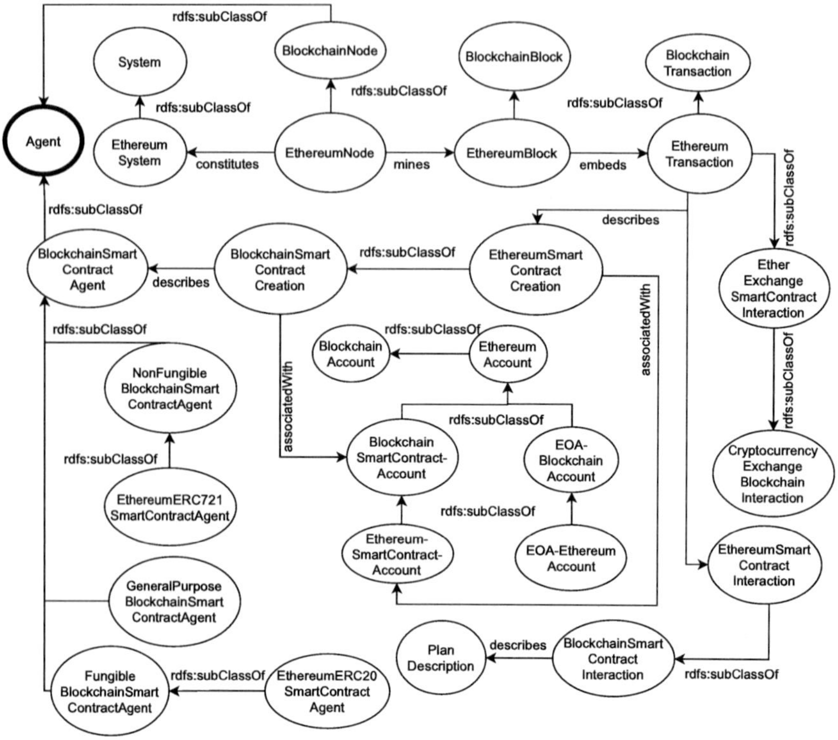

Fig. 2 Representing the Ethereum blockchain in OASIS

contained in them by means of the object-property embeds. Each Ethereum transaction is identified by an instance of the class *EthereumTransaction* (subclass of *BlockchainTransaction*), encapsulating all the transaction information.

Block miners are identified by instances of the class *EthereumNode* (subclass of *BlockchainNode*) and linked to the mined block through the object-property mines: instances of *BlockchainNode* are also instances of the class *Agent*, representing agents and provided with a behaviour as in the OASIS fashion. Moreover, instances of *BlockchainNode* (resp., *EthereumNode*) are connected to instances of the class *System* (resp., *EthereumSystem*) by means of the object-property constitutes. Such a characterization of nodes, blocks, and transactions is the main difference with analogous approaches such as Ethon and BLONDiE, since it allows one to describe activities carried out by both in-chain and out-of-chain agents, thus providing a higher-level model of the two ecosystems and a means to unify them. Specifically, Ethereum activities are classified into two main categories, namely, deployments of smart contracts, represented by instances of the class *EthereumSmartContratCreation* (subclass of *BlockchainSmartContractCreation*), and interactions with smart contracts, represented by instances of the class *EthereumSmartContractInteraction* (subclass of the *BlockchainSmartContractInteraction*).

In OASIS, smart contracts deployed into the Ethereum blockchain correspond to agents with well-defined behaviours: interactions with smart contracts are represented by OASIS *plan executions* and linked to the behaviour that induced the action. Specifically, a smart contract creation is represented by an instance of the class *BlockchainSmartContractCreation*, which is related with the description of the agent describing its behaviour by the object-property describes, the latter represented by an instance of the class *BlockchainSmartContractAgent* (subclass of the class *Agent*). Instances of *BlockchainSmartContractCreation* are also associated by means of the object-property associatedWith with the related Ethereum accounts, represented by instances of the class *EthereumSmartContractAccounts* (sublcass of the class *BlockchainSmartContractAccount*, which, in turn, is a subclass of *BlockchainAccount*). Users are instead associated with *Ethereum externally owned accounts* (EOA) represented by instances of the class *EOA-EthereumAccount* (sublcass of *EOA-BlockchainAccount*, which is a subclass of *BlockchainAccount*).

OASIS identifies four main general categories of smart contract agents: (a) smart contracts providing non-fungible token exchange mechanisms compatible with the Ethereum standard ERC721, which are represented by instances of the class *EthereumERC721SmartContractAgent* (subclass of the class *NonFungible-BlockchainSmartContractAgent*); (b) smart contracts providing fungible token exchange mechanisms compatible with the Ethereum standard ERC20, which are represented by instances of the class *EthereumERC20SmartContractAgent* (subclass of the class *FungibleBlockchainSmartContractAgent*); (c) agents responsible for exchanging Ether cryptocurrency, which are represented by instances of the class *EtherExchangeSmartContractAgent* (subclass of the class *CryptocurrencyExchangeBlockchainSmartContractAgent*); (d) general purpose and user-defined smart contract agents that do not enjoy the characteristics of the aforementioned smart contracts, which are introduced by instances of the class *GeneralPurposeBlockchainSmartContractAgent*.

In OASIS, agents may perform actions autonomously or as response to requests of executing some operations submitted by a peer. Concerning the blockchain ecosystem, we mainly limit ourselves to take into account only requests that modify the state of the chain (both internal and external), and hence induce transactions, even through *view functions*, namely functions that do not modify the state of the chain may be represented as well.

The transactions (represented by instances of the class *EthereumTransaction*) induced by interaction requests submitted to smart contracts are related with instances of the class *EthereumContractInteraction* (subclass of the class *SmartContractInteraction*) by means of the object-property describes. Instances of *EthereumContractInteraction* introduce plan descriptions as in the OASIS fashion by means of the object-property describes. Plan descriptions are ways of characterizing requests of actions and the related actions performed by agents. The most notable subclass of *EthereumContractInteraction* is the class *EtherExchangeSmartContractInteraction*, representing the transferring of Ether cryptocurrency from a wallet to another. The class *EtherExchangeSmartContractInteraction* is also a subclass of the class *Cryp-*

tocurrencyExchangeBlockchainSmartContractInteraction, which in turn is a subclass of the class *BlockchainSmartContractInteraction*.

An example of representing Ethereum transactions in OASIS is illustrated in [2], Appendix A.

5 The ERC721 Standard in OASIS

In this section we show how OASIS represents the main ERC721 standard functions for managing non-fungible tokens on the Ethereum blockchain. For space limitations, we illustrate how the ERC721 standard for minting non-fungible tokens is modelled in OASIS, whereas the ERC721 functions for transferring tokens, burning tokens, granting ownership of single tokens, of all the tokens stored in the wallet, and to verify the ownership of tokens, as represented in OASIS, are described in [2], Appendices B, C, D, E, and F, respectively.

As shown in the previous section, the smart contracts defined for the ERC721 token management are introduced by means of instances of the class *EthereumERC721 SmartContractAgent* (subclass of the class *Agent*). In OASIS, agents having similar behaviours inherit the representation of their behaviour from a common template providing general descriptions that may be customized by single agents. For this purpose, OASIS provides a template for the ERC721 standard introduced by the individual *ethereum_ERC721_smart_contract_behavior_template*, which describes the behaviours of agents minting, burning, and transferring Ethereum NFTs according to the guidelines of the standard. Other functions admitted by the standard ERC721, namely the *approve* function (delegating wallets for managing single tokens), the *setApprovalForAll* function (delegating wallets for managing all the tokens owned), and *ownerOf* (for retrieving the owner of a given token), are also represented in OASIS.

Figure 3 illustrates the behaviour template provided by OASIS describing the ERC721 function for generating new tokens. The behaviour template consists of a single goal, which in turn consists of a single task. Tasks comprise four elements. The first element is the task operator, providing the description of the action to be performed, namely minting. The latter is introduced by means of the individual *mint* (instance of the class *Action*), through the the object-property refersExactlyTo. We recall that in OASIS the object-properties refersExacltyTo and refersAsNewTo are introduced to describe the way how constituting elements of agent behaviours must be matched when a verification of compatible behaviours occurs. The object-property refersExactlyTo introduces well-known entities whose IRIs must correspond to the IRIs of the matched entities or for which the OWL object-property sameAs has been expressed. On the contrary, the object-property refersAsNewTo introduces entities (instances of the class *ReferenceTemplate*) that are used as general descriptions encapsulating the features that the matching entities must satisfy.

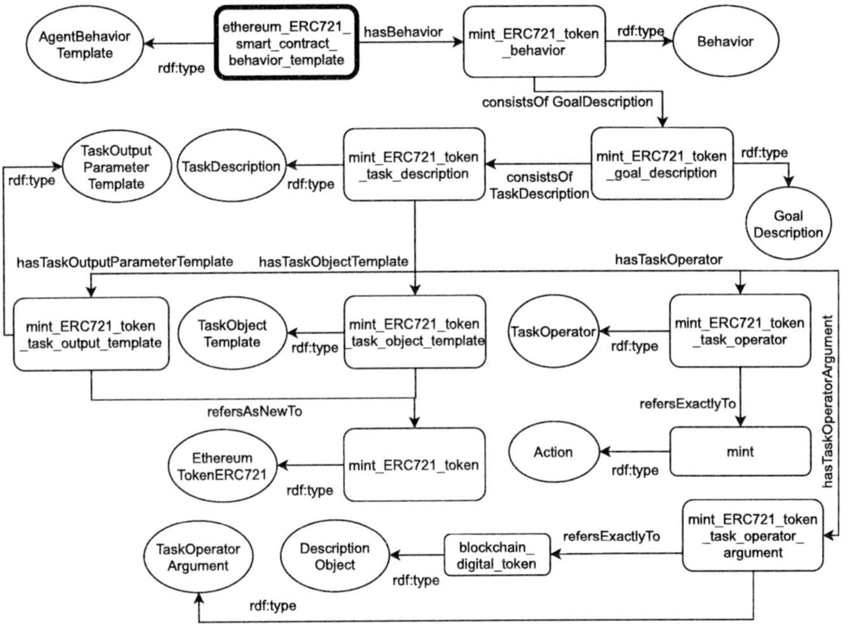

Fig. 3 OASIS behaviour template for the ERC721 minting function

The second element of the ERC721 token minting task is the operator argument introducing the individual *blockchain_digital_token* by means of the object-property refersExactlyTo. Operators and operator arguments identify unambiguously that the referred operation consists in the generation of (digital) tokens on the blockchain. The third and the fourth elements represent the recipient and the outcome of the operation, respectively. The recipient is introduced by means of a template of the task object, whereas the outcome is introduced by means of a template of the output parameter of the task. The object template and the output parameter are both connected through the object-property refersAsNewTo to the entity *mint_ERC721_token*, which describes the features that the recipient of the action must have, i.e., being an instance of the class *EthereumTokenERC721*.

6 Conclusions

This paper leveraged the OASIS ontology towards the representation of the Ethereum blockchain and the smart contracts deployed on it. Specific focus was on those that comply with the ERC721 standard for NFTs management. OASIS is exploited as a means of semantically representing Ethereum with the aim of probing the blockchain for locating smart contracts and related NFTs by specifying the desired features. In particular, the ontological representation of OASIS allows one to find smart contracts and related NFTs by inspecting their behavioural descriptions through purposely crafted SPARQL queries. It was already clear that the OASIS approach to semantic representation had the power of generality but our findings demonstrate it at an applied level.

Future work is dense. The very next step is to extend OASIS so as to model the standard protocols ERC20 and ERC1155 for fungible and semi-fungible tokens, respectively, and to represent different blockchains such as Stellar Lumens. Moreover, we intend to take up the design of a search engine exploiting OASIS to find desired smart contracts and tokens using a mechanism of auto-generating parametric ad-hoc SPARQL queries that could be borrowed from sibling applications [5]. The present work supports the claim that the potential of the semantic representation of blockchains has much to be unveiled in the near future.

References

1. Baqa, H., Truong, N., Crespi, N., Lee, G., Le Gall, F.: Semantic smart contracts for blockchain-based services in the internet of things. In: 2019 IEEE 18th International Symposium on Network Computing and Applications (NCA). pp. 1–5 (2019). https://doi.org/10.1109/NCA.2019.8935016
2. Bella, G., Cantone, D., Longo, C., Nicolosi-Asmundo, M., Santamaria, D.F.: Blockchains through ontologies: the case study of the ethereum ERC721 standard in OASIS (extended version) (2021). arXiv:2109.02899
3. Bresciani, P., Perini, A., Giorgini, P., Giunchiglia, F., Mylopoulos, J.: Tropos: an agent-oriented software development methodology. Auton. Agents Multi Agent Syst. **8**(3), 203–236 (2004)
4. Cano-Benito, J., Cimmino, A., García-Castro, R.: Towards blockchain and semantic web. In: Abramowicz, W., Corchuelo, R. (eds.) Business Information Systems Workshops, pp. 220–231. Springer International Publishing, Cham (2019)
5. Cantone, D., Longo, C.F., Nicolosi-Asmundo, M., Santamaria, D.F., Santoro, C.: Towards an Ontology-Based Framework for a Behavior-Oriented Integration of the IoT. In: Proceedings of the 20th Workshop From Objects to Agents, 26-28 June, 2019, Parma, Italy, CEUR Workshop Proceeding, vol. 2404. pp. 119–126 (2019)
6. Cantone, D., Longo, C.F., Nicolosi-Asmundo, M., Santamaria, D.F., Santoro, C.: Ontological smart contracts in OASIS: Ontology for agents, systems, and integration of services. In: To app. In: Proceedings of IDC 2021, The 14th International Symposium on Intelligent Distributed Computing, 16–18 September, on-line (2021)
7. Christidis, K., Devetsikiotis, M.: Blockchains and smart contracts for the internet of things. IEEE Access **4**, 2292–2303 (2016)
8. English, M.D., Auer, S., Domingue, J.: Blockchain technologies & the semantic web: a framework for symbiotic development. In: Lehmann, J., Thakkar, H., Halilaj, L., Asmat, R. (eds.) Computer Science Conference for University of Bonn Students. pp. 47–61 (2016)
9. Fill, H.: Applying the concept of knowledge blockchains to ontologies. In: AAAI Spring Symposium: Combining Machine Learning with Knowledge Engineering (2019)
10. Kim, H., Laskowski, M.: Toward an ontology-driven blockchain design for supply-chain provenance. Int. Syst. Acc. Finance Manag. **25**(1), 18–27 (2018)
11. de Kruijff, J., Weigand, H.: Understanding the blockchain using enterprise ontology. In: CAiSE (2017)
12. Pfeffer, J., Beregszazi, A., Li, S.: Ethon - an ethereum ontology (2016). https://ethon.consensys.net/index.html

13. Protocol Labs: The Interplanetary File System (IPFS), https://ipfs.io/
14. Ruta, M., Scioscia, F., Ieva, S., Capurso, G., Pinto, A., Di Sciascio, E.: A blockchain infrastructure for the semantic web of things. In: 26th Italian Symposium on Advanced Database Systems (SEBD 2018) (2018)
15. Szabo, N.: Formalizing and securing relationships on public networks. First Monday **2**(9) (1997)
16. Ugarte Rojas, H.E.: A more pragmatic web 3.0: Linked blockchain data. In: Google Scholar (2017)
17. World Wide Web Consortium: SWRL: A Semantic Web Rule Language Combining OWL and RuleML (2004). http://www.w3.org/Submission/SWRL/
18. Xu, X., Weber, I., Staples, M., Zhu, L., Bosch, J., Bass, L., Pautasso, C., Rimba, P.: A taxonomy of blockchain-based systems for architecture design. In: 2017 IEEE International Conference on Software Architecture (ICSA), pp. 243–252. IEEE (2017). http://design.inf.usi.ch/sites/default/files/biblio/icsa2017-blockchain.pdf

Preselection Based Leader Election in Distributed Systems

Amit Biswas and Anil Kumar Tripathi

Abstract This work presents a self-stabilizing leader election algorithm for arbitrary network topology. Here, we introduce the concept of the provisional leader that helps to improve the system performance. The proposed algorithm identifies r comparatively higher potential leader capable nodes in the system and designates the highest potential node among them as the leader. To reduce the time complexity and the message complexity of the election process, based on the eccentricity of the nodes, we divide a distributed system into two layers (i.e., inner-layer and outer-layer). Only the inner-layer nodes take part to identify the list of potential nodes. We also simulate the proposed algorithm on several arbitrary network topologies and compare the results with existing well-known algorithms.

1 Introduction

The distributed systems [7, 10] provide many advantages, including reliability, high performance, resource sharing, and higher scalability. With the ever-growing technological expansion of the world, distributed systems are becoming popular and widespread. The goal of a distributed system is to provide a high-performance and robust computing environment. In a distributed system, we simultaneously execute the different parts of a task on multiple nodes to complete the task faster. So coordination among these nodes is essential. The overall system performance and efficient system management of a distributed system depend a lot on the coordination of the

A. Biswas (✉) · A. K. Tripathi
Department of Computer Science and Engineering, Indian Institute of Technology (BHU), Varanasi, India
e-mail: amitbiswas.rs.cse17@iitbhu.ac.in

A. K. Tripathi
e-mail: aktripathi.cse@iitbhu.ac.in

D. Camacho et al. (eds.), *Intelligent Distributed Computing XIV*, Studies in Computational Intelligence 1026, https://doi.org/10.1007/978-3-030-96627-0_24

nodes. The concept of leader election [10] helps to efficiently control and coordinate the nodes by electing a node as the system leader, improving the system performance, and decreasing the system management complexity. A distributed system may suffer from inconsistency, inefficiency, and performance degradation in the absence of a leader. So whenever the leader fails, another node must be selected as the leader as soon as possible. If the leader election process takes a long time to select the leader, the system's collaborative works also get halted for a long time. On the other hand, a potential node can manage a system effectively. So a system always expects an election algorithm that takes less time, exchanges fewer messages, and elects a potential node as the system leader.

By considering various aspect of the distributed system, many leader election algorithms [1–3, 11] have been proposed so far. Among these, very few algorithms [4, 5, 8, 9, 11] are designed considering the arbitrary network topology. Gallager et al. and Nicola Santoro proposed two universal protocols known as the Mega-Merger protocol [5, 9] and the Yo-yo protocol [9] respectively for leader election. These two algorithms aim at minimum finding and leader election in a generic connected undirected graph. Vasudevan et al. [11] proposed a weak self-stabilizing algorithm for arbitrary network topology. Electing a node as the system leader in arbitrary network topology is more challenging than in a regular network topology. This work considers all the above-said issues and introduces a self-stabilizing leader election algorithm for arbitrary network topology. The significant contributions of this work are summarized as follows.

- We introduce the primary leader and provisional leader concept to reduce the system performance degradation during the election. Our algorithm not only elects a leader but also makes a list of r best leader capable nodes.
- We reduce the message and time complexity of the election process by dividing a distributed system into two layers (inner-layer and outer-layer) based on the eccentricity of the nodes. We introduce the leader factor concept to identify the potential nodes and elect a good quality leader for the system.
- The proposed algorithm is designed considering an arbitrary network topology, and it satisfies three necessary conditions i.e., termination, uniqueness, and agreement [2, 3, 11] that make the algorithm a self-stabilizing one.

The rest of the manuscript proceeds as follows. Section 2 introduces the leader election problem and the system model. Section 3 presents the proposed leader election algorithm and analyzes its complexity. The experimental analysis and discussion are presented in Sect. 4 and the last section concludes our work.

2 System Model and Problem Definition

We consider a distributed system comprised of n nodes connected through an arbitrary network topology, where n is an integer and $n \geq 2$. Every node has a unique Id and

Table 1 Variables and their meaning

Symbols	Meaning	Symbols	Meaning
NId_i	Node Id of a node i	LId_i	Leader Id stored by a node i
Ect_i	Eccentricity of a node i	Deg_i	Degree of a node i
D	Diameter of the system	R	Radius of the system
PL_i	It is a binary variable. When a provisional leader is selected, a node i sets $PL_i = 1$, otherwise $PL_i = 0$	$PNList_i$	List of potential nodes stored by a node i. A row of this list contains a node Id and its leader factor
Lf_i	Leader factor of node i	ILN	Set of all inner-layer nodes
Lf_c	Leader factor of the election message creator node	$NPNList$	List of new potential nodes of the system
OLN	Set of all outer-layer nodes	W_{in}	Width of the inner-layer of the system
$EmgId$	Election message generator's Id	$CLId$	Crashed leader Id
T_{ind}	A message takes T_{ind} time to travel between two farthest inner layer nodes	T_d	A message takes T_d time unit to travel between two farthest nodes in the system
TL	Type of leader. $TL = 1$ and $TL = 0$ indicate the provisional leader and the primary leader respectively	$NLId$	Newly elected leader Id
Mc_i	Memory capacity of a node i	Pc_i	Processing capacity of a node i

the links that connect the nodes are bidirectional. The nodes communicate through message exchanging, and message exchanging follows an upper bound of time. Every node of the system has a local clock, and the clocks of the nodes need not be synchronized. A graph $\chi = (\Gamma, \Delta)$ represents this system; where Γ is the set of nodes ($|\Gamma| = n$) and Δ is the set of all the links between the nodes. Here, we assume that every node knows its eccentricity. The details of the used variables and symbols in our algorithm are given in Table 1 and all the definitions and function regarding the proposed algorithm are as follows.

Width of the inner layer (W_{in}): Making use of the eccentricity [6] of the nodes, we divide the system into two layers i.e., the inner-layer and the outer-layer. W_{in} is the width of the inner layer, and we represent it as a mathematical function of the system's diameter. In this work, we consider $W_{in} = \lfloor\sqrt{D}\rfloor$.

Inner-layer nodes (ILN): Ect_i is the eccentricity of a node i. If $Ect_i < R + W_{in}$, the node i belongs to the inner-layer of the system.

Outer-layer nodes (OLN): If $Ect_i \geq R + W_{in}$, the node i belongs to the outer-layer of the system.

Leader factor (Lf_i): Depending on this leader factor, we identify the potential nodes in the system. In this work, we consider four properties (Processing capacity (Pc), Memory capacity (Mc), Eccentricity (Ect) and Degree (Deg)) of a node to calculate its Lf_i. A node with higher processing capacity, higher memory, higher degree, and lower eccentricity is considered a higher potential node. We calculate the leader factor of a node using Eq. (1). Here, we assume that every node knows the possible highest and lowest values of every property. If X is a property, then $max(X)$ and $min(X)$ refer to the possible highest and lowest values of X, respectively and w_X is the weight of X. The weight is used to prioritize the properties. Here, $w_{Pc}, w_{Mc}, w_{Deg}, w_{Ect} \geq 0$ and $w_{Pc} + w_{Mc} + w_{Deg} + w_{Ect} = 1$.

$$Lf_i = w_{Pc}\frac{Pc_i - min(Pc)}{max(Pc) - min(Pc)} + w_{Mc}\frac{Mc_i - min(Mc)}{max(Mc) - min(Mc)} + w_{Deg}\frac{Deg_i - min(Deg)}{max(Deg) - min(Deg)} + w_{Ect}\frac{max(Ect) - Ect_i}{max(Ect) - min(Ect)} \quad (1)$$

Type of Messages: In this work, we use three types of messages i.e., leader crashed information message ($Leader\_crash[CLId]$), election message ($Emsg[EmgId, CLId, Lf_c]$), and new-leader declaration message ($New\_leader[NLId, NPNList]$) and one function ($Create\_Emsg()$). When an outer layer node realizes the leader failure, the node creates a $Leader\_crash[CLId]$ and sends it to a nearest inner layer node to inform that the leader has crashed. The $Emsg[EmgId, CLId, Lf_c]$ is used to initiate the election and made the list of potential nodes. The $New\_leader[NLId, TL, NPNList]$ is used to declare the new leader.

$\boldsymbol{Create\_Emsg()}$

1. $EmgId \leftarrow NId_i$
2. $Lf_c \leftarrow Lf_i$
3. Create an $Emsg[EmgId, CLId, Lf_c]$ message, and send it to all the adjacent inner layer nodes.

3 Proposed Leader Election Algorithm

Based on the eccentricity of the nodes, we divide the network into two layers, i.e., inner-layer and outer-layer. Unlike the existing algorithms, instead of electing one node, the proposed algorithm identifies r potential leader capable nodes and designates the highest potential node as the leader. Only the inner-layer nodes participate in identifying the r potential nodes. Later on, when the system leader crashes, rather than running the whole election algorithm, instantly, the system can choose the next leading alive node from the list of potential leader capable nodes as the provisional leader. Afterward, the system elects the primary leader invoking the entire election algorithm as per its convenience. Here leader factor is used to choose the potential nodes. The higher leader factor indicates the higher potential node. The algorithm has three phases i.e., Election initiation, Finding potential nodes, and Leader declaration.

Election initiation: Algorithm 1 describes the election initiation steps. The leader's failure can be detected by the outer-layer or inner-layer node(s). When a

Algorithm 1: Election Initiation

```
// When a node i perceives the leader's failure and initiates the election.
if (the node i realizes the leader's failure) then
    Remove CLId from the PNList_i list.
    if (PNList_i == Empty) then
        if (node i ∈ OLN) then
            Create a Leader_crash[CLId] message, send it to a nearest inner layer node and start timer for 2T_d unit time.
        else
            Create_Emsg().
            Append NId_i and Lf_i into the PNList_i, start the timer for 2T_ind unit time
        end
    else
        Choose the highest potential alive node from the PNList_i list as the provisional system leader, NLId ← the new system leader Id,
        LId_i ← NLId
        TL ← 1, PL_i ← 1, NPNList ← PNList_i, create a New_leader[NLId, TL, NPNList] message and broadcast it over the network.
    end
end
```

node i perceives the leader's failure, it removes the crashed leader's information from its $PNList_i$. Then it checks its $PNList_i$ and performs the following action.

1. If the $PNList_i$ is empty, and the node i is an outer-layer node, the node i creates a $Leader\_crash[CLId]$ message and sends it to the nearest inner-layer node to inform the leader's failure and starts its timer for $2T_d$ time.
2. If the $PNList_i$ is empty, and the node i is an inner layer node or it gets a $Leader\_crash[CLId]$ message, the node i creates an $Emsg[EmgId, CLId, Lf_c]$ message and initiates an election by sending it to all the adjacent inner layer nodes. After that It appends self information (NId_i and Lf_i) into the $PNList_i$ and starts timer for $2T_{ind}$ time.
3. If the $PNList_i$ is not empty, the node i selects the highest potential alive node from the $PNList_i$ list and declares it as the provisional leader by creating and broadcasting the $New\_leader[NLId, TL, NPNList]$ message over the entire network.

Finding potential nodes: This phase explains how the proposed election method identifies the r potential nodes and lists them. Algorithm 2 describes the steps of finding the higher leader factor nodes. Suppose a node j gets a $Emsg[EmgId, CLId, Lf_c]$ message, then it performs the following actions.

1. If a node j gets the election messages for the first time, the node j makes its $PNList_j$ empty. Then it puts the node Id and leader factor received through the message in its $PNList_j$ and sends the received message to all the adjacent inner layer nodes except the message sender node. If node j is not an election initiating node, it places self Id and self leader factor in the $PNList_j$ in such a way so that the entries in the $PNList_j$ get arranged in descending order according to the leader factor. It creates an election message with self information, sends it to all the adjacent inner-layer nodes and starts its timer for $2T_{ind}$ time.
2. When node j gets further election messages, it checks whether it has already got the same election message. If it already got the same election message, it discards the message. Otherwise it checks whether its $PNList_j$ is full. If $PNList_j$ is full, node j compares the Lf_c (received by the message) with the lowest leader factor

Algorithm 2: Potential node finding

// When a node j receives an election messages ($Emsg[EmgId, CLId, Lfc]$).
if *(Node j gets an election messages for the first time)* **then**
  Make the $PNList_j$ empty.
  Put the node Id and leader factor received through the message in the $PNList_j$.
  Send the received message to all the adjacent inner layer nodes except its sender.
  if *(Node j is not an election initiating node)* **then**
    Place the self Id and self leader factor in the $PNList_j$ in such a way so that the $PNList_j$ gets arranged in descending order according to the leader factor.
    $Create\_Emsg()$, start the timer for $2T_{ind}$ unit time.
  end
else
  if *(Node j already got the same election message.)* **then**
    Discard the received election message ($Emsg[EmgId, CLId, Lfc]$)
  else
    if $(PNList_j == Full)$ **then**
      if *(Lfc is less than the lowest leader factor in $PNList_j$)* **then**
        Discard the received election message ($Emsg[EmgId, CLId, Lfc]$)
      else
        Remove the entry with the lowest leader factor from the $PNList_j$.
        Place the node Id and leader factor in the correct position in the $PNList_j$ so that the order gets maintained.
        Send the received election message to all the adjacent inner layer nodes except its sender.
      end
    else
      Place the node Id and leader factor received through the message in the correct position in the $PNList_j$ so that the order gets maintained.
    end
  end
end

in the $PNList_j$. If Lf_c is less than the lowest leader factor in the $PNList_j$, the node j discards the message. Otherwise, it removes the entry with the lowest leader factor from the $PNList_j$, places the node Id and leader factor received through the message in the correct position in the $PNList_j$ so that the order gets maintained. Then it sends the received message to all the adjacent inner layer nodes except the sender of the message. On the other hand, if $PNList_j$ is not full, node j places the node Id and leader factor received through the message in the correct position in the $PNList_j$. In the case of any tie or symmetry, node Id is considered to break the tie or symmetry, and the higher node Id gets the preference.

Leader declaration: In this phase, the inner layer nodes declare the newly elected leader by creating a $New\_leader[NLId, TL, NPNList]$ message and sends it to all other nodes of the system (cf. Algorithm 3).

1. After the timer's timeout, an inner layer node i selects the node with the highest leader factor from its $PNList_i$ as the new system leader. Next, the node i creates a $New\_leader[NLId, TL, NPNList]$ message (where, $NLID$ contains the newly elected leader Id, $TL = 0$ and $NPNList$ contains the r potential node Ids), and sends it to all the adjacent outer layer nodes.
2. When a node j gets a $New\_leader[NLId, TL, NPNList]$ message, it considers $NLId$ as the new leader, copies $NPNList$ into its $PNList_j$ and sends the received message to all the adjacent outer layer nodes. If node j further gets the $New\_leader[NLId, TL, NPNList]$ message it discards the message.

Algorithm 3: Elected leader declaration

```
// After the time out of a node i.
if (timer of node i == timeout) then
    if (node i ∈ ILN) then
        Choose the node with the highest leader factor from the PNList_i as the new system leader, NLId ← the new system leader Id.
        LId_i ← NLId, TL ← 0, NPNList ← PNList_i, create a New_leader[NLId, TL, NPNList] message and send it to all
        the adjacent outer layer nodes.
    else
        if (node i did not get any new-leader declaration message) then
            Re-initiate the election.
        end
    end
end
// Suppose, a node j gets an New_leader[NLId, TL, NPNList].
if (node j gets a New_leader[NLId, TL, NPNList] message for the first time) then
    LId_i ← NLId, PL_j ← TL, PNList_j ← NPNList, sends the New_leader[NLId, TL, NPNList] message to all the
    adjacent outer layer nodes except its sender
else
    Discard the message
end
```

3.1 Complexity Analysis

In this section, we calculate the time and message complexity of our algorithm.

Message complexity: According to the system model, the nodes communicate through message exchanging, so the message complexity of the algorithm depends on the total number of exchanged messages during the election.

Best case: When the primary leader crashes and a node is selected as the provisional leader from the list of potential nodes. It is the best-case scenario of the proposed algorithm. In the best case, only one node realizes the leader's failure and selects the highest potential alive node from the list as the provisional leader, and declares the newly selected leader by creating and broadcasting a new-leader declaration message. Hence, in the best case, the message complexity of the proposed algorithm is $O(|\Delta|)$, where Δ is the set of all network connectivity.

Worst case: When the list of r potential nodes needs to be identified through the election, it becomes the worst case of our algorithm. In this case, at a time, all nodes realize the leader's failure and initiate the election. So the outer layer nodes create $O(|OLN|)$ leader crashed messages. Furthermore, the inner layer nodes exchange $O(r \cdot |E_{in}|)$ (where E_{in} is the set of all network links in the inner layer.) election messages to identify the new r higher potential nodes of the system. After that, the highest prominent node from the list is selected and declared as the new leader. The algorithm exchanges $O(|E_{out}|)$ (where E_{out} is the set of all network links of the outer-layer.) messages for the declaration of the new leader. Hence, in the worst case, the message complexity is $O(|OLN| + r \cdot |E_{in}| + |E_{out}|)$.

Time complexity: In leader election, time complexity refers to the time steps required to elect a new leader. Here, we assume that a message takes c unit time to travel from one node to its adjacent node, where c is a constant.

Best Case: In the best case, the algorithm takes a maximum of $D \cdot c$ unit time to select and declare the provisional leader. Hence, the time complexity is $O(D)$.

Worst case: In the worst case, the algorithm takes a maximum of $2D \cdot c$ unit time to identify the r higher potential nodes and declare the highest potential node as the system leader. Hence, in this case, the time complexity is also $O(D)$.

3.2 Proof of Termination, Uniqueness, and Agreement Conditions

In this section, we show that our algorithm satisfies termination, uniqueness, and agreement conditions.

Termination: The propounded algorithm takes a maximum of $2D$ hops time to select the r most prominent nodes and elects a leader. According to the system model, a message gets exchanged between two nodes within a bounded time. Suppose a message takes c unit of time to get exchanged from one node to its adjacent node, where c is constant. After the time out of the timer, a node gets to know about the selected list of r number of potential nodes and the newly elected leader. So a node terminates the execution of the election process by $2 \cdot D \cdot c$ unit of time. The system consists of a finite number of nodes (n), so the diameter D of the network is also finite. Hence the proposed algorithm gets terminated in a finite time and elects a leader for the system.

Uniqueness: When an election is conducted to identify the r potential inner layer nodes, the election messages created by the r nodes with the comparatively higher leader factors get broadcast in the inner layer of the network. Thus all the inner layer nodes get to know about the r nodes with comparatively higher leader factors and elect the highest leader factor node as the leader. If every inner layer node has a distinct leader factor, then the node with the highest leader factor among them is elected. On the other hand, if multiple nodes have the highest leader factor, then the node Id is used to break the symmetry, and the node with the highest Id among them gets the preference to become the system leader. Every node has a unique Id, so the algorithm always selects only one node as the system leader.

Agreement: During the election, the election messages created by the r nodes with the comparatively higher leader factors get broadcast in the inner layer of the network. Thus all the inner layer nodes get to know about the r nodes with comparatively higher leader factors and the elected leader. The last phase of our algorithm is the leader declaration phase. In this phase, the leader declaration message informs all the outer layer nodes about the newly elected leader. So at the end of the election, all the nodes must know about the elected system leader.

Table 2 Details of the different arbitrary networks considered for the simulation

Sl. no.	Total no. of nodes	No. of ILN	No. of OLN	Total no. of edges	Diameter
Network 1	12	6	6	15	6
Network 2	20	8	12	29	8
Network 3	40	14	26	56	12
Network 4	60	21	39	81	17
Network 5	80	28	52	106	21
Network 6	100	32	68	127	25

4 Experimental Analysis and Discussion

The network layering concept of the proposed election method helps to reduce the message complexity and time complexity of the election process. Instead of all the nodes, only the inner layer's nodes directly participate in the election in identifying the list of potential nodes. That is why during the election, the required number of exchanged messages and time steps get reduced. The existing algorithms elect one node as the leader, whereas the proposed algorithm identifies the list of r most potential leader capable nodes and designates the highest potential node as the primary leader. Generally, when the system leader crashes, for the lack of a coordinator, the coordination-based tasks of the system get halted until a new leader is elected. According to the proposed election method, whenever a primary leader fails, instantly, the system can choose a provisional leader from the list so that due to the lack of coordination, the system's work does not get halted for a long time. This provisional leader controls and coordinates the system until the primary leader is being elected by the next election. The provisional leader makes the system more flexible because the system can wait for a suitable time to conduct a new election without hampering the system's performance. Thus the identification of r potential nodes helps to improve the overall system performance. Identification of the potential nodes and selecting one of them as the leader is made based on the leader factor. The leader factor of a node is calculated considering the processing capacity, memory capacity, eccentricity, and degree of the node. In the proposed method, a node gets elected as the leader with comparatively higher processing capacity, higher memory capacity, higher degree, and lower eccentricity. A higher processing and memory capacity leader can perform its leadership tasks faster, which improves the overall system performance. On the other hand, the leader's higher degree and lower eccentricity reduce network traffic and help in faster communication between the leader and other nodes. In the proposed method, the width of the inner layer (W_{in}) is changeable. It is favorable to keep the size of the inner layer small, but if required, we can change the W_{in} by taking a different mathematical function of D so that the election algorithm can always select a good quality leader.

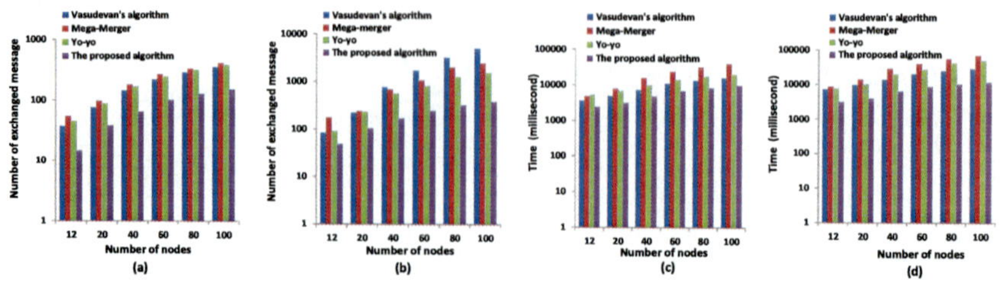

Fig. 1 **a** and **b** represent the number of exchanged messages in the best case and worst case, respectively. **c** and **d** represent the required time in the best case and worst case, respectively

We simulate the proposed leader election algorithm on six different arbitrary network topologies. We get to know the number of exchanged messages and time steps to elect the leader in the best-case and worst-case scenarios through this simulation. These arbitrary networks were constructed by the different number of nodes and links. In our experiment, we consider $r = 5$. The networks details are given in Table 2 and the simulation results are shown in Fig. 1. All experiments were performed on a single machine. The machine was decorated with Intel core(TM)i5-2410M processor (2.3 GHz, 4 MB cache, 2.9 GHz Turbo Boost), 8 GB RAM, 1 TB HDD, NVIDIA GeForce graphics, running Ubuntu Linux Release 16.04 (xenial kernel 4.4). We used *python 3.6* as the programming language and *mpi4py* as the message passing interface (MPI).

We compare the proposed algorithm with three well-known and highly cited leader election algorithms known as Mega-Merger [5, 9], Yo-yo [9], and Vasudevan's algorithm [11]. This comparison is done in terms of the requirement of the total exchanged messages and total time to complete the election process, and shown in Fig. 1. Figure 1a and b show the number of exchanged messages in the best case and worst case, respectively. These two graphs show that the proposed algorithm exchanged fewer messages than Mega-Merger, Yo-yo, and Vasudevan's algorithm to elect the leader. On the other hand, Fig. 1c and d show the time to elect the leader in the best case and worst case scenario, respectively. Here it is clear that the proposed algorithm takes less time than Mega-Merger, Yo-yo, and Vasudevan's algorithm to elect the leader. After the simulation, we get to know a limitation of the proposed algorithm. In some cases, it may happen that after the primary leader crashes, the next chosen provisional leader from the list might be slightly inferior because of the loss associated with the primary leader crash in the form of links failure.

5 Conclusion

This paper presented a new self-stabilizing leader election method where the concept of a provisional leader is introduced to reduce system performance degradation during the election. Instead of electing a single leader, the algorithm identifies a list of potential leader capable nodes and initially designate the best one from the list as the primary leader. After that, when the primary system leader crashes, the highest prominent node of the list is instantly selected as the provisional leader. This provisional leader controls and coordinates the system until the next election is electing the primary leader. Thus the provisional leader makes the system more flexible because the system can wait for a suitable time to conduct a new election without hampering system performance. The concept of layering the system and electing the leader based on the leader factor help to reduce the required number of exchanged messages and time steps to elect a good quality leader for the system.

References

1. Biswas, A., Dutta, A.: A timer based leader election algorithm. In: 2016 International IEEE Conferences on Ubiquitous Intelligence & Computing, Advanced and Trusted Computing, Scalable Computing and Communications, Cloud and Big Data Computing, Internet of People, and Smart World Congress (UIC/ATC/ScalCom/CBDCom/IoP/SmartWorld), pp. 432–439. IEEE (2016)
2. Biswas, A., Maurya, A.K., Tripathi, A.K., Aknine, S.: Frlle: a failure rate and load-based leader election algorithm for a bidirectional ring in distributed systems. J. Supercomput. **77**(1), 751–779 (2021)
3. Biswas, A., Tripathi, A.K., Aknine, S.: Lea-tn: leader election algorithm considering node and link failures in a torus network. J. Supercomput. 1–38 (2021)
4. Dagdeviren, O., Erciyes, K.: A hierarchical leader election protocol for mobile ad hoc networks. In: International Conference on Computational Science, pp. 509–518. Springer, Berlin (2008)
5. Gallager, R.G., Humblet, P.A., Spira, P.M.: A distributed algorithm for minimum-weight spanning trees. ACM Trans. Prog. Lang. Syst. (TOPLAS) **5**(1), 66–77 (1983)
6. Krnc, M., Sereni, J.S., Škrekovski, R., Yilma, Z.B.: Eccentricity of networks with structural constraints. Discussiones Mathematicae Graph Theory (2018)
7. Maurya, A.K., Tripathi, D., Biswas, A., Tripathi, A.K.: Design issues in distributed software. In: 2018 Fifth International Conference on Parallel, Distributed and Grid Computing (PDGC), pp. 563–567. IEEE (2018)
8. Raychoudhury, V., Cao, J., Niyogi, R., Wu, W., Lai, Y.: Top k-leader election in mobile ad hoc networks. Pervasive Mobile Comput. **13**, 181–202 (2014)
9. Santoro, N.: Design and Analysis of Distributed Algorithms, vol. 56. Wiley, New York (2006)
10. Tanenbaum, A.S., Van Steen, M.: Distributed Systems: Principles and Paradigms. Prentice-Hall, Hoboken (2007)
11. Vasudevan, S., Kurose, J., Towsley, D.: Design and analysis of a leader election algorithm for mobile ad hoc networks. In: Proceedings of the 12th IEEE International Conference on Network Protocols, 2004. ICNP 2004, pp. 350–360. IEEE (2004)

Physical Assets Tokenization for Blockchain Market

Giuseppe Ferrara, Fabrizio Messina, Massimiliano De Benedetti, and Corrado Santoro

Abstract Asset tokenization is a technique widely adopted in blockchain-based platforms that enables vendors to sell their assets in competitive markets characterized by direct participation and accessibility. Digital assets are natively based on information as, for instance, licenses, tickets, certificates or financial products. As a consequence, in order to perform asset tokenization, all the information must be previously stored, such that each single asset can be correctly identified. Tokens of digital assets are indeed defined as digital objects in which the relevant information are stored to guarantee a single, unique matching between asset and token. The benefit of this approach is represented by the public matching verification. In case of physical assets related to physical goods, the store and verification of all the relevant information to bind each asset to a specific token is a complex process. Moreover, unlike digital assets which only operate on digital domain, physical assets are subject to the regulation of the country where they are located and this asymmetry limits its use. Starting from this scenario, this work presents an architecture to support physical assets tokenization for digital markets. This is achieved through the definition of a process architecture with a specific trade-off between off-chain and on-chain tokenization.

G. Ferrara (✉) · F. Messina · C. Santoro
Department of DMI, University of Catania, Catania (CT), Italy
e-mail: giuseppe.ferrara@phd.unict.it

F. Messina
e-mail: messina@dmi.unict.it

C. Santoro
e-mail: santoro@dmi.unict.it

M. De Benedetti
Department of DEIB, Politecnico di Milan, Milano (MI), Italy
e-mail: massimilianomaurizio.debenedetti@polimi.it

D. Camacho et al. (eds.), *Intelligent Distributed Computing XIV*, Studies in Computational Intelligence 1026, https://doi.org/10.1007/978-3-030-96627-0_25

1 Introduction

Due to their nature, remote markets are able to offer 24/7 services to a wider audience than traditional physical markets—also known as "brick and mortar" market—that require in-place presence of buyers and sellers. Hence, digital markets represent an attractive alternative to the traditional ones. On the other hand, physical markets allow potential buyers to verify the assets to purchase, while on digital markets they have to trust a third party who provides some guarantees. This is why users still prefer to buy on brick and mortal market instead of e-commerce [1, 26].

The growth of e-commerce penetration in U.S., who is continuously increasing since its appearance, has boosted due to COVID-19 pandemic by doubling in just 8-weeks as much as the last 10 years, passing from 16% of 2019 to 27% of Q1 2020. Moreover, it's estimated that 95% of purchases will be facilitated by e-commerce by 2040 [8]. Since digital market operate by exchanging coded information, they perfectly fits on information based assets such as digital licenses, tickets or financial services but, on the the other hand, it cannot natively manage material assets, such as chickens, diamonds or sacks of wheat.

In order to extend the advantages of the remote digital market to the physical assets, two main processes have to be taken in consideration: the (i) tokenization process and the (ii) legitimation of the ownership. Tokenization is the process needed to make material assets available on digital market. It is obtained through the creation of a new digital entity called token, which is the representation of the physical asset. In order to work properly, a token must contain the description of all the relevant features that are useful to identify the physical asset. The tokenization process is, in principle, a simple and intuitive process, however the identification and the description of all the characteristics needed in order to identify a physical asset can actually be a very complex process which is subjected to the same challenge of the authentication problem, i.e. recognizing the right entity. Although physical asset tokenization is an high debated problem [13] due to his economic implications, according to [23] *users need to trust the correctness and accuracy of the operating algorithms and the provision of information about the asset by trusted third parties is still an important prerequisite.*

A Legitimate token is a token recognized by the market as the true single representation that detains the full ownership of the physical asset on the remote market [6, 20]. By taking its place on the digital market, a legitimate token can be traded on the digital market to transfer the ownership of the physical asset from the seller to the buyer (like the physical asset on the physical market), and this is the main feature of this kind of token. In this scenario, for each transaction of a legitimate token, a form of guarantee must be transferred from the seller to the buyer, in this way a trusted trading transactions of legitimate tokens can take in place. The main problem is related to the methodology/technique or algorithm used to guarantee the unique correspondence with the digital representation. Indeed the token and the physical asset must be associated with a specific approach, once that the related legitimate token is on sale on the digital market. Therefore, an important question arises in this approach: how can token correspondence and legitimacy be ensured? An additional level of confidence is therefore needed to enable physical assets trading on remote digital markets. In order to give an answer to the above question, in this work we present an architecture capable to enable the sales of physical assets into remote market with the lowest possible engagement of a third party. This *hybrid* architecture combines some promising features from different existing approaches. In particular, we first consider the state of the art of both in-presence and remote market in terms of threat and opportunity, then we analyze the current available approaches by pointing out the advantages and disadvantages. As we discuss later in this work, the hybrid solution here presented can ensure physical asset binding with the digital token counterpart by leaving in-between peer-to-peer trading work without any possible intermediary interference. The paper is structured as follow: Sect. 2 provides an overview of the related work. Section 3 reports the state of the art in terms of available features of the existing approaches, Sect. 4 describes the contribution of this paper by introducing a new hybrid architecture representing the best trade-off between reputation and distributed reply systems. Finally in Sect. 5 we provides the conclusions by highlighting benefits and limitation about the proposed solution, as well as future works.

2 Related Work

A common approach to the problem of assets tokenization is the label based solution [14, 17]. It consists on the application of a label with a printed content, such as a QR-code, that represents a "link" to the token itself [9]. However, since labels can be moved or cloned, this approach is only useful to guarantee the existence of the token; it does not provide any kind of guarantee about the association of the tangible asset represented by the token. There are a few more sophisticated application that use tamper-proof adhesives labels [5]. This type of labels have the characteristic to being irreparably damaged once removed, so that they cannot be removed without compromising the label itself. A more advanced application of this approach use NFC tags or RFID labels instead of printed code to achieve better anti-forgery feature [3], that can be considered as tamper-proof labels that can solve the relabeling problem. However, the production of these labels is more complex than a simple print of a QR-code and requires specific equipment that cannot be purchased and used by all the potential sellers, that must rely on an third party (intermediary) for the production of these labels. Furthermore, these labels do not solve the multiple labeling issue, and, although more sophisticated, they do not even totally solve the problem of being cloned by third parties without rely on on a third party certificate authority. Another approach to face this problem is to rely on IoT devices to read physical features and AI as decision engine to avoid hacks [25] but, since IoT devices cannot provide high collateral value as guarantee of their behaviour, this approach doesn't work for valuable assets. Other studies [15] presented solutions on this field of research about how to quantify, map and authenticate real or virtual asset while concluding that "trading and exchanging of Asset-backed token is still open question". With this work we want to put the foundation stone on this open question.

3 State of the Art

Let's suppose to assign a token to an artwork such as a painting or a sculpture. We could describe all the physical characteristics, such as dimensions, weight, volume, take several photographs [16] and we could even build an accurate 3D model [24]. Nevertheless the falsification of the original artwork is still possible, especially in case of a high valuable ones.

The example related to the artwork tokenization explains the need to put an additional level of confidence in order to sell physical assets through remote markets, for example by means of a trusted third party, i.e. an intermediary. This approach is called reputation-based. On the other hand the *best trade* is obtained when no interference occurs between the seller and the buyer. Therefore the best scenario is known as peer-to-peer (P2P) [10, 22], including no intermediaries. A typical example of this scenario is represented by the crypto-currencies, such as bitcoin [18]. In the following of this section we discuss the two different approaches, providing a brief comparison among them, in order to introduce our approach. In order to work properly, a reputation-based system is subject to the following conditions [19]: (i) the value of the reputation used as collateral is more than the value of transaction; (ii) it must be always possible to publicly demonstrate any incorrect behavior of the guarantor; in this case it must be possible to compromise the reputation of the guarantor; (iii) the guarantor gets enough incentive to provide the service with profit; (iv) all the information required to complete the transaction are shared with the guarantor.

For convenience we call the alternative architecture—the architecture based on blockchain [7] with no intermediaries—*Distributed Replay*. In this case the involved agents sign and spread the transactions among a network where each other agent is responsible to maintain a personal copy of the same shared ledger. Once received, the transaction terms are locally verified and stored within each local copy of the shared ledger. In order to work properly, each agent must behave in the same way so that each local copy of the ledger can be synchronized with the others. Distributed Replay takes its name from the fact that all involved nodes locally replay the same operations in order to verify and store every incoming transaction. Compared with the reputation based systems, the distributed replay architecture provide a new form of guarantee which can be trusted by all the other agents/nodes without the involvement of other intermediaries. This means that a replay architecture leverages only on the *health* of the network itself (size, decentralization and effort exposed to maintain it).

In order to work properly, this approach is subject to the following conditions: (i) the shared ledger is maintained by a relative majority of honest agents [2]; (ii) the community gets enough incentive to store and verify each transaction within the shared ledger; (iii) the information required for the transaction are shared among all agents of the network. Now we compare blockchain transactions that rely on

Table 1 On-chain versus Off-chain transactions

Parameters	Transaction type	
	On-chain	*Off-chain*
Speed	Low	High
Trust	High	Low

distributed replay systems, also known as on-chain transactions, with transaction that rely on reputation systems, also known as off-chain transaction [12]. Once signed, a blockchain transaction is sent to all the miners in order to perform validation, then it is *packed* together with all the other pending transactions to create the new last block. This process is highly redundant because there is no trusted part who indicates the right final state, thus each miner must run the same set of instructions at the same time to obtaining the same result. Although recent solutions have shown great improvements [4], this process is known to be somehow mandatory to ensure integrity within a trustless environment. This is the reason why transactions on public blockchains are expensive and needs to be optimized, as much as possible, in terms of stored data and computed data. One of the main approach in order to minimize the effort required by the transaction validation on a public blockchain, is based mainly on keeping only on the essential data on-chain while leave the non-essential information *off-chain* (but still linked with the essential one stored on-chain). This idea is the base of "layer-2" solutions like lightning network on Bitcoin [18], or rollups on Ethereum [11]. Indeed, while on-chain transactions are slow and expensive due to the trustless environment used, off-chain transaction are faster and cheaper because they don't need to achieve distributed consensus and can use legacy central technologies such as reputation systems. On the other hand, off-chain transaction are quite unsuitable for P2P purpose due to the fact that they lack of non-repudiation assurance. These different characteristics are reported in Table 1. In order to put physical asset on sale through the digital market, the first step consists in creating his digital representation. It is therefore necessary to identify the most suitable agent for this job, i.e. the one that has no advantage from interfering although able to do so. For obvious reasons neither of seller and buyer are good candidates as guarantor of the one-to-one association between token and physical asset, we can thus infer that at least one additional agent (or third party) must be introduced for this particular purpose. Based on the premises above, the only form of architecture directly applicable to grant binding association through physical asset and his corresponding token is the one based on reputation, and that this requires the involvement of an auxiliary trusted intermediary. The reputation system is the best compromise in order to achieve our goal and limit, at the same time, as much as possible the potential interference of the intermediary. In the next section we deal with this aspect by providing a hybrid solution that represents, in our view, a good compromise between distributed re-play and reputation-based systems to solve the problem we have set above.

4 Proposed Approach

The solution we present in this section has been designed to satisfy the main requirements discussed in the previous section: (i) limiting, as much as possible, the interference of any third party and (ii) to guarantee the minimum level of confidence on asset-to-token binding.

To this end we have performed an analysis of the tokenization process for a physical asset. After the analysis we designed a specific solution organized into three phases: phase 1 and phase 3, that represent the beginning and the end of the process itself, are designed to be reputation-based, while phase no. 2 leverages on distributed replay. As we discuss later in this section, in our approach the intermediary's domain is limited to those activities related to the transaction, acting as a provider of a fiduciary and logistic service. For the remaining of this section we denote, for convenience, as WT (*Warehouse-Tokenizator*) the intermediary of the transaction whose role is to make business by providing token legitimacy as exclusive representative of the ownership of the physical asset. In our approach WT will be responsible of safeguarding the physical asset until the token is back to him (after that the asset has been sold), or the agreed terms have expired. In this way, as we discuss later in this section, the token can work as a proof of asset availability. We also observe that, unlike most of the digital assets, such as cryptocurrencies, each physical asset is unique, i.e. non-fungible [21]. For this reason, the tokenization process of a physical asset must also produce non-fungible token. The schema of the proposed solution is depicted in Fig. 1, and detailed in Sects. 4.1, 4.2, and 4.3.

4.1 Phase 1

In the first phase (Fig. 1), the vendor agrees a set of terms of the service and, thereafter, it moves the asset into the WT location. Once the asset has been received, the WT performs a check of all its characteristics and, if they are compliant with the agreed terms, it mints a unique digital token. The ownership of the token is set to the identity of the vendor. Now the vendor can use the new digital token in place of the asset.

The key aspect, in this phase, is represented by the fact that the vendor trusts the WT, which acts as "safeguard" of the asset in its own warehouse. In this scenario, the vendor trusts the WT thanks to the reputation of the WT, that, in turn, assumes a crucial role in the transaction. Since the main interest of the WT is to make business and, in turn, to not incur in loss of reputation, it is highly desirable that the WT reputation has a value comparable with the asset value. This mechanism should guarantee to legitimate the digital token as exclusive representative of the ownership of the tangible asset.

Moreover, by moving the physical asset into the WT warehouse, the vendor agrees with the WT a set of terms for asset storage and tokenization. Anyway, while the physical movement of deposit is achieved off-chain for obvious reasons, the agree-

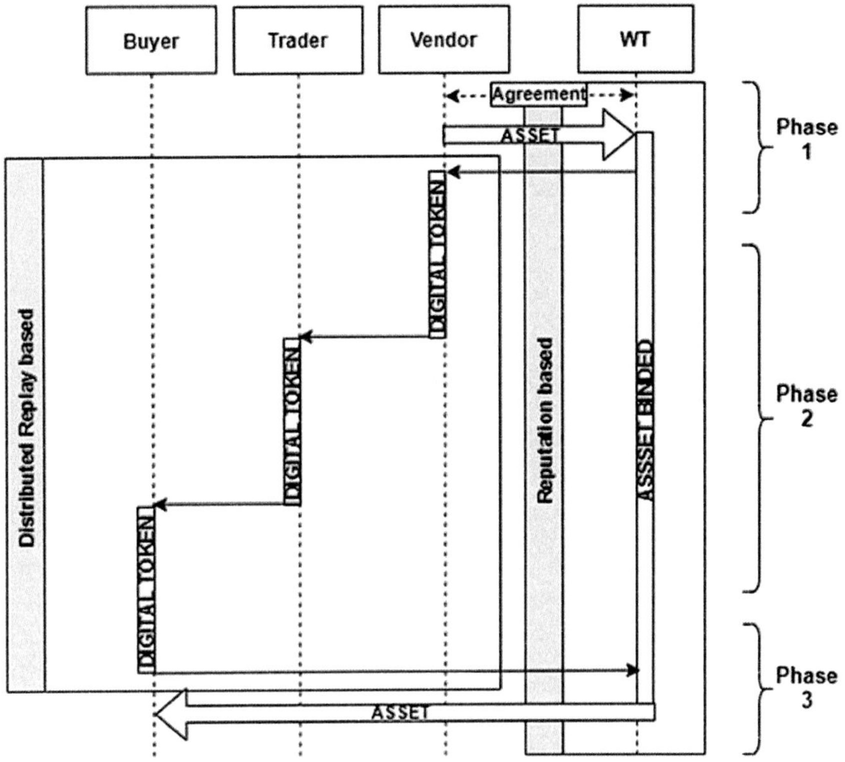

Fig. 1 The three phases of the hybrid solution with the application domains

ment can be made in either ways on-chain or off-chain as long as vendor can publicly leverage on the reputation of the WT.

Feedback. This phase includes the exchange of two different feedbacks. The former is left by the vendor to the WT once the vendor itself ha received a digital token that is compliant with the agreed terms. This first feedback will be auto-generated, because the compliance of the digital token can be checked automatically. If there is a complete match between agreement terms and token characteristics, a positive feedback will be released. In this manner the vendor cannot act in a dishonest manner by leaving a negative feedback to spoil the reputation of the WT. Finally the vendor must pay the WT for the service (the digital token) and, as a consequence, the WT will (automatically) leave a (positive) feedback once the payment has been performed. Moreover, in order to keep the record of the feedback history and apply these rules with transparency, this solution is applicable only in case the agreement is on-chain, i.e. managed by a smart contract on a blockchain that leverage on distributed replay system.

4.2 Phase 2

In the second phase (Fig. 1) the vendor is legitimated to trade the digital token obtained during the previous phase in the digital market by leveraging on WT reputation collateral. In this case it will leverage on a distributed replay system, i.e. excluding WT and any other intermediary from the the trading phase. In other words, phase two is a pure peer-to-peer trading process, without intermediaries.

4.3 Phase 3

In phase no. 3 the buyer, who has acquired the token on the digital market, must exchange the token with the its corresponding physical asset (Fig. 1). In order to achieve this final task, since the buyer was not known at the time of the tokenization agreement, the transferring of the physical asset to the buyer location must be organized and performed by the buyer itself. Once token ownership is set to his minter, buyer is automatically qualified to claim the physical asset and to leave a feedback to WT about the token received and feed his reputation. Moreover, buyer can always give a proof of being the last owner because the entire history of the token is publicly available and verifiable on the reference blockchain that is an append-only data structure. Moreover, once the token ownership came back to the minter (the WT), the token-asset binding is lost and token will be automatically destroyed.

Feedback. In this phase other two feedback are exchanged. The former is released by the WT to the buyer once the token has been sents back to him by the buyer itself. Also this (positive) feedback is automatically generated. The second feedback is released by the buyer to the WT once the buyer itself has acquired the asset. Since the process of transferring the asset is performed off-chain, this feedback cannot be generated automatically. Therefore, the buyer must send the feedback to the WT manually. In this case, in order to avoid malicious behaviours of buyers, aimed at spoiling the reputation of the WT with negative feedbacks associated with low-valued transactions, the feedback must be weighted with the value of the asset and and the reputation of the buyer itself.

5 Conclusions

In this work we have presented an architecture to enable peer-to-peer physical asset trading. The proposed solution is based on the combined usage of a reputation system and the distributed replay. By leveraging on reputation-based mechanisms the architecture ensures that the digital representation is always bonded with the physical asset, i.e. it is legitimate, while indeed, by leveraging on distributed replay it keeps the digital representation valid for p2p trading on blockchain. We have shown how a reputation system part is needed for the physical asset tokenization and how it can be coupled with p2p system in order to enable the assets trading without third party interference. We have discussed the details of this architecture and how it can avoid some of the issues that afflict monolithic solutions. We also discussed the most relevant risks related to the use of this system and we described how these risks could be mitigated through the proposed approach. As future work we aim to implement a prototype of the proposed architecture, in order to identify the most relevant applications that can take benefit of the proposed solution. To this end we will work on a specific case study such that the approach can be evaluated experimentally.

Acknowledgements This work has been supported by IDC 2021.

References

1. Agency, G.A.: The growth of ecommerce sales in the us
2. Alsinet, T., Puyol-Gruart, J., Torras, C.: Artificial intelligence research and development. In: Proceedings of the 11th International Conference of the Catalan Association for Artificial Intelligence, vol. 184. IOs Press (2008)
3. Alzahrani, N., Bulusu, N.: Securing pharmaceutical and high-value products against tag reapplication attacks using nfc tags. In: 2016 IEEE International Conference on Smart Computing (SMARTCOMP), pp. 1–6. IEEE (2016)
4. Ben-Sasson, E., Bentov, I., Horesh, Y., Riabzev, M.: Scalable, transparent, and post-quantum secure computational integrity. IACR Cryptol. ePrint Arch. **2018**, 46 (2018)
5. Olivier, D., Chiffelle, J., Casey, E., Bourquenoud, J.: Tamperproof timestamped provenance ledger using blockchain technology. Foren. Sci. Int.: Dig. Inv. **33**, 300977 (2020)
6. Buterin, V.: The most important scarce resource is legitimacy (2021)
7. Cachin, C., Vukolić, M.: Blockchain consensus protocols in the wild (2017)
8. Coppola, D.: E-commerce share of uk retail sales 2019-2024 (2021)
9. Thomasson, E.: Carrefour says blockchain tracking boosting sales of some products (2019)
10. Einav, L., Farronato, C., Levin, J.: Peer-to-peer markets. Ann. Rev. Econ. **8**(1), 615–635 (2016)
11. 2021
12. Gudgeon, L., Moreno-Sanchez, P., Roos, S., McCorry, P., Gervais, A.: Sok: Off the chain transactions. IACR Cryptol. ePrint Arch. **2019**, 360 (2019)
13. Ross, O., Jensen, J.: Assets under tokenization: Can blockchain technology improve post-trade processing? In: ICIS 2019 Proceedings (2019)
14. Kim, M., Hilton, B., Burks, Z., Reyes, J.: Integrating blockchain, smart contract-tokens, and iot to design a food traceability solution. In: 2018 IEEE 9th Annual Information Technology, Electronics and Mobile Communication Conference (IEMCON), pp. 335–340. IEEE (2018)
15. Li, X., Wu, X., Pei, X., Yao, Z.:. Tokenization: Open asset protocol on blockchain. In: IEEE 2nd International Conference on Information and Computer Technologies (ICICT), pp. 204–209. IEEE (2019)
16. Magnus: Shazam for art
17. Ahmed, S., Broek, N.: Blockchain could boost food security. Nature **550**(43) (2017)
18. Nakamoto, S.: Bitcoin: a peer-to-peer electronic cash system (2009)
19. Ooi, B.C., Liau, C.Y., Tan, K.-L.: Managing trust in peer-to-peer systems using reputation-based techniques. In: International Conference on Web-Age Information Management, pp. 2–12. Springer (2003)
20. Pontikes, E.G.: Two sides of the same coin: How ambiguous classification affects multiple audiences' evaluations. Admin. Sci. Q. **57**(1), 81–118 (2012)

21. Regner, F., Urbach, N., Schweizer, A.: Nfts in practice–non-fungible tokens as core component of a blockchain-based event ticketing application (2019)
22. Schoder, D., Fischbach, K.: Peer-to-peer prospects. Commun. ACM **46**(2), 27–29 (2003)
23. Notheisen, B., Cholewa, J.B., Shanmugam, A.P.: Trading real-world assets on blockchain. Bus. & Inf. Syst. Eng. **59**(6), 425–440 (2017)
24. Stanco, F., Battiato, S., Gallo, G.: Digital Imaging for Cultural Heritage Preservation: Analysis, Restoration, and Reconstruction of Ancient Artworks. CRC Press, Boca Raton (2017)
25. Weingärtner, T.: Tokenization of physical assets and the impact of iot and ai. In: European Union Blockchain Observatory and Forum, vol. 10, pp. 1–16 (2019)
26. Ysabel, K.: E-commerce vs. brick & mortar: the state of retail in 2021

MAS & ALGORITHMS (Main Track)—Chair: Fabrizio Messina

Intelligent Agents for Diffused Cyber-Physical Museums

Angelo Ambrisi, Rocco Aversa, Dario Branco, Massimo Ficco, Salvatore Venticinque, Giuseppina Renda, and Sabrina Mataluna

The Cleopatra project investigates cyber-physical systems, where the social interaction among human users and intelligent agents leverages a collaborative fruition of cultural contents and distributed services. The social interaction is distributed among physical and virtual locations. Conversational and Proactive Agents communicate and interact with users in a credible human-like manner, as well as can simply deliver cultural information on demand, carrying out interactive storytelling. In particular, Proactive Agents implement gateways between multi-user conversations, as well as the diadic interaction with conversational agents, which uses the Natural Language Processing technology to provide cyber-intelligence to the sys-

A. Ambrisi · R. Aversa · D. Branco (✉) · M. Ficco · S. Venticinque
Department of Engineering, University of Campania Luigi Vanvitelli, Caserta, Italy
e-mail: dario.branco@unicampania.it

A. Ambrisi
e-mail: angelo.ambrisi@unicampania.it

R. Aversa
e-mail: rocco.aversa@unicampania.it

M. Ficco
e-mail: massimo.ficco@unicampania.it

S. Venticinque
e-mail: salvatore.venticinque@unicampania.it

G. Renda · S. Mataluna
Dipartimento di Lettere e Beni Culturali, University of Campania Luigi Vanvitelli, Caserta, Italy
e-mail: giuseppina.renda@unicampania.it

S. Mataluna
e-mail: sabrina.mataluna@unicampania.it

D. Camacho et al. (eds.), *Intelligent Distributed Computing XIV*, Studies in Computational Intelligence 1026, https://doi.org/10.1007/978-3-030-96627-0_26

tem. A 'Diffused Museum' case study is implemented to demonstrate the proposed model that integrates digital collection and live tours of archaeological sites.

1 Introduction

The common experiences in the field of the cultural heritage is usually limited to the most known cultural sites. Because of time constraints, long distances and lack of communication, it cuts off minor attractions, which may be very relevant even if considered secondary locations out of touristic itineraries. This issue affects both the cultural experience and the self-sustaining capability of minor museums or archaeological sites. Innovative technologies can be used to allow visitors to share their cultural experience interacting in real time with other users, exchanging pictures, videos and opinions, for example, while visiting the same wide archaeological sites or different museums (that are geographically distributed). The visitors may even be connected from home enjoying virtually a live experience of other visitors. Moreover, intelligent agents can foster the delivery of semantically related content from secondary sites or digital libraries to augment the cultural value and to recommend alternative itineraries. In such scenarios, the dissemination of knowledge, based on the digital exchange of cultural contents among museums and cultural sites could be also regulated by well defined innovative business models. Collaborative models and gamification based on well known formats, such as escape room or Easter eggs, can be designed to foster the user participation. Finally, the same approach, implemented at a larger scale, could be used to build one continuously growing diffused museum with user-generated contents consisting of self-made pictures of accidental discovery and share of private collections.

The Cleopatra project aims at building a cyber-physical system, where the social interaction among human users and intelligent agents leverages a collaborative fruition of contents and distributed services. The 'Diffused Museum' represents the project case study in the field of the cultural heritage application domain. The users are distributed among physical and virtual distributed locations and can interact remotely with other human users and intelligent agents, using multi-modal interaction models (voice, video, text, voice) and legacy social channels and applications. Roles of intelligent agents may be different. Specifically, they can deliver cultural information on demand, leverage the collaboration stimulating the users, carry out interactive storytelling in a multi-user environment, control Internet of Things devices (such as cameras or drones), and try to foster an emergent behaviour conditioning the social interaction according to specific optimization criteria.

2 Related Work

Virtual guides represent a widely investigated research topic, that includes the deployment of avatars in virtual heritage. Avatars help users to navigate the virtual world that either preserves a given culture in a traditional form of a museum or re-creates the actual site as a virtual environment. In particular, intelligent conversational agents have been used in different application contexts to promote the cultural heritage, such as smart cities, archaeological sites and museums [6]. In [2] a computer-generated intelligent agent represented with human-like body can interact with other agents within a virtual environment and with users, using speech, facial expressions, and gestures. Users can interact with the agent to inquire about requested actions, virtual objects they should manipulate, and assistance. Also Cari et al. [3] presents a framework of rules and best practices for developing of the virtual environment by autonomous agents and multi-agent systems needed for supporting cultural heritage applications. By combining gamification and artificial intelligence, sub-domains are identified that can interact to automatically create interactive personalized virtual environments. In [5] an agent is employed as guide in a public computer museum, where he engages with visitors in natural face-to-face communication, providing them with information about the museum or the exhibition, and conducts natural small talk conversations. Nevertheless of all widely used technologies support only a dyadic interaction between human user and agent [11]. Our contribution aims at developing agents, which are able to tell a story while they interact with multiple users that control in such a way their experience. It means that the distributed system must guarantee both coherence, that is the user's ability to comprehend the relationships between the events in the story, and the degree of engagement by the user herself. On one hand, interactive storytelling concerns the management of interaction of human users with computer-controlled agents in an interactive narrative-oriented virtual environment balancing these two, often conflicting, objectives [5, 10]. On the other hand, in our cyber-physical scenario, the exploitation of digital content aims at augmenting the live experience [4]. For this reason, agents driven coordination must either guarantee the coherence of the story told and the emergent behaviour that maximizes the enjoyment of the social experience.

3 The Cleopatra Framework

Figure 1 shows a layered architecture of the Cleopatra framework. Virtual Agents coordinate a collaborative user-experiences by either one-to-one or one-to-many interactions with human users. Like a human guide the virtual agents can propose and illustrate arguments or can answer questions asked by any visitors. Moreover, they can be at the same time on multiple distributed physical and virtual sites, and can start private conversations with single users, if necessary, as a personal guide.

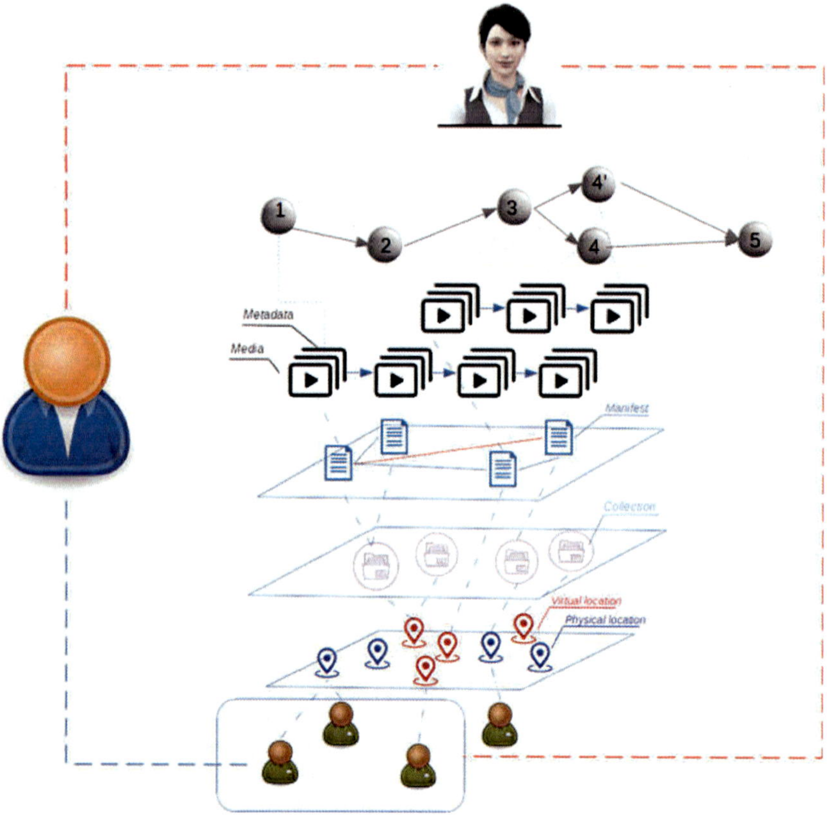

Fig. 1 Conceptual representation of Cleopatra framework

At the lower level, we can see blue-physical points of interest, which are geographical distributed, and red-virtual locations, which can have e geographical attribute or not (a virtual access to a physical museum or multi-media information about a physical location). At the middle level, we have the meta-representation of locations, by collections of *manifests*. A manifest is a representation of either a concrete artifact or an abstract cultural content. It can be the virtual representation of a building, a monument, a masterpiece, a natural landscape, the room of a museum or any other self-consistent information, trough a sequence of semantically related multimedia objects with their meta-data. The Cleopatra agent will be able to build a cultural itinerary proactively, or answering the users' questions, complementing an interactive storytelling with the presentation of semantically related multi-media.

3.1 Cyber-Physical Collaboration

The Use Case diagram shown in Fig. 2 provides a high level representation of the cyber-physical collaboration among software agents and human users, enjoying a diffused museum. Each user belongs to a primary site (the one she is visiting live or she has joined virtually). In such an environment, she can communicate with other human users or with *Proactive Agents* located in that site. One Proactive Agent per site is in charge of leading the interactive storytelling and of coordinating the group of users, whose are visiting that site. It plays its role sending messages or presenting digital contents, but also suggesting the interaction with other agents and users located in secondary sites. Even tough, the social interaction between remote users is free, the interaction among local users is monitored and controlled by the local Proactive Agent to maximize the utility of the social experience, leveraging an emergent social behaviour. On the other hand, any secondary Proactive Agent can answer to primary users, or can be stimulated by the primary Proactive Agent to participate to the conversation, but cannot start autonomously any action in order to not affect the local coordination. We assume that each Proactive Agent has its own knowledge about the digital library, describing its own site and eventually is aware about semantic link to cultural content belonging to other sites. In this way, it can recommend virtual or physical itinerary across other sites or can invite, during its storytelling, other Proactive Agents to present related cultural contents.

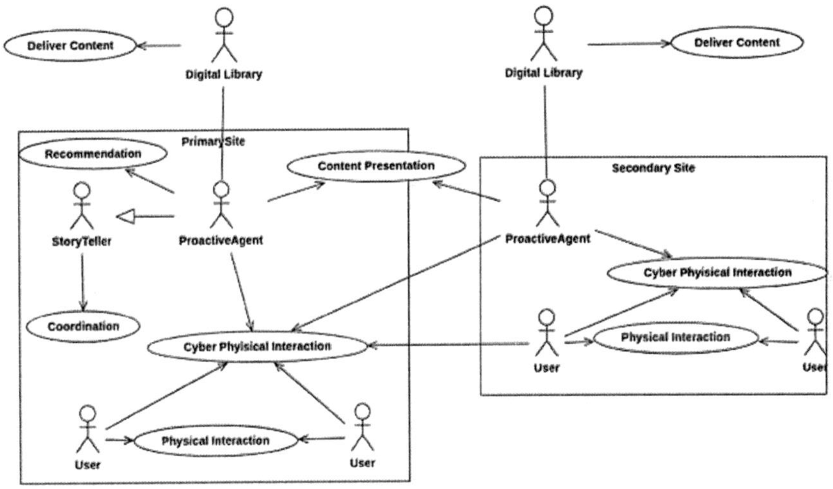

Fig. 2 Cyber-physical collaboration in a diffused museum

3.2 Software Architecture

The component diagram of the Cleopatra platform is shown in Fig. 3. The main functionalities implemented have been designed to support multi-user communication, proactivity and standard protocols. *Multi-users communication* aims at advancing state of art technologies for development of NLP conversation agents, which usually support only diadic interactions. It means that an agent can run multiple conversations at same time, but with only one user each. Moreover, usually conversational agents send a message only for responding to one user's request. On one side, the Proactive Agent has been designed to implement the autonomous and proactive behavior of the cyber side of the Cleopatra platform. On the other side, it implements a gateway between a multi-user conversation that is built upon the XMPP protocol and the diadic interaction with a conversational agent, which uses the NLP technology to provide cyber intelligence to the system [1]. The XMPP protocol is the main choice that allows to exploit all the provided functionalities, but other legacy communication channels will be supported to grant the access by the most widespread messenger applications such as Telegram and Facebook.

The Proactive Agent will be able to receive messages from users and forward them to three conversational agents in charge of handling different class of message, according to its recipient. *One-To-All* messages, whose recipient is the multi-user chat, are public messages without a '@tag'. They are addressed to everyone in the multi-user conversation. The Proactive Agent can decide whether to process that messages, forwarding them to the conversational agents or to ignore them based on well defined criteria. *One-To-Cleopatra* messages are sent to the multi-user chat with the '@cleopatra' tag. They are still public, but must be forwarded by the Proactive Agent to a different conversational agent, that will always respond with tagged message visible by all users. *One-To-One* messages, whose recipient is the Cleopatra agent itself, have been sent privately by the user. The Proactive Agent will forward the message

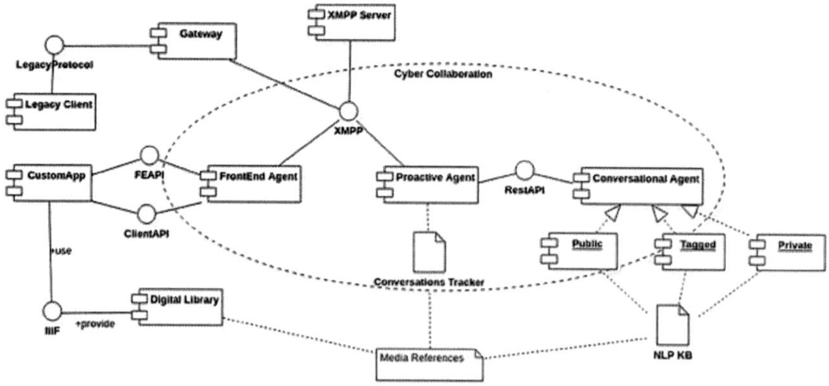

Fig. 3 Component diagram

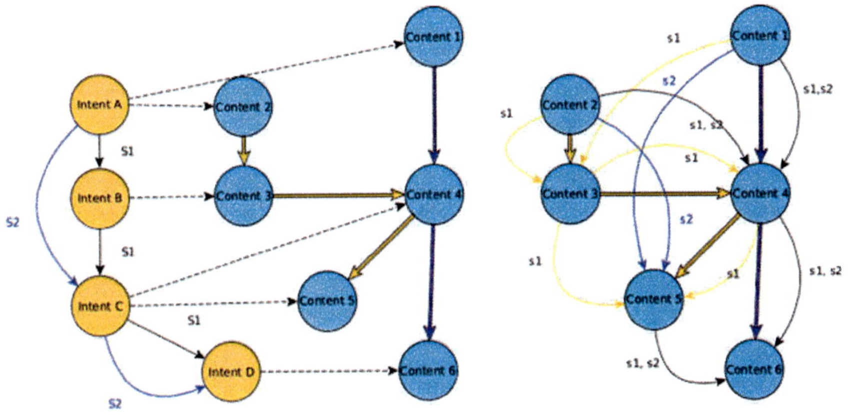

Fig. 4 Story-line development

to the last conversational agent and will return back a private response. A distributed shared memory is exploited by the Proactive Agent and the Conversational agents to build a collective knowledge about the multi-user session. *Multi-media delivery and presentation* is another task of the Proactive Agent that supports the delivery of cultural information carrying on an interactive storytelling. The storyline will include the presentation of digital contents according to the planned plot and semantically related to the current conversation. The utilization of the IIIF (International Image Interoperability Framework) for the standard delivery and presentation of digital objects allow us to maximize both the cultural and the technological impact of the research activity. IIIF defines several application programming interfaces for describing and delivering images over the web, as well as "presentation based metadata" [12]. It supports the definition of structured sequences of canvas, which can include mainly images, but also annotations and other kind of media contents. The digital library, which is an outcome of the Cleopatra project, will be made available to external applications by such an interoperable interface. On the other hand, the agents will be able to access and delivery any other compliant IIIF digital library, which has available in the network. At the stat of art, explicit references link each intent, which is recognized by the Natural Language Understanding unit of the conversational agents, to objects of the digital library.

Multi-users coordination is handled by the Proactive Agent pursuing the fulfillment of well defined objectives, which aim at leveraging emergent behaviours by the community of human users.

As an example let us consider what it is shown in Fig. 4. On left side, the yellow nodes represent user's intents. The user's intention is detected by the NLU capability of the NLP agent on each received message. In a set of pre-defined conversations flows (stories), the NLP agent knows which one is going on with each user. In our toy example, two stories $s1$ and $s2$ are available, which correspond respectively by the black and the blue tiny edges. The dashed arrows represent annotations that link an intent to some relevant cultural contents, which may be presented to the users. We assume here that only one relevant content can be chosen once an intent has been detected. The blue and yellow marked arrows represent two available story-lines, which can be told by the agent to two users who are following respectively conversations $s1$ and $s2$. In this context, the decision problem of planning the story-line for all users consists of finding the best paths in the graph on the right side of Fig. 4, composed of those edges that are tagged with the user's conversation $s1$ or $s2$. The two story-lines in the example have different length and share one content, which is presented in a different step of the conversation. Of course, the length of $s1$ and $s2$ may increase, as well as in the same story the intent X could be followed by an intent Y or Z according to the user's input. The decision problem could depend on multiple parameters (e.g., the user's position and time) and may aim at optimizing multiple local or global optima. The main goals of a Proactive Agent are defined to foster emergent behaviours by group of visitors, which improve their cultural experience or allow to leverage an effective utilization of available facilities. Some examples are the exploitation of unmissable artworks, the path of a complete and coherent itinerary trough the available spots, or liven up the attention and interest of with riddles or any gamification of the cultural experience. With these aims the agents can plan actions to leverage the achievement of their objectives. Examples are prompt users to take action within a predefined time, or inform about relevant artworks or findings in order to increase curiosity, engaging discussion with users to make them curious about the visit or recommending to change location within a predefined time. The achievement of the objectives is measured in different ways. Examples are the percentage of completion of the recommended itinerary, or the sentiment analysis of specific responses.

4 Case Study and Software Prototype

The archaeological scenario is focused on the Samnites, a proud civilization who settled—from the end of the 6th c. BC—in central-southern Italy.

In particular, the case study revolves around the fortified centre of Monte Pugliano [9]. The Samnites built monumental fortifications (often with polygonal walls) on the top of the hills to visually control the access routes to the inner Samnium. In addition to Monte Pugliano, three major sites are within a few minutes' drive: San Pasquale di Faicchio, Monte Acero, Rocca di San Salvatore Telesino [8]. The proposed solution seems to work even better considering the absence of a (virtual) info point or signs that might serve to lead the visitor in this area. It is for this reason that the displacement of multimedia contents and follow-this-route signs becomes crucial to the visitors' experience.

In Fig. 5 the UML Component Diagram of the implemented prototype illustrates the baseline open source technologies integrated both at client and server side. The conversational agents have been built by the RASA open source platform (https://rasa.com), but also Google dialogflow has been evaluated as alternative solution.

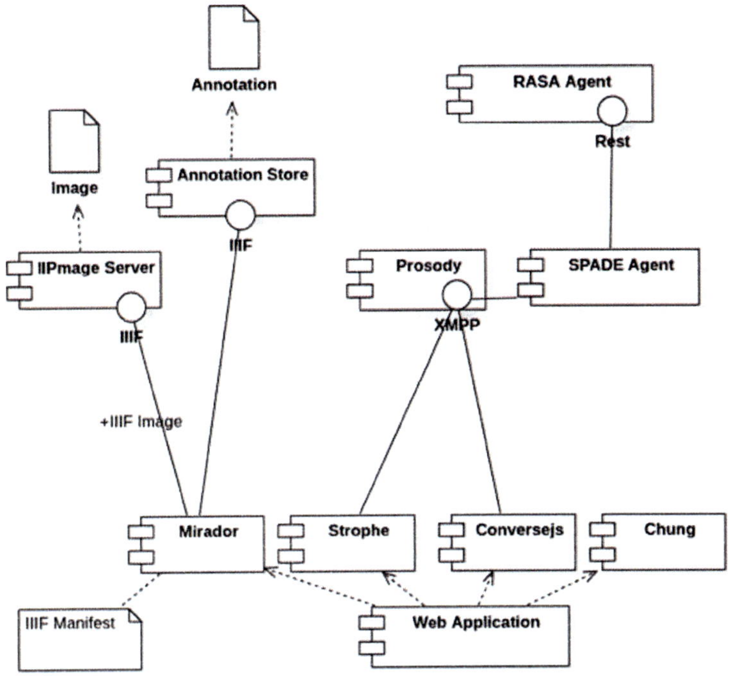

Fig. 5 Technology stack of prototype implementation

The Spade technology has been used to implement the proactive agent [7]. Such a library supports natively the communication via the XMPP protocol. Prosody (https://prosody.im) is the XMPP server technology, which has been chosen to provide a peer to peer overlay for the cyber-physical interaction. In addition to to the legacy channels (such as Facebook, Slack and Telegram), we support and advance multi-modal interaction with the embodied conversational agent integrated into the Mirador IIIF viewer (https://projectmirador.org). The IIPImage server is used to deliver high resolution images of our self-hosted digital libraries. The result of integration of the conversational interface with the Mirador viewer is shown in Fig. 6. An embodied agent shows to the user his position on the map and provides information view about the main archaeological find in the area. The embodied agent is implemented by an open source javascript library (https://sourceforge.net/projects/chatbotchung) that allows for a real time rendering in a web browser and a real time text to speech conversion. The communication is occurs in a chatroom by the conversejs client (https://conversejs.org), while control messages, which allows to the agent for controlling the mirador viewer (e.g: change image, zoom, select annotation) are exchanged by an hidden parallel XMPP session, handled by the strophe.js library.

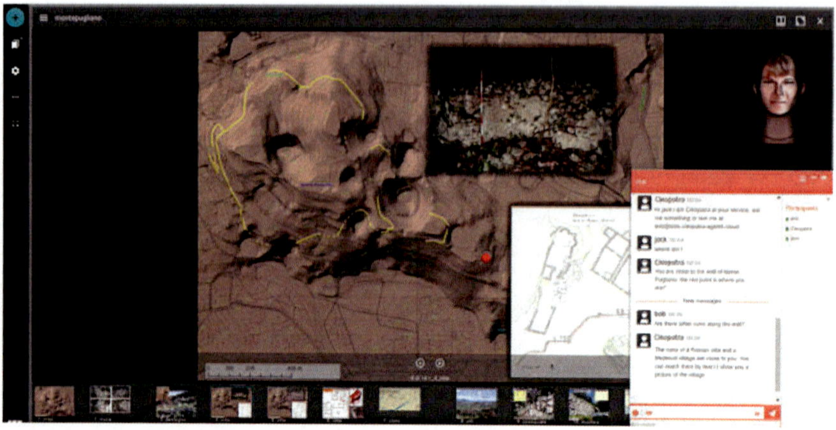

Fig. 6 Converse Web Chat and Mirador integration with javascript embodied agent

5 Conclusion

The proposed platform aims at exploiting distributed intelligent agents as a way to create and reinforce a dense network of experience-sharing, which is eventually able to promote and disseminate information about many overlooked naturalistic and archaeological sites, as well as museums, in the entire area. The presented prototype is going to be tested in controlled environments. A first trial will be tested by 12 students attending a summer school. A second trial will be held at the next exhibition at the "Museum of Capua". In the last case, a kiosk will allow for a touch-less interaction between visitors the Cleopatra agent by one voice assistant. We plan to collect users' feedback and experimental measures for preliminary performance evaluation. Collected data will be used later to build synthetic workloads and to evaluate the scalability of the framework in simulated scenarios, before testing larger deployments. Sentimental analysis of users' response to specific agent's questions will be used to evaluate their feedback. Performance metrics will include: average response time measured at client side, XMPP and web server statistics at server side. Duration of user-sessions, number and frequency of interactions, hot questions and percentage of fallback responses will be used to analyse users behaviours.

Acknowledgements The Cleopatra Project has been funded by the University of Campania "Luigi Vanvitelli" through the VALERE 2019 research program.

References

1. Ambrisi, A., Aversa, R., Ficco, M., Cacace, D., Venticinque, S.: Intelligent cloud agents in multi-participant conversations for cyber-physical exploitation of cultural heritage. Lecture Notes in Networks and Systems, vol. 227, pp. 97–106 (2021). https://doi.org/10.1007/978-3-030-75078-7_11
2. Bilal, N., Bilal, N., Ronan, Q.: Cognitive embodied conversational agents in virtual learning environment. In: Proceedings of the 29th International Conference on Microelectronics (ICM), pp. 1–4 (2017)
3. Chairi, K., George, P., Anestis, K., Dimitis, K.: Multi-agents based virtual environments for cultural heritage. In: Proceedings of the 26th International Conference on Information, Communication and Automation Technologies (ICAT), pp. 1–6 (2017)
4. Škola, F., Rizvić, S., Cozza, M., Barbieri, L., Bruno, F., Skarlatos, D., Liarokapis, F.: Virtual reality with 360-video storytelling in cultural heritage: Study of presence, engagement, and immersion. Sensors **20**(20) (2020). https://doi.org/10.3390/s20205851
5. Kopp, S., Gesellensetter, L., Krämer, N.C., Wachsmuth, I.: A conversational agent as museum guide - design and evaluation of a real-world application. In: Panayiotopoulos, T., Gratch, J., Aylett, R., Ballin, D., Olivier, P., Rist, T. (eds.) Intelligent Virtual Agents, pp. 329–343. Springer, Berlin (2005)
6. Machidon, O.M., Tavčar, A., Gams, M., Duguleană, M.: Culturalerica: A conversational agent improving the exploration of European cultural heritage. J. Cult. Herit. **41**, 152–165 (2020). https://doi.org/10.1016/j.culher.2019.07.010
7. Palanca, J., Terrasa, A., Vincete, J., Carrascosa, C.: Spade 3: supporting the new generation of multi-agent systems. IEEE Access **8**, 182537–182549 (2020)
8. Renda, G.: Carta archeologica e ricerche in campania. comuni di melizzano e frasso telesino. comuni di cerreto sannita, guardia sanframondi, san lorenzello. Atlante Tematico di Topografia Antica **XV** (2010)
9. Renda, G.: Il centro fortificato di monte pugliano e le nuove ricerche sull'altura. Atlante Tematico di Topografia Antica **27**, 125–141 (2017)
10. Riedl, M., Saretto, C.J., Young, R.M.: Managing interaction between users and agents in a multi-agent storytelling environment. In: Proceedings of the Second International Joint Conference on Autonomous Agents and Multiagent Systems, AAMAS '03, pp. 741–748. Association for Computing Machinery, New York, NY, USA (2003). https://doi.org/10.1145/860575.860694
11. Seering, J., Luria, M., Kaufman, G., Hammer, J.: Beyond Dyadic Interactions: Considering Chatbots as Community Members, pp. 1–13. Association for Computing Machinery, New York, NY, USA (2019). https://doi.org/10.1145/3290605.3300680
12. Snydman, S., Sanderson, R., Cramer, T.: (iiif): A community & technology approach for web-based images. In: Archiving Conference, Archiving 2015, pp. 6–21(6). Society for Imaging Science and Technology (2015)

T-Patterns: A Way to Model Complex Trustworthiness in a Social Multi-agent Community

Domenico Rosaci and Giuseppe M. L. Sarné

Abstract A key issue to realize community of agents involving social aspects is that of modeling trustworthiness between the actors of the society and, to this purpose, many trust and reputation models have been proposed in the past. Most of these proposals focused on representing trust dimensions using apposite scalar measures, and integrating these measures in synthetic indicators of trustworthiness, possibly collected into a trust vector. We highlight as this choice is an evident limitation and to overcome it, we propose a new model of trust and reputation for a community of social agents, where the trust perceived by an agent about another agent is modeled by a directed, weighted graph whose nodes and edges represent trust dimensions and their relationships, respectively. This way, we can represent also those situations in which an agent does not knows a given trust dimension, e.g. the honesty, but it is capable to derive it from other correlated dimensions, e.g. the reliability. Our model, called *T-pattern*, has been specifically designed to represent any situation in which many different trust dimensions are mutually dependent. We also introduce the notion of T-Pattern Network (TPN) as an integrated framework to represent both the trust and reputation values as well as the dependencies between trust dimensions for all the pairs of agents.

D. Rosaci
DIIES, University Mediterranea of Reggio Calabria, Reggio Calabria, Italy
e-mail: domenico.rosaci@unirc.it

G. M. L. Sarné (✉)
Department of Psychology, University of Milano Bicocca, Milano, Italy
e-mail: giuseppe.sarne@unimib.it

D. Camacho et al. (eds.), *Intelligent Distributed Computing XIV*, Studies in Computational Intelligence 1026, https://doi.org/10.1007/978-3-030-96627-0_27

1 Introduction

Nowadays, the paradigm of multi-agent systems is rapidly evolving including social aspects, deriving from the necessity that agents communicate with each other also to perform complex interactive tasks, as requesting and providing support and services, negotiating and forming groups [5, 11, 17]. This need to consider social interactions naturally emerged in the field of cooperative agents [18, 19] to promote collaborative behaviors. In this context, a remarkable example is that of IoT [12], where many proposals exist in the literature to promote collaboration into society of smart objects to improve both effectiveness and efficiency of agents in achieving common goals. Moreover, also in the field of competitive agents, when the actors are selfish as in e-business activities, several approaches have been proposed [8, 10] to model and implement social behaviors providing the agents with sophisticated engines to better realize their ends. A key issue to realize a community of agents involving social aspects is that of modeling trustworthiness between the actors. In the last years, trust-based approaches in MAS has been recognized as a good solution to improve system performances, especially in those contexts where software agents are distributed on large-scale networks and interact with each other. When two agents must interact for services, one of them (called the *trustor*) acts as a service requester while the other (called the *trustee*) acts as an e-service provider. It is important to note as agent trust relationships can involve multiple dimensions, depending on the adopted viewpoint, like competence, honesty, security, reliability, expertise, etc. The dimensions above are considered as subjective measure, but often it is necessary to consider a trust dimension in which the whole community associates a trust measure with a given trustee. This dimension is called *reputation*, and assumes a key role in those situation in which an agent a does not have a sufficient direct knowledge about another agent b and, therefore, it has the necessity to exploit the b's reputation. Many trust and reputation models have been proposed in the past in the context of multi-agent community [1, 13]. Most of these proposals focused on representing trust dimensions using apposite scalar measures, and integrating these measures in synthetic indicators of trustworthiness. We here highlight that an evident limitation in this approach is that of assuming trust and reputation as simple scalar values, possibly collected into a trust vector. As an example, the trust of an agent a about another agent b could be represented by the vector $trust_{a,b} = [honesty, reliability, expertise..]$. However, often the individual elements of the trust vector are perceived by the agent as independent of each other. It might be possible for a trust value, e.g., honesty, to be perceived as derivable from another measure, e.g., the reliability, since an agent might think that a reliable partner should also be regarded as honest, with a given degree of probability.

In this paper, to cover this situation we propose a new trust and reputation model conceived for a community of social agents, in which the trust perceived by an agent a about another agent b is modeled not as a vector of trust values, but using a directed, weighted graph whose nodes are trust dimensions (which we denote as *trust aspects*) and whose edges represent relationships between trust dimensions. In this way, we have the advantage of making representable even situations in which an agent does not know one dimension of trust, e.g., honesty, but is able to derive it from another dimension, e.g., the reliability. Our model, called *T-pattern*, can represent any situation of the type illustrated above, in which many different trust dimensions are mutually dependent. This is achieved by using a formalism that invokes logical rules: for example, to represent the fact that honesty is derived from reliability, we use the rule: $reliability \xrightarrow{z} honesty$. However, in such a formalism, $reliability$ and $honesty$ are not logical literals, but real variables that take values in the range [0, 1], where 1 (resp. 0) means maximum (resp. minimum) trustworthiness in the variable, and z is a real value that allows us to quantify the value of the derived variable. In this way, it is possible to derive the same trust variable simultaneously from many different trust variables, without the difficulty of solving a logic program, but simply by tuning the values of z. We also introduce the notion of a T-Pattern Network (TPN) as an integrated framework to represent both trust and reputation values and dependencies between trust dimensions for all pairs of agents.

The paper is organized as follows. In Sect. 2 we discuss some related work. In Sect. 3, we introduce the scenario we deal with and the architecture of our multi-agent community. In Sect. 4, we describe our proposal in detail, also providing an explanatory example of our T-pattern, while Sect. 5 proposes a distributed architecture, called TPN, to automatically manage a Trust Pattern Network. Finally, in Sect. 6, we draw some conclusions and discuss our ongoing research.

2 Related Work

Several different approaches have been proposed in the past to model trust and reputation in multi-agent communities. From the viewpoint of logics, at a high level of abstraction, in [4] the authors propose a logical language, called Trust Computation Tree logic (TCTL), that extends the Computation Tree Logic (CTL) with a new modality to represent trust. More recently, [3] constructs in the logical language $TCTL^G$ can represent the quantitative aspect of trust and develop a symbolic model checking algorithm for quantifying the relationships among the interacting agents. With respect to this type of logical approaches, our T-pattern model proposal exploits a formalism that recall the deductive mechanism of logical programs, but that it is not based on logical predicates but uses a sort of *quantification* of the strength of the rules, particularly suitable in a trust evaluation perspective.

In the Machine Learning area, [20] reports a thorough survey on trust evaluation approaches. Also in our paper, we will present a general multi-agent architecture to automatically learn the knowledge encapsulated in the T-patterns, that can be adapted to several different learning mechanisms already known in the literature. In particular, as for our ongoing research, in order to automatically build T-patterns we are planning to apply the connectionist approach, based on logical neural networks capable to extract logical rules, presented in [15]. As for practical applications of trust-based multi-agent systems approaches, many proposals have been presented in the field of collaborative agents, especially in IoT environments [7, 9, 16]. For example, in [6], a model of competitive MAS scenario is presented in the framework of non cooperative games, and a group formation algorithm able to asymptotically maximize the social capital, in terms of trustworthiness of the agents, is proposed, where the computed solution is a Nash equilibrium. Moreover, in [14], the authors deal with a Social IoT environment and focus on the possible types of trust attacks that can affect the IoT, to this aim they propose a trust management model able to overcome all the analyzed attacks, effectively isolating malicious nodes in the network.

Other applications of trust-based mechanisms for multi-agent systems have been proposed, for instance, (*i*) to improve the security and safety of intersections crossing by the group of autonomous vehicles [2]; (*ii*) to analyse the influence of trust on the functioning of a market for perishable goods, where there not exists quality signal and quantities can be scarce; (*iii*) to create a trust and reputation based framework for social networks that takes into consideration the users relationships, the historic evolution of their reputations and their profile similarity for developing a tamper resilient network to guarantee trustworthiness of both communications and transactions. In all the aforementioned scenarios, our novel proposal of modeling trust and reputation using T-patterns could be fruitfully applied, with the advantage of providing a mechanism to represent complex situations with respect to other approaches modeled by simple trust vectors, instead of considering also the possibility to causal implications between trust dimensions.

3 The Multi-agent Social Scenario

In our scenario, the actors are *agents* assumed as entities operating some tasks. An agent can be either a human being or a software application acting on behalf of a human being. We suppose that in our scenario we have a set A of agents, capable of interact with each other, thus constituting a social community. In such a community, the ways by which the trustworthiness of each agent is perceived by the other agents are modeled by some *trust aspects*. As an example of multidimensional trust, we can consider the well known trust measures used in social communities, as: (*i*) **expertise**: The expertise of an agent represents how the agent is capable of giving expert opinions in a particular domain. Therefore, many different expertise related to different domains can be used as trust dimensions. For instance, in an e-commerce scenario, we could denote as $e_{hardware}$ and $e_{software}$ the expertise of an agent to give opinions about hardware and software, respectively; (*ii*) **honesty**: The honesty of an agent represents in what measure that agent shows a truthful behavior, and it is not fraudulent or misleading; (*iii*) **security**: The security of an agent describes in what measure that agent confidentially manages private data and does not allow unauthorized access to them; (*iv*) **reliability**: A reliable agent provides reliable services. In other words, reliability represents the degree of reliance that can be placed on the services provided by the agent, including effectiveness and efficiency.

In our scenario, we denote as $T = \langle t_1, t_2, \ldots, t_n \rangle$ a set of n trust dimensions associated with the set of agents A. Note that the elements of T are simply strings representing the names of the trust dimensions, and they are not variables containing quantitative values. Indeed, in order to represent how the agents quantitatively evaluate the trust dimensions contained in T, we introduce the notion *confidence* in a trust dimension, and we characterize it with respect to a *group of agents* g, which as particular case can coincide with the whole agent community.

The confidence that an agent a has in a trust dimension x of an agent b with respect to the group g, that we denote as $conf_{a,b,g}(x)$, represents the trust that the agent a assigns to the trust dimension x of the agent b in the context of the group g, and it is either (*i*) a real value belonging to the interval $[0, 1]$, where $conf_{a,b,g}(x) = 1$ (resp. $conf_{a,b,g}(x) = 0$) means maximum (resp. minimum) trust or (*ii*) the value *null* if that confidence has not yet been evaluated. The confidence that group g (which can coincide with the whole social community) has in a trust dimension x of an agent a, that we denote as $conf_{a,a,g}(x)$, represents (*i*) the *group reputation* that the group g assigns to the trust dimension x of the agent a, and it is either a real value belonging to the interval $[0, 1]$, where $conf_{a,a,g}(x) = 1$ (resp. $conf_{a,a,g}(x) = 0$) means maximum (resp. minimum) reputation or (*ii*) the value *null* if that confidence has not been yet evaluated. If the group g coincides with the whole agent community A, we call $conf_{a,a,A}(x)$ *community reputation* of the agent a on the trust dimension x.

4 The T-Pattern Model

A *T-pattern* on T associated with an ordered triplet (a, b, g), where T is a set of trust dimensions and a, b are two agents, is a tuple $\langle a, b, g, N_T \rangle$, where (*i*) N_T is a network (i.e., a weighted directed graph) representing some trust relationships, associated with the ordered pair $\langle a, b \rangle$, involving the trust dimensions contained in T and (*ii*) g is a group, i.e. a set of agents, in the context of which the trust dimensions are evaluated. As a particular case, the group g could be the whole agent community. Note that we consider also the possibility to have a T-pattern associated with the triplet $\langle a, a, g \rangle$. In this case we say that such a T-pattern is a *global T-pattern*, characterizing a single agent by means of global evaluations expressed by the whole group g (as a special case, by the whole agent community), differently where $a \neq b$, we have a *peer-to-peer T-Pattern* (*P2P T-pattern* for short), characterizing how an agent a trusts the agent b.

The N_T network contained in a T-pattern $p = \langle a, b, g, N_T \rangle$ represents how the agent a perceives the relationships existing among the trust dimensions with respect to the agent b and in the context of a group g. More in particular, $N_T = \langle TA, E \rangle$ is a network composed by a set of trust dimensions TA and a set of edges E, where each edge $e = \langle x, y, g, z \rangle$ is an ordered tuple such that $x, y \in N$ are two trust dimensions, g is a group and z is a real number ranging in $[0, 1]$.

In the case $b \neq a$, i.e. in the case of a P2P T-pattern, and in the case $z \neq y$, the value z represents how the agent a perceives the relationship between the trust dimensions x and y, with respect to the agent b. More in particular, the value of z represents the ratio $\frac{conf_{a,b,g}(y)}{conf_{a,b,g}(x)}$ and therefore we can derive $conf_{a,b,g}(y)$ as $z \cdot conf_{a,b,g}(x)$. For instance, if x is the trust dimension *reliability* and y is the trust dimension *honesty*, the arc $\langle x, y, 0.7\rangle$ means that the confidence in the honesty of b is evaluated by a as the 70% of the confidence assigned of the reliability of b. In the case $b = a$, i.e. in the case of a global T-pattern, z represents how the whole group (as a particular case, the whole social community) perceives the relationship between the trust dimensions x and y, with respect to the agent a. The value of z is also in this case equal to the ratio $\frac{conf_{a,a,g}(y)}{conf_{a,a,g}(x)}$.

Therefore, each arc $e = \langle x, y, g, z\rangle$ of the N_T network of a P2P T-pattern means that $conf_{a,b,g}(y) = z \cdot conf_{a,b,g}(x)$. We call such a type of rule a *derivation rule* and we also denote it by the following formalism:

$$x \xrightarrow{z} y \tag{1}$$

Moreover, we define another type of rule, called *assignment rule*, that allows to assign to the confidence $conf_{a,b,g}(y)$ a given value $v \in [0, 1] \cup \{null\}$ and we denote it by the following formalism:

$$v \rightarrow y \tag{2}$$

Finally, we define a third type of rule, called *z-assignment*, that allows to assign to the value z of an edge e a value c belonging to the interval [0, 1], denoting it by:

$$c \rightarrow z \tag{3}$$

For the global T-Pattern, we have a completely identical formalism to indicate that $conf_{a,a,g}(y) = z \cdot conf_{a,a,g}(x)$.

4.1 Automatic Rules Application

In our model, the rules above can be applied either manually or automatically, as side effect of other rule applications. In particular, each time the confidence $conf_{a,b,g}(x)$ (resp. $conf_{a,a,g}(x)$ in the case the T-pattern is associated with an auto-edge) of a trust dimension x has been updated (in consequence of the application of the rule (2) or by side effect of other updates), automatically the rule (1) is applied to each edge outgoing from the node associated with x in the network N_T, and the propagation of the update is applied in cascade to each other nodes outgoing from now updated nodes, excepting for the node x. Moreover, for all the edges $e = \langle k, x, , g, z\rangle$ incoming in the node x, it is automatically applied the rule (3) updating z by using a value

$c = \frac{conf_{a,b,g}(x)}{conf_{a,b,g}(k)}$. Finally, each time the rule (3) is applied to update the value z of an edge e, it is automatically applied the rule (1) to the final node of the edge e, and the propagation of the update is applied in cascade to each other nodes outgoing from now updated nodes, excepting the initial node of the arc e.

4.2 T-Pattern Network

A *T-Pattern Network* (TPN for short) is defined as a triplet $\langle A, G, T, conf, P\rangle$, where A is a set of agents, G is a set of groups, T is a set of trust dimensions, $conf$ is a mapping that maps each confidence $c_{a,b,g}$ with a value belonging in $[0, 1] \cup null$, and P is a set of T-patterns on T, such that there do not exist two T-patterns belonging to P associated with the same ordered triplet (a, b, g). Since a T-pattern can be viewed as an edge between two agents, a TPN can be considered as a network where the nodes are agents and the edges are defined by the T-patterns of P, each edge weighted by the pair (g, G_T) associated with the corresponding T-pattern. The mechanisms of automatic rule activation described above guarantees the consistence of all the T-patterns with the $conf$ mapping.

4.3 An Explicative Example

In order to practically understand the formalism introduced in the above section, let us consider the example graphically represented in Fig. 1.

In this example, we have two agents, denoted as a and b, respectively, that live in a multi-agent community represented by the tuple $\langle A, T, conf, G, P\rangle$, where:

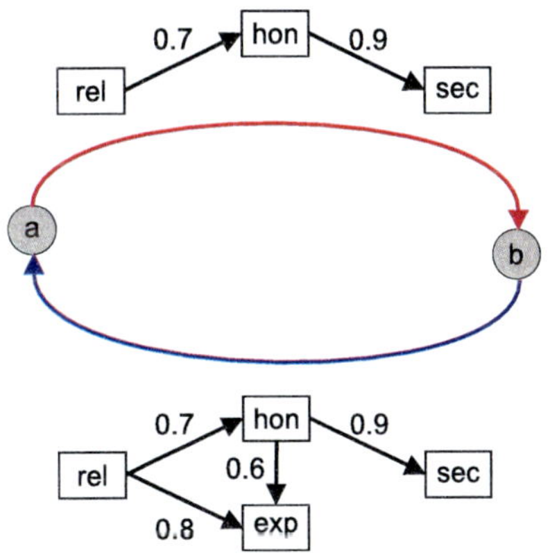

Fig. 1 An example of TPN with two agents

- $A = \{a, b\}$ is the set of agents;
- $G = \emptyset$; in this example we do not consider any group;
- $T = \{rel, hon, sec, exp\}$ is a set of trust dimensions representing reliability (rel), honesty (hon), security (sec) and expertise (exp);
- $conf$ is a mapping containing the confidence values of each agent with respect to the other agent. More in particular, we suppose that in this example all the values contained in $conf$ are equal to null, representing an initial situation in which we do not yet have any knowledge about the mutual trustworthiness of the agents. For simplicity, we have omitted to indicate the group g, since in this case the unique group is the whole community.
- P is a set containing two T-patterns, graphically depicted by the two arcs $< a, b, N_{a,b} >$ and $< b, a, N_{b,a} >$, respectively, representing the a-priori knowledge we have about how the two agents perceive the trustworthiness. In particular, the network $N_{a,b}$ informs us that the agent a derives the honesty of b from the reliability by using the derivation rule $rel \overset{0.7}{\rightarrow} hon$. In other words, if a knows the reliability of b, it automatically trusts the honesty of b as a percentage of the reliability equal to the 70%. Therefore the agent a sufficiently believes to the honesty of b also in those cases it did not directly experiment this honesty, provided that the reliability of b has been verified. Analogously, the network $N_{a,b}$ also informs us that the agent a derives the security of b from the honesty by using the derivation rule $hon \overset{0.9}{\rightarrow} sec$. In other words, when a has verified the honesty of b, it is also almost sure of its security. Note that in the initial situation of the confidence values represented by the mapping $conf$, no information is available and therefore no information can be derived from the derivation rules, that will can be applied only when some confidence values will be directly derived from the agents. The meaning of the the network $N_{b,a}$ is analogously represented by the following derivation rules:

$$rel \overset{0.7}{\rightarrow} hon; \qquad hon \overset{0.9}{\rightarrow} sec$$

$$hon \overset{0.6}{\rightarrow} exp; \qquad exp \overset{0.8}{\rightarrow} rel$$

5 The T-Pattern Architecture (TPA) to Automatically Manage a TPN

In this section, we describe a multi-agent architecture, called T-Pattern Architecture (TPA) capable of automatically managing a T-Pattern Network $NET = \langle A, G, T, conf, P \rangle$, deriving the information necessary to update the T-patterns from the direct observation of the agent behavior. The TPA is distributed on three logical levels, namely:

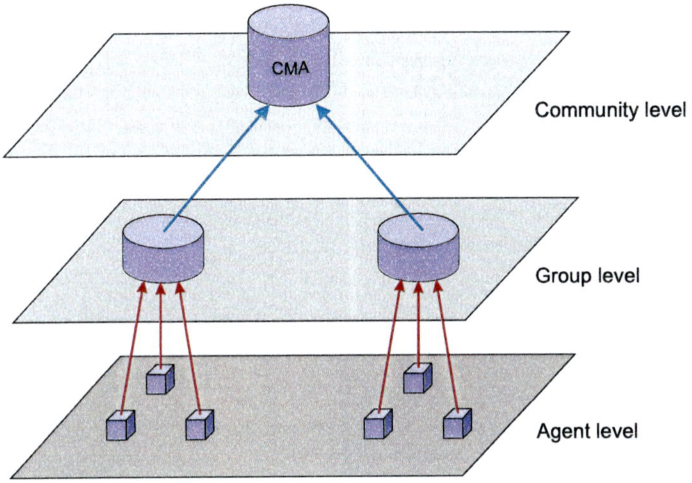

Fig. 2 The three-layers TPA architecture

- an *agent level*, composed of n *trust manager agents* $tm_1, tm_2, \ldots, tm_n$, where each tm_i is associated with the corresponding agent $a_i \in A$ of NET and it is capable of updating the trust patterns associated with all the edges outgoing from a_i in NET.
- a *group level*, composed of l *group manager agents* $g_1, g_2, \ldots, g_l$, where each gm_i is associated with the corresponding group $g_i \in G$ of NET and it is capable of computing the group reputation $conf_{a,a,g_i}(t)$ for all the agents $a \in g_i$ and for all the trust aspects $t \in T$.
- a *community level*, composed from a *community manager agent* CMA, capable of computing the community reputation $conf_{a,a,A}(t)$ for all the agents $a \in A$ and for all the trust aspects $t \in T$.

In this paper we do not deal with the mechanism used by the trust manager agents to build trust patters, and we do not face the issue of computing group level reputation and community reputation starting by the trust patterns. Many different mechanisms could be proposed, and finding effective and efficient proposals in this setting is the purpose of our ongoing research (Fig. 2).

6 Conclusion

In this paper we have proposed a novel model of trust and reputation for a community of agents, having the particular feature of modeling trust by a weighted graph whose nodes are trust dimensions and whose edges represent relationships between trust dimensions. We have argued that our proposal introduces the advantages of making representable those situations in which an agent does not knows a trust dimension, but it is capable to derive it from another dimension.

We have also introduced the notion of T-Pattern Network (TPN), as an integrated framework to represent both the trust and reputation values as well as the dependencies between trust dimensions for all the pairs of agents. As for the future developments of this work, our ongoing research is mainly focused on finding suitable mechanisms to build trust patterns, as well as to compute group level reputation and community reputation starting by the trust patterns.

References

1. Brogan, C., Smith, J.: Trust Agents: Using the Web to Build Influence, Improve Reputation, and Earn Trust. Wiley, Berlin (2020)
2. Chuprov, S., Viksnin, I., Kim, I., Reznikand, L., Khokhlov, I.: Reputation and trust models with data quality metrics for improving autonomous vehicles traffic security and safety. In: 2020 IEEE Systems Security Symposium (SSS), pp. 1–8. IEEE (2020)
3. Drawel, N., Bentahar, J., Qu, H.: Computationally grounded quantitative trust with time. In: Proceedings of the 19th International Conference on Autonomous Agents and MultiAgent Systems, pp. 1837–1839 (2020)
4. Drawel, N., Bentahar, J., Shakshuki, E.: Reasoning about trust and time in a system of agents. Procedia Comput. Sci. **109**, 632–639 (2017)
5. Esmaeili, A., Mozayani, N., Motlagh, M., Matson, E.: A socially-based distributed self-organizing algorithm for holonic multi-agent systems: case study in a task environment. Cognit. Syst. Res. **43**, 21–44 (2017)
6. Fortino, G., Fotia, L., Messina, F., Rosaci, D., Sarné, G.M.L.: A meritocratic trust-based group formation in an iot environment for smart cities. Future Gener. Comput. Syst. **108**, 34–45 (2020)
7. Fortino, G., Fotia, L., Messina, F., Rosaci, D., Sarné, G.M.L.: Trust and reputation in the internet of things: state-of-the-art and research challenges. IEEE Access **8**, 60117–60125 (2020)
8. He, Z., Han, G., Cheng, T., Fan, B., Dong, J.: Evolutionary food quality and location strategies for restaurants in competitive online-to-offline food ordering and delivery markets: An agent-based approach. Int. J. Prod. Econ. **215**, 61–72 (2019)
9. Hussain, Y., Zhiqiu, H., Akbar, M., Alsanad, A., Alsanad, A., Nawaz, A., Khan, I., Khan, Z.: Context-aware trust and reputation model for fog-based iot. IEEE Access **8**, 31622–31632 (2020)
10. Jafari, S., Navidi, H.: A game-theoretic approach for modeling competitive diffusion over social networks. Games **9**(1), 8 (2018)
11. Jaques, N., Lazaridou, A., Hughes, E., Gulcehre, C., Ortega, P., Strouse, D., Leibo, J., De Freitas, N.: Social influence as intrinsic motivation for multi-agent deep reinforcement learning. In: International Conference on Machine Learning, pp. 3040–3049. PMLR (2019)
12. Khan, W., Aalsalem, M., Khan, M., Arshad, Q.: When social objects collaborate: concepts, processing elements, attacks and challenges. Comput. & Electr. Eng. **58**, 397–411 (2017)
13. Kowshalya, A., Valarmathi, M.: Trust management for reliable decision making among social objects in the social internet of things. IET Netw. **6**(4), 75–80 (2017)
14. Marche, C., Nitti, M.: Trust-related attacks and their detection: a trust management model for the social iot. IEEE Trans. Netw. Serv. Manag. (2020)
15. Rosaci, D.: Cilios: Connectionist inductive learning and inter-ontology similarities for recommending information agents. Inf. Syst. **32**(6), 793–825 (2007)
16. Sharma, A., Pilli, E., Mazumdar, A., Gera, P.: Towards trustworthy internet of things: a survey on trust management applications and schemes. Comput. Commun. (2020)

17. Telang, P., Singh, M., Yorke-Smith, N.: Maintenance of social commitments in multiagent systems. In: Proceedings of the AAAI Conference on Artificial Intelligence, vol. 35, pp. 11369–11377 (2021)
18. Torreño, A.I., Onaindia, E., Komenda, A., Štolba, M.: Cooperative multi-agent planning: a survey. ACM Comput. Surv. **50**(6), 1–32 (2017)
19. Walczak, S.: Society of agents: a framework for multi-agent collaborative problem solving. In: Natural Language Processing: Concepts, Methodologies, Tools, and Applications, pp. 160–183. IGI Global (2020)
20. Wang, J., Jing, X., Yan, Z., Fu, Y., Pedrycz, W., Yang, L.: A survey on trust evaluation based on machine learning. ACM Comput. Surv. **53**(5), 1–36 (2020)

An Algorithmic Proposal for the Qualitative Comparison Between the Chen-Yeh EIT Mixture Model and the Susan-Resiga GSMs Experimental Model for MR Fluids

Mario Versaci and Annunziata Palumbo

Abstract In the framework of extended irreversible thermodynamics, the Chen-Yeh theoretical "mixture" model treats magnetorheological (MR) fluids by evolutionary constitutive equations that, in the pre-yield region, manifest the co-presence of elastic, viscoelastic and viscoplastic behaviors. However, such model is characterized by high computational complexity; so, it is hoped to find possible qualitative correspondences between the evolutionary equations of this theoretical model and the evolutionary equations of experimental ones. Therefore, in this paper, we present an innovative sequential algorithm to achieve the qualitative correspondence between the Chen-Yeh model with the Susan-Resiga experimental one that, here, under the same operating conditions and in the framework of generalized standard materials (in shear thinning flow), has been proved to be characterized by acceptable computational complexity compatible with the most important industrial applications.

1 Introduction

Magnetorheological (MR) fluids are made up of polarizable fine particles of ferromagnetic material suspended in mineral oil with rheological characteristics that can change suddenly and reversibly when a magnetic field **H** is externally impressed [1–4]. Without **H**, these fluids become liquids whose viscosity is equal to that mineral oils. Conversely, if **H** is impressed, the induced magnetic moment aggregates

Supported by the Italian National Group of Mathematical Physics (GNFM-INdAM) and the University of Messina through FFO 2019.

M. Versaci (✉)
DICEAM Department, Mediterranea University, Cittadella Universitaria,
89122 Reggio, Calabria, Italy
e-mail: mario.versaci@unirc.it

A. Palumbo
MIFT Department, University of Messina, Via F. Stagno d'Alcontres 31, 98166 Messina, Italy
e-mail: apalumbo@unime.it

D. Camacho et al. (eds.), *Intelligent Distributed Computing XIV*, Studies in Computational Intelligence 1026, https://doi.org/10.1007/978-3-030-96627-0_28

the dipoles into chains parallel to **H** forming poorly movable columns with high viscosity [5]. Then, to overcome the strong friction between the columns, it is necessary to overcome a shear stress threshold, τ_0 (yield stress), which depends on $|\mathbf{H}|$. If $|\mathbf{H}|$ varies in a controlled manner, τ_0 will also vary, controlling the ability of the fluid to transmit forces [6, 7] up to the pre-yield region, beyond which the correspondence between τ_0 and $|\mathbf{H}|$ cannot be easily modeled [7]. Recently, theoretical research has produced excellent complete but computationally noncompetitive models [7] such as the "mixture" models elaborated by Chen and Yeh at National Chi-Nan University (Taiwan) that, in the extended irreversible thermodynamics (EIT) framework in the sense of Jou et al. [8], consider the MR fluids as a mixture of a fluid continuum and an equivalent solid continuum to obtain evolutionary constitutive equations which, in the pre-yield region,[1] manifests the co-presence of elastic-visco-plastic components [9]. In parallel, the experimental research has produced good models with low computational load respectful of the macroscopic behaviors of MR fluids with variable shear rate, $\dot{\gamma}$ [10]. Among them, the Susan-Resiga [11] experimental model, developed at the University of Timisoara (Romania), selects the electric current I by adapting the experimental measurements of the shear stress, τ, depending both on $\dot{\gamma}$ and on the absolute temperature θ, producing the well-known flow curves, $\tau(\dot{\gamma}, \theta)$, defining the magnetorheological behavior of the fluid. Particularly, $\tau(\dot{\gamma}, \theta)$ exhibits quasi-linear behavior below a particular shear rate value, $\dot{\gamma}^*$. Moreover, if $\dot{\gamma} < \dot{\gamma}^*$, Newtonian formulations are used; otherwise if $\dot{\gamma} > \dot{\gamma}^*$, the model shows plastic behavior. It follows that $\tau(\dot{\gamma}, \theta)$ is expressed in terms of the weighted sum of two addends: the first is of a Newtonian nature, while the second is distinctly non-linear ($\dot{\gamma} > \dot{\gamma}^*$) in which viscoplastic phenomena prevail. Mathematically, the Susan-Resiga model is structured in terms of the weighted sum of two addends: the first determines the Newtonian behavior of the MR fluid while the second, modeled by Herschel-Bulkley formulations in the framework of generalized standard material, GSMs (according to the theory of Halphen-Nguyen [6, 7]), manifests the co-presence of viscoplastic components. So, it appears clear the need to have an effective algorithm for the qualitative comparison, in the pre-yield region, between the Chen-Yeh theoretical model and the Susan-Resiga experimental one in order to have a sort of "low computational load translation" of a model complete but industrially unappetizing. Therefore, in this paper, we present an algorithm proposal (structured in three fundamental steps and four results) for the qualitative comparison between the Chen-Yeh model and the Susan-Resiga model, in the pre-yield region, which highlights how such a qualitative comparison is possible.

The reminder of the paper is structured as flows. Sections 2 and 3 briefly describe the Che-Yeh and Susan-Resiga models, respectively; while Sects. 4 and 5 details the steps of the proposed algorithm and the main results achieved, respectively. Finally, some conclusions and future perspectives are drawn.

[1] Industrially, thepre-yieldregion, for an MR fluid, is very important because it represents a sort of upper limit of theme chanical resistance of the MR fluid above which it loses its rheological characteristics to assume the properties of an ordinary fluid.

2 The Chen-Yeh Model

Point Materials and Positions
The Chen-Yeh model treats MR fluids as a mixture composed of a fluid continuum, F, and an equivalent solid continuum, S, that simultaneously occupy the same space [9, 10]. We indicate S and F continua by the material points ξ^F and ξ^S so that, $\forall t$, $x_i^F = x_i^F(\xi_\alpha^F, t)$ and $x_i^S = x_i^S(\xi_\alpha^S, t)$ are their positions, respectively.

Velocity Vectors and Material Derivatives Since F and S simultaneously the same space, it follows that $x_i = x_i^F = x_i^S$, from which we achieve the velocity vectors, $v_i^F = \frac{d_F x_i^F}{dt}$, $v_i^S = \frac{d_S x_i^S}{dt}$, where $\frac{d_S}{dt}$ and $\frac{d_F}{dt}$ are the material derivatives for fixed ξ_α^F and ξ_α^S, respectively.

"Intermediate" States in the Sense of Lee [9, 10]
S is a set of magnetizable particles; therefore, the model exploits the concept of "intermediate state"[2] to decompose the deformation of S. Let us denote the position of the material point in S in the reference, released intermediate, and final positions as ξ_α^S, X_I^S, and x_i^S, respectively. Moreover, to consider the displacements among two different states, the model introduces three vectors, $\mathbf{u}^S$, $\mathbf{u}^{SP}$, and $\mathbf{u}^{SE}$,[3] such that $u_i^S(x_i, t) = x_i - \delta_{i\alpha}\xi_\alpha^S$, $u_i^{SP} = u_i^{SE} + \delta_{iK}u_K^{SP}$ and $u_i^{SE} = x_i - \delta_{iK}X_K^S$.

Velocity Gradient Tensor and Elastic Strain Tensors
Let L_{ij}^S, L_{ij}^{SE} and L_{ij}^{SP} be the elastic and plastic components of the velocity gradient tensors, respectively such that $L_{ij}^S = L_{ij}^{SE} + L_{ij}^{SP}$. Moreover, decomposing L_{ij}^S in its symmetric, $L_{(ij)}^S$, and anti-symmetric components, $L_{[ij]}^S$, one obtains $L_{(ij)}^S = \underbrace{D_{ij}^{SE} + D_{ij}^{SP}}_{\text{elongations and distortions}}$ and $L_{[ij]}^S = \underbrace{\Omega_{ij}^{SE} + \Omega_{ij}^{SP}}_{\text{rotations}}$. Finally, taking into account both Lagrangian and Eulerian strain tensors, E_{IJ}^{SE} and e_{ij}^{SE} respectively, and their derivatives with respect the time, one achieves $e_{ij}^{SE} = \frac{1}{2}(u_{i,j}^{SE} + u_{j,i}^{SE} - u_{k,i}^{SE}u_{k,j}^{SE})$[4] that represents the link between the elastic deformation and the displacement gradient [9, 10].

Stress Tensors and Magnetization
Under the assumption that the fluid stress tensor is a symmetric one, it follows that $\tau_{ij}^F = \tau_{ji}^F$; concerning the solid stress tensor (due to the external $\mathbf{H}$), it follows that $\tau_{ij}^S \neq \tau_{ji}^S$. Therefore, $\tau_{ij}^F = \tau_{ji}^F = -p^F\delta_{ij} + \tau^{FB}\delta_{ij} + \tau_{ij}^{FS}$, $\tau_{ii}^{FS} = 0$, $\tau_{ij}^S = \tau_{ij}^{SR} + \tau_{ij}^{SD} = \tau_{(ij)}^{SR} + \tau_{[ij]}^{SR} + \tau_{ij}^{SD}$ and $\tau_{ij}^{SD} = -\tau_{ji}^{SD}$ where FB and FS indicate the bulk of the fluid and the shear fluid; SR and SD indicate the reversible part and dissipative part of the solid continuum, respectively. Moreover, p^F is the equilibrium pressure of the fluid, while τ^{FB} and τ_{ij}^{FS} are the quantities induced in the non-equilibrium state due to the dissipative viscous mechanical forces. Applying the

[2] Artificial state to separate plastic and elastic deformations.

[3] $\mathbf{u}^{SP} + \mathbf{u}^{SE} = \mathbf{u}^S$, $\xi^S + \mathbf{u}^{SP} = X^S$ and $\xi^S + \mathbf{u}^S = x$.

[4] $_{,i} \equiv \frac{\partial}{\partial x_i}$ stands for the partial derivatives by coordinates.

angular momentum balance, the anti-symmetric component of the stress is related to the magnetization M_i^S, $\tau_{[ij]}^S = -M_{[i}^S B_{j]}^S$, where B_j^S is the magnetic induction. Finally, by decomposing the magnetization into a reversible part, M_i^{SR}, and a dissipative part M_i^{SD}, that is $M_i^S = M_i^{SR} + M_i^{SD}$, it follows that $\tau_{[ij]}^{SR} = -M_{[i}^{SR} B_{j]}^S$ and $\tau_{[ij]}^{SD} = -M_{[i}^{SD} B_{j]}^S$. It is worth noting how the link between mechanical and magnetic tensor quantities is only explicit for S and not for F.

Field Equations

Since MR fluids are magnetizable particles suspended in mineral oil, it is assumed that the F does not undergo the action of **H** while the continuous solid is sensitive to it. So, the following operational steps make sense:

1. the starting point is represented by the writing of Maxwell's equations only for the continuous solid;
2. the mass balance, linear momentum, and energy equations for both F and S are written;
3. by combining the balances referred to in point 2., one obtains the balance equations where, in each of which, both S and Fappear;
4. the local versions of the second law of the thermodynamics (depending entropy densities and fluxes), for both F and S, are written at the same temperature, θ;
5. therefore, introducing the entropy density, η, and the entropy flux of the mixture, S_i, and exploiting the Helmotz free energy of the mixture, H, we obtain the following interesting inequality:

$$-\rho\frac{dH}{dt} - \rho\eta\frac{d\theta}{dt} + \theta S_{i,i} + \tau_{ji}^F v_{i,j}^F + \tau_{ji}^S v_{i,j}^S - q_{i,i} - M_i^S\frac{d_S B_i^S}{dt} + P_i v_i^R \geq 0. \tag{1}$$

To obtain evolutionary constitutive equations it is necessary to consider the problem in the framework of the EIT which exploits the dependence of the free Helmotz energy on a set of conserved variables and some flows. Therefore, exploiting the material derivatives and defining the generalized entropy density η, the equilibrium pressure of the fluid, p^F, the reversible stress of the solid, τ_{ij}^{SR}, and the reversible part of the magnetization, M_i^{SR}, after calculations, the following equations are obtained:

$$\begin{cases} \rho\frac{\partial F}{\partial q_i} = \lambda_{11}q_i + \lambda_{12}M_i^{SD} + \lambda_{13}v_i^R; & \rho\frac{\partial F}{\partial M_i^{SD}} = \lambda_{21}q_i + \lambda_{22}M_i^{SD} + \lambda_{23}v_i^R \\ \rho\frac{\partial F}{\partial v_i^R} = \lambda_{31}q_i + \lambda_{32}M_i^{SD} + \lambda_{33}v_i^R; & \rho\frac{\partial F}{\partial \tau_{ij}^{FS}} = \lambda_4\tau_{ij}^{FS}; \quad \rho\frac{\partial F}{\partial \tau^{FB}} = \lambda_5\tau^{FB} \end{cases} \tag{2}$$

where $\lambda_{ij} = \lambda_{ji}$ are scalars depending on the conserved variables; λ_4 and λ_5 do not explicitly depends on B_i^S. Finally, assuming the linear approximation, λ_{ij} will depend on the conserved variable, so that, after some calculations, (1) becomes

$$\tau_{ji}^{SR}L_{ij}^{SP} + q_i X_i^q + M_i^{SD}X_i^M + v_i^R X_i^v + \tau_{ij}^{FS}X_{ij}^{FS} + \tau^{FB}X^{FB} \geq 0 \tag{3}$$

in which X_i^q, X_i^M, X_i^v, X_{ij}^{FS}, and X^{FB} are the thermodynamic force.

Evolutionary Constitutive Equations in the Pre-Yield Region

To achieve evolutionary constitutive equations, we need to fix the link between the generalized flux and generalized forces. The simplest link to model, as highlighted in [9], is the linear one. Therefore, after some steps of calculations, we obtain a system of evolutionary constitutive equations that, in pre-yield region, is simplified as follows:

$$\begin{cases} \lambda_4 \mu^S \frac{d\tau_{11}^{FS}}{dt} = -\tau_{11}^{FS} + \mu^S D_{11}^F \\ \lambda_4 \mu^S \frac{d\tau_{22}^{FS}}{dt} = -\tau_{22}^{FS} + \mu^S D_{22}^F \\ \lambda_5 \mu^B \frac{d\tau^{FB}}{dt} = -\tau^{FB} + \mu^B v_{1,1}^F + \mu^B v_{2,2}^F \end{cases} \tag{4}$$

(in which both μ^S and μ^B are viscosities) so that inequality (3) becomes

$$\tau_{ji}^{FS} X_{ij}^{FS} + \tau^{FB} X^{FB} \geq 0. \tag{5}$$

Therefore, the MR material, in the pre-yield region, behaves like a viscous fluid or as a viscoelastic body, while in the post-yield region it behaves like a viscoplastic fluid. It is worth noting that (4) only makes sense in the EIT framework; otherwise, if H depended only on conserved variables (i.e., in the framework of classical thermodynamics) one would achieve constitutive equations instead of evolutionary constitutive equations for the dissipative fluxes.

3 The Susan-Resiga Model

Structure of the Experimental Model

This model is mathematically structured in such a way that $\tau(\dot{\gamma}, \theta)$ results as a weighted sum of two addends as follows:

$$\tau(\dot{\gamma}, \theta) = W_1(\cdot) \cdot f_1(\cdot) + W_2(\cdot) \cdot f_2(\cdot) \tag{6}$$

where $f_1(\cdot)$ describes the Newtonian behavior while $f_2(\cdot)$ is formulated according to the Herschel-Bulkley theory with shear-thinning conditions, where the functional dependence on both the viscosity and the yield stress is explicit [6, 7, 11], as it takes into account any co-presence of visco-plastic behaviors (for details, see [5–7, 10]), but other formulation can be taken into account [7]. Moreover, both the absolute temperature, θ_0, and the activation energy, E_a, for the viscous flow are present; therefore, the model is sensitive both to temperature variations (induced or accidental) and to eventual changes in viscous behavior. Obviously, the model does not lack the presence of $\dot{\gamma}$, which represents the shear rate around the transition from Newtonian to plastic behavior. Both addends are weighted by two functions, $W_1(\dot{\gamma})$ and $W_2(\dot{\gamma})$, respectively, such that their sum equals unity([5]). If $W_1(\dot{\gamma}) \gg W_2(\dot{\gamma})$, then Newtonian behavior is dominant and the shear rate is low; on the other hand, if $W_1(\dot{\gamma}) \gg W_2(\dot{\gamma})$, the plastic behavior takes place and the shear rate is high. Obviously, $W_1(\cdot)$ and $W_2(\cdot)$ weigh the individual contributions so that, as one approaches the pre-yield region, the Newtonian contribution reduces in favor of the second addend showing contributions of a different nature.

The Experimental Model in the Pre-Yield Region
At the pre-yield region, the Newtonian contribution is minimized while the other contributions highlighted by $W_2(\cdot)$ and weighted by $f_2(\cdot)$ are preponderant. However, the power-law formulation of $W_2(\cdot)$ (Herschel-Bulkley formulation for the Susan-Resiga experimental model) is not able to highlight the co-presence of elastic, viscous and plastic components for which, in the pre-yield region, it is not possible to algorithmically structure the comparison between the Chen-Yeh with the Susan-Resiga one. Then, in the GSMs framework, it is necessary to generalize $W_2(\cdot)$ to obtain a $3D$ version that highlights the aforementioned co-presence in evolutionary terms [7].

4 The Proposed Algorithm

In this Section we present the proposed algorithm which is structured in three fundamental steps and some important results for which we omit, for obvious reasons of space, the proofs.

4.1 *Step 1: Elast-Visco-Plastic Model and GSM and Simple Shear Flow*

It is worth noting that the constitutive equations for MR fluids must be both objective (i.e., invariant with respect to the observer's changes) and satisfy the II law of thermodynamics. Such requests are satisfied if we introduce GSMs as a framework for the $f_2(\cdot)$ component of the experimental model. With this goal in mind, we split the total deformation tensor ϵ_{ij} into its elastic component, ϵ_{ij}^E and its plastic one, ϵ_{ij}^P, such that $\epsilon_{ij} = \epsilon_{ij}^E + \epsilon_{ij}^P$. GSMs are special materials whose behavior is describable, according to Halphen-Nguyen's theory [7], by a free energy of the system, $H(\epsilon_{ij}, \epsilon_{ij}^E)$, that represents the parameter of elasticity and a dissipation potential, $D(\dot{\epsilon}_{ij}, \dot{\epsilon}_{ij}^E) = \phi(\dot{\epsilon}_{ij}) + \phi_p(\dot{\epsilon}_{ij}^P)$,[6] in which $\phi(\dot{\epsilon}_{ij})$ models the incompressible viscoelastic behavior and $\phi_p(\dot{\epsilon}_{ij}^P)$ expresses the viscoelastic behavior (for analytical details, see [7]). It is worth observing that the formulations of $\phi(\dot{\epsilon}_{ij})$ and $\phi_p(\dot{\epsilon}_{ij}^P)$

[5] $W_1(\dot{\gamma})$ and $W_2(\dot{\gamma})$ must be continuous together their first derivative to ensure smooth transitions between the two different behaviors.
[6] $\dot{(\cdot)} = \frac{d}{dt}(\cdot)$, so that the time variable appeared.

represent important theoretical studies on complex fluids that have been reflected in many practical laboratory experiences. Furthermore, $D(\dot{\epsilon}_{ij}, \dot{\epsilon}_{ij}^P)$ is a convex functional satisfying the II law of thermodynamics, as proved in [7].

4.2 Step 2: GSMs Framework and The Constitutive Laws

We observe that the fluid under study has a relevant plastic component; so, it does not allow the calculation of the differential of $D(\dot{\epsilon}_{ij}, \dot{\epsilon}_{ij}^P)$ in the ordinary sense (due to the non-linearity and non-differentiability of both $\phi(\dot{\epsilon}_{ij})$ and $\phi_p(\dot{\epsilon}_{ij}^P)$). For this reason, we need to consider the sub-differential of $D(\dot{\epsilon}_{ij}, \dot{\epsilon}_{ij}^P)$ which allows us to exploit an important result of the theory of complex fluids according to which a complex functional admits a minimum, δ_0, *iff* the straight line with no slope (indicated by 0) belong to boundary of $D(\dot{\epsilon}_{ij}, \dot{\epsilon}_{ij}^P)$, $\partial D(\dot{\epsilon}_{ij}, \dot{\epsilon}_{ij}^P)$, that is $0 \in \partial D(\delta_0)$. This important result allows us to write, after some calculations, the constitutive laws as $\tau_{ij} \in \partial\phi(\dot{\epsilon}_{ij}) + \partial\phi_p(\dot{\epsilon}_{ij}^P)$ and $0 \in 2\omega\epsilon_{ij}^E - \partial\phi_p(\dot{\epsilon}_{ij}^P)$. In parallel, τ_{ij} is decomposed as $\tau_{ij} = -p\delta_{ij} + 2\mu\dot{\epsilon}_{ij}$ (where p is the pressure and $|2\mu\dot{\epsilon}_{ij}|$ is indicated by $|T_d|$), so that $\partial\phi(\dot{\epsilon}_{ij})$ and $\partial\phi_p(\dot{\epsilon}_{ij}^P)$ become those exploited in [7, 10]. In Fig. 1 are depicted the steps 1 and 2 of the proposed algorithm.

4.3 Step3: GSMs Framework and Evolutionary Constitutive Equations

Starting from the constitutive laws, after some calculations, we obtain the following differential problem

$$\begin{cases} \rho\left(\frac{\partial v_i}{\partial t} + v_k v_{i,k}\right) - (-p\delta_{ik} + 2D_{ik} + \tau_{ik})_{,k} = \rho g_i \\ \frac{1}{\omega}\left(\frac{\mathcal{D}_a \tau_{ij}}{\mathcal{D}t}\right)_d + \max\left(0, \frac{|\mathrm{T}_d| - \tau_0}{K|v_i|^n}\right)^{\frac{1}{n}} \tau_{ij} - 2D_{ij} = 0 \\ v_{i,i} = 0 \ \text{ in } (0,T)\times\Omega \\ \tau_{ij}(0) = (\tau_{ij})_0 \quad \wedge \quad v_i(t=0) = (v_i)_0 \ \text{ in } \Omega \\ \tau_{ij} = (\tau_{ij})_\Gamma \quad \wedge \quad v_i = (v_i)_\Gamma \ \text{ on } (0,T)\times\partial\Omega \end{cases} \tag{7}$$

in which $\left(\frac{\mathcal{D}_a \tau_{ij}}{\mathcal{D}t}\right)_d$ is the Gordon-Schowalter derivative [7] and $a \in [-1, 1]$ is parameter characteristic for the MR fluid. Obviously, given the presence of $\left(\frac{\mathcal{D}_a \tau_{ij}}{\mathcal{D}t}\right)_d$, we need to dimensionless the differential model (7). Therefore, introducing the Weis-

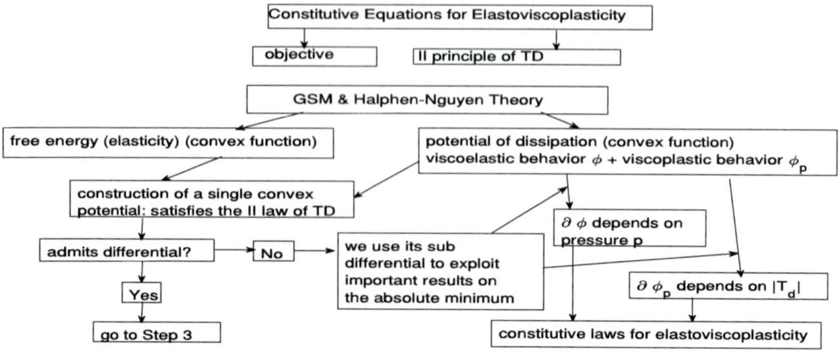

Fig. 1 Steps 1 and 2 of the proposed algorithm

senberg numbers, We, Bingham, Bi, and Reynolds, Re, and establishing the *simple shear flow condition*,[7] i.e. $\nabla v_i = ([0, 1]; [0, 0])$ and $s = |\mathrm{T_d}|$, we obtain:

$$\begin{cases} \zeta\dot{\gamma}\frac{d\tau_{11}}{dt} + k_n(|\mathrm{T}_d|)\tau_{11} - \zeta\dot{\gamma}(1+a)\tau_{12} = 0 \\ \zeta\dot{\gamma}\frac{d\tau_{22}}{dt} + k_n(|\mathrm{T}_d|)\tau_{22} + \zeta\dot{\gamma}(1-a)\tau_{12} = 0 \\ \zeta\dot{\gamma}\frac{d\tau_{12}}{dt} + \zeta\dot{\gamma}\frac{1-a}{2}\tau_{11} - \zeta\dot{\gamma}\frac{1+a}{2}\tau_{22} + k_n(|\mathrm{T}_d|)\tau_{12} = \Pi \\ \tau_{ij}(0) = 0. \end{cases} \tag{8}$$

where ζ is the relaxation time and Pi is the slowdown parameter (for details, see [7]).

Remark 1 We observe that the solution in (8), as $t \to +\infty$, tends to a constant value with a significant decrease of τ_{12} (see [6]). Moreover, in our case, $s = |\mathrm{T_d}|$ as proved in [10].

Therefore, taking into account Remark 1 and by separating the contributions due to FS and FB we finally obtain:

$$\begin{cases} \zeta\dot{\gamma}\frac{d\tau_{11}^{FS}}{dt} = -k_1(|\mathrm{T}_d|)\tau_{11}^{FS} + k_1(|\mathrm{T}_d|)p^F \\ \zeta\dot{\gamma}\frac{d\tau_{22}^{FS}}{dt} = -k_1(|\mathrm{T}_d|)\tau_{22}^{FS} + k_1(|\mathrm{T}_d|)p^F \\ \zeta\dot{\gamma}\frac{d\tau^{FB}}{dt} = -k_1(|\mathrm{T}_d|)\tau^{FB}; \quad \frac{\zeta\dot{\gamma}}{\Pi}\frac{1-a}{2}\tau_{11}^F - \frac{\zeta\dot{\gamma}}{\Pi}\frac{1+a}{2}\tau_{22}^F = 1 \\ \tau_{ij}(0) = 0, \end{cases} \tag{9}$$

where $k_1(|\mathrm{T}_d|)\tau_{ij} = \tau_{ij} - \frac{Bi}{|\mathrm{T}_d|}\tau_{ij}$ [10], so that models (4) and (9) are ready to be compared. In Fig. 2 is depicted the step 3 of the proposed algorithm.

[7] This hypothesis is plausible since in the usual rheometers it is the operating condition more frequent.

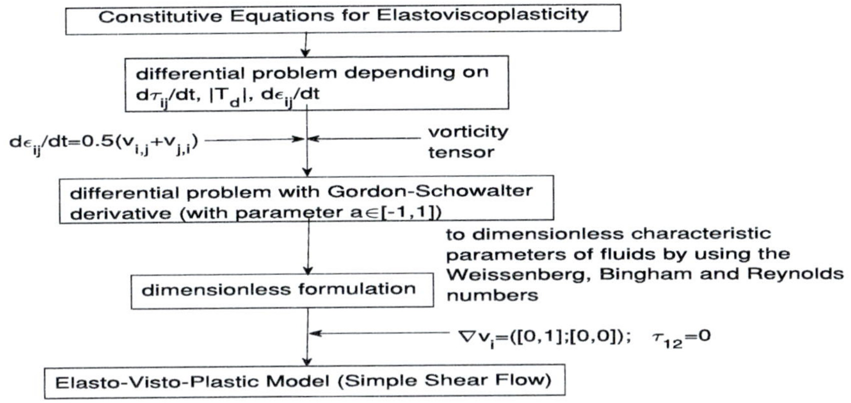

Fig. 2 Step 3 of the proposed algorithm

5 Some Important Results on the Qualitative Correspondence Between the Models

The following results prove the qualitative correspondence between the theoretical model and the experimental model.

Theorem 1 *$\zeta\dot{\gamma}\frac{d\tau_{11}^{FS}}{dt}$, $\zeta\dot{\gamma}\frac{d\tau_{22}^{FS}}{dt}$, and $\zeta\dot{\gamma}\frac{d\tau^{FB}}{dt}$ in (9) correspond, qualitatively, to $\lambda_4\mu^S\frac{d\tau_{11}^{FS}}{dt}$, $\lambda_4\mu^S\frac{d\tau_{22}^{FS}}{dt}$ and $\lambda_5\mu^B\frac{d\tau^{FB}}{dt}$ in (4), respectively.*

Theorem 2 *$-k_1(|\mathrm{T}_d|)\tau^{FB}$ in (9) corresponds, qualitatively, to $-\tau^{FB}+\mu^B v_{1,1}^F+\mu^B v_{2,2}^F$ in (4).*

Theorem 3 *$-k_1(|\mathrm{T}_d|)\tau_{11}^{FS}+k_1(|\mathrm{T}_d|)p^F$ and $-k_1(|\mathrm{T}_d|)\tau_{22}^{FS}+k_1(|\mathrm{T}_d|)p^F$ in (9) correspond, qualitatively, to $-\tau_{11}^{FS}+\mu^S D_{11}^F$ and $-\tau_{22}^{FS}+\mu^S D_{22}^F$ in (4), respectively.*

Theorem 4 *The information content present in the fourth equation of (9) is contained, in the pre-yield region, in the dissipative inequality of (4) (i.e., (5)).*

6 Conclusions and Perspectives

The proposed algorithm is able to qualitatively prove the correspondence, in the pre-yield region, of the theoretical model mixture of modeling of MR fluids performed by Chen-Yeh with the elasto-visco-plastic generalization of the Susan-Resiga model (in simple shear flow). This allows us to state that the Susan-Resiga model, which is easier to implement, qualitatively represents a "simplified version" of the theoretical "mixture" model which is characterized by prohibitive computational costs, in order to get an implementable version of the mixture model very interesting for industrial applications such as automotive industry. Obviously, the algorithm proposed in this paper is to be considered as the starting point of a research line that aims at obtaining an efficient algorithm for this important qualitative comparison. Moreover, the link between τ_0 and the induction field, **B** in the modeling of MR fluids in the pre-yield region to obtain the mathematical link between the impressed electric current and τ_0, is still poorly modeled in literature. This is mainly due to the fact that, in the theoretical model, in the pre-yield region, the action of **H** (and, then, **B**) is "perceived" only by S. Therefore, in the near future, it is our intention to act on the theoretical model so that F also "perceives" the action due to **H**.

References

1. Brigadnov, I.A., Dorfmann, A.: A mathematical modeling of magnetorheological fluids. Continuum Mech. Thermodyn. **17**, 29–42 (2000)
2. Drouot, R., Napoli, G., Racineux, G.: Continuum modeling and testing of magnetorheological fluid in low shear rate sweezing flows. Int. J. Mod. Phys. B. **21**, 1–6 (2001)
3. Morabito, F.C.: Independent component analysis and feature extraction techniques for NDT data. Mater. Eval. **58**(1), 85–92 (2009)
4. Versaci, M., Cutrupi, A., Palumbo, A.: A magneto-thermo-static study of a magneto-rheological fluid damper: a finite element analysis. IEEE Trans. Magn. **57**(1), 9234465 (2021)
5. Hajalilou, A., Mazlan, S.A., Lavvafi, H., Shameli, K.: Field Responsive Fluids as Smart Materials. Springer Nature, Singapore (2016)
6. Saramito, P.: A new elastoviscoplastic model based on hesrschel-bulkley viscoplastic model. J. Non-Newtonian Fluid Mech. **158**(1), 154–161 (2009)
7. Saramito, P.: Complex Fluids, Modeling and Algorithms. Springer International Publishing, Switzerland (2016)
8. Jou, D., et al.: Extended Irreversible Thermodynamics. Springer Nature, Singapore (2010)
9. Chen, K.C., Yeh, C.S.: Extended irreversible thermodynamics approach to magnetorheological fluids. J. Non-Equilib. Thermodyn. **15**(1), 495–510 (2002)
10. Versaci, M., Palumbo, A.: Magnetorheological fluids: qualitative comparison between a mixture model in the extended irreversible thermodynamics framework and an Herschel-Bulkley experimental elastoviscoplastic model. Int. J. Non-linear Mech. **118**(103288), 1–13 (2020)
11. Susan-Resiga, D.: A rheological model for magneto-rheological fluids. J. Intell. Mater. Syst. Struct **20**, 1001–1010 (2009)

DataX: A System for Data eXchange and Transformation of Streams

Giuseppe Coviello, Kunal Rao, Murugan Sankaradas, and Srimat Chakradhar

Abstract The exponential growth in smart sensors and rapid progress in 5G networks is creating a world awash with data streams. However, a key barrier to building performant multi-sensor, distributed stream processing applications is high programming complexity. We propose DataX, a novel platform that improves programmer productivity by enabling easy exchange, transformations, and fusion of data streams. DataX abstraction simplifies the application's specification and exposes parallelism and dependencies among the application functions (microservices). DataX runtime automatically sets up appropriate data communication mechanisms, enables effortless reuse of microservices and data streams across applications, and leverages serverless computing to transform, fuse, and auto-scale microservices. DataX makes it easy to write, deploy and reliably operate distributed applications at scale. Synthesizing these capabilities into a single platform is substantially more transformative than any available stream processing system.

1 Introduction

Smart sensors today sense, digitize and communicate over the internet [15]. The exponential growth in such sensors [27], and progress in 5G networks [33], is creating a world awash with digital streams. In the future, this exponential growth in smart sensors will result in billions of digital data streams, each describing an increasingly smaller aspect of the physical or digital worlds in greater detail. A rich understanding

G. Coviello (✉) · K. Rao · M. Sankaradas · S. Chakradhar
NEC Laboratories America, Inc., Princeton, NJ 08540, USA
e-mail: giuseppe.coviello@nec-labs.com

K. Rao
e-mail: kunal@nec-labs.com

M. Sankaradas
e-mail: murugs@nec-labs.com

S. Chakradhar
e-mail: chak@nec-labs.com

D. Camacho et al. (eds.), *Intelligent Distributed Computing XIV*, Studies in Computational Intelligence 1026, https://doi.org/10.1007/978-3-030-96627-0_29

of these complex worlds, which will be impossible to do by using information from any single sensor, will inevitably require the fusion of information in various data streams.

A key barrier to building performant multi-sensor distributed applications is high programming complexity, which results in the low productivity of application developers. *First*, they have to express the functionality of an application as a collection of interacting microservices [12, 16] that form an efficient processing pipeline. There is no standard way to do this for arbitrary applications. *Second*, they have to design and specify appropriate data communication mechanisms among the distributed microservices, and this has a big impact on the performance of the application. *Third*, they have to worry about the actual hardware on which the microservices will run and therefore, manage the underlying computing infrastructure as well. *Fourth*, if any of the microservices need to maintain state, then developers are forced to manage appropriate databases.

Scaling multi-sensor stream processing applications, and ensuring reliable operation in the face of software and hardware failures, is yet another problem that application developers have to address. Furthermore, if the developers want to extend their application pipeline by reusing microservices from another application pipeline, then there is no easy way to do so, other than obtaining the microservice, understanding the interfaces, implementing appropriate data communication, and then embedding the microservice within the desired application pipeline. This is very tedious and cumbersome. Removing such barriers and improving application developer productivity is the focus of this paper.

In this paper, we present DataX, a platform that improves programmer productivity by enabling easy exchange, transformation and management of data streams in complex, multi-sensor distributed stream processing applications. We propose simple programming abstractions and SDKs to specify multi-sensor applications. Developers define and register objects like sensors, drivers, streams, analytics units, actuators, and gadgets, all of which enable succint specification of the overall application pipeline (Sect. 2).

DataX runtime (Sects. 4 and 5) automatically determines appropriate data communication mechanisms among the application's objects, including network connections, serialization and deserialization of data streams. DataX runtime provides serverless computing, and the developers only provide the business logic for different types of analytics processing. Application-specific allocation, scheduling and execution on the underlying distributed computing resources, as well as auto-scaling and reliable operation, is automatically managed by DataX. For stateful applications, DataX manages and exposes a database, which can readily be used by the developer. DataX also makes it easy for developers to effortlessly reuse any registered microservice in any application pipeline by just subscribing to the service's output streams.

The synthesis of all of these capabilities into a single platform like DataX is substantially more transformative than any known stream processing system. To the best of our knowledge, there is no other system that addresses all these developer concerns and simplifies multi-sensor application development, deployment and reliable operation on different distributed computing fabrics.

2 DataX Abstraction

DataX offers a simple abstraction for building complex, multi-sensors applications. It allows developers to focus on the business logic rather than the complex intricacies of writing, deploying, or operating containerized applications on distributed systems. Application developers use DataX abstractions like *sensors*, *gadgets*, *streams*, *analytics units* and *actuators* to build distributed stream processing applications.

A *sensor* is a device that produces raw data. Examples of sensors include cameras, location sensors, temperature and pressure sensors, LIDAR, and radar. Applications process and analyze raw data from sensors to generate insights. Sensors may have wired or wireless networking capability, or they may be physically attached to a computing device through interfaces like USB. Sensors are the starting point of a DataX application.

A *gadget* is a (physical or virtual) device that can be controlled by using insights derived from data analysis. Examples of gadgets include entry/exit gates, displays, LED arrays, or graphical user interfaces. Like sensors, some gadgets may have networking capability, while others may be attached to computing devices. Gadgets are ideally the ending point of a DataX application. Not all DataX applications may have gadgets. Also, a physical or virtual device can be both a sensor and a gadget if the device produces data, and the device can also be controlled. For example, a camera can produce a video stream; and it also allows remote management to configure camera parameters.

A data *stream* is a continuous flow of homogeneous discrete messages. Some streams have only data produced by the sensors, while others can contain the insights obtained by analyzing and fusing the data from sensors. An *analytics unit* (AU) processes on one or more input streams to mine insights (i.e., analytical processing), or transforms the input streams for future analysis, and produces output streams with rich insights obtained from analytical processing.

For example, DataX produces a stream containing the frames (images) captured for each registered camera. An object detector, which is an example of an analytics unit, can subscribe to the stream, process the stream, and generate an output *augmented* stream that has all the objects detected in each frame. The augmented stream can be used as input to other analytics units or *actuators*, which control various gadgets by using the information in the input streams.

Figure 1 shows the analytics processing pipeline in a generic DataX application. Application developer provides the business logic (or full docker image) for generating a stream from a sensor (i.e., a driver), analyzing a stream or fusing multiple streams (i.e., an analytics unit), or controlling a gadget (i.e., actuator). Application developers can also re-use business logic functions already available in DataX like drivers for cameras, actuators for entry/exit turnstiles, and analytics units for

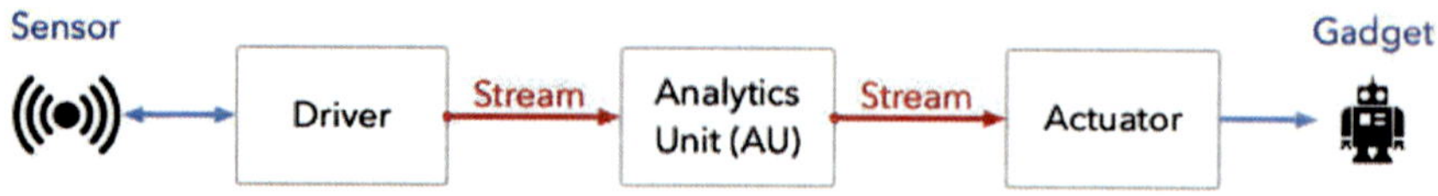

Fig. 1 A Generic DataX Application

recognizing faces or detecting objects. DataX also provides a simple SDK (software development kit) for writing drivers, analytics units, and actuators. DataX supports *serverless* computing. Application developers only provide business logic to implement drivers, AUs, and actuators. For stateful applications, DataX provides an abstraction for creating databases and attaching them to drivers, analytics units, and actuators. DataX installs and maintains the databases, while applications are responsible for the content in the databases (like creating a schema or storing data).

3 Benefits

Simple abstraction. DataX abstractions and SDK are intuitive and easy to use. Analytics applications typically receive input data from one or more sensors using drivers. The sensor data from drivers then undergoes a series of transformations and processing by analytics units, which produce output streams that can activate a gadget through an actuator. Sensors, drivers, analytics units, actuators, and gadgets are all simple DataX abstractions, which make it easier to write, as well as deploy and reliably operate distributed applications at scale.

Automated data communication. DataX automatically chooses appropriate communication mechanisms to exchange data among drivers, analytics units, and actuators to accomplish application-specific goals. In addition, DataX handles the network connections as well as manages serialization and deserialization of data when data is being transferred among drivers, analytics units and actuators. By handling all aspects of data communication, DataX relieves application developers from additional effort to design application-specific data communication.

Serverless stream processing. In addition to writing the software components, application developers typically also have to worry about the computing resources necessary to execute these components. This is often not so straightforward in a distributed system, where multiple applications are running simultaneously on a shared computing infrastructure that is dynamic and heterogeneous. This daunting task of specifying and managing execution on the underlying compute resources is automatically handled by DataX. This is similar to serverless computing, where developers only provide the business logic and actual execution is handled transparently.

Easy state management. AUs sometimes need to manage state to provide specific functionality. Developers usually have to handle state by themselves, and there is little to no support from the underlying platform. DataX makes this state management easy by exposing in-built database management systems and the associated databases. Developers can choose the specific database, create the desired schema, and manage the desired content/state. This DataX platform-provided state management makes it easy for developers to manage any state within and across AUs.

Effortless data streams reuse. One of the most beautiful features about DataX is that it allows developers to reuse data streams produced by third-party applications effortlessly. This fosters incremental development and there is no need to reinvent the wheel; developers already know what sensors are registered with DataX and what kind of augmented streams are produced from various sensor data streams. Using this knowledge from DataX, developers can quickly build new and more exciting sensor-fusion applications by reusing existing raw or augmented data streams.

4 Design and Implementation

The target of DataX is programmers, and the focus is on programming abstractions for application developers. These abstractions simplify application development, and make distributed stream processing applications easier to write. Furthermore, data scientists who are users of DataX are likely not familiar with Kubernetes or other container management systems. Enabling scientists to rapidly get their work done, and focus on their domain-specific streaming applications rather than distributed systems, is the primary goal of DataX.

Accordingly, the implementation of DataX ensures that developers and programmers who use DataX platform are largely unaware that they are running on top of Kubernetes, the de-facto standard for cloud-native development. Kubernetes is an open-source orchestrator for deploying containerized applications. It is a complex system that makes applications faster to deploy and more reliable to operate at scale. Although Kubernetes simplifies the deployment and operations of distributed applications, it does little to make the development of distributed applications easy. With DataX as a complete abstraction on top of Kubernetes, we can now not only deploy and reliably operate containerized applications at scale, but we can also simplify application development and make stream processing applications easier to write.

There are at least two different options for realizing DataX on top of Kubernetes. *First* option is to use stock Kubernetes API server out of the box. Applications developers use Kubernetes's resources like Pods, StatefulSets, and ReplicaSets, to describe the application's desired state, and Kubernetes deploys the application and strives to ensure that the current state matches the declarative description of desired state [7]. However, in this option, Kubernetes does not understand the workload running in Kubernetes, objects like StatefulSets or ReplicaSets, or the application's domain-specific knowledge of running a specific application. *Second*, and more performant option is to extend the Kubernetes API server by adding new functionality using operators [11], which allow encapsulation of domain-specific knowledge of running a specific stream processing application. DataX extends Kubernetes API by defining operators, allowing DataX to take advantage of the vast Kubernetes ecosystem of tools.

DataX defines *custom resources* as extensions of Kubernetes API that are not available in a default Kubernetes installation. A resource is an endpoint in Kubernetes API that stores a collection of API objects of a certain kind; for example, the built-in Pods resource contains a collection of Pod objects. Once a custom resource is installed, we can create and access its objects using kubectl, just as we do for built-in resources like Pods. DataX makes components like driver, AU, actuator, sensor, gadget, stream, and databases as first-class citizens of Kubernetes by installing them as custom resources.

DataX also extends Kubernetes API by defining a new operator, which is a software extension to Kubernetes that makes use of custom resources to manage DataX applications and their components. The Operator pattern aims to capture the key aim of a human operator managing a service or set of services. Human operators who look after specific applications and services have deep knowledge of how the system should behave, how to deploy it, and how to react if there are problems. The Operator pattern [11] captures how you can write code to automate a task beyond what Kubernetes itself provides. DataX Operator adds one more custom resource to Kubernetes. The Operator provides the logic for monitoring and maintaining the resources it defines, which means that the Operator will take actions based on the resource's state defined by DataX.

Extending Kubernetes API server leads to multiple benefits. DataX administrators can interact with Kubernetes using familiar and standard interface (YAML resource descriptions, kubectl, etc.). Also, DataX components like driver, AU, actuator, sensor, gadget, stream, and database become Kubernetes-managed objects, just like a pod, for example. This way, DataX does not have to implement new, custom software modules to auto-scale, store or update application domain-specific components.

Figure 2 shows (in yellow) the components of DataX: a Kubernetes Operator defines and manages DataX's custom resources, a message bus facilitates commu-

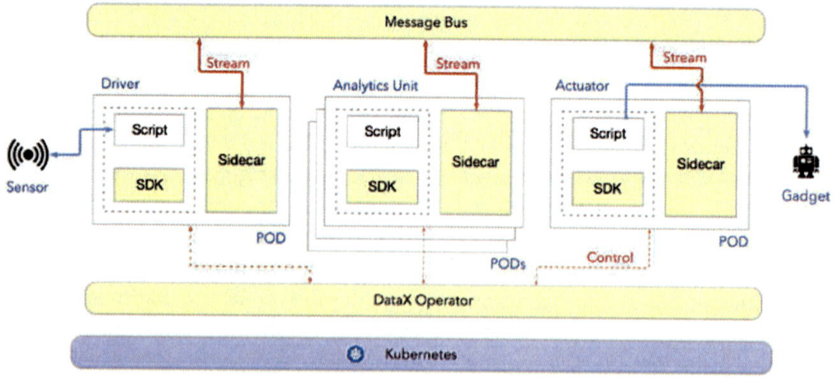

Fig. 2 DataX Components

nication among the microservices of the application, a sidecar container in each Pod handles data communication among various DataX entities (drivers, AUs or actuators), and language-specific SDKs enable easy use of DataX APIs.

DataX Operator. The DataX Operator extends Kubernetes to support the entities defined in Sect. 2; it provides the CRDs (custom resource definition) and logic for the driver, AU, actuator, sensor, gadget, and stream. The DataX Operator takes necessary actions to ensure that all DataX applications are in a coherent state at all times. It also protects the system from user's actions that might bring the system into an unrecoverable incoherent state. For example, uninstalling a driver while a sensor is being used can bring the system into an incoherent state, which will be detected and avoided.

Drivers, AUs, and actuators. To register a driver, an AU, or an actuator, the user provides the name, either a script (pure serverless) or a docker image containing the business logic and (optionally) the configuration schema. When the user provides just a script, DataX will automatically create a docker image for executing the script. When the user requests an upgrade of drivers, AUs, and actuators, DataX Operator will automatically cascade the upgrade to running instances. DataX Operator ensures the coherency of the upgrade by enforcing that new configuration schemas are compatible with the schemas of the running instances. When performing an upgrade, optionally, the user can provide a script to convert the configuration schemas; in such case, DataX Operator will accept the upgrade only if the script can be executed successfully for all the running instances (compatible configurations are generated for all instances). Also, if a user requests the deletion of a driver, AU, or actuator, then DataX Operator will check if the entity is currently in use and refuse the operation if there is already a running instance for that entity.

Sensors, streams, and gadgets. When registering a sensor, DataX Operator ensures that (a) the required driver is installed, and (b) the driver configuration schema provided by the user is compatible with the configuration schema expected by the installed driver. DataX Operator will also maintain the driver's running instance (on appropriate computing resources) as long the sensor is registered. For example, if the sensor is physically attached to a computing node through a USB interface, then DataX Operator will maintain a running instance on the same computing node. A registered sensor always generates an output stream that has the same name as the sensor. The user requests to create augmented streams by providing: (a) AU that generates the stream, (b) the input streams, and (c) the AU configuration. DataX Operator checks that the AU is available, the configuration is compatible, and the input streams are registered; DataX Operator, unless the user requests a fixed number of instances, auto-scales the number of instances of the AU. DataX Operator performs similar operations when the user requests to register a new gadget. Before deleting any sensors or streams, DataX Operator ensures that they are not input to produce other streams.

Message bus. Communication among microservices in a distributed application is typically accomplished by using either (some variant of) REST/HTTP or a messages bus/distributed queue [5]. DataX uses NATS, which is a lightweight, distributed, and scalable message queue [26], for automatically setting up communication mechanisms among various entities like drivers, AUs, and actuators. DataX Operator manages the deployment and configuration of NATS (the DataX Operator uses Kubernetes DaemonSet construct to deploy a NATS cluster). The NATS cluster uses authentication and authorization, and only services deployed on DataX will be able to connect to the NATS server (they will be able to subscribe and publish only on the defined and registered streams).

DataX Sidecar. DataX Sidecar is a containerized application that DataX Operator runs alongside each instance of a user-provided driver, AU, or actuator (in the same Pod). The main role of the DataX Sidecar is to automatically manage data communication (it manages the connection, subscriptions, and publishing to the messages bus). Also, DataX Sidecar monitors the health of the user's application; it exposes, using REST API, the metrics such as the systems resources utilization and the number of messages received, dropped, and published. DataX Operator and Kubernetes use those metrics to ensure that all the components are working correctly, and these metrics also drive the auto-scaling process.

DataX SDKs. DataX provides SDKs for several programming languages (C, Go, JavaScript, Python). SDKs provide APIs for retrieving configuration, receiving data, and publishing data. These APIs are implemented using the idiom of the programming language; for example, SDK for Python exposes a class `DataX` having three public methods: `get_configuration()`, to retrieve the configuration in the form of a dictionary (pairs of key-values); `next()` to obtain a message from one of the input streams, this method returns a tuple of two values, the name of the stream and the message in the form of a dictionary where keys are strings; `emit(message)` for publishing a new message, which is a dictionary where keys are strings. DataX SDKs are simple and lightweight. DataX Sidecar does most of the work in managing data communication, and the SDKs provide an interface between DataX Sidecar and the business logic (SDKs and DataX Sidecar use shared memory for communicating with each other).

5 Distributed Application Development Using DataX

We designed a popular fever screening application [28] pipeline, shown in Fig. 3 on DataX. This complex DataX application has two sensors, two drivers, one actuator, one gadget, five analytics units, and one database. Application developer only provides the business logic for these entities, and DataX takes care of stream data communication, serverless computing, state management, and stream reuse. Also, DataX leverages Kubernetes to deploy and reliably operate the application at scale. The synthesis of these capabilities into a single product like DataX is substantially more transformative than any available stream processing system.

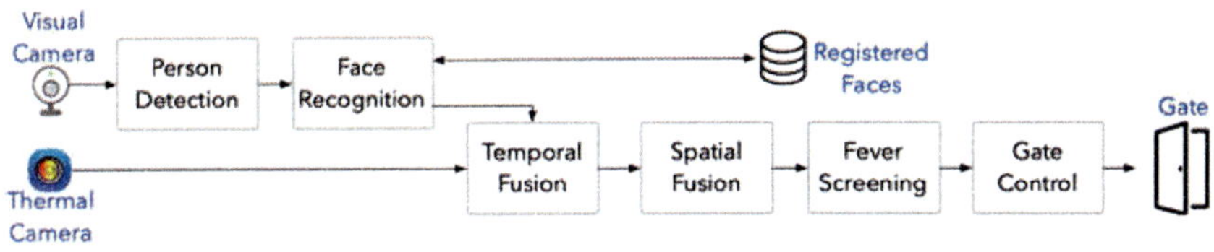

Fig. 3 Fever screening application on DataX

6 Related Work

There are several open source and proprietary stream processing systems like Flink [14], Spark Streaming [9], Storm [30], Heron [18], ksqlDB [23], Esper [13], Hazelcast Jet [17], Hitachi Streaming Data Platform [19], IBM Streams [21], Microsoft Azure Stream Analytics [24], Oracle Stream Analytics [25], SAS Event Streaming [29], TIBCO Streaming [34], etc. that provide capabilities to handle data streams. However, these systems have full-fledged and specific programming models and supporting runtime systems, which restricts the way applications are developed in order to be deployed on these systems. In contrast, DataX provides flexibility in choice of programming language and simplifies programming using simple abstractions.

There are systems for stream data integration like Apache Airflow [1], Cloudera DataFlow [8], Data Stream Manager [10], IBM DataStage [20], Informatica Data Engineering Streaming [22], StreamSets Data Collector [31], Striim Platform [32], etc. that deal with ingestion and storage of data streams. Such storage of data streams is not included in DataX to avoid additional storage overhead.

AWS Lambda [3] is a service provided by Amazon for running serverless computing, wherein the developer provides just a "Lambda" function that operates on data. These functions can be combined and integrated with Amazon Kinesis [2] to create a pipeline for processing streaming data in real-time [4, 6]. These services i.e. Kinesis and Lambda operate as separate services. DataX, on the other hand, combines functionality of Amazon Kinesis and AWS Lambda into a middleware product that can be deployed anywhere, including resource constrained edge environments.

7 Conclusion

A key barrier to building multi-sensor distributed stream processing applications is high programming complexity. DataX makes it easy to write, deploy, and reliably operate distributed stream processing applications at scale. Although DataX uses Kubernetes and NATS internally, DataX can be implemented on other container management or message queuing systems. DataX can be used as an independent platform or embedded in any streaming application. Also, DataX applications can run on-premise, in the cloud, or any edge-cloud computing fabric.

References

1. Airflow, A.: Apache Airflow. https://airflow.apache.org/
2. Amazon: Amazon Kinesis Data Analytics. https://aws.amazon.com/kinesis/data-analytics/
3. Amazon: AWS Lambda. https://aws.amazon.com/lambda/
4. Amazon: Serverless Reference Architecture: Real-time Stream Processing. https://github.com/aws-samples/lambda-refarch-streamprocessing
5. Bakshi, K.: Microservices-based software architecture and approaches. In: 2017 IEEE Aerospace Conference, pp. 1–8. IEEE (2017)
6. Beswick, J.: Using AWS Lambda as a consumer for Amazon Kinesis, 30 Sept. 2020. https://aws.amazon.com/blogs/compute/using-aws-lambda-as-a-consumer-for-amazon-kinesis/
7. Burns, B., Beda, J., Hightower, K.: Kubernetes: up and running: dive into the future of infrastructure. O'Reilly Media (2019)
8. Cloudera: Cloudera DataFlow. https://www.cloudera.com/products/cdf.html
9. Databricks: Spark Streaming. https://databricks.com/glossary/what-is-spark-streaming
10. datastreams.io: Data Stream Manager. https://datastreams.io/wp-content/uploads/2018/12/User_Manual_Data_Stream_Manager.pdf
11. Dobies, J., Wood, J.: Kubernetes Operators: Automating the Container Orchestration Platform. O'Reilly Media, Inc. (2020)
12. Dragoni, N., Giallorenzo, S., Lafuente, A.L., Mazzara, M., Montesi, F., Mustafin, R., Safina, L.: Microservices: Yesterday, Today, and Tomorrow (2017)
13. EsperTech: Esper. https://www.espertech.com/esper/
14. Flink, A.: Apache Flink. https://flink.apache.org/
15. Fortino, G., Trunfio, P. (eds.): Internet of Things Based on Smart Objects. Internet of Things. Springer, Cham (2014). https://doi.org/10.1007/978-3-319-00491-4
16. Fowler, M., Lewis, J.: Microservices (2014). https://martinfowler.com/articles/microservices.html
17. Hazelcast: Hazelcast Jet. https://hazelcast.com/products/stream-processing/
18. Heron, A.: Apache Heron. https://heron.apache.org/
19. Hitachi: Hitachi Streaming Data Platform. https://download.hitachivantara.com/download/epcra/hsdp0034.pdf
20. IBM: IBM DataStage. https://www.ibm.com/products/infosphere-datastage?cm_sp=Scheduler-_-CopyChng2-_-C
21. IBM: IBM Streams. https://www.ibm.com/downloads/cas/NAX7DDYQ
22. Informatica: Informatica Data Engineering Streaming. https://www.informatica.com/content/dam/informatica-com/en/collateral/data-sheet/informatica-big-data-streaming_data-sheet_3236en.pdf
23. ksqlDB: ksqlDB. https://ksqldb.io/
24. Microsoft: Azure Stream Analytics. https://azure.microsoft.com/en-us/services/stream-analytics/
25. Oracle: Oracle Stream Analytics. https://www.oracle.com/middleware/technologies/stream-processing.html
26. Quevedo, W.: Introduction to NATS. In: Practical NATS, pp. 1–18. Springer (2018)
27. Ranger, S.: What is the IoT? Everything you need to know about the Internet of Things right now, Feb. 2020. https://www.zdnet.com/article/what-is-the-internet-of-things-everything-you-need-to-know-about-the-iot-right-now/
28. Rao, K., Coviello, G., Feng, M., Debnath, B., Hsiung, W., Sankaradas, M., Yang, Y., Po, O., Drolia, U., Chakradhar, S.: F^3S: free flow fever screening. In: 7th IEEE International Conference on Smart Computing (SMARTCOMP 2021), pp. 276–285 (2021)
29. SAS: SAS Event Stream Processing. https://www.sas.com/en_us/software/event-stream-processing.html
30. Storm, A.: Apache Storm. https://storm.apache.org/
31. StreamSets: StreamSets Data Collector. https://streamsets.com/products/dataops-platform/data-collector/

32. Striim: Striim Platform. https://www.striim.com/
33. Tamsons, A.: How 5G and the Internet of Things can create a winning business (Jan. 2020). https://www.ericsson.com/en/blog/2020/1/how-5g-and-the-internet-of-things-can-create-a-winning-business
34. TIBCO: TIBCO streaming. https://community.tibco.com/products/tibco-streaming

Monitoring Hierarchical Systems for Safety Assurance

Franz Wotawa and Horst Lewitschnig

Abstract Assuring safety for autonomous safety-critical systems like cars equipped with autonomous driving functionality seems to be hard if even impossible to achieve. Checking the behavior of the system online during operation regarding its degree of fulfillment of given safety requirements provides an alternative countermeasure for hazards. In this paper, we discuss the concept of monitoring devices that implement run time verification based on safety and functional requirements. We introduce a hierarchical approach where monitoring information is passed from one lower level to other higher-levels in order to finally come up with verification results that would not have been able to achieve at the lower level. Besides presenting the principles, we use a potential example from the automotive industry for illustrating the approach.

1 Introduction

Safety critical systems, like ordinary cars, are systems where faults or malfunctions may seriously harm people, properties, or the environment. For such systems it is inevitable to assure that faults or malfunctions do not arise during operation or to define appropriate mitigation mechanisms that prevent from harm. Because of increased functionality including the introduction of automated or autonomous (sub-) systems, the challenge of safety assurance becomes more and more difficult. For example, in autonomous driving ordinary mitigation mechanisms like passing control to the driver may not be possible anymore.

Even worst in autonomous driving, the vehicle has to cope with a huge number of situations where only a part has been maybe considered during verification before

F. Wotawa (✉)
Graz University of Technology, Institute for Software Technology, Inffeldgasse 16b/2, 8010 Graz, Austria
e-mail: wotawa@ist.tugraz.at

H. Lewitschnig
Infineon Technologies Austria AG, Siemensstrasse 2, 9500 Villach, Austria
e-mail: Horst.Lewitschnig@infineon.com

D. Camacho et al. (eds.), *Intelligent Distributed Computing XIV*, Studies in Computational Intelligence 1026, https://doi.org/10.1007/978-3-030-96627-0_30

deployment. This is due to the fact that the environment an autonomous car has to drive provides a huge number of different potential combinations of environmental elements. For example, a street can have different shapes ranging from straight lines, to curves, crossings of various sizes, and roundabouts. The marking of the street may vary as well. There might be one or several lanes in every direction. We may see a lot of obstacles passing the street including pedestrians, but also animals or even plastic bags moved by the wind. When considering only 100 different parameters each may take 5 values only, there are $5^{100} \approx 10^{70}$ different situations. Hence, when verifying such systems, we have to focus on the most critical cases or combinations.

In order to test autonomous and automated driving systems, there has been a lot of work already done. For a general overview of challenges we refer to [1] and more recently [2]. In the latter paper, the author summarized work in the context of combinatorial testing [3], and genetic programming [4]. In [5] a more detailed comparison of these two methods was provided. However, in all the previous work, test case generation was limited coping only with some parts of the environment that are most appropriate for the automated driving function to be tested. It is worth noting that even random testing was able to provide critical scenarios, which may not be the case for a real autonomous car.

When considering the huge search space, we have to deal with for verifying autonomous driving, we may ask ourselves whether alternative ways of assuring safety exists. In this paper, we contribute answering this question and provide an approach that checks for validity or safety during operation of the system, i.e., at run time. The underlying idea is to monitor the system state and the values of important variables of the system, introduce constraints representing safety requirements or the validity of systems, and to check whether the system fulfills these requirements over time. In case of contradictions arising during this check, mitigation procedures have to be triggered aiming at avoiding disasters. Note that we will not focus on the latter part in this paper, but assume that the checking part can be handled. Hence, future research has to focus on providing decision systems that take the outputs of monitoring into account for coming up with appropriate mitigation procedures.

Monitoring of systems and in particular condition monitoring is a concept that is already reflected in several standards, like [6–8]. In a wider sense, monitoring can be seen as part of a prognostics and health management (PHM) system, for example to detect failures occurring during operation. Another typical application of monitoring is predictive maintenance. The monitoring system gives a warning that triggers the maintenance operation (see, e.g., [9]).

For automotive applications, a concept how to support PHM is given in [10]. This standard describes the principles how to exchange health status information in an automotive application. In the context of this standard, health status information can be seen in a wide range, from warning to the reporting of failing functionalities or devices.

Condition monitoring, in our case with the mean of a monitoring system, gets more and more relevance for automotive applications. On one hand, if it is known when a device will fail, proper action can be taken in advance and a failure can be avoided. If this concept works in a reliable way, redundancy in the design can be saved. Condition monitoring contributes to cost efficiency, reduced complexity, and safe operation. It turns a failure into a maintenance action. This has also a major impact to functional safety analysis, like for semiconductors ([11]). On the other hand, components can be highly reliable. In classical stress tests, there are hardly any failures seen. Condition monitoring tracks the health status of a component itself and can, therefore, support to calculate the remaining useful lifetime (RUL) of this component ([12]). This is based on the specific usage conditions and strength of a device.

In this paper, we restrict our view on hierarchical systems. Such systems comprise other systems as their components are highly interconnected. Ordinary but also autonomous cars fall into such a category of systems. A car comprises various parts. Those parts themselves have sub-parts like sensors or actuators. In Fig. 1, we depict such an hierarchical structure. There we include the different levels of the hierarchy including examples of for all level instances like sensors used to comprise a LiDAR that is used for object detection in autonomous driving. Afterwards, we show the different stages of monitoring, where we assume a monitoring device (or short MonDev) in every level, and an example of how information is passed from one level to the other. It is worth noting that this communication of errors, values, and other data can be within monitoring devices of immediately following levels or go up to higher-levels. At each of the levels the monitoring device may use different sets of safety or validity constraints to check the overall health of the system. In this paper, we will see that passing information over different monitoring levels is required to identify critical situations that cannot be handled alone at a certain level.

The proposed hierarchic monitoring approach seems to be appropriate for our purpose, because it allows for checking safety (and other requirements) during operation at different stages of the hierarchy. A distribution independently from the system itself seems to be possible. Moreover, because of identifying critical situations at run time, risks regarding faults or malfunctions might be reduced which leads to more safety. However, this can only be achieved when appropriate mitigation mechanisms are installed as well (which we assume to exist).

This paper is organized as follows. In the next section, we formally introduce the proposed monitoring concept for hierarchical systems, which may serve as a specification for a concrete implementation. Afterwards, we outline one use case from the automotive industry showing that the concept is able to cope with critical scenarios (i.e., interactions between the behavior of different sub components) to assure safety. Finally, we discuss future research activities.

2 Monitoring of Hierarchical Systems

Hierarchical systems are systems that are tree-structured comprising interconnected components as nodes. There is exactly one root node, i.e., a node that has no predecessors, and a set of leave nodes, i.e., nodes having no successors. We further assume that all nodes are working independently, and that communication works in

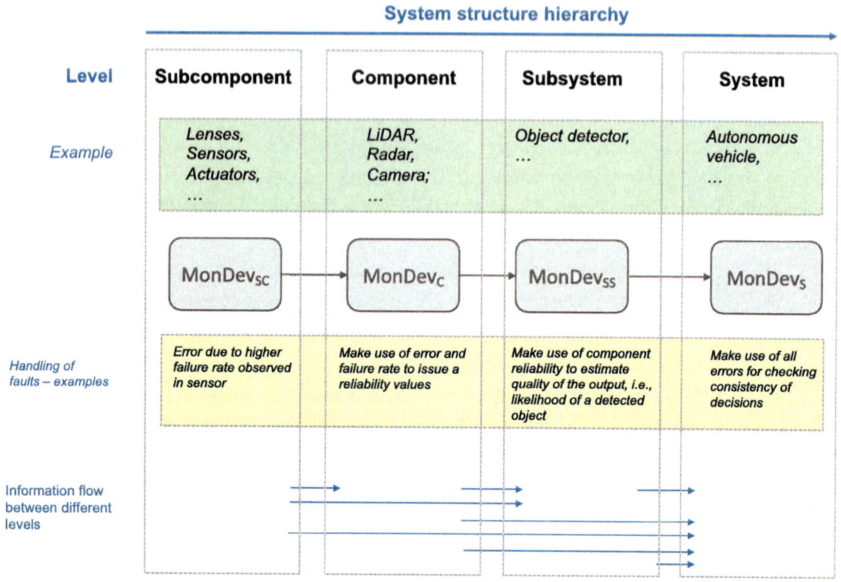

Fig. 1 System structure hierarchy

the direction from the leaves to the root. This is not a restriction, because we only consider this tree-structured systems for monitoring purposes. For implementing the real behavior (which we do not consider) there might be an additional communication network not being tree-structured. It is worth noting that the proposed monitoring system is a separate, add-on unit of a system.

In the following, we formally introduce the concepts behind the monitoring and the required information. These concepts may serve as specification for concrete implementations in practice. We start with defining a system.

Definition 1 (System) A tuple $(COMP, CONN)$ is a system where $COMP$ is a set of components, and $CONN$ a set of connections between components, i.e., a set of pairs from $COMP \times COMP$. For a system, the following must hold:

- There is only one node $C \in COMP$ for which there is no $(C', C) \in CONN$. This node C is called the root note.
- For all $(C, C') \in CONN$ $C \neq C'$ must hold, and there is no (C'', C') in $CONN$ with $C \neq C''$.
- For all $C \in COMP$ there must be a (C, C') or a (C', C) in $CONN$.

Because of simplicity, we further introduce the following functions:

- *Predecessor* of $C \in COMP$: $pred(C) = \{C' | (C', C) \in CONN\}$.
- *Successor* of $C \in COMP$: $succ(C) = \{C' | (C, C') \in CONN\}$.
- *Root* node: $root = C$ if and only if $pred(C) = \emptyset$.

- *Leaf* nodes: $leaves = \{C|(C', C) \in CONN\} \wedge succ(C) = \emptyset$.

Note that because of Definition 1 every node has to be connected with every other node in the tree. It is not allowed that a node has no interconnections with other nodes in $COMP$. What is missing is the monitoring and checking behavior of a system. For this purpose, we introduce three additional functions that have to be defined for each component in a system. We assume without restricting generality to use first order logic (FOL) for representing the information delivered by these functions.

- *Output*: The $out : COMP \mapsto FOL$ function delivers a set of predicates that has to hold when executing the component given in the parameter.
- *Constraints*: The $constr : COMP \mapsto FOL$ function delivers first order logic sentences that must hold in the context of a given component.
- *Behavior*: The $behav : COMP \mapsto FOL$ function returns a set of first order logic sentences that are used with the inputs of a component (i.e., the outputs of the successor nodes) to derive the outputs.

We use these functions to express the behavior of the system. For each component C_i its output can be derived from the inputs considering the defined behavior of the component. Formally, we can express this as follows:

$$behav(C_i) \cup (\bigcup_{C \in succ(C_i)} out(C)) \models out(C_i).$$

Moreover, for each component C_i, if working correctly, all constraints have to be fulfilled considering given inputs and computed outputs, i.e.:

$$constr(C_i) \cup out(C_i) \cup (\bigcup_{C \in succ(C_i)} out(C)) \not\models \bot.$$

We are able to summarize the monitoring and checking activities to be carried out. Algorithm 1 implements the necessary steps. The algorithm will be called using the root node and the system. First, the algorithm makes a depth-first traversal of the tree until the leaves are reached. In each step, the outputs are computed and the constraints are checked. Obviously, Algorithm 1 terminates because of considering finite tree structured systems. Because of constructions, the constraints are checked only if all information for a particular node are available, i.e., all successor nodes are already traversed. The run time of $monDev$ is linear in the number of nodes considering that the behavior as well as the constraints can be checked in one time unit. Otherwise, the run time would be mainly determined by the derivation and consistency checking function $\models$. Assuming that we are able to map the first order logic sentences to propositional logic via grounding, the run time of solving would be exponential in the size of the logic sentences. Because logic sentences are local with respect to one node, we do not expect larger logic sentences making reasoning feasible.

As already mentioned in practical applications, the implementation of the monitoring device may not rely on a logic representation but instead of ordinary procedural

Algorithm 1 $monDev(n, (COMP, CONN))$

Input: A node n of the system $(COMP, CONN)$ at the current state
Output: $\top$ if all constraints are fulfilled, and $\bot$ otherwise

Let in be the empty set
for C in $succ(n)$ **do**
  if $monDev(C, (COMP, CONN)) = \bot$ **then**
    return $\bot$
  end if
  Let in be $in \cup out(C)$
end for
Let $out(n)$ be logic sentences derived from the input and the behavior, i.e., $behav(n) \cup in \models out(n)$.
if $constr(n) \cup in \cup out(n) \models \bot$ **then**
  return $\bot$
else
  return $\top$
end if

programs that provide the behavior or the consistency checks. In the next section, we discuss how the proposed monitoring approach can be applied to an use case from the automotive industry.

3 Automotive Use Case

In the following, we discuss a use case from the automotive industry where the speed of a vehicle is measured. We assume that we have sensors that are capable of detecting faults, and ordinary sensors without such capabilities, and also virtual sensors. The latter do not directly measure a physical quantity but allow to derive such values. For example, we do have a function that allows to map current and voltage of an electric motor to its speed. Hence, when measuring the power consumption of the motor, we also get an indicator of its rotational speed from which we are able to derive the speed of the car when knowing the behavior of the whole powertrain of the car.

For sensors and other integrated circuits, a potential example for failure detection is a failure counter. Complex integrated circuits may have a function that whenever a failure occurs at a cycle, a failure counter increases its value by 1. If the next cycle is without a failure, the failure counter decreases its value by 1. A certain threshold is set by the application when the number of such counted failures is too high and a warning signal is given. E.g., such a threshold maybe set to 10 failures. Typically, the level of such failures below the threshold are not observed. Nevertheless, this can be useful information whether the failure level is, e.g., in the range of 0–1, or 1–9. Taking this information into account can potentially announce a warning in advance.

Furthermore, on application level this information can be fused with, e.g., monitoring data from a sensor. The individual devices see only deviations in their domains,

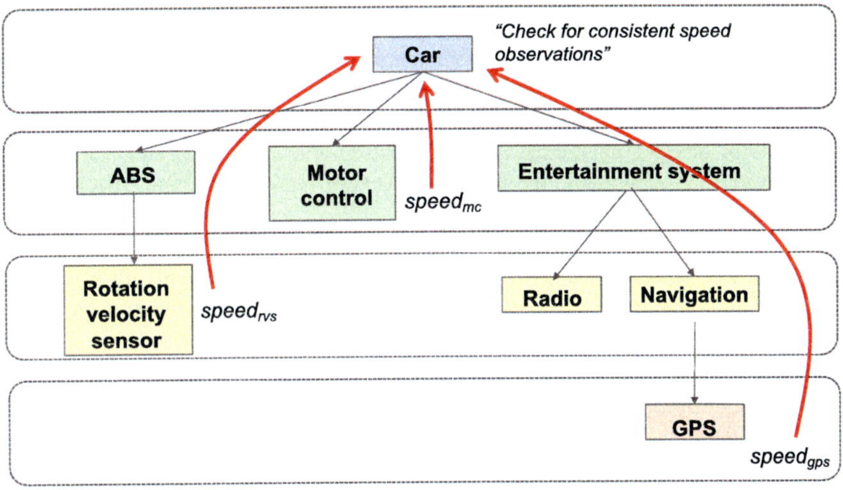

Fig. 2 Use case: Measuring speed

but on a higher-level monitoring information can be fused to identify degradation on, e.g., module or subsystem level. This is exactly what we want to capture with our proposal for a hierarchic monitoring approach.

Let us now make use of the suggested monitoring approach. For this purpose, let us have a look at the system depicted in Fig. 2, where we have a car comprising an anti-blocking system (ABS), a motor control, and an entertainment system (besides many others) as subsystems. The ABS makes use of a rotation velocity sensor to measure whether a wheel is rotating or not. For the motor controller we do not consider more components in this example. However, from the motor controller we assume also that speed measures are taken. The entertainment system comprises a radio and a navigation component where the latter is connected to a GPS from which we obtain also the speed of the vehicle.

Obviously, all speed measurements do have a certain fraction of error and comparing them directly may not be appropriate. However, in any case it has to be assured that all speed measures must be within a certain range. Otherwise, there is a failure and a fault isolation, and mitigation has to be triggered.

We start formalizing this use case with the definition of its corresponding system $(COMP, CONN)$:

$$COMP = \{Car, ABS, MC, ES, RVS, R, Nav, GPS\}$$

$$CONN = \left\{ \begin{array}{c} (Car, ABS), (Car, MC), (Car, ES), (ABS, RVS), \\ (ES, R), (ES, Nav), (Nav, GPS) \end{array} \right\}$$

Obviously, the root node is Car and the leaves of the system are $leaves = \{RVS, MC, R, GPS\}$. The predecessor node of the radio R is the entertainment system ES. The successor nodes of ES are the radio R and the navigation Nav.

In order to allow the system to be used for monitoring, we further have to specify the behavior of the components and the constraints used for checking the current state. We reduce our view to the necessary parts of our example, i.e., checking the speed sensor and ignore the rest.

For the leaf nodes RVS, MC, and GPS, we assume that they are measuring directly or indirectly the speed of the car. For the measured speed (and other physical or virtual quantities), we introduce a predicate $val(Q, V)$ having the two attributes name of the quantity Q and its value V. It is worth noting that the predicate is added to the monitoring device for a particular point in time if the value of the given quantity holds. This is important because it might be the case that sensors or other devices do only provide information if certain parameters are set. For example, the speed of a car using rotational speed might only be obtained with sufficient accuracy if the speed is higher than a certain threshold.

Formally, the behavior of RVS, MC, and GPS, can be formalized as follows, when assuming that they are providing any speed value v_i for $i = 1, 2, 3$:

$$\begin{aligned} behav(RVS) &= \{val(speed_{rvs}, v_1)\}, \\ behav(MC) &= \{val(speed_{mc}, v_2)\}, \\ behav(GPS) &= \{val(speed_{gps}, v_3)\}. \end{aligned}$$

Using the definition of the output function for a component, we are able to derive the following outputs:

$$\begin{aligned} out(RVS) &= \{val(speed_{rvs}, v_1)\}, \\ out(MC) &= \{val(speed_{mc}, v_2)\}, \\ out(GPS) &= \{val(speed_{gps}, v_3)\}. \end{aligned}$$

Hence, for predecessor components like ABS or Nav, their outputs are (ignoring other predicates that maybe derivable):

$$\begin{aligned} out(ABS) &= \{val(speed_{rvs}, v_1)\}, \\ out(Nav) &= \{val(speed_{mc}, v_2)\}. \end{aligned}$$

Finally, after processing all nodes, the output available at the root node Car is:

$$out(Car) = \{val(speed_{rvs}, v_1), val(speed_{mc}, v_2), val(speed_{gps}, v_3)\}.$$

In order to check the status of the whole system, we now introduce the constraints $constr$ for components. For our example, there are only constraints defined for Car:

$$constr(Car) = \left\{ \begin{array}{l} \neg(val(speed_{rvs}, v_1) \wedge val(speed_{mc}, v_2) \wedge |v_1 - v_2| > boundary) \\ \neg(val(speed_{rvs}, v_1) \wedge val(speed_{gps}, v_3) \wedge |v_1 - v_3| > boundary) \end{array} \right\}.$$

There the constraints are stating that the speed difference between two sensors are not allowed to be larger than a preset boundary value *boundary*. When assuming $v_1 = 1.00$, $v_2 = 1.11$, and $v_3 = 0.91$ for sensors RVS, MC, and GPS, respectively, and setting *boundary* to 0.1, we would be able to derive $\bot$ because $1.11 - 1.00 = 0.11$ which is larger than 0.1.

Hence, when using the proposed monitoring approach, it is possible not only to pass information from lower levels up to higher-levels, but also to check consistency with expectations. Note that in the given use case, the consistency can only be checked at the level of the whole system *Car* and not directly at the respective components like the motor controller.

The proposed concept of a monitoring device for hierarchical systems makes it easy to summarize information for the whole system, to check consistency at the different levels, and also to add new components. As already mentioned, the monitoring device can be seen separately from the system to be monitored. The structure of the system to be monitored may also be different.

4 Conclusions

In this paper, we discussed challenges arising when considering systems of increased autonomy. We argued that—in case of autonomous driving—safety can hardly be assured using testing. Hence, there is a need for coming up with monitoring where the system state is not only measured over time but also to check the degree of fulfillment of given constraints. The proposed concept relies on a hierarchic structure of systems from which information can be passed from the leaf components to the root. At each stage, not only information is transferred but also checked regarding given constraints. The constraints state safety and other requirements, like security, that must always hold.

It is worth noting that the system to be monitored may rely on a non-hierarchic structure. Although not explicitly discussed, the integration of time-related information into this concept should not cause any severe changes nor difficulties. There are strong relationships to PHM (as already discussed in the paper) but also to other approaches like passive testing [13] and run time verification [14]. With both approaches we share the concept of checking constraints that must hold over time. However, we allow separate checking of constraints at different levels of a given system.

Future research includes coming up with a concrete implementation of the proposed monitoring device concept. This includes coming up with data formats for specifying the behavior and the constraints, the communication of information between connected components, and a method for integrating monitoring devices easily into systems or systems under development, respectively. Moreover, the integration method has to be applied to concrete systems to check its usefulness for practical applications.

Acknowledgements ArchitectECA2030 receives funding within the Electronic Components and Systems for European Leadership Joint Undertaking (ESCEL JU) in collaboration with the European Union's Horizon2020 Framework Programme and National Authorities, under grant agreement n° 877539. All ArchitectECA2030 related communication reflects only the author's view and the Agency and the Commission are not responsible for any use that may be made of the information it contains. The work was partially funded by the Austrian Federal Ministry of Climate Action, Environment, Energy, Mobility, Innovation and Technology (BMK) under the program "ICT of the Future" project 877587.

References

1. Wotawa, F.: In: Watzenig, D., Horn, M. (eds.), Automated Driving. Springer International Publishing Switzerland (2017). https://doi.org/10.1007/978-3-319-31895-0_22
2. Wotawa, F.: In: SafeAI@AAAI, CEUR Workshop Proceedings, vol. 2808. CEUR-WS.org (2021)
3. Li, Y., Tao, J., Wotawa, F.: Inf. Softw. Technol. **117**, 2020 (2019). https://doi.org/10.1016/j.infsof.2019.106200
4. Klück, F., Zimmermann, M., Wotawa, F., Nica, M.: In: 2019 IEEE 19th International Conference on Software Quality, Reliability and Security (QRS), pp. 418–425 (2019). https://doi.org/10.1109/QRS.2019.00058
5. Klück, F., Zimmermann, M., Wotawa, F., Nica, M.: In: Gaston, C., Kosmatov, N., Le Gall, P. (eds.) Testing Software and Systems, pp. 140–156. Springer International Publishing, Cham (2019)
6. ISO17359. Condition monitoring and diagnostics of machines? General guidelines (2018)
7. ISO18129. Condition monitoring and diagnostics of machines ? Approaches for performance diagnosis (2015)
8. ISO20958. Condition monitoring and diagnostics of machine systems ? Electrical signature analysis of three-phase induction motors (2013)
9. Pecht, M., Wang, W.: IEEE Trans. Ind. Electron. **7**(58), 3077 (2011)
10. SAE-JA6268TM. Design and Run-Time Information Exchange for Health-Ready Components (2018)
11. ISO26262-11. Road vehicles—Functional Safety—Part 11: Guidelines on applications of ISO 26262 to semiconductors (2018)
12. Meeker, W., Escobar, L., Lu, J.: Technometrics **40**, 89 (1998)
13. Lee, D., Netravali, A., Sabnani, K., Sugla, B., John, A.: In: Proceedings 1997 International Conference on Network Protocols, pp. 113–122 (1997). https://doi.org/10.1109/ICNP.1997.643699
14. Bartocci, E., Falcone, Y.: Lectures on Runtime Verification—Introductory and Advanced Topics. Lecture Notes in Computer Science (LNCS), vol. 10457. Springer (2018). https://doi.org/10.1007/978-3-319-75632-5

DATA ANALYTICS (Main Track)—Chair: Michele Malgeri

Preliminary Characterization of Italian Academic Scholars by Their Bibliometrics

Vincenza Carchiolo, Marco Grassia, Michele Malgeri, and Giuseppe Mangioni

Abstract In this paper we analyze the bibliometrics of Italian academic researchers that belong to three "*academic disciplines*" (SSDs), topic-based groupings defined by law. In particular, we extract the data from Elsevier's Scopus profiles of the researchers and analyze the impact of the SSDs and Research areas (ASJC) they belong to on various productivity indices, such as the Herfindahl–Hirschman (H–index), that are widely used by universities to evaluate people, this paper presents a preliminary works aiming at characterizing the mentioned disciplines and create a dataset useful to investigate existing correlation, if any, between bibliometrics and scholar's habits.

1 Introduction

The evaluation of the quality of research become a topic of considerable importance in the last decades, indeed Universities are constantly under evaluation by both public and private entities. In the first case, evaluation is often used to determine the extent of public funding for each university while, in the second case, the evaluation is mainly used to draw up rankings with the aim of establishing which are the best universities because of predefined indices. For this reason, universities are interested in raising their indexes. Unfortunately, more often they cannot act transversely, but they must focus their efforts on areas of interest [1]. On the other hand, also individual

V. Carchiolo · M. Grassia · M. Malgeri (✉) · G. Mangioni
Dip. Ingegneria Elettrica, Elettronica e Informatica, Università degli Studi di Catania, viale Andrea Doria, 6, Catania, Italy
e-mail: michele.malgeri@unict.it

V. Carchiolo
e-mail: vincenza.carchiolo@unict.it

M. Grassia
e-mail: marco.grassia@unict.it

G. Mangioni
e-mail: giuseppe.mangioni@unict.it

D. Camacho et al. (eds.), *Intelligent Distributed Computing XIV*, Studies in Computational Intelligence 1026, https://doi.org/10.1007/978-3-030-96627-0_31

researchers are usually subject to evaluation of their research activity, and they pay more and more attention to the value of the indices used for the evaluation. Indeed, an improvement in the indices of individual researchers may also lead to an improvement in the rank of the universities to which they are affiliated.

While there are many factors influencing the score assigned to each researcher, since there is no general agreement on what is the best/minimal set of evaluation indexes, there are in the literature several works that tried to correlate performance indexes with some social aspects/elements, such as number of collaborators, the co–authorship network size [2], the use of social media for papers/projects advertising [3], and the impact of publication multidisciplinarity [4].

In this paper we investigate how some metrics used to score a researcher depend on the specific research area they belong to. More specifically, we quantify such a difference through an analysis of the data extracted from a real–world publication database.

A first attempt in this direction was the research published in [5], where the authors study the link between the Herfindahl-Hirschman index (H-index) and the degree of multidisciplinarity of a researcher, making a comparison between researchers from Italian universities in 6 different scientific areas. In carrying out this analysis, the distribution of Elsevier's Scopus[1] subject areas of authors' documents, their bibliographical references and their citing documents are taken into consideration. The limit of this research is twofold: (1) only an extremely limited part of the researchers is considered and (2) the chosen areas include a mixing of scientific and social/humanistic sectors that are not easily comparable.

This paper presents a study of a subset of scholars working into Italian University with the aim of comparing their habits and finding correlation, if any. The Italian academic researchers are grouped, by law, into 370 "*academic disciplines (SSDs)*" according to common themes and interests. We study three SSDs with comparable number of researchers: information processing system (ING-INF/05), which belongs to computer engineering, mathematical analysis (MAT/05) belonging to mathematics, and economics (SECS-P/01).

The paper is organized as follows: Sect. 2 introduces the structure of our dataset together with a brief discussion on how data are extracted, Sect. 3 presents a set of data analysis to compare the three considered groups of researchers, and, Finally, Sect. 4 discusses the result of the dataset analysis.

2 Data Set Description

The information about the three groups of researchers, analised in this paper, was extracted from Elsevier's Scopus database using the provided APIs. Scopus is one of the largest collections of scientific publications, covering about 78 million items from more than 25 thousand journals, conference proceedings, books, etc. Scopus is con-

---

[1] http://www.scopus.com

Table 1 Number of records in the dataset

SSD	Total records	Filtered
ING-INF/05	843	786
MAT/05	723	679
SECS-P/01	733	676
Total	2299	2141

stantly growing both in Journals and researchers referred by. However, Bordignon [6] notes that the growth in Journals during the last year was not evenly distributed over different areas.

Scopus provides accurate author information like bibliographic data and several quality and performance measures that are widely used to rank researchers in many European countries including Italy [7, 8]. Specifically, Scopus' provides three different search interfaces: *Affiliation* that collects all referred institutions, *Authors* that collects the referred authors' profiles and *Scopus* that covers abstracts and documents information.

The data analysed in our study was collected in May 2021 and was processed to obtain information concerning the bibliometrics and the subject areas of selected authors' documents. The list of scholars belonging to SSD under study was obtained by *MIUR (Ministero dell'Universitá e della Ricerca, the Italian Ministry of Education, Universities and Research)* [9], then, for each scholar, we performed several queries against SCOPUS databases until only one profile match else profile is discarded.

The resulting dataset contains 2141 records almost equally distributed among the chosen SSDs as reported in Table 1. All records contain the following bibliometrics: author's Herfindahl–Hirschman (H-index); the number of times the papers have been cited by other papers covered by Scopus (citation-count); the number of document that cited a scholar's paper (cited-by-count); the number of document presents on Scopus for every researcher (document-count); the year of the author's first publication and more recent publication (respectively pub-range-min and pub-range-max); number of co-authors for each document (coauthor-count) and the frequency of publication in each subject area and field classified according to "All Science Journal Classification (ASJC)" Codes [10] .

ASJC classifies the documents using 301 fields of interest that are organized in 26 Areas that, in turn, are grouped into the following four subject Area of Sciences: Life, Sciences, Physical and Health. Each field is coded by a 4 digits code where the first 2 digits identify the area.

Table 2 summarizes the classification into Area and Field used by ASJC and Table 3 introduces the description of the fields used in this paper.

Finally, to analyse the impact of the subject areas of authors' documents on the bibliometrics information, we collected, for each researcher, the field with higher frequency of publication, the number of different areas covered by author's publica-

Table 2 ASJC subject area and area

Subject Area	Area		#Fields
	Description	Code	
Life Sciences	Agricultural and Biological Sciences	(11)	11
	Biochemistry, Genetics and Molecular Biology	(13)	13
	Immunology and Microbiology	(24)	6
	Neuroscience	(28)	9
	Pharmacology, Toxicology and Pharmaceutics	(30)	5
	Psychology	(32)	7
Social Sciences	Arts and Humanities	(12)	11
	Business, Management and Accounting	(14)	10
	Decision Sciences	(18)	4
	Economics, Econometrics and Finance	(20)	3
	Social Sciences	(33)	22
Physical Sciences	Chemical Engineering	(15)	8
	Chemistry	(16)	6
	Computer Science	(17)	12
	Earth and Planetary Sciences	(19)	13
	Energy	(21)	5
	Engineering	(22)	16
	Environmental Science	(23)	12
	Materials Science	(25)	8
	Mathematics	(26)	14
	Physics and Astronomy	(31)	10
Health Sciences	Medicine	(27)	48
	Nursing	(29)	23
	Veterinary	(34)	4
	Dentistry	(35)	6
	Health Professions	(36)	15

Table 3 Excerpt of ASJC Code area related to scholars belonging to chosen SSDs

Code	Field description
1400	General Business, Management and Accounting
1702	Artificial Intelligence
1706	Computer Science Applications
1708	Hardware and Architecture
1712	Software
2002	Economics and Econometrics
2208	Electrical and Electronic Engineering
2603	Analysis
2610	Mathematical Physics
3305	Geography, Planning and Development
1700	General Computer Science
1705	Computer Networks and Communications
1707	Computer Vision and Pattern Recognition
1710	Information Systems
2000	General Economics, Econometrics and Finance
2207	Control and Systems Engineering
2600	General Mathematics
2604	Applied Mathematics
2614	Theoretical Computer Science

tion, the number of different fields covered by author's publication, the frequency of a publication in each area and the frequency of publication in each field.

3 Data Set Analysis

In this section we analyse some bibliometric performance indexes of the three previous mentioned groups of researchers, *information processing system—ING-INF/05*, *mathematical analysis—MAT/05* and *economics—SECS-P/01*. As in [5], we compare the distribution of Elsevier Scopus subject areas of authors' documents, their bibliographical references, and their citing documents.

3.1 Bibliometrics Analysis

In this subsection we explore the distribution of bibliometric indexes. The first bibliometric index we compare is the Herfindahl–Hirschman index (H–index) because

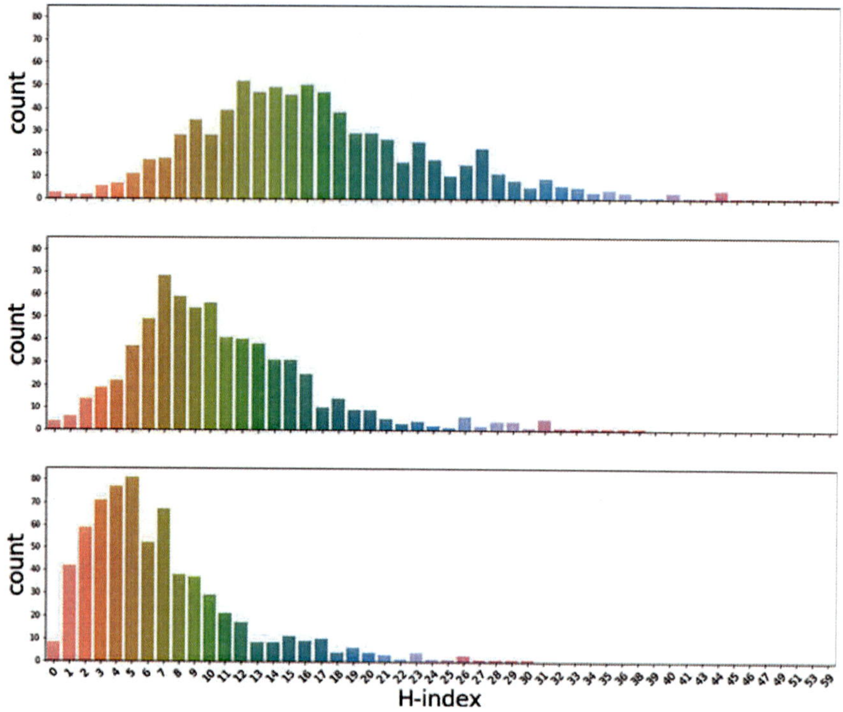

Fig. 1 Distribution of H–index of ING-INF/05 (top), MAT/05 (mid) and SECS-P/01 (bottom)

it is widely used by many countries to rank researchers. In Italy, it represents one of the three metrics used to define the threshold that allows researchers to be hired at university and to measure the quality of research, therefore most Italian researchers are carefully about it.

Figure 1 shows the distribution of the H-index over the three academics. It is worth mentioning that the distributions differ both in average and maximum. In particular, ING-INF/05 presents more scholars with high H–index and also most researchers have H–index in the range [14–24]. On the contrary, for SECS-P/01 only a few researchers have H–index values higher than 22 and most of them are in range [6–10]. The distribution of MAT/05 has an intermediate figure. The next step in our analysis concerns the distribution of each studied bibliometric index and the identification of any outliers. Figure 2 reports the distributions of the following indexes: coauthors-count, document-count, citation-count, and multiple citation. We cut on x-axis the first two indexes to compare the value for all three SSDs.

Diagrams in Fig. 2a highlight that people belonging to ING-INF/05 tend to collaborate much more than other two SSDs that, otherwise, exhibit similar behaviour. Specifically, ING-INF/05 has a great number of authors that have more than 500 co-authors that is the largest value drawn in the diagrams. Please, note that the co-authors in the dataset range over all people referenced in Scopus not only researchers working in Italian academics and/or belonging to the same academic discipline, as, for example, other countries authors, independent researcher, people working for companies, etc. Co-authors number, considering both author belonging to Italian academics and total present in Scopus, is different, but MAT/05 and SECS-P/01 show some similarity whilst ING-INF/05 has a completely different dynamic. This fact is also reflected by the lack of collaboration between researchers in the sector even within the SSD, in fact, only 66% of researchers of SECS-P/01 are not isolated nodes in the co-authorship network. This percentage rises to 90% for MAT/05 and 96% for ING-INF/05.

The ING-INF/05 also exposes a greater number of citations (citation-count) than other two SSDs (Fig. 2b). In these diagrams x-axis has been cropped to 10000 citation that are reached only by few scholars of ING-INF/05 and one on SECS-P/01. Figure 2c depicts the distribution of the author's number of documents, indeed, this has a similar pattern of number of citations through the analysed SSDs.

Finally, Fig. 2d reports the distribution of the percent of multiple citations by the same document (i.e., one document cites more works of the same author). This indicator is calculated as citation-count and citation-by-count ratio. The higher the value, the less multiple citation, thus SECS-P/01 authors are less used to multiple citation than others while it is quite common in MAT/05. This fact may be explained by the greater connection of the scholars inside the ING-INF/05 and MAT/05 SSDs.

The next step in our analysis aims at studying the difference among SSDs of the author's period of activity. Diagram 3a highlights that people belonging to MAT/05 have the shortest activity period while Diagram 3b shows that SECS-P/01 researchers have a lower left tails that means that after hiring diminish their publication rate.

A different perspective is obtained reading correlation matrix among the indexes. Figure 4 shows the heatmap where each field has the colour that corresponds to its value, the lighter the colour, the higher the correlation.

As shown in Fig. 4c, there is a strong correlation between citation-count and coauthor-count in SECS-P/01 (0.92), whilst the value is lower in MAT/05 (0.54) and ING-INF/05 (0.33). Vice-versa, the correlation between citation-count and document-count is meaningful in ING-INF/05 (0.68) and MAT/05 (0.77), but less in SECS-P/01 (0.31).

As shown in Fig. 4a, H–index's correlates in ING-INF/05 with citation-count (0.84), citation-by-count (0.79) and document count (0.80), while the correlation between H–index and number of co-authors is quite low (0.30). MAT/05 exhibits similar values, as illustrated in Fig. 4b: H–index is highly correlated with citation-count (0.86), citation-by-count (0.76) and document count (0.85) whilst the correlation between H–index and number of co-authors is low (0.22). On the contrary, as shown in Fig. 4c, only the document-count strongly correlates with H-index (0.85) in SECS-P/01.

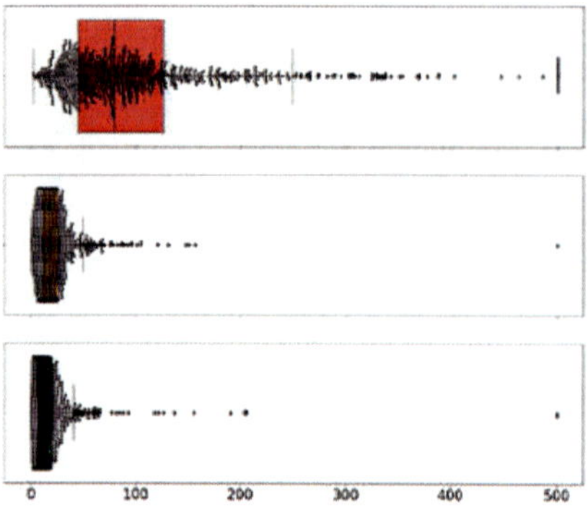

(a) co-authors count

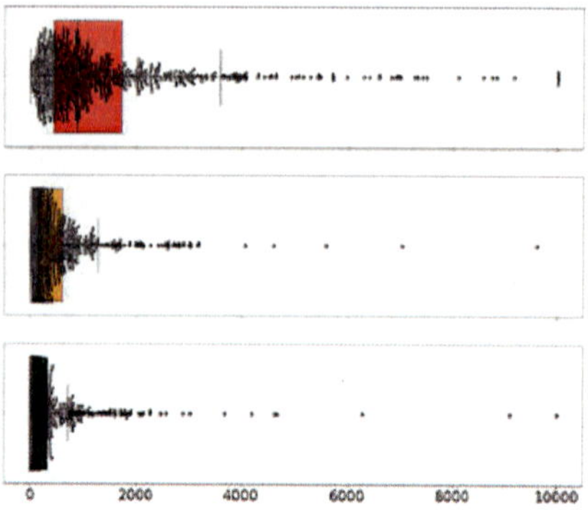

(b) citation count

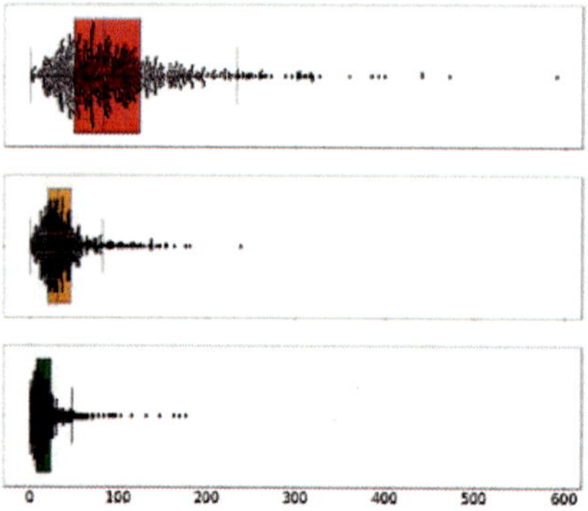

(c) document count

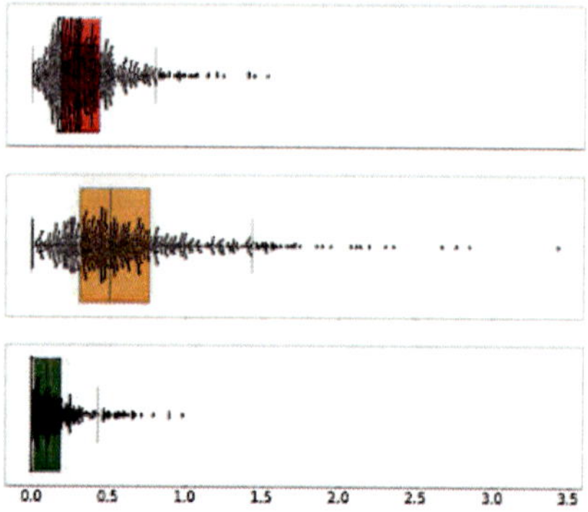

(d) percent of multiple citation

Fig. 2 Distribution of some indices. All diagrams have same scales, the values greater than maximum x-scale are aggregated and represented at a single coordinate which means "all values greater than this". Red colour (top) refers to ING-INF/05, yellow (mid) to SECS-P/01 and green (bottom) to MAT/05

3.2 Analysis of Publication Area and Fields

We also analyse the number, and the main areas and fields researchers publish in. In particular, the publications of researchers belonging to ING-INF/05 cover all 27 areas and almost of all the 301 fields (more precisely 285), also SECS-P/01 publish in all 27 areas but covers only 256 fields whilst MAT/05 researchers are wrapped up on few topics covering only 223 fields. For the full distributions of the number of areas and fields covered by each researcher, see Fig. 5a, b respectively. In particular, ING-INF/05 researchers are involved, on average, in more areas and fields which reflect multidisciplinary activity. By comparing the distribution of the H–index previously discussed (see Fig. 1) with the multidisciplinary nature of the publications it appears

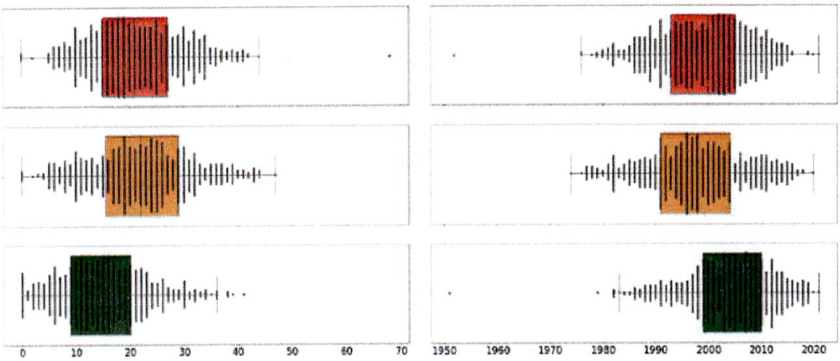

(a) Activity period **(b)** Date of the oldest publication

Fig. 3 Analysis of the period of author's activity, it is calculated by Scopus using the oldest and the more recent publication dates respectively. Figure **a** shown the years of activity and **b** the date of the oldest publication of each researcher

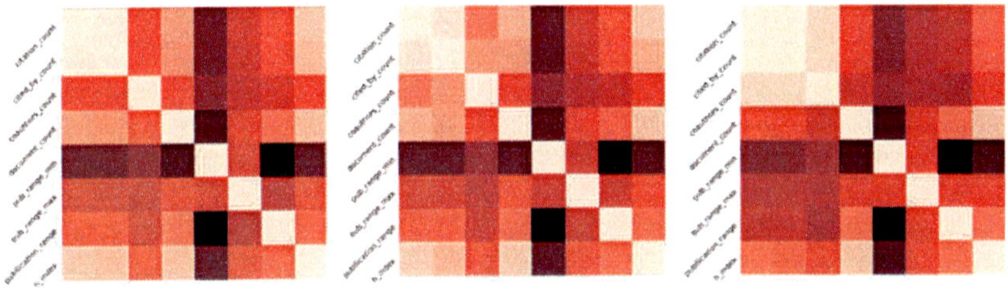

(a) ING-INF/05 **(b)** MAT/05 **(c)** SECS-P/01

Fig. 4 Correlation Matrix among some bibliometrics indices. The rows (columns), from top to bottom (left to right, respectively), are: citation-count, cited-by-count, coauthors-count, document-count, pub-range-min, pub-range-max, publication-range and H–index

that greater multidisciplinarity corresponds to a higher H–index value. To definitively assert such a result an analysis on a larger number of SSDs and further studies are needed.

To evaluate a kind of multidisciplinary degree, we evaluate for each researcher the field with higher frequency of publication, and we call this value "main field". Figure 6 shows the scatter plots of the "main field" where green dots correspond to SECS-P/01, orange to MAT/05 and red to ING-INF/05.

The most common main fields are 2002, 2604 and 1700 respectively per SECS-P/01, MAT/05 and ING-INF/05. The greater multidisciplinarity of ING-INF/05 can also be appreciated in Table 4 where are listed the "main field" of each SSD that bring together at least 10 researchers. We see that while for SECS-P/01 and MAT/05 are only 4 fields for ING-INF/05 there are 10 different fields that are common for more than 10 researchers. The description of each code can be found in Table 3.

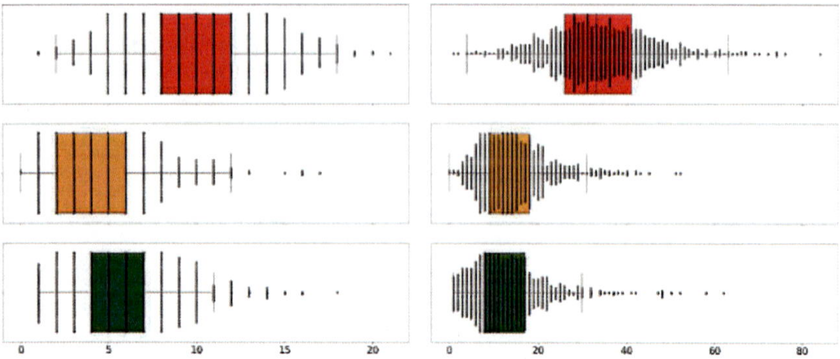

(a) Distribution of each scholar on number of area

(b) Distribution on number of fields for each scholar

Fig. 5 Number of areas (**a**) and fields (**b**) of author's publications, calculated by Scopus

Fig. 6 Main fields scattering of the three SSDs, where ING-INF/05 is red, MAT/05 is orange and SECS-P/01 is green

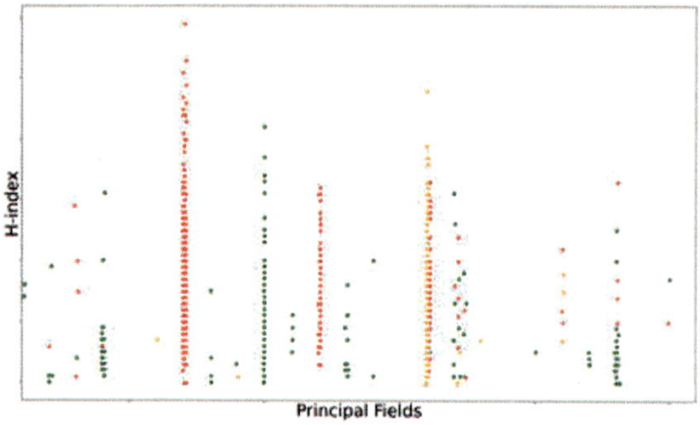

Table 4 Number of scholars in each Main field

SECS-P/01		MAT/05		ING-INF/05	
Field	Number	Field	Number	Field	Number
2002	499	2604	316	1700	226
2000	49	2600	196	1712	133
3305	15	2603	131	1705	111
1400	14	2610	12	1702	50
–	–	–	–	2208	46
–	–	–	–	1707	42
–	–	–	–	1708	33
–	–	–	–	1706	33
–	–	–	–	2614	31
–	–	–	–	2207	14
–	–	–	–	1710	14

(a) ING-INF/05 (b) MAT/05 (c) SECS-P/01

Fig. 7 Correlation Matrix among Areas of interest (rows from top upper to lower and columns from left to right) and H–index (bottom and rightmost): the lighter the colour, the greater the correlation

Our last analysis concerns the areas of interest and H–index. In Fig. 7 we show the correlation matrices between the areas of interest and the H–index for each SSDs. Specifically, the entries of the matrix represent the code of several areas of interest (two digits code ranging over 10 to 36) and the last one is the H–index. While there is no correlation for most, few pairs (specific for each SSD) show strong correlation. In particular, the H-index in SECS-P/01 is correlated to areas 20, 33 and 14, MAT/05 to area 26 only and, ING-INF/05 to areas 17, 26 and 22.

4 Conclusions

The paper presents a preliminary work aiming at investigating the correlation among the bibliometrics widely used to rank researcher in many countries studying groups of researchers divided by common topics or by law, as in the case of Italy. We selected three *academic disciplines*, i.e., grouping of Italian academic researchers imposed by law, that encompass almost the same number of people but different topics and habits.

The analysis highlights that some correlations exist, but usually they are inside the same *academic discipline*. It is quite difficult to generalize to groups that have different essentials uses, such us, number of co-authors, types of journals or book, etc. However, the results are interesting enough to promote further analysis and the extension of the dataset covering more *academic disciplines* or different approaches to group people, for instance using the ASJC areas and fields, studying the topological properties of co-authorship network, as performed in [11].

Acknowledgements This work has been partially supported by the project of University of Catania PIACERI, *PIAno di inCEntivi per la Ricerca di Ateneo*.

References

1. Haddawy, P., Hassan, S.U., Abbey, C.W., Lee, I.B.: Uncovering fine-grained research excellence: The global research benchmarking system. J. Inf. **11**(2), 389–406 (2017)
2. Guan, J., Yan, Y., Zhang, J.J.: The impact of collaboration and knowledge networks on citations. J. Inf. **11**(2), 407–422 (2017)
3. de Winter, J.C.F.: The relationship between tweets, citations, and article views for plos one articles. Scientometrics **102**(2), 1773–1779 (2015)
4. Hassan, S.U., Imran, M., Gillani, U., Aljohani, N.R., Bowman, T.D., Didegah, F.: Measuring social media activity of scientific literature: an exhaustive comparison of scopus and novel altmetrics big data. Scientometrics **113**(2), 1037–1057 (2017)
5. Moschini, U., Fenialdi, E., Daraio, C., Ruocco, G., Molinari, E.: A comparison of three multidisciplinarity indices based on the diversity of scopus subject areas of authors'documents, their bibliography and their citing papers. Scientometrics **125**(2), 1145–1158 (2020)
6. Bordignon, F.: Tracking content updates in scopus (2011-2018): a quantitative analysis of journals per subject category and subject categories per journal. In: ISSI (2019)
7. Elsevier, B.V.: Elsevier Developer—Academic Research (2021). https://dev.elsevier.com/api_service_agreement.html. Accessed July 2021
8. Elsevier, B.V.: Elsevier Developer—API Service Agreement (2021). https://dev.elsevier.com/academic_research_scopus.html. Accessed July 2021
9. Ministero dell'Universitá e della Ricerca (MIUR): Professori e Ricercatori (2021). Accessed May 2021
10. Elsevier, B.V.: Scopus: Access and use Support Center (2021). https://service.elsevier.com/app/answers/detail/a_id/15181/supporthub/scopus/. Accessed July 2021
11. Carchiolo, V., Grassia, M., Longheu, A., Malgeri, M., Mangioni, G.: A network-based analysis of a worksite canteen dataset. Big Data Cogn. Comput. **5**(1) (2021)

ERAKS—Interactive Interface to Annotate Entities, Co-references, and Relationships

Cătălina Mărănduc, Mihaela Colhon, and Anca-Diana Bibiri

Abstract This paper aims to build a tool/an interface for automatic recognition of entities in a literary text and the relationships between them. The result of the practical work of three series master students of natural language processing was the creation of a manually annotated corpus from the Romanian version of the novel Quo Vadis by Henryk Sienkiewicz. It was previously annotated with a POS-tagger and received the sentences and words id. Entities, human characters, or gods, as well as geographic locations, were then annotated in XML format, then co-references between them sent to a common trunk. Afterwords, affect, social, and kinship relationship between these entities has also been annotated. Entities participating in a relationship have been annotated along with triggers, words in the text that are markers of that type of relationship. A tool was developed, which automatically searched triggers in a text annotated with more types of entities and co-references. Next, we searched with the same tool affect triggers in a chat corpus, i.e. diagnostic words for affect relationships between real characters. Further, we switched to the development of the tool, transformed into an interactive interface which annotates entities, then searches for triggers, and annotates them if they are validated by the human assistant. This interface can be used to annotate any text with real or fictional characters.

1 Introduction

Quo Vadis: a Narrative of the Time of Nero, written by Henryk Sienkiewicz, was the basis for a project of annotating entities and relations between entities in order to develop an automatic tools for annotating different levels encountered in the book.

C. Mărănduc
"Alexandru-Ioan Cuza" University, Iaşi, Romania

M. Colhon (✉)
University of Craiova, Craiova, Romania
e-mail: mihaela.colhon@edu.ucv.ro

A.-D. Bibiri
Interdisciplinary Research Department, "Alexandru-Ioan Cuza" University, Iaşi, Romania

D. Camacho et al. (eds.), *Intelligent Distributed Computing XIV*, Studies in Computational Intelligence 1026, https://doi.org/10.1007/978-3-030-96627-0_32

Begun as a theme research, it was quickly developed into a project with many possibility of exploitation. It was envisaged to draw upon several directions: the first attempt was to identify characters of the book (entities) and the relations established among them (semantic links) [1]. The next step was to summarize the levels of annotation and to generate statistics from the corpus [2]. Subsequently, the group was concerned with establishing correspondences between book pages and real entities, people, event, locations. The corpus, consisting of 7,281 sentences, was manually annotated marking entities, persons and gods, then marking four types of semantic relations of these entities: *referential*, *affect*, *kinship* and *social*. Each of them has more sub-types [3, 4]. A further step was the study of the interpretation of semantic relationships among major characters of the literary text [5, 6].

The QV (*Quo Vadis*) experiment involving many master and Ph.D. students, led to several partial results. However, several inconsistencies were reported due to the large number of participants, (who did not devise a method for determining or measuring their agreement) to the changes in the annotated concept grid along the way, to the interface used, which required annotators to memorize ids and write them, something that was conducive to multiple possible errors.

On the basis of these enumerated achievements so far, the goal of this study is to exploit and validate the computer's capacity of assessing what a human annotator is able to do. We attempt to generate automatic programs capable to eliminate as many errors made by humans as possible and to approach human thinking as far as possible.

In 2014, a trigger detector was built, which memorized the triggers annotated by humans and which generated a list of suggested triggers in certain contexts. It returned 5,136 suggestions and a total of 3,839 were validated by humans. 74.74% of the suggestions were validated as triggers [7]. A chat corpus, consisting of 2,580 sentences has been also annotated and studied [8, 9].

On the one hand, the way the relations were organized led to statistically and objectively obtained information about the annotated novel *Quo Vadis*, they represented information which literary analysts identify intuitively: that there is an opposition between two systems, Nero's empire and the Christians' camp; that the passage of a character from one camp to another leads to a change in the type of relationship he establishes. The very complicated imaginary architecture of the human mind can consequently be modeled as computational data [10].

We have reached to some conclusions regarding the need to widen the corpus directly proportional with the increasing need of obtaining complex data, and also with the need for more rigorous annotation tools, which the computer contributes moreover by the systematization of the results.

Through this approach we make a contribution for creating a more refined conception of human-machine interaction. The computer must work in the same way as the human mind classifying objects and relationships, and must contribute to their storage and evaluation. By classifying the triggers and by computing their efficiency, we can increase the validation percent and we can approach the desideratum of a detector with unassisted operation.

2 Related Work

Lately, there are researches interested in semantics and in the level of textual organization study in various fields; the research attempts to surpass the only partial achievements at the level of morphology and syntax of natural language. The semantic orientation approaches usually exploit the term extraction methods to obtain concepts from texts, using links to Princeton Word Net (PWN) [11, 12].

In other research, the semantic relationships of words are identified by a dependency parsing process [13, 14]. Another research, [15] explored a new direction in the concept mining field by means of lexicon-syntactic patterns; while others explore the human behavior on various social networking platforms [16].

To study the fictional text from the point of view of its truth value in relation to the real world is incorrect because it does not aim to describe reality but to convey ideas through an isomorphic image. John Searle [17] has already shown, long ago, that the logical value of truth is suspended in the fictional text.

In terms of relations, which govern the choices of the real characters in the consumer society, the affect relations (sentiment analysis) were the subject of a lot of computational research, the opinion meaning [18] having applications in the field of business and in the electoral one, see [16]. Other researchers are interested in the study of public blogs or forums or product reviews [19, 20].

Sentiment analysis tasks are usually to search an already existing lexicon which "triggers" words that express sentiments in texts. For this reason, there have been created dedicated resources, consisting for example in lexicons of triggers [21, 22]. Other important procedures described in related papers included bag-of-word features, hash-tags, handling of negation, word shape and punctuation features, elongated words, emoticons, etc. [23].

Other studies deal with the classification of emotions. Here it is worth recalling Ekman's [24] six classes of emotions: *happiness*, *anger*, *sadness*, *fear*, *disgust*, and *surprise*. Using graphs made with the NodeXL program, a work related to ours is [25], which shows the similarity of the relationship structure in the fictional and real worlds.[1]

First of all, our research starts from the annotation of some words of texts as entities, because relationships and feelings cannot be established among words, but among the entities they designate. The problem of correct correlations between the annotated entities and pronouns or among more general common nouns is another corner stone of computational linguistics.

Secondly, we are not only interested in emotional relationships, but also in social ones, which are sometimes dominant in the text, we are also interested in the kinship relationships, real ones or figuratively used ones. We do not believe that only emotional relationships suffice for characterizing the real or fictive world in which the characters happen to move.

[1] Hansen analyzes the social media network and the relationships in Victor Hugo?s novel "Les Miserables".

Thirdly, we use trigger lists extracted from our previous annotations on various types of text, and we will describe below a method to evaluate their effectiveness, for adding or removing them from our trigger list.

3 General Architecture of the Annotation

The text to be annotated is in XML format. It is segmented into sentences and words, and then annotated with morphological, sometimes also syntactic, information. This experiment was performed on two texts. The first has 534 sentences and 7,446 tokens, punctuation included; the text is Mateiu Caragiale's short story *Remember*. The second experiment was performed on a part of a Chat corpus, containing 504 sentences, with 6,202 tokens, which represent the conversation between two persons. We had the intention of annotating two short texts, one fictional and the other representing the conversation of two real characters.

3.1 Annotation of Entities and Co-references

The annotation process has three steps. The first step is the annotation of entities and of simple co-references. Through these, we determine which nouns or pronouns represent people, places, groups of people, etc. By associating entities with common reference trunks designated by ids, we establish how many such entities there are in the text and which are the most important ones, for example the main characters of a novel will be designated by most references to the same common trunk, a big number of co-referent words in the text.

The annotation tool allows us to enter lists of entities from the text; we have sorted these into five types: Persons, Locations, Animals, Groups, and Parts (of the body). The number of types can be larger or smaller. These lists are open. If someone wants to annotate a word in the text, he/she selects it in the interface and then searches in the lists for an entity co-referential with it.

Proper nouns are marked in lists with the Np symbol and they are automatically annotated with the same id for all occurrences in the text, but the co-reference of other nouns (the women, the young man) cannot be established by the computer. Annotations entered in the text include the word or words selected by the user. There are entities made up of several words of which one may be a nested entity. Example:

```
<entity id="LO00019">
 <word chunk="" deprel="sbj." form="fantana" head="4" id="5"
  lemma="fantana" postag="Ncfsry"/>
  <word chunk="" deprel="det." form="lui" head="7" id="6"
  lemma="lui" postag="Tf-so"/>
 <entity id="PE00037">
  <word chunk="" deprel="a.subst." form="Roland" head="5" id="7"
```

```
  lemma="Roland" postag="Npmsrn"/></entity>
</entity>
```

Fântâna lui Roland (Roland's Fountain) is a location entity formed of three words, with id LO00019, and within it, the word Roland designates a person entity with id PE00037.

This first step of the annotation process establishes the simple entities and co-references that occur in the text.

3.2 Annotation of Triggered Relationships Between Entities

The annotation of triggered relationships contains three ids: two of the two entities that establish the relationship and the third one of the trigger that determines its detection and classification in a relationship type and sub-type. The annotation marker is also inside the sentence, among the marks of the entities, without including them; therefore the ids must contain the word id and the sentence id, like in the second type of annotation.

```
<entity id="LO0001">
  <word chunk="" deprel="prep." form="Berlin" head="2" id="3"
  lemma="Berlin" postag="Npmsrn"/></entity>
  <word id="4" form="mare" lemma="mare" postag="Afpfsrn" head="5"
  chunk="" deprel="a.adj."/>
  <trigger="T-AF0555">
  <word id="5" form="sl?biciune" lemma="slabiciune" postag=
  "Ncfsrn" head="1" chunk="" deprel="c.d."/><trigger/>
  <REL subtype="love" from="PE0001010012" to="LO0001010003"
  trigger="T-AF0555"/>
```

This annotation shows that Berlin is an entity type Location, and that an entity of the type Person has feelings for it, marked by the "affect" trigger (*slăbiciune*=weakness), and classified as belonging to the "love" sub-type.

The tool makes the annotation of the relation in the XML document, containing the three ids and not including words in the mark. Annotated relationships are in this research of the AFFECT and SOCIAL types (polarized relationships that may be positive or negative), as in the following list.

- AFFECT relations
  - POSITIVE: love, liking, joy, worship, bravery, happiness, friendship, gratitude
  - NEGATIVE: hate, disgust, sadness, disdain, fear, distress, enmity, ingratitude
- SOCIAL relations
  - POSITIVE: richness, cooperation, obedience, protection, compensation
  - NEGATIVE: poverty, rivalry, revolt, persecution, punishment

4 Automatic and Interactive Manual Annotation

Our interface is called E(nities), R(eferences) and AKS (Affect, Kinship, Social) annotator (ERAKS). For this step, the role of the computer is more important. It stores the list of the triggers which we found in previous annotations. We reject lists built or taken through dictionaries using various procedures to have a high degree of artificiality.

Then the program searches the entire list in the text to be annotated (what we call trigger detector, a program created in 2014 and now integrated into the interface), and returns a list of triggers suggestions, each with the context in which it appears.

The human assistant must first click Y for validating the trigger. At this moment, the word(s) proposed as trigger and validated will be automatically annotated in the XML. The annotation of triggers, like the annotation of entities, include the word(s) in the mark of the new annotation.

Other options are: click to the button **opposite** if the trigger is appropriate for the AFFECT type "hate", or click the button **Add trigger** if another trigger exists in the context or if the proposed trigger has a totally different type and subtype than the proposed one. The button **Delete** is for deleting an annotation, and the button **Repeat** is for annotating the reciprocal relationships, by selecting the arguments in inverse order. Then, the human assistant must choose the two entities that participate in the relation, from the annotated entities in the proposed context. The first choice will be annotated as the source of the relation, the second choice will be annotated as the target. In the example, "I have a very soft spot for Berlin, no very sad circumstances have prevented me from revisiting it with pleasure", **me** is the source (the author) and **Berlin** is the target. If the words which must be selected are not annotated as entities, then the human assistant must return to the first page (first step) and annotate them. (see Fig. 1).

The annotation, as shown in the Sect. 3.3, has three arguments (the source, the target, and the trigger), and is placed in the XML between the words, not including them. In addition to the inclusion of validated annotations in XML, the interactive annotator, has to fulfill two other important functions.

First, it memorizes data for trigger efficiency in-formation. The trigger list must be dynamic. Triggers can be easily added, either as new words or as words already included in the list, but illustrating another type and sub-type of relation (another interpretation). Triggers that have been proposed by the program many times and have not been validated are removed from the list.

In this way, the efficiency of our list will increase as the annotated body grows and when we reach a degree of adequacy of over 90%, we can assume that we have an independent trigger annotator. If the context has only two annotated entities, the program will annotate these entities as arguments according to a set of simple rules. If the context has more than two annotated entities, the program will make a suggestion and the human assistant will select the participants to the relationship.

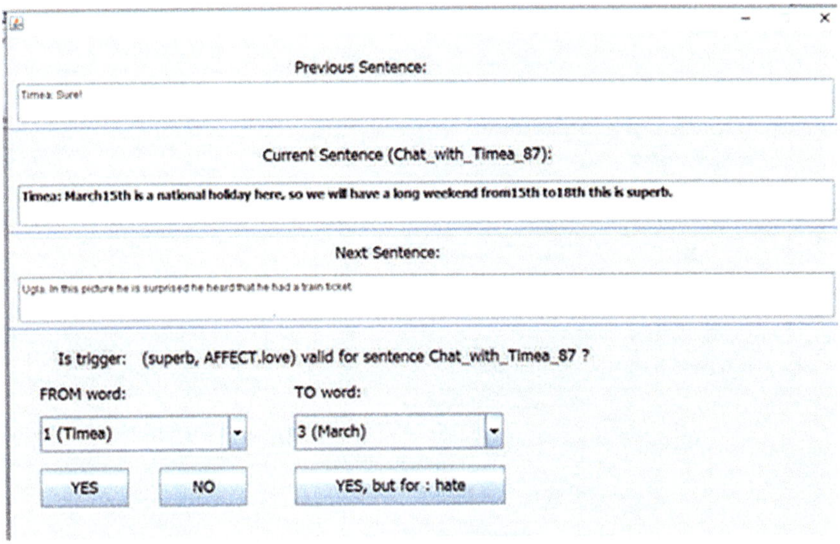

Fig. 1 The third page of the ERAKS interactive annotator

5 Graphic Interpretation of the Lists Extracted from the Annotated Documents

The lists with entities and their Affect, Kinship and Social relations can be introduced in a NodeXL or Gephi program that make graphs of relations between real or fictional Characters. We can also order our interactive tool to make special selections, for example to return a list (in CoNLLU format) with the positive and negative relations only.

The purpose is to understand more thoroughly the literary text. But *Quo Vadis* is a classic novel which has many socially organized characters, while the novel now analyzed is a modern text that has two characters one of which is the author, narrating at the first person singular (see Fig. 2).

Their relationships are established with abstract entities. Both are solitary, but become friends because their affect relationships focus on common points of interest: the past, the old liquor store, the old trees, the icon of old times, the darkness, the early paintings.

Another issue that we propose to answer is whether real-world relationships, as they appear in the *Chat* corpus, are similar to or different from those of fictional characters. The chat corpus of our treebank has 2580 sentences, and the communicators are four persons. For this research, we have selected the verbal exchanges of two persons, i.e. a part of the corpus, which has comparable dimensions with the novel *Remember*.

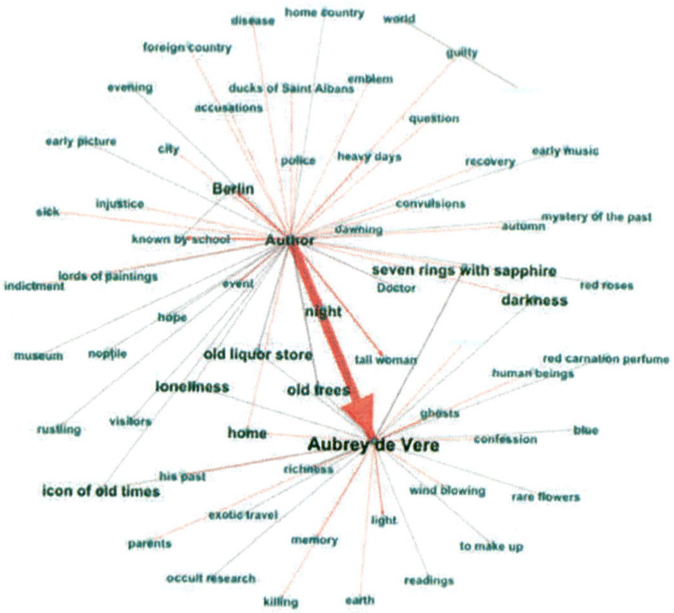

Fig. 2 The positive and negative relationships in the novel Remember

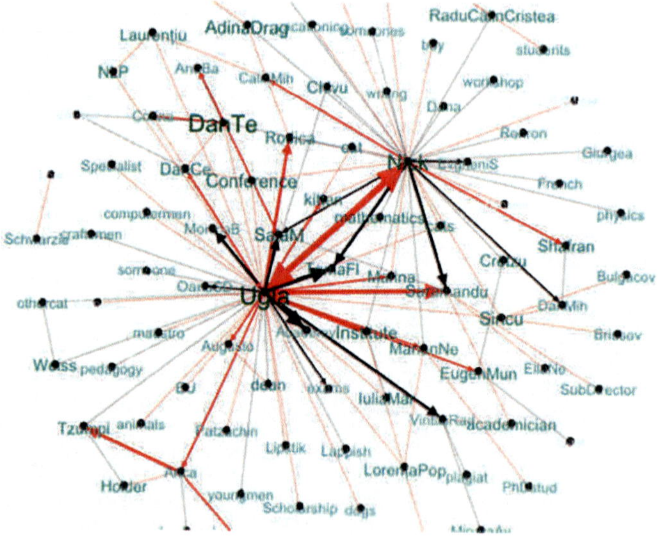

Fig. 3 The positive and negative relationships in the Chat corpus

In this corpus, there are also two characters who have been closely linked through friendship since their student period, and who share their (usually critical) views about common knowledge. Friendship is maintained because each of them gives up slightly positive opinions contrary to the negative ones of their friend, both being inclined towards the humorous style. They have common views about more characters. The views do not coincide about other characters, but the supporter does not hesitate to enhance their defects for the sake of consensus. Both interlocutors have animals and they refer to them as to human beings, designating each one by his name: Ronron, Pusha, Weiss, Schwarzie, Holder (see Fig. 3).

What can be remarked is that in modern literature the character that is spoken about does not have an active role in the relationship of friendship, the author presents only his own point of view, he does not consider himself entitled to speak in the name of the character. By contrast, the relationships between the Chat characters are reciprocal, the arrow between them has two peaks.

Another difference is that relationships in the fictional text are more efficiently organized even if they want to give a chaotic impression at times. We have noticed this when comparing *Quo Vadis* with the characters in the chat.

In the graph of relationships in Fig. 2 you can see a lot of sympathy or rivalry between the characters that are involved. The names were abbreviated to be included in the chart.

The exemplified graphs are for minimal sub-corpora, if the relations in larger dimensional corpora such as *Quo Vadis* are extracted by the same method, the resulting graphs will be much more complicated, and we should make separate graphs of the relation types and sub-types.

6 Conclusions and Future Work

Our goal has been to build an annotated corpus with affect, kinship and social relationships, after a prior annotation of the entities. The research began with the manual annotation of *Quo Vadis*. Only the experience gained then made future development possible. These annotations at a semantic and textual level represent a step superior to the simple logical-semantic annotation and have numerous applications in the extraction of textual meaning, summaries, opinion mining, the social configuration of groups, etc. Future works, are intended to broaden this corpus with both literary texts and communication between real characters on social media sites and platforms.

Our aim is to design a working interface capable to perform whatever is very difficult or impossible for the human mind, while it is something that comes easy to the computer and allows the human assistant to intervene wherever the computer cannot make distinctions that are obvious to humans.

We aim to improve this instrument in the future, to automatically annotate common nouns by searching for their lemma, to introduce a module of rules for the change of pronouns in conversation. The possibilities for data retrieval and extraction will be multiplied. The appropriate selection of triggers can lead to a greater share of the automatic annotation of AKS relationships but the validation of the automatic annotation of this type of relationship by the human annotator will probably not be completely eliminated.

References

1. Bibiri, A.D.: An Annotated Corpus of Entities and Semantic Relations, Disertation thesis, Faculty of Computer Science, Alexandru Ioan Cuza University of Iaşi (2014)
2. Bibiri, A.D., Colhon, M. Diac, P., Cristea, D.: Statistics over a corpus of semantic links: QuoVadis. In: Proceedings of the 10th International Conference Linguistic Resources and Tools for Processing the Romanian Language, pp. 33–44. Alexandru Ioan Cuza University Publishing House (2014). ISSN 1843–911X
3. Cristea, D., Dima G.E., Postolache O.D., Mitkov R.: Handling complex anaphora resolution cases. In: Proceedings of the Discourse Anaphora and Anaphor Resolution Colloquium (DAARC 2002) (2002). ISBN 972-772-350-0
4. Cristea, D., Ignat, E. (2013) Linking book characters. Toward a corpus encoding relations between entities. In: Proceedings of the 7th International Conference on Speech Technology and Human-Computer Dialogue, Spet. 2013, pp. 1–8
5. Cristea, D., Gîfu, D., Colhon, M., Diac, P., Bibiri, A.D., Mărănduc, C., Scutelnicu, L.A.: Quo Vadis: A Corpus of Entities and Relations, Language Production, Cognition, and the Lexicon, pp. 505–543. Springer International Publishing (2015). ISBN: 978-3-319-08043-7
6. Colhon, M., Gîfu, D., Cristea, D.: The QuoVadis storytelling. In: Proceedings of the the 11th International Conference Linguistic Resources and Tools for Processing the Romanian Language (ConsILR 2015), Alexandru Ioan Cuza University Publishing House (2015). ISSN 1843-911X, 93–108
7. Colhon, M., Diac, P., Mărănduc, C., Perez, C.A.: Quo Vadis research areas—text analysis. In: Proceedings of the 10th International Conference Linguistic Resources and Tools for Processing the Romanian Language, pp. 45–56. Alexandru Ioan Cuza University Publishing House (2014). ISSN 1843–911X
8. Perez, C.A., Mărănduc, C., Simionescu, R.: Social Media—Processing Romanian Chats and Discourse Analysis, Computación y Sistemas, 20. ISSN **404–414**(3), 1405–5546 (2016)
9. Perez, C. A., Mărănduc, C., Simionescu, R.: Including Social Media—A Very Dynamic Style—in the Corpora for Processing Romanian Language, Linguistic Linked Open Data: 12th EUROLAN 2015 Summer School and RUMOUR 2015 Workshop, Sibiu, Romania, July 13-25, 2015, Revised Selected Papers, pp. 139–153. Springer International Publishing (2016). ISBN 978-3-319-32942-0
10. Colhon, M. Cristea, D., Gîfu, D.: Discovering Semantic Relations Within Nominals, Linguistic Linked Open Data: 12th EUROLAN 2015 Summer School and RUMOUR 2015 Workshop, Revised Selected Papers, pp. 85–100. Springer International Publishing (2016). ISBN 978-3-319-32942-0
11. Tufiş, D., Ştefănescu, D.: Connotative meaning markup for wordnet. In: Proceedings of the Romanian Academy, Series A, Bucharest, vol. 12(2), pp. 143–155 (2011)
12. Esuli, A., Sebastiani, F.: SENTIWORDNET: a publicly available Lexical resource for opinion mining. In: Proceedings of the 5th Conference on Language Resources and Evaluation (LREC?06), pp. 417–422 (2006)
13. Poria, S., Agarwal, B., Gelbukh, A., Hussain, A., Howard, N.: Dependency-Based Semantic Parsing for Concept-Level Text Analysis, Computational Linguistics and Intelligent Text Processing. CICLing,: Lecture Notes in Computer Science, vol. 8403. Springer, Berlin, Heidelberg (2014)

14. Dozat, T., Manning C.D.: Simpler but more accurate semantic dependency parser. In: Proceedings of the 56th Annual Meeting of the Association for Computational Linguistics (Short Papers), pp. 484–490 (2018)
15. Hearst, M.A.: Automatic acquisition of hyponyms from large text corpora. In: Proceedings of the 14th Conference on Computational Linguistics, vol. 2. Association for Computational Linguistics, pp. 539–545 (1992)
16. Malandrakis, N., Potamianos, A., Iosif, E., Narayanan, S.: Distributional semantic models for affective text analysis. IEEE Trans. Audio Speech Lang. Process. **2379–2392**(11), 1558–7916 (2007)
17. Searle, J.R.: The Logical Status of Fictional Discourse, American Poetics. Current Orientations, Dacia Publisher, Cluj-Napoca, pp. 154–173 (1981)
18. Pang, B., Lee, L.: Opinion mining and sentiment analysis. In: Foundations and Trends in Information Retrieval, vol. 2(1–2), pp. 1–135 (2008)
19. Wiebe, J., Mihalcea, R.: Word sense and subjectivity, In: Proceedings of the 21st International Conference on Computational Linguistics and the 44th annual meeting of the Association for Computational Linguistics, Association for Computational Linguistics, pp. 1065–1072 (2006)
20. Hu, M., Liu, B.: Mining and summarizing customer reviews. In: Proceedings of the Tenth ACM SIGKDD International Conference on Knowledge Discovery and Data Mining, Association for Computational Linguistics, pp. 168–177 (2004)
21. Valitutti, R.: WordNet-Affect: an affective extension of WordNet. In: Proceedings of the 4th International Conference on Language Resources and Evaluation, pp. 1083–1086 (2004)
22. Bradley, M.M., Lang, P.J.: Affective norms for English words (ANEW): instruction manual and affective ratings, Technical report c-1, University of Florida, The Center for Research in Psychophysiology (1999)
23. Farzindar, A., Inkpen, D.: Natural Language Processing for Social Media. Morgan & Claypool Publishers (2015). ISBN 1627053883, 9781627053884
24. Ekman, P.: An argument for basic emotions. Cogn. Emot. **6**(3–4), 169–200 (1992)
25. Hansen, D.L., Shneiderman, B., Smith, M.A.: Analyzing Social Media Networks with NodeXL. Morgan Kaufmann (2011). ISBN 978-0-12-382229-1
26. Barnes, J., Kurtz, R., Oepen, S., Oevrelid, L., Velldal E.: Structured sentiment analysis as dependency graph parsing. In: Proceedings of the 59th Annual Meeting of the Association for Computational Linguistics and the 11th International Joint Conference on Natural Language Processing, pp. 3387–3402 (2021)

Graph Theory Applied to Brain Network Analysis in Alzheimer's Disease

Serena Dattola and Fabio La Foresta

Abstract Alzheimer's disease (AD) is an incurable brain disorder which affects especially elderly. Over the years, the analysis of the brain functional connectivity from EEG signals has been exploited for promoting an early diagnosis of AD. Graph theory provides helpful tools to describe complex brain networks. In this work, starting from High-Density EEGs, we estimated the functional connectivity by the Lagged Linear Connectivity (LLC) parameter, for 84 Regions of Interest (ROIs), and analyzed the brain networks properties for three groups of subjects: control subjects (CNT), Mild Cognitive Impairment patients (MCI) and AD patients. We computed three network parameters: the Clustering Coefficient, the Characteristic Path Length and the Randić Index. The results showed that the functional connectivity of MCI and even more of AD patients declines in comparison to healthy people. Moreover, the results deriving from the Randić Index about robustness of brain networks outperform those deriving from the Connection Density Index, commonly used for brain network analysis.

1 Introduction

Alzheimer's Disease (AD) is a degenerative brain disease that is the most widely recognized reason for dementia among elderly. The first stage, called Mild Cognitive Impairment (MCI), is mainly characterized by memory issues, which however do not alter significantly the quality of life of the patients. Every year, 10–15% of MCI patients converts to AD. As the disease progresses, cognitive and motor abilities are lost, finally leading to death. A detection of the disease in its prodromal stage may contribute to an early intervention which can slow the progression of the disease.

S. Dattola (✉) · F. La Foresta
DICEAM Department, Mediterranea University of Reggio Calabria, Via Graziella Feo di Vito, 89060 Reggio Calabria, Italy
e-mail: serena.dattola@unirc.it

F. La Foresta
e-mail: fabio.laforesta@unirc.it

D. Camacho et al. (eds.), *Intelligent Distributed Computing XIV*, Studies in Computational Intelligence 1026, https://doi.org/10.1007/978-3-030-96627-0_33

Electroencephalography (EEG) has been used for years for studying and monitoring AD [9, 10]. EEG allows to record the brain electric activity in a non-invasive way, with electrodes conveniently placed on the scalp. The simultaneous activity of cortical neurons is detected by each electrode as a function of the electric signals' direction and distance from each sensor. Moreover, the intensity of the signals is affected by the conductivity of the scalp and the brain tissues. In this respect, the choice of an adequate head model is required for a correct interpretation of the EEG recordings [23]. The main strengths of EEG are represented by its affordability and ease of use. EEG provides an excellent temporal resolution but a low spatial one, which represents its main drawback. In order to address this issue, High-Density EEG (HD-EEG) montages were introduced, including up to 256 electrodes. In fact, several studies showed that the scalp potential can be correctly sampled when the space between electrodes ranges from 1 to 3 cm [5, 20].

Graph theory provides helpful tools to describe complex brain networks. In particular, the nodes represent the brain areas and the edges that connect them denote the structural or functional connectivity. In this paper we focused on functional connectivity, which quantifies the relationship between brain regions through coherency measurements. Many techniques have been used for estimating the functional connectivity during the last two decades, as reviewed in [2]. As for AD, several studies revealed a decrease in functional connectivity, reported in [17, 22]. Recent studies showed that healthy brain networks have small-world properties [1, 18]. In particular, brain networks are characterized by a combination of segregation and integration properties: the first denotes the ability to form interconnected clusters of brain regions, the latter refers to the tendency to rapidly exchange information from different brain regions. Segregation and integration can be quantified by the Clustering Coefficient (CC) [18] and the Characteristic Path Length (λ), respectively [18]. According to literature, it is expected that the small-world properties of AD patients decline as compared with healthy subjects [21, 24]. Moreover, it could be useful to analyze aspects relative to the robustness of brain networks, in order to provide a better interpretation of the above-mentioned indexes. As reported in literature, it is detected a decrease in graph robustness as the disease progresses, due to a reduction in functional connections between regions for AD patients [7, 19]. In order to evaluate the functional connectivity, the localization of the brain active sources is necessary. One of the most promising method used for the so-called EEG inverse problem is LORETA (Low Resolution Electromagnetic Tomography) [13]. It is a computational method which provides a three-dimensional reconstruction of the cortical electrical activity from the potentials measured on the scalp (EEG). The two new LORETA family algorithms later introduced, sLORETA and even more eLORETA, showed a significant improvement in source localization accuracy [11, 14]. Starting from the reconstructed brain electrical activity, it is possible to quantify the functional connectivity by a tool implemented in the LORETA-KEY software. Differently from previous works, which applied this tool to low-density EEGs, this study focused on HD-EEGs for exploring the small-world properties of brain networks for three groups of subjects: control subjects (CNT), Mild Cognitive Impairment patients (MCI) and Alzheimer's Disease patients (AD). In addition to the estimation of the Clustering

Coefficient and the Characteristic Path Length, we computed the Randić Index for quantifying the robustness of the brain networks. So, this work can be regarded as a further development of our previous study described in [8]. The results we obtained showed that the small-world properties of MCI and even more of AD patients are weakened in comparison to healthy people. Moreover, the information deriving from the Randić Index about the robustness of brain networks are more significant than those deriving from the Connection Density index. In fact, as reported in [3], the Randić Index showed a better performance in discriminating between healthy people and MCI/AD patients as compared with the Connection Density index, commonly used in functional connectivity analysis. So, the analysis we carried out confirms the promising results obtained by means of the Randić Index and suggests that it can be successfully employed also in this field.

2 Materials and Methodology

This section provides information about the EEG recordings and the data preprocessing (Sect. 2.1); then, Lagged Linear Connectivity and the analyzed complex network parameters are described (Sect. 2.2).

2.1 *HD-EEG Acquisition System and Data Preprocessing*

EEG signals, acquired at "IRCCS–Centro Neurolesi Bonino-Pulejo" of Messina (Italy), belong to three groups of subjects: 10 CNT, 21 MCI patients, and 9 AD patients. The research was conducted following a protocol approved by the local Ethics Committee (Approved N. 003/17) and all the subjects or their caregivers signed an informed consent form. The EEG acquisition system is the 256-channel HydroCel Geodesic Sensor Net, belonging to the Geodesic EEG System. The electrode impedance was maintained $<50\,\text{k}\Omega$, according to EGI guidelines (https://www.egi.com/research-division/eeg-systems/geodesic-eeg-systems). The reference electrode was Cz, positioned in the middle of the scalp. EEGs were recorded in resting state with eyes-closed, the sampling rate was 250 Hz and the signals were band-pass filtered between 1 and 40 Hz. After filtering, artifacts were visually identified and removed. Nevertheless, only 173 channels were taken into account in this study because the potentials deriving from the electrodes placed on the face and the neck were affected by too much noise. For each subject we considered two minutes of cleaned EEGs, which were finally transformed to a common average reference montage. All the preprocessing operations were performed by MATLAB (R2018a) of Mathworks.

Algorithm 1 eLORETA algorithm

Input: **K**; **H**; α; ϵ.
1: Set $error > \epsilon$
2: $Initialize\_Weight\_Matrix$ (**W**)
3: **while** $error > \epsilon$ **do**
4: $\mathbf{M} = (\mathbf{K}\mathbf{W}^{-1}\mathbf{K}^T + \alpha\mathbf{H})^+$
5: **for** $j = 1 \cdots N_V$ **do**
6: $W_j = [K_j^T \mathbf{M} K_j]^{SySqrt}$
7: **end for**
8: **end while**
9: $\mathbf{T} = \mathbf{W}^{-1}\mathbf{K}^T(\mathbf{K}\mathbf{W}^{-1}\mathbf{K}^T + \alpha\mathbf{H})^+$

2.2 *eLORETA Lagged Linear Connectivity Toolbox and Complex Network Parameters*

The first step for a correct functional connectivity analysis is represented by the localization of the active sources. The algorithm employed in this study is eLORETA, which provides zero error localization in the presence of measurement and structured biological noise and considers the deeper sources more adequately in comparison with the previous LORETA family algorithms [11]. The brain volume is divided into 6239 grid points (voxels) at 5 mm resolution and the brain sources are placed on each voxel. The algorithm provides the computation of the current density vector for each source.

The mathematical expression of the estimated current density is given by:

$$\hat{\mathbf{J}} = \mathbf{T}\mathbf{\Phi} \tag{1}$$

where **T** is the pseudoinverse and $\Phi \in R^{N_E x1}$ is the vector of the electric potential differences measured at N_E electrodes with respect to a single common reference electrode.

The pseudocode of the eLORETA algorithm is reported in Algorithm 1.
The eLORETA algorithm takes as input the average reference lead field **K** and the average reference operator **H**, the regularization parameter $\alpha > 0$, and the parameter ϵ representing the maximum admissible *error* calculated according to the variations occurring between two successive iterations in the weight matrix **W** (lines 3–8).

First, a function initializes the weight matrix **W** (line 1) and then the matrix **M** is computed (line 2). Until the value of the parameter *error* is greater than ϵ (lines 3–8), for each of the N_V voxels, the corresponding value of W (lines 5–7) is calculated. Note that, at line 6, the term $[K_j^T \mathbf{M} K_j]^{SySqrt}$ denotes the symmetric square root of the matrix $[K_j^T \mathbf{M} K_j]$. Finally, when the value of the parameter error is negligible (i.e., $error \leq \epsilon$), the matrix **T** is computed and the algorithm ends.

After the active source reconstruction performed by the eLORETA algorithm, the brain functional connectivity was quantified by means of the Connectivity toolbox implemented in the LORETA-KEY software (v20201109). In particular, the authors of LORETA proposed a new parameter for the evaluation of functional connectivity [12]. They considered the total connectivity as the sum of an instantaneous and a lagged contribution: the first one considers the potentials generated by distant cortical regions and conducted across the head volume (the so-called "volume conduction"), whereas the lagged component provides almost pure physiological information.

The linear dependence between time series X and Y at frequency ω is given by:

$$F_{X,Y}(\omega) = \ln \frac{\left| \begin{pmatrix} \mathbf{S}_{Y,Y}(\omega) & \mathbf{0} \\ \mathbf{0}^T & \mathbf{S}_{X,X}(\omega) \end{pmatrix} \right|}{\left| \begin{pmatrix} \mathbf{S}_{Y,Y}(\omega) & \mathbf{S}_{Y,X}(\omega) \\ \mathbf{S}_{X,Y}(\omega) & \mathbf{S}_{X,X}(\omega) \end{pmatrix} \right|} \quad (2)$$

where $\mathbf{S}_{X,X}$, $\mathbf{S}_{Y,Y}$, $\mathbf{S}_{X,Y}$, $\mathbf{S}_{Y,X}$ denote the complex valued covariance matrices and $|\mathbf{M}|$ is the determinant of $\mathbf{M}$.

Linear dependence can be considered as the sum of the lagged linear dependence $F_{X \rightleftarrows Y}(\omega)$ and the instantaneous linear dependence $F_{X \cdot Y}(\omega)$:

$$F_{X,Y}(\omega) = F_{X \rightleftarrows Y}(\omega) + F_{X \cdot Y}(\omega) \quad (3)$$

The instantaneous linear dependence is defined as:

$$F_{X \cdot Y}(\omega) = \ln \frac{\left| \mathrm{Re} \begin{pmatrix} \mathbf{S}_{Y,Y}(\omega) & \mathbf{0} \\ \mathbf{0}^T & \mathbf{S}_{X,X}(\omega) \end{pmatrix} \right|}{\left| \mathrm{Re} \begin{pmatrix} \mathbf{S}_{Y,Y}(\omega) & \mathbf{S}_{Y,X}(\omega) \\ \mathbf{S}_{X,Y}(\omega) & \mathbf{S}_{X,X}(\omega) \end{pmatrix} \right|} \quad (4)$$

where Re($\mathbf{M}$) denotes the real part of $\mathbf{M}$.

Finally, the measure of lagged linear dependence is:

$$F_{X \rightleftarrows Y}(\omega) = F_{X,Y}(\omega) - F_{X \cdot Y}(\omega) = \ln \frac{\left\{ \left| \mathrm{Re} \begin{pmatrix} \mathbf{S}_{Y,Y}(\omega) & \mathbf{S}_{Y,X}(\omega) \\ \mathbf{S}_{X,Y}(\omega) & \mathbf{S}_{X,X}(\omega) \end{pmatrix} \right| \Big/ \left| \mathrm{Re} \begin{pmatrix} \mathbf{S}_{Y,Y}(\omega) & \mathbf{0} \\ \mathbf{0}^T & \mathbf{S}_{X,X}(\omega) \end{pmatrix} \right| \right\}}{\left\{ \left| \begin{pmatrix} \mathbf{S}_{Y,Y}(\omega) & \mathbf{S}_{Y,X}(\omega) \\ \mathbf{S}_{X,Y}(\omega) & \mathbf{S}_{X,X}(\omega) \end{pmatrix} \right| \Big/ \left| \begin{pmatrix} \mathbf{S}_{Y,Y}(\omega) & \mathbf{0} \\ \mathbf{0}^T & \mathbf{S}_{X,X}(\omega) \end{pmatrix} \right| \right\}} \quad (5)$$

All these measures are non-negative. Their value is zero only when there is independence (lagged, instantaneous, or both). They can also be transformed into a "squared coherence", with values in the [0 …1] range.

So, the Lagged Linear Connectivity (LLC) parameter, computed by Eq. 5, estimates the statistical dependence among active sources for the all 84 possible Regions of Interest (ROIs) defined by the LORETA-KEY software, for the frequency range 1–40 Hz. The ROIs denote distinct Brodmann areas, 42 for each hemisphere. EEGs were divided into artifact-free non-overlapping epochs of 250 samples (1 s). LLC was computed for windows of three epochs, resulting in 40 connection matrices for each subject. In this way, we gained 40 values of each connectivity parameter for each subject.

According to graph theory, brain networks can be represented by a collection of nodes, which denotes the brain regions, and the links between each pair of nodes, which can be directed or undirected, weighted or not. In this study, we considered a weighted undirected graph, where each node denotes a ROI. Weighted graphs provide information about connection strengths.

Functional integration measures the ease of communication between brain regions and can be quantified by the Characteristic Path Length (λ). λ is defined as the average shortest path length between all pairs of nodes [18]:

$$\lambda = \frac{1}{n}\sum\nolimits_{i\in N}\frac{\sum\nolimits_{j\in N, j\neq i} d_{ij}^{w}}{n-1} \tag{6}$$

where n is the number of nodes and d_{ij}^{w} is the shortest weighted path length between the nodes i and j.

Functional segregation is the capacity to create interconnected groups of brain regions (clusters) and can be measured by the Clustering Coefficient (CC). It gives an estimation of the probability that the neighbors of one node become neighbors each other. The weighted Clustering Coefficient has the following expression [18]:

$$CC^{w} = \frac{1}{n}\sum_{i\in N} CC_{i}^{w} = \frac{1}{n}\sum_{i\in N}\frac{2t_{i}^{w}}{k_{i}(k_{i}-1)} \tag{7}$$

where CC_i is the Clustering Coefficient of node i, t_i is the number of triangles around the node i and k_i is the number of links connected to a node i, so that $k_i(k_i - 1)/2$ is the maximum number of possible edges can exist between neighbors of node i.

A complex network has small-world properties if the Clustering Coefficient is high and the average path length is short.

The Randić Index of a graph is a topological index used to describe the molecular structure and widely employed in chemistry [4, 16]. The Randić Index associated with a graph G is defined as:

$$R(G) = \sum_{\langle u,v\rangle\in E}\frac{1}{d_{u}d_{v}} \tag{8}$$

where u is a node belonging to G and d_u is the degree of u, defined as the number of edges incident on u.

For a graph G of size m, the computational complexity of $R(G)$ increases linearly with m. So, the Randić Index can be easily recomputed (in an incremental way) when the graph topology changes in terms of added/deleted links (or nodes).

The Randić Index reaches its minimum value in case of a complete graph, so the lower the $R(G)$, the more connected (i.e., robust) the G.

In our study, we used a different formulation of the Randić Index, already proposed in [15]. It is defined as:

$$R(G) = \sum_{(u,v)\in E} d_u d_v \tag{9}$$

According to Eq. (9), the higher the $R(G)$, the more connected (i.e., robust) the G.

Data processing and network analysis were performed in a MATLAB environment (R2018a). In particular, the Clustering Coefficient and the Characteristic Path Length were computed by means of the MATLAB toolbox "Brain Connectivity Toolbox" (https://sites.google.com/site/bctnet/).

3 Experimental Results

The methodology described in Sect. 2 was applied to the HD-EEG dataset. Figure 1 shows the boxplots of the Characteristic Path Length and the Clustering Coefficient of the three groups of subjects. Each boxplot consists of 1600 parameters, 40 for each patient. From Fig. 1a arises that λ median value increases for MCI and even more for AD in comparison to CNT. Figure 1b shows the opposite trend, that is the median CC decreases for MCI and even more for AD compared with CNT. The trend of both λ and CC reflects the expected behavior. A statistical analysis was carried out for validating the results. Specifically, the Wilcoxon rank sum test [6] was performed to determine if the data belonging to two different groups are samples from continuous distributions with equal medians (null hypothesis). The significance level was set at 5%, so the differences between groups are statistically significant when $p < 0.05$. The results of the statistical analysis are reported in Table 1. As for λ, there is a significant difference between all the three groups. In regard to CC, only the difference between CNT and AD is statistically significant.

The computation of the Randić Index was performed as a function of a threshold applied to the weighted connection matrices of CNT, MCI, and AD at step of 5%. The network will be totally connected when the threshold is set to 0 and disconnected if

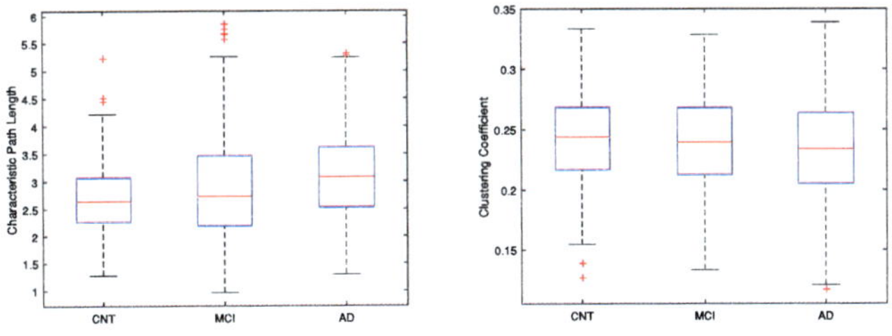

(a) (b)

Fig. 1 **a** Boxplot of the characteristic path length of CNT, MCI, and AD; **b** boxplot of the clustering coefficient of CNT, MCI, and AD

Table 1 Results of Wilcoxon rank sum test for the characterisic path length and the clustering coefficient

Subjects	Characteristic path length p-value	Clustering coefficient p-value
CNT-MCI	0.01[a]	0.14
MCI-AD	2.54 e-07[a]	0.16
CNT-AD	1.72 e-15[a]	0.01[a]

[a]Denotes statistically significant differences between groups ($p < 0.05$)

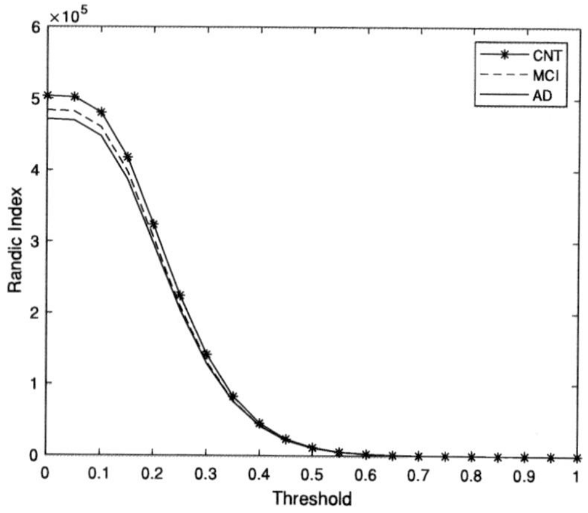

Fig. 2 Mean values of the Randić Index ($R(G)$) computed by thresholding the connection matrices of CNT, MCI, and AD

the threshold is set to 1. Note that the mean Randić Index of each group of subjects is given by the average of the Randić Index of all subjects belonging to the given group. Figure 2 shows that, as expected, the network robustness is greater for CNT, followed by MCI and finally by AD. Once again, the Wilcoxon rank sum test was used for the statistical analysis. Note that the analysis was carried out for the threshold range [0%, 25%] because most of the links of the graph are removed outside this range. In this case, most of the existing information is lost and the related results cannot be reliable. The results, reported in Table 2, show a significant difference between CNT and AD groups, for all considered threshold values.

Table 2 p-values resulting from the Wilcoxon rank sum test for the Randić Index

Threshold	Randić Index p-value		
	CNT-MCI	MCI-AD	CNT-AD
0	0.1084	0.2508	0.0181[a]
0.05	0.1083	0.2514	0.0179[a]
0.1	0.1086	0.2535	0.0184[a]
0.15	0.1029	0.2556	0.0177[a]
0.2	0.0832	0.2751	0.0168[a]
0.25	0.0686	0.3150	0.0156[a]

[a]Denotes statistically significant differences between groups ($p < 0.05$)

4 Discussion and Conclusions

Alzheimer's Disease is an incurable brain disorder which affects especially old people. Over the years EEG has been exploited for studying the brain electrical activity and functional connectivity in AD, in order to provide useful information for an early diagnosis. Moreover, brain functional connectivity has recently been explored using graph theory. In this work we analyzed the brain networks of 10 CNT, 21 MCI, and 9 AD patients. Starting from the active source reconstruction by eLORETA, the brain functional connectivity was evaluated by means of the Lagged Linear Connectivity parameter, computed through a toolbox implemented in the LORETA-KEY software. The complex network analysis was carried out by the computation of three parameters: the Characteristic Path Length, the Clustering Coefficient and the Randić Index. All the three parameters showed a trend consistent with the expected behaviour of AD brain networks. In particular, for the Characteristic Path Length there is a significant difference between all groups, whereas for the Clustering Coefficient the difference is significant between CNT and AD. Also the Randić Index shows statistically significant difference in network robustness between CNT and AD. These outcomes outperform those obtained using the Connection Density Index, reported in our previous study [8], and can be conveniently exploited for further investigations. In a future research we will perform a complex network analysis in a longitudinal study, in order to evaluate the network parameters and robustness during the evolution of the disease for each patient monitored over time, with a more sizeable dataset.

In conclusion, this study confirms that the brain functional connectivity of MCI and, even more, AD patients is weakened in comparison to healthy people. It also provides interesting results about the Randić Index, proving that it can be conveniently exploited beyond the purpose for which it was originally intended, and so employed for graph robustness measurements in case of brain networks.

References

1. Bassett, D.S., Bullmore, E.: Small-world brain networks. Neuroscientist **12**(6), 512–523 (2006)
2. Bastos, A.M., Schoffelen, J.M.: A tutorial review of functional connectivity analysis methods and their interpretational pitfalls. Front. Syst. Neurosc. **9**, 175 (2016)
3. Dattola, S., Mammone, N., Morabito, F.C., Rosaci, D., Sarné, G.M.L., La Foresta, F.: Testing graph robustness indexes for EEG analysis in alzheimer's disease diagnosis. Electronics **10**(12), 1440 (2021)
4. De Meo, P., Messina, F., Rosaci, D., Sarné, G.M., Vasilakos, A.V.: Estimating graph robustness through the randic index. IEEE Trans. Cybern. **48**(11), 3232–3242 (2017)
5. Freeman, W.J., Holmes, M.D., Burke, B.C., Vanhatalo, S.: Spatial spectra of scalp EEG and EMG from awake humans. Clin. Neurophysiol. **114**(6), 1053–1068 (2003)
6. Gibbons, J., Chakraborti, S.: Nonparametric Statistical Inference Springer (2011)
7. König, T., Prichep, L., Dierks, T., Hubl, D., Wahlund, L., John, E., Jelic, V.: Decreased eeg synchronization in alzheimer's disease and mild cognitive impairment. Neurobiol. Aging **26**(2), 165–171 (2005)

8. La Foresta, F., Morabito, F.C., Marino, S., Dattola, S.: High-density EEG signal processing based on active-source reconstruction for brain network analysis in alzheimer's disease. Electronics **8**(9), 1031 (2019)
9. Labate, D., La Foresta, F., Palamara, I., Morabito, G., Bramanti, A., Zhang, Z., Morabito, F.C.: EEG complexity modifications and altered compressibility in mild cognitive impairment and alzheimer's disease. In: Recent Advances of Neural Network Models and Applications, pp. 163–173. Springer (2014)
10. Mammone, N., De Salvo, S., Bonanno, L., Ieracitano, C., Marino, S., Marra, A., Bramanti, A., Morabito, F.C.: Brain network analysis of compressive sensed high-density EEG signals in AD and MCI subjects. IEEE Trans. Ind. Inf. **15**(1), 527–536 (2019)
11. Pascual-Marqui, R.D.: Discrete, 3d distributed, linear imaging methods of electric neuronal activity. Part 1: exact, zero error localization (2007). arXiv:0710.3341
12. Pascual-Marqui, R.D.: Instantaneous and lagged measurements of linear and nonlinear dependence between groups of multivariate time series: frequency decomposition (2007). arXiv:0711.1455
13. Pascual-Marqui, R.D., Michel, C.M., Lehmann, D.: Low resolution electromagnetic tomography: a new method for localizing electrical activity in the brain. Int. J. Psychophysiol. **18**(1), 49–65 (1994)
14. Pascual-Marqui, R.D., et al.: Standardized low-resolution brain electromagnetic tomography (sloreta): technical details. Methods Find Exp. Clin. Pharmacol. **24**(Suppl D), 5–12 (2002)
15. Phillips, D.J., McGlaughlin, A., Ruth, D., Jager, L.R., Soldan, A., Initiative, A.D.N., et al.: Graph theoretic analysis of structural connectivity across the spectrum of alzheimer's disease: the importance of graph creation methods. NeuroImage: Clin. **7**, 377–390 (2015)
16. Randic, M.: Characterization of molecular branching. J. Am. Chem. Soc. **97**(23), 6609–6615 (1975)
17. Rossini, P.M., Di Iorio, R., Vecchio, F., Anfossi, M., Babiloni, C., Bozzali, M., Bruni, A.C., Cappa, S.F., Escudero, J., Fraga, F.J., et al.: Early diagnosis of alzheimer's disease: the role of biomarkers including advanced EEG signal analysis. Report from the IFCN-sponsored panel of experts. Clin. Neurophysiol.**131**(6), 1287–1310 (2020)
18. Rubinov, M., Sporns, O.: Complex network measures of brain connectivity: uses and interpretations. Neuroimage **52**(3), 1059–1069 (2010)
19. Sankari, Z., Adeli, H., Adeli, A.: Intrahemispheric, interhemispheric, and distal EEG coherence in alzheimer's disease. Clin. Neurophysiol. **122**(5), 897–906 (2011)
20. Spitzer, A.R., Cohen, L.G., Fabrikant, J., Hallett, M.: A method for determining optimal interelectrode spacing for cerebral topographic mapping. Electroencephalogr. Clin. Neurophysiol. **72**(4), 355–361 (1989)
21. Stam, C.J., Jones, B., Nolte, G., Breakspear, M., Scheltens, P.: Small-world networks and functional connectivity in alzheimer's disease. Cereb. Cortex **17**(1), 92–99 (2006)
22. Tijms, B.M., Wink, A.M., de Haan, W., van der Flier, W.M., Stam, C.J., Scheltens, P., Barkhof, F.: Alzheimer's disease: connecting findings from graph theoretical studies of brain networks. Neurobiol. Aging **34**(8), 2023–2036 (2013)
23. Wang, G., Ren, D.: Effect of brain-to-skull conductivity ratio on EEG source localization accuracy. BioMed Res. Int. **2013** (2013)
24. Wang, R., Wang, J., Yu, H., Wei, X., Yang, C., Deng, B.: Decreased coherence and functional connectivity of electroencephalograph in alzheimer's disease. Chaos: Interdiscip. J. Nonlinear Sci. **24**(3), 033136 (2014)

Detection of Postpartum Depression-Related Posts: An Analysis for Serbian

Ulfeta Marovac and Aldina Avdić

Abstract This paper presents a model of an intelligent distributed support system for pregnant women and mothers with an emphasis on the detection of postpartum depression. There are psychological tests that can be used to determine the degree of postpartum depression, but it is very important to detect the existence of this depression in women who are not aware of it. A lot of women use social networks to share information about their problems. Using tools for natural language processing and machine learning, social network posts that describing problems similar to postpartum depression can be detected. The presented system was tested on a data set of posts from two social media forums (one Serbian and one English). It has been shown that posts indicating postpartum depression can be labeled with an accuracy of up to 89%. This system aims to identify the potential existence of problems and refer to the system for detection and professional assistance. It can contribute to improving the lives of mothers and thus babies.

1 Introduction

The use of ICT technologies in informing, detecting, and treating various medical problems is becoming more common. Problems with the advent of COVID have made many health services inaccessible. Sometimes it is not easy to recognize that a problem exists, so medical help is not given at the right time. Seeking help online is very often the first step in healing. It is very important to recognize that type of action on the Internet and direct it in the right direction. The current statistics show that stressful situations in postpartum period accelerate the current psychological state of women, which in 10% of cases leads to postpartum depression.

U. Marovac (✉) · A. Avdić
Department for Technical Sciences, State University of Novi Pazar, Novi Pazar, Serbia
e-mail: umarovac@np.ac.rs

A. Avdić
e-mail: apljaskovic@np.ac.rs

D. Camacho et al. (eds.), *Intelligent Distributed Computing XIV*, Studies in Computational Intelligence 1026, https://doi.org/10.1007/978-3-030-96627-0_34

How to recognize the state of postpartum depression? This question should be answered by psychiatrists and psychologists. Postpartum depression signs and symptoms may include: depressed mood or severe mood swings, excessive crying, difficulty bonding with the baby, withdrawing from family and friends, loss of appetite or eating much more than usual, inability to sleep (insomnia) or sleeping too much, overwhelming fatigue or loss of energy, reduced interest and pleasure in usual activities, intense irritability and anger, fear of unsuccessful parenting, hopelessness, feelings of worthlessness, shame, guilt or inadequacy, etc.

To determine postpartum depression, the Edinburgh scale is used, with a score greater than 12 indicating the postpartum depression. EPDS is one of the most commonly used scales for assessing depressive symptoms for women who have given birth. The respondent estimates the weight for ten different depressive symptoms in the past seven days on a scale of 0 to 3 [1]. The possible range of results is from 0 to 30, where higher scores refer to more difficult symptomatic. The items on the scale refer to the assessment of the possibility of experiencing joy, concern, guilt, sorrow, then self-harming, etc.

In the postpartum period, when a woman is most needed by the baby and herself, each health problem is more difficult to solve due to the attachment of the mother and the baby. Women often seek their first advice on social networks and specialized forums. It is very important to recognize forum posts related to postpartum depression and refer the participants to platforms that can give them more information and help. The aim of this paper is to mark non-English posts that indicate postpartum depression using resources in English. This paper proposes a system for labeling postpartum depression-related posts on Serbian and refers them to an information service with the possibility of detecting the degree of depression, and expert advice on the topic of postpartum depression (PPD).

The rest of the paper is organized as follows. The second section presents a description of similar research. The third section presents the properties of used data sets. The methodology and presentation of the system model are given in the fourth section. The fifth section shows the results of the machine learning techniques for postpartum depression detection. The last section contains the conclusion and directions for further research.

2 Related Work

Many studies link mental state with the use of language. There are also indications that the writer's feelings and condition can be studied through linguistic errors. In the era of the Internet and the social network of Facebook, Twitter, and Reddit, conditions have been created to collect large amounts of textual data through which it is possible to monitor personality behavior. Applications based on crowdsourcing are used in the implementation of voting systems (e.g. Amazon Mechanical Turk, [2]), information sharing systems (Quora, [3]), games (Google Image Labeler), and creative systems (TopCoder, [4]), etc.

Social networks have become another source of support for mental health professionals in drawing conclusions and finding indications of mental disorders, such as depression. Social networks, forums are used to analyze the mental state of the user through the information they leave in the form of text, images, user information, and additional tools such as lexicons NRC Word-Emoticon Association Lexicon, WordNet-Affect, Anew, and LIWC tool [5]. The paper [6] presents a study of tweets in which users declared themselves depressed. Social networks are a place where people communicate with each other on common topics. By applying machine learning techniques, depression can be detected with a high level of accuracy from textual data collected, for example, on Facebook [7]. Changes in the mood of mothers before and after childbirth on Twitter were monitored in [8] and based on them, models were made that can predict what the mood will be after childbirth. In [9], a marker for marking fathers with possible postpartum depression was proposed, which was examined on fathers' publications given in the period of 6 months from the birth of the child. The possibility of applying machine learning to predict postpartum depression was investigated over the corpus created from Reddit posts in [10]. What is characteristic is that most of this research is related to the English language, non-English social media are quite unexplored. A model for diagnosing postpartum depression via the crowdsourcing platform has been proposed for Serbian, which also includes an automated test for determining the degree of postpartum depression using the Edinburgh scale [11]. This paper presents a system for labeling postpartum depression-related posts in social media forums in Serbian which referred users to appropriate support systems such as the one mentioned above.

3 Evaluation Dataset

For this research, two data sets were collected. One contains posts from the Reddit [12] forum in English and the other set of data is from the ana.rs [13] forum in Serbian.

The first data set (DS1) consists of Reddit posts from the depression section (community about: "Peer support for anyone struggling with a depressive disorder.") and happy sections (community about: "Too many depressing things on the main page, so post about what makes you warm and fuzzy inside!") which containing the word "pregnancy" data are in English. They were collected using the Pushshift API. 131 posts from the depression group were singled out, and 128 posts from the happy group.

The second set of data (DS2) is from the forum ana.rs, which is one of the largest women's forums in Serbian. This forum contains the topic of postpartum depression from which 98 posts have been selected. Also, 95 posts unrelated to postpartum depression were singled out from the sections in "Pregnancy and Weight"and "Pregnancy Chat Corner". Posts are marked according to the section to which they belong to the group of PPD-related posts and group of posts not related to PPD. Both sets were manually checked and posts related to postpartum depression are marked.

The research aims to identify publications related to postpartum depression, so based on these data, a model that classifies publications into the group of postpartum depression or something else was made.

4 Methodology

Most postpartum depression forums are in English. Therefore most researches on PPD-related posts detection are for English. To include a larger population, languages other than English must also be used. The following is an example of processing posts in the Serbian language that does not contain the specifics of the Serbian language (language-dependent resources) and can be easily transferred to other languages.

Figure 1 shows a postpartum depression-related post tagging system that redirects participants to a crowdsourcing platform that support pregnant women and mothers in stressful situations. This platform is described in [11] and contains part of the automated determination of the degree of postpartum depression based on the Edinburgh scale as well as the possibility of obtaining professional help. To label posts as potentially associated with postpartum depression, a classification model was developed using machine learning methods. The PPD labeling system goes through the following steps:

- detection of the language of post
- translation of the post into English

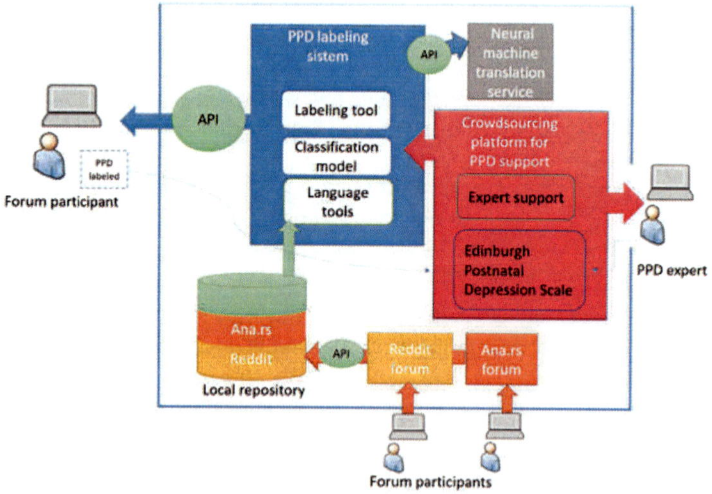

Fig. 1 System for tagging postpartum depression-related posts, PPD level detection, and expert support

- marking the post based on the classification model in English or corresponding language (for example Serbian)
- if the post is related to PPD, a link to the PPD support platforms is sent to the user in the appropriate language.

A neural machine translation service that is part of the Azure Cognitive Services family of REST APIs [14] was used for language detection and translation into English.

5 Results and Discussion

The goal of the system is to detect posts that may be associated with postpartum depression. That is why a classification model was made. Weka was used to build the model. Tokenization was performed on the textual data and n-grams of length 1 to 4 tokens were extracted. SMO, J48, and RandomForest machine learning methods were applied. The first model (M1) was made over a set of posts in English DS1 and the second (M2) on a set of posts in Serbian DS2. The Lovins stemmer was also used for the English language. The results are shown in Table 1.

To estimate the skill of a machine learning model on both data set 10 fold cross-validation is used. The best results were obtained by applying the SMO method on the set DS1 and by applying the Random Forest method on the set DS2. The obtained results for the English are comparable to similar research on Reddit data for depression where the accuracy ranges from 66 to 90% using different attributes and methods [15]. For the Serbian, even it was done on a smaller sample and without the application of normalization, which is necessary for highly inflected languages, satisfactory results were obtained.

The classification of the English set was done to show that in case of lack of resources in a certain language, marking can be done quite well by translating into English. A set of 52 posts from ana.rs from the sections "Postpartum depression", "Pregnancy and Weight" and "Pregnancy Chat Corner" were taken for verification. The same set was translated into English using the MS Azure translator so that it was tested on it. The obtained results are shown in Table 2.

Both models justified their accuracy at the test set. Again, better results are obtained for English, which can be explained by the lack of normalization of words

Table 1 Accuracy for different classification methods

Classification model	Data set	SMO	RandomForest	J48
M1	DS1	**89, 58**	85, 33	72, 97
M2	DS2	84, 46	**86, 53**	75, 65

Table 2 Accuracy for different classification methods on test data set

Classification model	Data set	SMO	RandomForest	J48
M1	DS1	90, 20	**98, 04**	86, 27
M2	DS2	86, 27	**88, 24**	74, 51

in Serbian. Such results provide a good basis for conducting more detailed research for multilingual systems in this field.

6 Conclusions and Future Work

This system has shown that even without the use of language-specific tools, the existence of postpartum depression can be detected or indicated. Adequate application of a language-specific tool can make this detection more accurate. The goal of further research is to create tools specific to Serbian.

Acknowledgements This paper is partially supported by the Ministry of Education, Science and Technological Development of the Republic of Serbia under project III44007.

References

1. Cox, J. L., Holden, J. M., Sagovsky, R.: Detection of postnatal depression. Development of the 10-item edinburgh postnatal depression scale. Br. J. Psychiatr. 150:782–786 (1987). https://doi.org/10.1192/bjp.150.6.782.PMID:3651732
2. Amazon Mechanical Turk. https://www.mturk.com/. Accessed 15 July 2021
3. Quora (2021). https://www.quora.com/. Accessed 15 July 2021
4. Topcoder (2021). http://topcoder.com/. Accessed 15 July 2021
5. Giuntini, F.T., Cazzolato, M.T., dos Reis, M.D.J.D., Campbell, A.T., Traina, A.J., Ueyama, J.: A review on recognizing depression in social networks: challenges and opportunities. J. Ambient Intell. Hum. Comput. **11**(11), 4713–4729 (2020)
6. Coppersmith, G., Dredze, M., Harman, C., Hollingshead, K. : From ADHD to SAD: Analyzing the language of mental health on Twitter through self-reported diagnoses. In: Proceedings of the 2nd Workshop on Computational Linguistics and Clinical Psychology: from Linguistic Signal to Clinical Reality, pp. 1–10 (2015)
7. Islam, M.R., Kabir, M.A., Ahmed, A., Kamal, A.R.M., Wang, H., Ulhaq, A.: Depression detection from social network data using machine learning techniques. Health Inf. Sci. Syst. **6**(1), 1–12 (2018)
8. De Choudhury, M., Counts, S., Horvitz, E.: Predicting postpartum changes in emotion and behavior via social media. In: Proceedings of the SIGCHI Conference on Human Factors in Computing Systems, pp. 3267–3276 (2013)
9. Shatte, A.B.R., Hutchinson, D.M., Fuller-Tyszkiewicz, M., Teague, S.J.: Social Media Markers to Identify Fathers at Risk of Postpartum Depression: A Machine Learning Approach,

Cyberpsychology, Behavior, and Social Networking (2020). https://doi.org/10.1089/cyber.2019.074
10. A., Semeraro, D., Drake, J., Bukowski, R., Oliveira, J.L.: Social media mining for postpartum depression prediction. In: Digital Personalized Health and Medicine, pp. 1391–1392. IOS Press (2020)
11. Marovac, U., Ljajić, A., Avdić, A., Fazlagić, A.: Automation of psychological testing of stressful situations in the Serbian, ICIST. In: 2019 Proceedings, pp. 102–106 (2019)
12. Reddit (2021). https://www.reddit.com/. Accessed 15 July 2021
13. Ana.rs (2021). https://www.ana.rs/forum/. Accessed 15 Jul 2021
14. Translator (2021). https://api.cognitive.microsofttranslator.com. Accessed 15 July 2021
15. Tadesse, M.M., Lin, H., Xu, B., Yang, L.: Detection of depression-related posts in reddit social media forum. IEEE Access **7**, 44883–44893 (2019)

SmartTeV—Leveraging eVoting with Smart Voting Ballots

Ioanna Dionysiou and Harald Gjermundrød

Abstract Research and commercial approaches on electronic voting (e-Voting) exist and the recently emerged blockchain technology has given this research area yet another boost. However, one may claim that the actual deployment of e-Voting is at its infancy, mainly due to election integrity issues such as vote non-tampering, voter anonymity, and vote verifiability. The integration of cryptography and blockchain technology in a single framework could provide a solid foundation for secure, anonymous and still verifiable electronic voting. The *TrustedEVoting* (`TeV`) framework is a blockchain-based eVoting approach, which wraps not only the required key elements for secure and verifiable e-Voting but also additional features such as support for re-voting and post-election vote check, among others. This paper presents a novel feature to be integrated in `TeV`, the smart voting ballot, that implements a ranked-choice voting by allowing the voter to cast two votes: a primary vote and an alternative vote. The latter one is utilized in the case the primary vote went to an election party that didn't reach the threshold. In this way, a more flexible voting scheme is available to a voter, mitigating the voter reluctancy to vote for smaller parties. In order to protect the voter's privacy by not revealing trends/preferences/statistics regarding the two votes, an intelligent mechanism is in place that provides a privacy shield for the alternative votes not being used in the voting tallying.

1 Introduction

The proliferation of technology has been undoubtedly the driving force behind the disruption of the public sector by digital transformation. There are numerous exemplary digital services supported by governmental virtual one-stop front desks, but

I. Dionysiou · H. Gjermundrød (✉)
Department of Computer Science, School of Sciences and Engineering, University of Nicosia, Nicosia, Cyprus
e-mail: gjermundrod.h@unic.ac.cy

I. Dionysiou
e-mail: dionysiou.i@unic.ac.cy

D. Camacho et al. (eds.), *Intelligent Distributed Computing XIV*, Studies in Computational Intelligence 1026, https://doi.org/10.1007/978-3-030-96627-0_35

at the same time there are services still at the beginning of their digitization journey. Electronic voting (e-Voting) is one of the digital services still at its infancy, despite the fact that research and commercial approaches on electronic voting do exist. Indeed, a study on the reform of the EU electoral law [7] finds that only three countries are currently deploying e-Voting nationwide or in selected parts of the country (Estonia, France, Belgium) and a mere few are running (or had run but discontinued) pilot programs (Germany, Ireland, UK, among others). On the bright side, two recent initiatives (both in March 2019) are promising indicators of the long overdue turn towards e-Voting: Switzerland conducting a public intrusion test (PIT) on its open source Internet-based voting solution [12] and US DARPA (Defense Advanced Research Projects Agency) announcing the development of a multi-million open source e-Voting system [10].

Any practical deployment and usage of an e-Voting solution must overcome one major challenge: persuading the political establishment as well as the voters to accept and support it. The reluctance of the political establishment to throw support to this voting reform is mainly due to the lack of confidence that the integrity of the election process and results can be guaranteed by a paperless system. Zero-day vulnerabilities due to design and implementation flaws on any layer of the network stack, including application flaws [3, 4, 11] and network/transport protocol flaws [8], are difficult to mitigate and any exploitation during an ongoing election could have a catastrophic impact on the democratic system itself. Voters, on the other hard, show little confidence in paperless voting. For example, even though e-Voting in Estonia has been operational since 2005, the number of voters that use the electronic voting alternative has been growing in a rather slow pace. Only 1 out of 3 voters favored the e-Voting option in the elections of 2014 and 2015 [9].

In order to alleviate the valid concerns that hinder the practical deployment of e-Voting, the digital version of voting must provide assurances that a transparent, secure and rigid process is followed, supporting end-to-end vote verifiability, while at the same time protecting the voter's identity. Verifiability and anonymity may seem conflicting at first, but these are key properties expected from an e-Voting system. *Trusted E-Voting* (`TeV`) [13], a novel theoretical framework that uses blockchain technology as its backbone building block does provide secure and anonymous e-Voting, compliant with the key requirements of e-Voting as stated in [2]. Additional features are incorporated in `TeV` that leverage the flexibility, transparency and accountability of the overall electoral process. For example, re-voting is supported, external auditing as well as post-election vote verifiability by the voter himself/herself.

The digitization of the voting process could now facilitate alternative voting systems, traditionally deemed impractical for nationwide elections deploying paper-based voting. The logistics of an election method (vote casting and vote tallying) could be encapsulated in a pluggable programmable module, allowing the integration of multiple election methods into a single voting framework. This paper investigates a customized version of the *ranked-choice* voting, applicable to multi-party parliament elections that require a minimum vote threshold to be met in order to compete for parliament seats (usually 3–5%). The voter casts two votes, a primary and an alternative, with the latter being considered in the case the primary vote went for a party that didn't meet the required threshold. Several policies could be configured on the alternative vote tallying. This smart ballot feature will advance the framework to `smartTeV`.

The rest of the paper is organized as follows: Sect. 2 gives a brief overview of the design principles and baseline modules of `TeV` within an electoral cycle. Section 3 discusses the concept of smart voting ballots in `TeV`. The intelligent vote tallying process is presented in Sect. 4. Section 5 concludes.

2 **`TeV`** Framework: Design Principles and Functionality Overview

There are several election-specific elements that differentiate national elections. `TeV` [6, 13] takes a holistic approach to e-Voting with configurable modules supporting multiple variations (e.,g support of multiple elections in an election cycle, configurable length of voting cycle, support of multiple ways to cast a vote). Even though there is no standardized election process as this is dictated by national legal regulations and policies, still a general election cycle could be divided into three main phases: the pre-electoral period, the electoral period and the post-electoral period. `TeV` spans the entire election cycle phases.

The backbone network platform of `TeV` is a consortium blockchain, a shared permission blockchain administered by a federation of organizations that constitute an e-Voting ecosystem. As an example, one could consider a consortium blockchain with eight nodes, six being controlled by government authorities and two managed by third parties, possibly election oversight authorities acting as independent observers that verify the election results (national election oversight committee, Office for Democratic Institutions and Human Rights (ODIHR), EU Election Observation Missions organization (EUEOM), to just name a few).

Two blockchain instances, *Voting Status Register* and *Cast Vote Register*, run on each consortium node, one per `TeV` record type (*Voting Status* type, *Vote* type). The added value in maintaining two different repositories is to allow post-election vote checking without disclosing the voter's identity. The correct operation of the two blockchain instances relies on their real-time synchronization, requiring that a *Voting Status* record and its corresponding *Vote* record to be stored simultaneously in their perspective repositories.

The *Voting Status Register* instance is essentially a repository of *Voting Status* records, where voter-specific metadata on the status of his/her voting activity is kept. The second blockchain instance is the *Cast Vote Register* that stores *Vote* records. Data related to a cast vote includes, among others, the actual vote that an eligible voter cast along with an anonymized identifier (*AV-ID*), both in an encrypted form. After the election, the two quantities are decrypted for vote tallying and cast vote checking. Due to the fact that re-voting is supported, multiple records for the same voter could exist but only the last vote counts. Figure 1 illustrates the structure of both record types (more details could be found in [6]).

Once the vote casting is completed, vote tallying commences. Only one blockchain instance is needed to determine the preliminary election results, which are stored

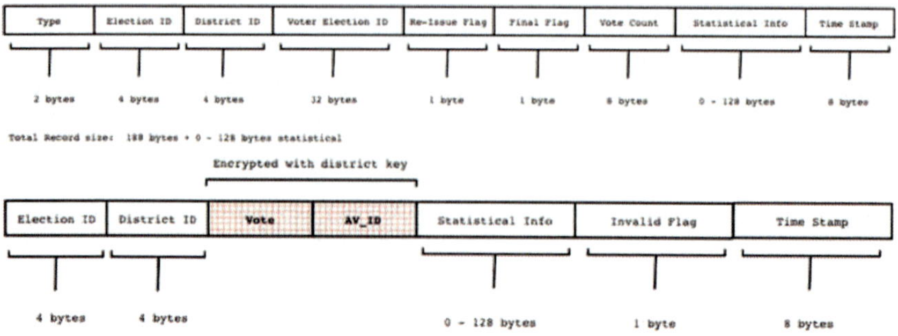

Fig. 1 TeV record types: voting status record and vote record

in the *Election Results Register*. Verification of the results could be performed by repeating the vote tallying process using a different blockchain node, with a minimum of 3 nodes to ensure the outcome correctness. The external nodes could verify the election results independently by following the same procedure. Any discrepancies would prompt for an investigation according to the national electoral legislation. An *Election Results Register* record has the same information as the *Vote* record, but in its decrypted format. In order to determine the final vote for a voter, the procedure shown in Algorithm 1 is followed.

Algorithm 1 Determine Final Vote

function DETERMINE FINAL VOTE(**Vector<AVID>** $AV-IDs$)
  for each $avID$ in $AV-IDs$ **do**
    Retrieve all vote records for $avID$ from *Cast Vote Registry*
    Check Retrieved Records Integrity
    Find latest added record based on the time stamp, and mark as the final vote
    Check if the invalid flag is set. If the flag is set dismiss the record
    Add 1 to that candidate/party (Invalid, Blank if thats the case)
  end for
end function

3 Motivation for Smart Ballot

It has been shown that the effect of Internet-based voting on voter turnout is minimal [1, 5], raising questions whether or not the use of electronic voting could be a countermeasure for the low voter turnout. One possible explanation for the low voter turnout, despite an electronic voting option being available, could be the perceived impact of one's vote on the overall election. Consider the *ranked-choice* voting, applicable to multi-party parliament elections that require a minimum vote threshold to be met in order to compete for parliament seats (usually 3–5%). A voter supporting a party that, according to the pre-election polls, is not expected to reach the threshold may not exercise his/her voting right or be reluctant to vote for a small party as the vote is going to be considered a *thrown-away* vote.

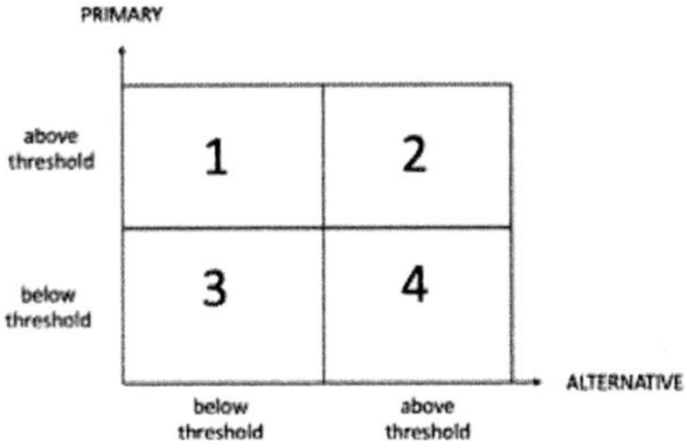

Fig. 2 Vote configurations after primary vote tallying

With the introduction of e-voting frameworks like TeV, features could be added to mitigate the hesitation to vote for smaller parties. Such a feature is the *smart ballot* that supplements the primary vote with an alternative/backup vote. The voter still casts his/her primary vote for a specific party. However, if that party does not meet the electoral threshold after the vote tallying, then the alternative vote will be considered in a second vote tallying. Figure 2 shows the four possible configurations for the primary and alternative votes after the first round of vote counting:

Configurations 1 and 2: The primary vote is cast for a party above the electoral threshold, thus the alternative vote is not considered in either case. This is how current election schemes work.

Configuration 3: Both primary and alternative votes are cast for parties not meeting the electoral threshold. One could suggest that the alternative vote should count even though the party is below the threshold. The implications of this decision though are not simple and indeed this configuration is rather complicated as demonstrated in the scenarios 1 and 2 below. It is possible that the alternative vote counting will give leverage to a small party and meet the electoral threshold but at the same time borderline parties may pushed down the threshold.

- Scenario 1-Voter V1 has the ballot (Party A, Party B) and voter V2 has the ballot (Party B, Party A). If both Party A and Party B are below the threshold then the alternative votes will cancel each other.
- Scenario 2-Voter V1 has the ballot (Party A, Party B), voter V2 has the ballot (Party A, Party B), voter V3 has the ballot (Party C, Party A). Both Party A and B are below the electoral threshold after the first round, but Party C is above. However, in the second round Party B will actually meet the threshold, but at the expense of Party C, which will then be pushed below the threshold.

Configuration 4: The primary vote is cast for a party not meeting the electoral threshold. In the traditional voting systems, this vote is a *thrown-away* vote. The alternative vote though is cast for a party that is already eligible to compete for seats. Counting the alternative vote could impact the number of seats that the party could eventually get allocated. Similarly, it is also possible that a party that will loose some seats if none of the alternative votes are given to it, hence its overall percentage (with relation to the other parties) will be decreased. Furthermore, there is a case that a party that indeed met the electoral threshold after the first round will find itself below the threshold after the second round vote counting. A possible solution to address this scenario is the following: A party that met the threshold after the first round, will at a minimum get at least one seat after the reshuffling of the second round. The specifics of how this *loss* will be distributed among all the other parties that met the threshold would need to be set as an election policy and/or law before the start of the election.

As explained earlier, one of `TeV`'s strong attributes is the transparency of its operation within the electoral lifecycle and as a result all its blockchain instances are available to third parties. For example, the *Election Results Register* hosts the unencrypted ballots that were counted during the vote tallying. If the existing ballot were to support a primary-alternative vote pair, under the current vote tallying procedure, both votes would have to be decrypted regardless of the alternative vote eligibility to be counted (e.g. Configurations 1–3). We claim that the alternative vote should only be decrypted when needed (e.g. Configuration 4) and this is in no violation of the transparency policy of `TeV`. On the contrary, it is aligned with its privacy policy that prevents a vote to be traced back to the actual voter via its anonymization mechanism. In a sense, the privacy policy is extended to avert attempts to correlate primary–alternative votes per voter that may hinder the democratic process of an election:

- A voter optionally casts an alternative vote, thus decrypting the entire vote pair violates the voter's privacy by revealing whether or not he/she exercised the option of alternative vote
- A voter casts an alternative vote under the assumption that it will be used only if needed, thus decrypting both votes when only the primary one is used violates the voter's rights
- Unless a voter explicitly waives his/her right to vote privacy, then no correlation between the primary–alternative votes should be allowed as it may be used by the political establishment to determine alliances and/or close competitors, etc.

The next section discusses the protocol that builds the *vote-to-be-disclosed* intelligence into the smart ballot and the overall vote tallying process.

4 Intelligent Vote Tallying

In order to incorporate *vote-to-be-disclosed* intelligence in a two-vote ballot, `TeV` needs to be augmented as follows:

- Support of a public-private pair per candidate party (PU_{Party}, PR_{Party})
- Replacement of the *Vote* field in the *Vote Record* by two fields, *Primary* P and *Alternative* A
- Use of two intermediary blockchain instances, *Vote Tallying R1* and *Vote Tallying R2* to temporarily store the results from the two rounds of vote counting
- Update of the vote tallying algorithm

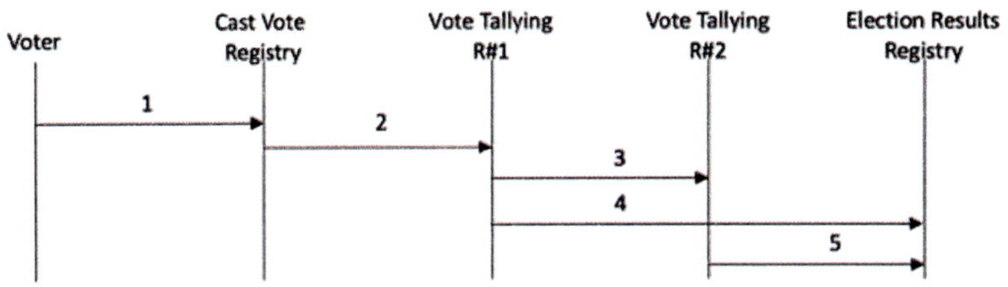

Fig. 3 Intelligent vote tallying

Figure 3 illustrates the communication protocol among the main `TeV` entities after the introduction of the two intermediary blockchain instances. The message exchanges[1] communicate the *Vote* record or parts of it as shown below:

`Exchange` 0—Voter: $V[AVID|P|E_{PU_{PartyP}}(A)]$
`Exchange` 1—Voter→ Cast Vote Registry: $E_{PU_{district}}(V[AVID|P|E_{PU_{PartyP}}(A)])$
`Exchange` 2—Cast Vote Registry → Vote Tallying R1: $AVID|P|E_{PU_{PartyP}}(A)$
`Exchange` 3—Vote Tallying R1 → Vote Tallying R2: $AVID|A$
`Exchange` 4—Vote Tallying R1 → Election Results Registry: $AVID|P$
`Exchange` 5—Vote Tallying R2 → Election Results Registry: $AVID|A$

The vote record gets constructed locally at the voter (*Exchange 0*) using the processes and algorithms in [6, 13]. A voter could utilize the option of casting an alternative vote, which it will be only considered if the requirements of Configuration 4 hold. In order to safeguard the privacy of the alternative vote, the following mechanisms are in place:

- Suppose that the Alternative vote field is X bits. The first bit of the Alternative vote indicates whether or not the voter exercised his/her right to cast an alternative vote. In case that the value is 0, then a nonce is used for the remaining X-1 bits. In the case that the value is 1, then R bits are reserved for the actual vote, followed by a X-1-R bit nonce. This technique is deployed to avert attempts to deduce whether an alternative vote is cast.
- The Alternative vote field is encrypted using the PU_{partyP} of the Party that the primary vote was cast. The decryption of this field will only take place during vote tallying round 2, if and only if, the primary vote party did not meet the threshold. In this way, possible correlation between primary-alternative votes gets limited.

Once the vote record is received by the entity that hosts the *Cast Vote Registry* (see *Exchange 1*), it gets added to the corresponding blockchain instance. After the voting process completes, the vote tallying process commences. Primary votes get decrypted and stored in the intermediary storage *Vote Tallying R1*, which could reside on the same entity or on a different one (*Exchange 2*). This constitutes the first round of vote counting, very similar to existing voting systems. For the parties that didn't meet the threshold, the alternative votes are decrypted and stored in *Vote Tallying*

[1] For brevity reasons, only the encrypted part (with the $PU_{district}$) of the Vote Record is shown.

R2 (*Exchange 3*). This constitutes the second round of vote counting. The last two exchanges (*Exchange 4*, *Exchange 5*) accommodate the merging of the primary and alternative results into a final blockchain instance (*Election Results Registry*). Note that for each AV-ID, at-most one vote is present in this final registry: either the primary or the alternative one, or none in case both the primary and alternative vote were casted for parties below the electoral threshold.

`smartTeV` provides the technical infrastructure to support voting based on two vote preferences, where the alternative vote is protected under the *vote-to-be-disclosed* policy. The way that these two votes are going to be utilized by the electoral system is beyond the scope of this paper and `smartTeV` itself. The Vote Tallying module is built as a configurable module, properly adjusted to meet a particular election's rules and policies.

Assume that the vote tallying procedure follows the specifications of Configuration 4, as explained earlier. To be more specific, the alternative vote replaces the primary vote if and only if:

1. The primary vote was cast for a party that didn't reach the electoral threshold
2. The alternative vote is cast for a party that met the electoral threshold after the primary vote counting

Algorithm 2 describes the mechanism that determines the final vote for a specific voter, by examining the *Cast Vote Registry* along with the two intermediary blockchain instances, and deciding which vote is going to count towards the final election result.

Algorithm 2 Determine Final Vote Updated

```
function TallyingRoundOne(Vector<AVID> AV − IDs)
    for each avID in AV − IDs do
        Retrieve all vote records for avID from Cast Vote Registry
        Check Retrieved Records Integrity
        Determine the latest added record based on the timestamp
        Add [avID|P|E_{PU_{PartyP}}(A)|] to Vote Tallying R1
        Update count for Party P
    end for
    Update threshhold for all Parties
end function
function TallyingRoundTwo(void)
    for each record in Vote Tallying R1 do
        if P in record was cast for a party that didn't reach the threshold then
            Retrieve PR_{PartyP}
            Decrypt D_{PR_{PartyP}}(A) to get A
            Check the alternative vote A
            if A in record was cast for a party that reached the threshold then
                Add avID|A| to Vote Tallying R2
                Add avID|A| to Election Results Registry
            end if
        else
            Add avID|P| to Election Results Registry
        end if
    end for
end function
function Determine Final Vote(Vector<AVID> AV − IDs)
    TallyingRoundOne(AV − IDs)
    TallyingRoundTwo()
end function
```

5 Concluding Remarks and Future Directions

All national elections are governed by the nation's electoral laws. The digital transformation of voting could be leveraged with alternative voting systems. The eVoting approach empowers the electoral law to implement voting policies not feasible in a paper-based voting system, such as a ballot with an alternative vote that will increase the voter turnout. This kind of feature would be very difficult to introduce in a traditional election system especially when it comes to the post-election validation.

This paper presented a novel feature, the smart voting ballot, that allows the voter to cast two votes, a primary vote and an alternative vote, where the alternative vote replaces the primary one under certain conditions. The capability is integrated into `TeV`, a blockchain-based eVoting approach that provides for secure, verifiable and transparent digital voting, leveraging it into `smartTeV`. An ongoing investigation is underway to explore the utlilization of smart contracts as the building block for the smart ballot, something that will allow its seamless integration in `TeV`.

References

1. Germann, M., Serdült, U.: Internet voting and turnout: evidence from switzerland. Elector. Stud. **47**, 1–12 (2017). https://doi.org/10.1016/j.electstud.2017.03.001 http://www.sciencedirect.com/science/article/pii/S026137941630453X
2. Gritzalis, D.A.: Principles and requirements for a secure e-voting system. Comput. Secur. **21**(6), 539–556 (2002). https://doi.org/10.1016/S0167-4048(02)01014-3. http://www.sciencedirect.com/science/article/pii/S0167404802010143
3. Kubjas, I., Pikma, T., Willemson, J.: Estonian voting verification mechanism revisited again. In: Krimmer, R., Volkamer, M., Braun Binder, N., Kersting, N., Pereira, O., Schürmann, C. (eds.), Electronic Voting, pp. 306–317. Springer International Publishing, Cham (2017)
4. Lewis, S.J., Pereira, O., Teague, V.: Ceci n'est pas une preuve, the use of trapdoor commitments in bayer-groth proofs and the implications for the verifiabilty of the scytl-swisspost internet voting system. (2019). https://people.eng.unimelb.edu.au/vjteague/UniversalVerifiabilitySwissPost.pdf. Accessed Apr. 2019
5. Lust, A.: I-vote, therefore i am? internet voting in Switzerland and Estonia. SAIS Rev. Int. Affairs **38**(1), 65 (2018). https://search.proquest.com/docview/2131929175?accountid=29916. Copyright Johns Hopkins University Press Winter/Spring 2018
6. Verwer, M.: Prev framework: a practical e-voting frameowrk,: Master Thesis. University of Nicosia, Department of Computer Science (2019)
7. Nogaj, M., Poptcheva, E.M.: The Reform of the Electoral Law of the European Union. The European Parliament (2015). https://doi.org/10.2861/26011.http://www.europarl.eu/thinktank
8. OpenSSL: The heartbleed bug (2014). https://www.heartbleed.com. Accessed July 2020
9. Solvak, M.V.K.: E-voting in Estonia: technological diffusion and other developments over ten years (2005–2015). http://skytte.ut.ee/. Accessed Apr. 2018
10. Naked Security by Sophos, D.B.: Darpa is working on an open source, secure e-voting system (2019). https://nakedsecurity.sophos.com/2019/03/18/us-government-works-to-secure-electronic-voting/. Accessed Mar. 2019
11. Springall, D., Finkenauer, T., Durumeric, Z., Kitcat, J., Hursti, H., MacAlpine, M., Halderman, J.A.: Security analysis of the estonian internet voting system. In: Proceedings of the 2014 ACM SIGSAC Conference on Computer and Communications Security, CCS '14, pp. 703–715. ACM, New York, NY, USA (2014). https://doi.org/10.1145/2660267.2660315. http://doi.acm.org/10.1145/2660267.2660315

12. Swisspost: E-voting website (2019). https://www.evoting.ch/en. Accessed Mar. 2019
13. Verwer, M.B., Dionysiou, I., Gjermundrød, H.: Trustedevoting (tev) a secure, anonymous and verifiable blockchain-based e-voting framework. In: Katsikas, S., Zorkadis, V. (eds.) E-Democracy—Safeguarding Democracy and Human Rights in the Digital Age, pp. 129–143. Springer International Publishing, Cham (2020)

IDCCPSS 2 (Special Session)—Chair: Vasily Desnitsky

A Technique for the Design of Abstract Models of Microcontroller-Based Physical Security Systems

Dmitry Levshun, Andrey Chechulin, and Igor Kotenko

Abstract The paper describes a new technique for the design of abstract models of microcontroller-based physical security systems. Under the abstract model of the system, this work considers an abstract representation of the system, the design of which is carried out based on abstract elements such as electronic components, microcontrollers and single-board computers. Physical security systems were chosen as an area of application due to the need to ensure not only the functionality of the designed systems, but also to ensure their security against cyber-physical attacks. The novelty of the technique is in taking into account complex dependencies between system elements, namely, their hierarchy, nesting, communications and requirements. Experimental evaluation of the technique was carried out by the design of a secure system of mobile robots for perimeter monitoring. During the experiment, the design time was measured depending on the parameters of the attacker against which system security must be ensured.

1 Introduction

Modern microcontroller-based systems are an integral part of any sphere of human activity. They collect data from multiple sensors, set complex mechanisms in motion, analyze and display data. And because a failure of such systems leads not only to financial and reputational losses but can also entail a threat to human life and health, the importance of ensuring their security is critical [8].

There are many approaches to ensure the information security of microcontroller-based systems. As a rule, these approaches are associated with individual stages

D. Levshun (✉) · A. Chechulin · I. Kotenko
SPC RAS, 39, 14th Liniya, St. Petersburg, Russia
e-mail: levshun@comsec.spb.ru

A. Chechulin
e-mail: chechulin@comsec.spb.ru

I. Kotenko
e-mail: ivkote@comsec.spb.ru

D. Camacho et al. (eds.), *Intelligent Distributed Computing XIV*, Studies in Computational Intelligence 1026, https://doi.org/10.1007/978-3-030-96627-0_36

of such systems life cycle: planning, analysis, design, development, integration and testing, implementation, maintenance [10]. The technique suggested in this paper technique is used at the system design stage.

The design of secure microcontroller-based systems is an important task, that is why various design techniques have been developed and embedded into practice. The disadvantages of existing solutions include the lack of consideration of the physical layer features, impossibility of working with an abstract representation of the system, as well as the absence of a compromise between security and expended resources. Most of the existing solutions are focused on specialized areas of the application and taking into account only certain aspects of security if any [9].

This work describes a new technique for the design of an abstract model of microcontroller-based physical security systems. Under the abstract model of the system, an abstract representation of the system is considered, which design is carried out based on abstract elements such as electronic components, microcontrollers and single-board computers without specifying their platforms and architectures. Physical security systems were chosen as an area of application due to the need to ensure not only the functionality of the designed systems, but also to ensure their security against cyber-physical attacks [2].

The novelty of the developed technique is in taking into account complex dependencies between system elements, namely, their hierarchy, nesting, communications, conflicts and requirements. The contribution to the scientific field lies in the possibility to design secure microcontroller-based systems without limitations to specific platforms and architectures. Such an approach reduces the number of parameters to be searched, thereby significantly increasing the work speed.

The paper is organized as follows. Section 2 considers techniques for secure microcontroller-based systems design. Section 3 presents the new technique for the design of abstract models of microcontroller-based physical security systems. Section 4 contains an example of the technique application, where it is used to design an abstract model of a secure system of mobile robots for perimeter monitoring. Section 5 contains a discussion of the work done and a comparison of the developed technique with existing solutions. Section 6 contains the main conclusions and the future work directions.

2 Related Work

Recent work in this area can be divided into manual system design approaches that are based on the knowledge of experts in the field of building microcontroller-based systems and automated approaches that take on routine tasks while helping an expert. Let us consider them in more detail.

Authors design and implement different microcontroller-based systems manually, for example, plant irrigation system [7], smart home system [4], temperature monitoring system [11], automatic power factor correction system [12], wireless humidity monitor [6] and computer numerical control plotter [5]. As a rule, the approach to design such a system contains (1) design of the block diagram, which divided the system into the transmission and receiving sections; (2) design of flow charts for each section of the system; (3) development of the software for each part of the system; (4) implementation of the system. Authors built such a system manually, using their own knowledge.

In [3] a technique is proposed for identifying the list of possible harmful impacts, to which the microcontroller-based device may be subject in accordance with the selected model of the intruder, and also by used hardware and software components. In this approach, the means of protection are an internal part of the device, ensuring its security. The main steps of the suggested design approach are (1) definition of functional and non-functional requirements; (2) identification of the set of alternatives of component composition of the device in accordance with the functional requirements; (3) choice of the optimal component composition of the device from the point of view of non-functional requirements; (4) identification of the list of possible harmful impacts on the device based on the static testing. The disadvantage of the proposed approach is the need for a complete enumeration of possible device components to form system alternatives that meet the requirements.

In [1] an approach for the design of secure communication between devices based on microcontrollers was presented. The key idea of the solution is to use a topological approach to build secure communication channels between devices. The task is solved by calculating the security index for the path between two points of the network graph on the basis of the numerical security values assigned to the nodes. This index serves as a basis for changing the requirements for devices. One of the disadvantages of the proposed approach is the fact that this approach does not take into account the interaction of devices with external systems. This limitation restricts the scope of application of this approach when designing secure and reliable data transfer environments.

Thus, modern approaches to the design of secure systems are either manual or initially operate with specific devices and perform a full enumeration of possible alternatives, which limits their scope of application. Alternatively, in this paper, a new technique is proposed that works with an abstract representation of the system, which will be discussed in more detail in the next section.

3 Technique Description

A technique for the design of abstract models of microcontroller-based physical security systems is used to construct an abstract representation of a secure system based on its devices list, their abilities, elements and sub-elements as well as security recommendations. This technique represents the system as an abstract hierarchical model that takes into account connections between system devices, their elemental composition, dependencies between device elements and requirements to them. The work process of the technique is automatic, the operator is not required. Its overview is presented in Fig. 1.

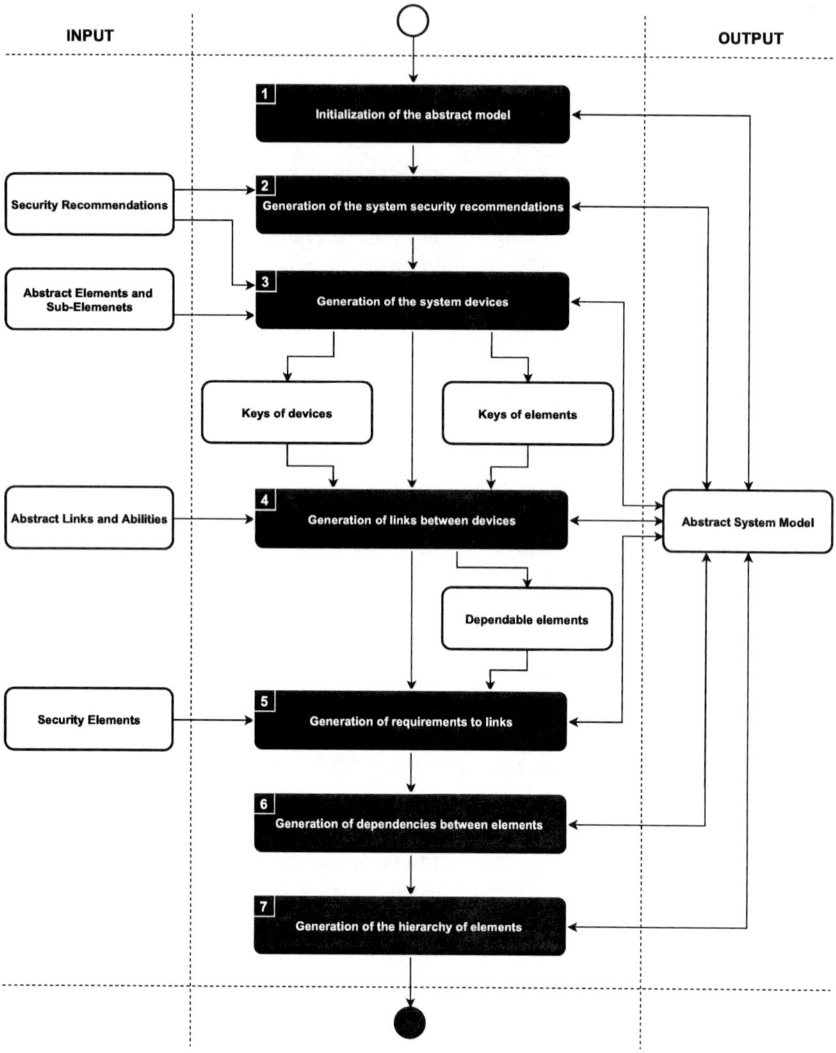

Fig. 1 Overview of the technique for the design of abstract models

As **input data**, the technique takes:

- *security recommendations:* abstract security recommendations to the system implementation as a whole as well as for each of its devices separately that are extracted based on security elements and can't be interpreted as abstract elements or sub-elements;
- *abstract elements and sub-elements:* abstract component compositions of the system devices, where abstract elements are extracted based on requirements to the

device and possible attack actions and represent controllers, electronic components, software and firmware, while abstract sub-elements are extracted based on elements and represent techniques, settings and requirements;
- *abstract links and abilities:* abstract types of communications that are possible between devices of the system with corresponding devices abilities that are related to their interaction;
- *security elements:* abstract methods and means of protection that are required to make the designed system secure against attackers with certain parameters, interpretable as security recommendations, abstract elements and sub-elements.

As **output data**, the technique provides the abstract system model that contains abstract system representation. The structure of the abstract model of the system is JSON-based and contains the next fields:

- *devices:* data about each device of the system, including its unique key, id, name, components and recommendations;
- *recommendations:* data about recommendations to the implementation of the system to ensure its security against attackers with certain parameters, including unique key, id and name (description);
- *links:* data about links between devices of the system, including its unique key, id, type, parties, dependencies and requirements.

Each element of the device from the "components" field has its unique key and id as well as data about its own components (sub-elements), links, requirements and dependencies. The work of the technique contains **7** main stages. Let us consider them in more detail.

Stage 1: Initialization of the abstract model. This stage defines the data structure for storing the abstract model of the system. At the end of the stage, the abstract model consists of fields for data about devices, links between them and recommendations for the implementation of the system related to ensuring its security.

Stage 2: Generation of the system security recommendations. At this stage, the abstract model of the system is filled with data on the recommendations for the implementation of the system related to ensuring its security. Each of the recommendations has a unique key by which its id and text description are available.

Stage 3: Generation of the system devices. This stage is aimed at filling the abstract model of the system with data about its devices. For each device, data is generated about its unique identifier, name and components composition. Data on recommendations related to ensuring the security of devices after their implementation is also generated.

The main part of this stage is the generation of the device components composition. This part contains the initialization of abstract electronic components of each device as well as the generation of their requirements based on each component sub-elements (including security ones). For example, depending on the component of the device, the technique assumes how much flash memory of the firmware it needs to work correctly.

After this stage is done, each device of the abstract system model is filled with a number of elements in their "components" field. Each element represents an abstract electronic component of the microcontroller-based system—operating system, firmware, sensor, receiver, transmitter, single-board computer, database, microcontroller, etc. Each element in the abstract model has its own key that is unique only inside each device. By using this key data about its unique identifier, name, components, links and requirements can be extracted. It is important to note that data about each element's components and links during this stage is empty and would be filled only during stage 7.

Stage 4: Generation of links between devices. At this stage, the abstract model is filled with data on links between devices of the system. Firstly, the technique detects all links that are possible between each pair of devices according to their abilities. And if the link is detected, its generation starts. In the abstract model, each link has its own unique key, by which data about its unique identifier, type, parties, dependencies and requirements can be extracted. For example, the "dependencies" field is filled with data about abstract elements, the selection of a specific implementation of which directly depends on the selection of the interface and protocol of this link between devices. As an output of this stage, unique keys of links with unique identifiers of elements the selection of which depends on the selection of a specific interface and protocol of the link.

Stage 5: Generation of requirements to links. This stage is aimed at filling the abstract model with data about requirements to links between devices of the system. This field was empty after stage 4 and now is filled with data generated based on security elements that are required to design a secure system. Generated during this stage requirement defines if a link is wired or wireless, transfers data, signal or charge, requires encryption, authentication or access point and so on.

Stage 6: Generation of dependencies between elements. At this stage, the abstract model is filled with data on requirements to elements of devices as well as with data about dependencies between them. For example, for each microcontroller data about dependencies between their selection and the subsequent selection of sensors that will be connected to them would be generated. It is done to ensure the compatibility of the elements of the device after the transmission from the abstract model to the implementation of the system. Also, for each microcontroller the number of required digital and analogue pins is calculated.

Stage 7: Generation of the hierarchy of elements. This stage is aimed at the reconstruction of the "components" field of each device of the system. The technique generates hierarchical elements composition instead of their enumeration. The transmission to the hierarchical structure is based on a graph representation of the components of each device and recursive conversions. Firstly, graph nodes are generated based on unique identifiers and keys of elements. After that elements of each device are checked pair by pair in terms of the possibility to connect one element to another. For example, a sensor can be connected to a microcontroller if they are compatible, while compatibility can be checked according to their parameters. And if two elements can be connected to each other then the edge between nodes that are representing them is generated.

After the graph structure for each device is generated, the process of hierarchy building starts. Firstly, the root node of the graph is obtained based on topological sorting. After that, the child node of the lowest level is obtained together with its parent node. It is required for the technique to encapsulate data about the obtained child into the "components" field of its parent as well as for the generation of a link between them. After it is done, the data about the encapsulated child is deleted from the abstract model (this data is in the "components" field of its parent now) as well as the corresponding to this child node is deleted from the graph representation of the device. This process continues until all graph nodes are not deleted.

It is important to note that during stages 3, 4, 6 and 7 the technique relies on the database for making decisions:

- *stage 3:* to decide if the abstract element of the device is an electronic component;
- *stage 4:* to decide if devices are linked based on their abilities; to decide what abstract elements selection will depend on the selection of interface and protocol of the abstract link between devices;
- *stage 6:* to decide if one abstract element selection will depend on the selection of another abstract element; to decide how many digital or analogue pins are required to connect the abstract element to the microcontroller;
- *stage 7:* to decide if one abstract element is compatible with another one; to decide what link will be between two abstract elements when they are connected.

This technique can be used for the intelligent design of abstract system representation based on information about its devices, their elements and sub-elements. The output data is well-structured, while the technique takes into account the hierarchy of elements and dependencies between them as well as generates requirements for them. The experimental evaluation of the technique is presented in the next section.

4 Experimental Evaluation

For the experiment, it was decided to design a microcontroller-based physical security system that provides perimeter monitoring based on mobile robots. This system contains a server as well as multiple mobile robots and charging stations with different electronic components. Robots (mobile objects) and stations (static objects) are monitoring the perimeter via different sensors based on server instructions. And if the battery of one of the robots is low, it moves to the nearest free charging station. The information about the perimeter map, locations of robots and stations as well as the charge state of robots and occupancy of stations is stored on the server.

Such a system was chosen due to the presence of several types of devices, multiple communications between them, as well as the need to use many different elements for each device in the system. Also, during the design of such a system it is necessary to ensure not only its functionality (perimeter monitoring), but also to ensure that the system is secure against attacks on it.

Software implementation of the technique is an application that consists of Python script and PostgreSQL database. The database was filled with data manually, while

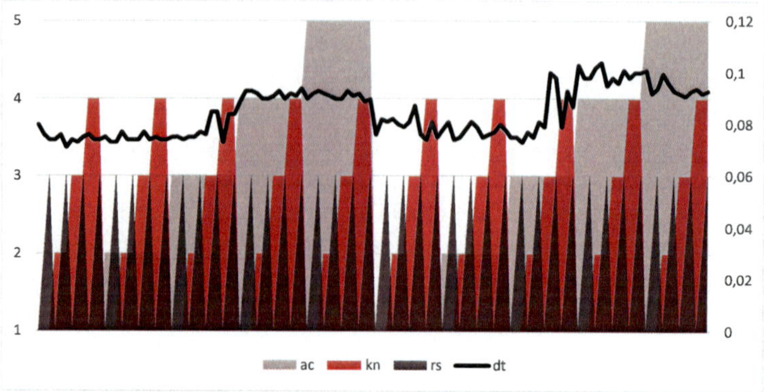

Fig. 2 Relationship between the attacker's parameters and the system design time

input data for the technique was generated based on the database content. The technique was executed 1000 times on a computer with Windows 10 × 64 operating system, Intel Core i7-4790 CPU 3.60 GHz processor and 32 GB of RAM.

The average time of design of the system was **0.1109** seconds. In the designed system, the server consists of 8 elements with sub-elements, each station consists of 12 elements with sub-elements and each robot consists of 17 elements with sub-elements. Moreover, there are links between devices of the system and elements of devices, security recommendations to the system and devices implementation, requirements to links and elements as well as dependencies between them.

Note that during the system design process, the required level of security is ensured in accordance with the parameters of the attacker model. These parameters, together with the system element composition, determine possible attack actions, which in turn determine the methods and means of protection required for use.

The attacker's model characterizes his capabilities in accordance with three parameters: *ac*—type of access (from 1 to 5), *kn*—type of knowledge (from 1 to 4) and *rs*—type of resources (from 1 to 3). The relationship between the attacker's parameters and *dt*—system design time is shown in Fig. 2. The scale on the left from 0 to 5 reflects changes in the attacker's parameters (*ac*, *kn* and *rs* as area charts), while the scale on the right from 0 to 0.12 reflects the design time (*dt* as black line).

The minimum design time was **0.07098** s, while maximum was **0.10397**. The input and output data of the technique was saved as files and can be downloaded: https://github.com/levshun/IDC-abstract_model.

Table 1 Comparison of design techniques

		[7]	[4]	[11]	[12]	[6]	[5]	[3]	[1]	Developed
Parameters	Abstract	−	−	−	−	−	−	±	±	+
	Hierarchy	−	−	−	−	−	−	−	±	+
	Nesting	−	−	−	−	−	−	±	±	+
	Links	+	+	+	+	+	+	−	+	+
	Conflicts	−	−	−	−	−	−	+	−	+
	Requirements	+	+	+	+	+	+	+	+	+
	Recommendations	−	−	−	−	−	−	+	+	+
	Security	−	−	−	−	−	−	+	+	+

5 Discussion

To the best of our knowledge, there is no data available about the average time of design of the microcontroller-based system by techniques from the related work analysis. Moreover, even if this parameter would be available, it is difficult to compare design techniques when different systems with different amounts of devices that contain different amounts of elements are designed. That is why it was decided to compare the techniques with respect to the parameters of the microcontroller-based systems taken into account during their design, see Table 1.

It is also important to note that although this technique is useful to an expert in the design of secure systems, its full potential is revealed when interacting with other techniques within the framework of the design methodology. In such a methodology other techniques will provide input data, while the abstract model—output of this technique—would be detailed by replacing abstract elements with their concrete implementations (taking into account indicated requirements and dependencies). Such a methodology is currently under development and was verified on physical objects—LEGO 9797 Mindstorms NXT mobile robot [8].

6 Conclusion

The new technique for the design of microcontroller-based physical security systems was presented in this paper. The key difference from analogs is the operation of the technique with an abstract representation of the system, which is based on abstract elements such as electronic components, microcontrollers and single-board computers as well as their firmware and software. The novelty of the technique is in taking into account complex dependencies between system elements, namely, their hierarchy, nesting, communications and requirements. The output data is well-structured and presented in JSON format. Experimental evaluation of the technique was carried out by the design of a secure system of mobile robots for perimeter monitoring. The technique was executed 1000 times on a computer with Windows 10 × 64 operating system, Intel Core i7-4790 CPU 3.60 GHz processor and 32 GB of RAM. The average system design time was 0.1109 s. Because of lack of data about the average design time of other techniques, the developed one was compared with them based on parameters that are taken into account during the design process.

Acknowledgements The reported study was funded by RFBR, project number 19-37-90082.

References

1. Chechulin, A., Kotenko, I., Desnitsky, V.: An approach for network information flow analysis for systems of embedded components. In: International Conference on Mathematical Methods, Models, and Architectures for Computer Network Security, pp. 146–155. Springer (2012)
2. Costin, A.: Security of cctv and video surveillance systems: threats, vulnerabilities, attacks, and mitigations. In: Proceedings of the 6th International Workshop on Trustworthy Embedded Devices, pp. 45–54 (2016)
3. Desnitsky, V., Kotenko, I., Chechulin, A.: Configuration-based approach to embedded device security. In: International Conference on Mathematical Methods, Models, and Architectures for Computer Network Security, pp. 270–285. Springer (2012)
4. Hasan, M., Anik, M.H., Islam, S.: Microcontroller based smart home system with enhanced appliance switching capacity. In: 2018 Fifth HCT Information Technology Trends (ITT), pp. 364–367. IEEE (2018)
5. Hasan, M.M., Khan, M.R., Noman, A.T., Rashid, H., Ahmed, N., Reza, S.T.: Design and implementation of a microcontroller based low cost computer numerical control (cnc) plotter using motor driver controller. In: 2019 International Conference on Electrical, Computer and Communication Engineering (ECCE), pp. 1–5. IEEE (2019)
6. Islam, J., Habiba, U., Kabir, H., Martuza, K.G., Akter, F., Hafiz, F., Haque, M.A.S., Hoq, M., Mannan, M.A.: Design and development of microcontroller based wireless humidity monitor. IOSR J. Electr. Electron. Eng. **13**(2), 41–46 (2018)
7. Kumar, B.D., Srivastava, P., Agrawal, R., Tiwari, V.: Microcontroller based automatic plant irrigation system. Int. Res. J. Eng. Technol. **4**(5), 1436–1439 (2017)
8. Levshun, D., Chevalier, Y., Kotenko, I., Chechulin, A.: Design and verification of a mobile robot based on the integrated model of cyber-physical systems. Simul. Model. Pract. Theory **105**, 102151 (2020)
9. Levshun, D., Kotenko, I., Chechulin, A.: The application of the methodology for secure cyber–physical systems design to improve the semi-natural model of the railway infrastructure. In: Microprocessors and Microsystems, pp. 103482 (2020)
10. Levshun, D.S., Gaifulina, D.A., Chechulin, A.A., Kotenko, I.V.: Problematic issues of information security of cyber-physical systems. Inf. Autom. **19**(5), 1050–1088 (2020)
11. Tayab, U.B., Yuen, C.W.: Design and implementation of microcontroller based temperature monitoring system. Adv. Res. Des. **30**, 1–11 (2017)
12. Uddin, M.M., Al Mahmud, A., Islam, N.: Design & implementation of a microcontroller based automatic power factor rectification system for different loads. In: 2019 1st International Conference on Advances in Science, Engineering and Robotics Technology (ICASERT), pp. 1–6. IEEE (2019)

An Approach to the Synthesis of a Neural Network System for Diagnosing Computer Incidents

Igor Kotenko, Vladimir Avramenko, Albert Malikov, and Igor Saenko

Abstract To diagnose computer incidents based on a neural network model, it is necessary to determine the optimal number of neurons of both the hidden and the input layer. An increase in the frequency of feature collection entails an increase in the dimension of the input layer, which, with a limited set of training examples, leads to a deterioration in the quality indicators of the artificial neural network. In addition, the training time of the neural network increases. The paper discusses a combined artificial neural network as the basis of the neural network system for diagnosing computer incidents. Through the use of deep learning, the disadvantage of the multilayer perceptron, associated with the need to form a massive base of training examples for a specific structure of the neural network, is eliminated. In the experiments, the dependences of the dimension of the neural network input layer on the diagnosis result was revealed. This made it possible to minimize the time for collecting features while maintaining the quality of the neural network at the required level.

I. Kotenko (✉) · V. Avramenko · A. Malikov · I. Saenko
St. Petersburg Federal Research Center of the Russian Academy of Sciences (SPC RAS), 39, 14th Liniya, St. Petersburg, Russia
e-mail: ivkote@comsec.spb.ru

V. Avramenko
e-mail: vsavr@yandex.ru

A. Malikov
e-mail: mkv.vas@yandex.ru

I. Saenko
e-mail: ibsaen@comsec.spb.ru

D. Camacho et al. (eds.), *Intelligent Distributed Computing XIV*, Studies in Computational Intelligence 1026, https://doi.org/10.1007/978-3-030-96627-0_37

1 Introduction

To adapt modern diagnostics systems to the improving technologies for implementing security violations, it is advisable to use expert knowledge in addition to the long and widely used automation tools for collection and processing of information security events (which mainly use preset rules) [9–11].

The accumulated experience allows us to apply machine learning methods to diagnosing computer incidents. Based on the values of the characteristics of the security violations that have occurred for a certain set of events, a database of training examples is formed on the basis of the values of the characteristics of the occurred security violations. Further, on its basis, neural networks of direct propagation are trained, the use of which can significantly increase the efficiency and reliability of diagnosing computer incidents in comparison with existing approaches [14, 18]. The effectiveness of diagnosing computer incidents based on artificial neural networks depends on the structure of the created network. The structure of the input and output layers are established at the initial stage of solving the problem, and the structure of the hidden layer is determined empirically during the experiment.

The dimension of the input data set to cover events that affecting the outcome of diagnostics is large enough and the number of neurons in the input layer can be several tens of thousands. To train such a neural network, a many training examples are required. It is necessary to optimize the structure of the neural network to prevent the negative effect of retraining the network using a limited number of examples. For this purpose, it is proposed to build a combined neural network consisting of a coding part of an autoencoder and a multilayer perceptron [14].

This paper presents the results of a further study of the influence of the number of training examples on the structure of the combined artificial neural network previously proposed by the authors [2]. The study of the dependence of the precision and recall indicators on the number of diagnostic features, accumulated during feature collection, was carried out.

The main theoretical contribution of this work lies in the formulation and solution of a new problem of diagnosing computer incidents based on an optimized combined neural network that provides high-quality training of the artificial neural network on a limited set of training examples. The advantages and merits of this approach are the gain in training time for the neural network while maintaining the required values of the precision and recall indicators.

The paper is structured as follows. Section 2 provides an overview of the relevant work. Section 3 defines the formulation of the problem of synthesizing the structure of the neural network system for diagnosing computer incidents. Sections 4 and 5 consider the parameters and structure of the combined artificial neural network used to diagnose computer incidents. Section 6 discusses the experimental results. In Sect. 7, conclusions are given, and directions for further research are determined.

2 Related Work

The result of diagnosing computer incidents is largely determined by the preparatory operations associated with the processing of initial data from event logs and from other sources. The papers [4, 5, 12, 19] indicate the main approaches to the selection and processing of information security events. It is emphasized that traditional approach-es to manually examining event logs by a security administrator are currently extremely ineffective.

Examples of the successful use of artificial neural networks for diagnosing out-of-service equipment of information systems are given in [15, 20]. Event logs contain data about system failures, which are preprocessed and transmitted for troubleshooting to the input of artificial neural networks. In [1, 6, 16, 17], artificial neural networks are used to search for attacks and anomalous actions. The high accuracy of the obtained experimental results and the ability of search system to function in a time mode close to real are noted. In [13], effective optimization procedures using unsupervised learning are presented. They have demonstrated high performance for deep neural architectures. Despite the fact that artificial neural networks are a powerful tool for data processing, the issues of automatic diagnostics of computer incidents based on artificial neural networks have not been properly elaborated in the known literature. In [2], a combined artificial neural network is proposed, which allows determining the values of the characteristics of a security violation with high reliability and close to real time. Currently, the issue of choosing the structure and determining the values of the parameters of the formed combined neural network remains relevant. The structure is influenced by the number of training examples, as well as the size of the time interval for collecting security events. The parameters of the network are set during the training procedure in the form of the weight coefficients of neural connections.

The paper presents the results of studying the dependence of the values of quality indicators of the formed combined artificial neural network precision and recall on the number of diagnostic features (on the size of the time interval for collecting features).

3 Statement of the Problem of Synthesizing a Neural Network System for Diagnosing Computer Incidents

To detect a computer incident, the information protection system collects data from various sources, which include event logs of information protection tools, servers and workstations. In other words, the initial set of events X registered in the information system is converted into a set of informative diagnostic features X', $F : X \rightarrow X'$, where $X = \{x_{tsob}^{ist}\}$, $ist = \overline{1, N_s}$, $tsob = \overline{1, N_{ts}}$, N_s is the number of information sources, N_{ts} is the number of event types; $X' = \{x'_{tpr}\}$, $tpr = \overline{1, N_{tp}}$, N_{tp} is the number of types of diagnostic features.

Table 1 The investigated security violation characteristics

Description of characteristics	Characteristics values	Characteristics type
The nature of the violation	Intentional Unintentional	Secondary
Effects	Major damage Minor damage	Secondary

Next, it is necessary to determine the values of the characteristics of a security violation based on the available set of diagnostic features $hn_i{}^j \in HN$, $i = \overline{1, N_{ch}}$, $j = \overline{1, N_{zn}}$, where N_{ch} is the number of security violation characteristics, the values of which need to be determined in the course of diagnostics, N_{zn} is the number of possible values of the ith security violation characteristic. Thus, the solution to the problem of diagnostics is to find a mapping of the set of informative events $X^{'}$ to the set of values of the security violation characteristics HN, $F : X^{'} \to HN$. The set of security breach characteristics HN is actually a detailed description of a security breach. In this work, we investigate the dependences of the structure of the diagnostics system for two security violation characteristics given in Table 1.

According to the method of determining the values of security violation characteristics, we will divide the characteristics into primary and secondary.

Primary are those characteristics whose values are determined by direct measurement or calculation (for example, user IDs, time, and others). Secondary characteristics include those characteristics for determining the values of which it is required to construct functional dependencies on the values of diagnostic features.

To preserve the original sequence of the appearance of diagnostic features, the time interval Δt for collecting diagnostic features is divided into uniform sections $\Delta t^{'}$ with a certain frequency k, $\Delta t = k \cdot \Delta t^{'}$. Then the objective function for the developed diagnostic system can be represented as follows:

$$\Delta t^{'} \to max, Q_n(nn) \geq Q_f(nf) \tag{1}$$

where $Q_n(nn)$ is the value of the quality indicator of the artificial neural network containing nn neurons after retraining, $Q_f(nf)$ is the value of the quality indicator of the artificial neural network containing nf neurons obtained during the initial training.

4 Parameters of the Combined Artificial Neural Network

As applied to the problem of diagnosing computer incidents, the multilayer perceptron generalizes diagnostic features followed by finding the values of security violation characteristics. However, in order the neural network could generalize the available examples, and not just memorize them, it is necessary to maintain a balance

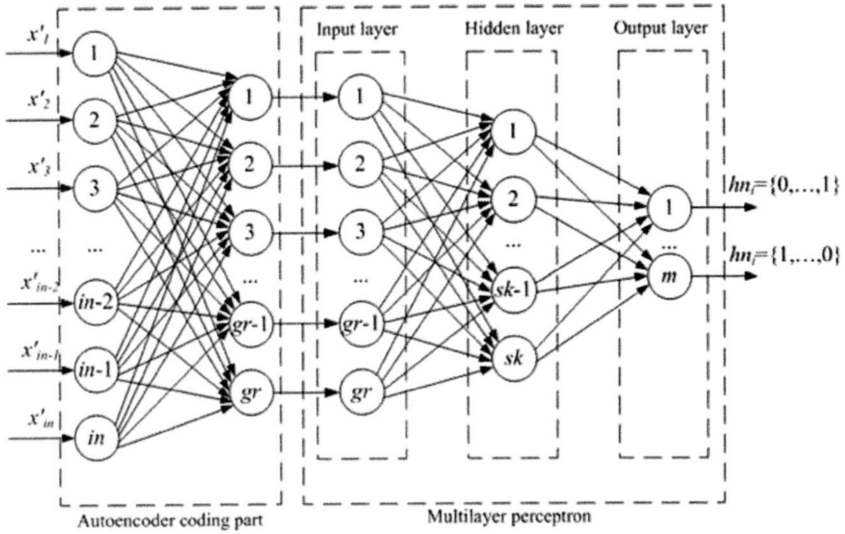

Fig. 1 The structure of a combined artificial neural network for diagnosing a computer incident

between the volume of the training sample and the number of neurons in the structure of the neural network. To ensure that the quality requirements of the created neural network are met, it is proposed to combine the multilayer perceptron with the coding part of the autoencoder, that is, to use a combined artificial neural network.

A model of a combined artificial neural network for diagnosing computer incidents is shown in Fig. 1, where in is the size of the input layer, gr is the size of the hidden layer of the autoencoder, sk is the size of the hidden layer of the perceptron, m is the size of the output layer of the combined neural network. The ratio between the number of neurons is as follows: $in \gg gr > sk > m$.

An autoencoder reduces the dimension of the vector of the initial data X, entering the input of the combined artificial neural network [3]. Thus, the first stage is carried out to generalize the input data (compression is performed), $F_a : X \to X'_g$. Group diagnostic features X'_g are formed in the hidden layer of the autoencoder. Then they enter the input of a multilayer perceptron, three-layer in this case, which carries out the procedure for processing diagnostic features and outputs a set of values of the m-ary characteristic of a security violation, i.e. $F_p : X'_g \to HN$.

Thus, the output of the considered model is an estimate of the degree of compliance with the values of the security violation characteristic for the set of diagnostic features presented at the input.

By training a combined artificial neural network, the weighting coefficients of neural connections are determined, $w_{ij} \in W, i = \overline{1, N_{prev}}, j = \overline{1, N_{cur}}$, where N_{prev} is the number of neurons in the previous layer, N_{cur} is the number of neurons in the layer under consideration. These coefficients make it possible to record in general form the dependence of the value of the i-th characteristic of a security violation on the input vector of diagnostic features X' : $hn_i = f_3(f_2(f_1(W_1 \cdot X') \cdot W_2) \cdot W_3)$, where $W_k \in W, k = \overline{1,3}$, is the vector of weight coefficients of the k-th layer; f_k is a function of activation of neurons of the k-th layer.

5 The Structure of a Neural Network System for Diagnosing Computer Incidents

The implementation of the procedure for diagnosing computer incidents using a combined artificial neural network involves the following sequence of actions:

- collection and processing of data from information sources;
- preparation of an input dataset for an artificial neural network;
- formation of group diagnostic features;
- classification of the obtained group diagnostic features.

The structure of a neural network system for diagnosing a computer incident is shown in Fig. 2.

The order of operation of the diagnostic system is as follows. Upon detecting a computer incident, information protection tools give out the values of the primary characteristics of a security breach. Along with this, the determination of the secondary characteristics of a security breach begins. For this, information sources transmit data to the feature preparation unit, where they are processed and presented to the required form before being fed to the input of combined artificial neural networks. It is necessary to select the time interval Δt within which the collection of diagnostic features in the event logs is carried out. In the course of experiments on a detailed study of the contents of a large number of event logs of a typical computer network, it was found that the number of diagnostic features is limited and amounts

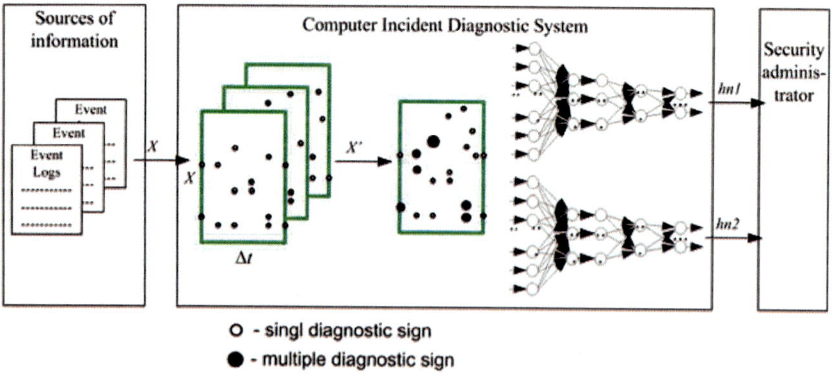

Fig. 2 The structure of a neural network system for diagnosing a computer incident

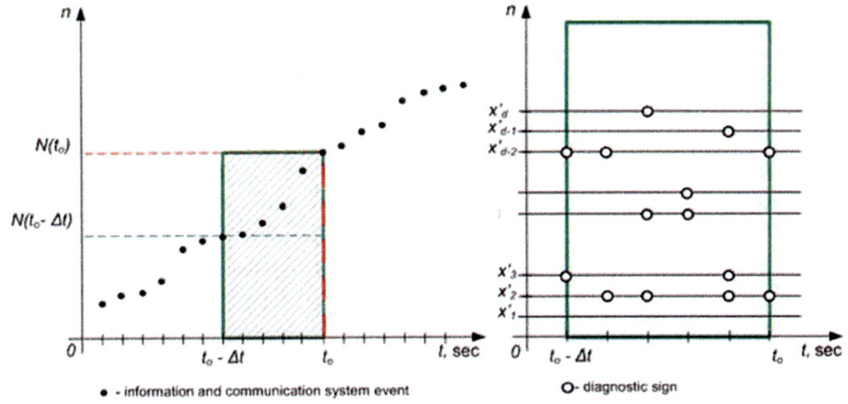

Fig. 3 The principle of forming a matrix of diagnostic features

to about 100, which coincides with research in the field of event analysis for technical diagnostics [20]. Diagnostic features are selected by experts.

For each event log, its own matrix of features is formed, the rows of which are diagnostic features, and the columns are temporary reports with a given frequency. The principle of forming a matrix of features is shown in Fig. 3.

Examples of diagnostic features are: (1) exceeding the number of connection attempts using single ICMP requests from one IP address (if more than 7 per hour, then in the matrix of features we indicate the value 1, if less, then 0); (2) the value of the last 4th byte for ICMP requests (if 0 or 255, which indicates broadcasting in the net-work, then in the attribute matrix we indicate the value 1, if another value, then 0); (3) starting a process as administrator or root; (4) increased processor load during detection of a security violation; (5) increasing the intensity of access to RAM; (6) opening ports and others. The generated matrix serves as the basis for the input set of values for the applied artificial neural networks.

Having entered the analysis unit, the normalized values of diagnostic features are processed in parallel by several neural networks at once. At the final stage of diagnostics, the final set of values of the security violation characteristics is formed, which includes the values of the primary and secondary characteristics.

The set of initial data and diagnostic results can, if necessary, be used as new training examples. The number of examples in the training set N_{ex} affects the number of neurons in the hidden layer N_{ck} of an artificial neural network, according to the relationship established by R. Hecht-Nielsen [7]:

$$N_{ck} = \frac{N_w}{N_x + N_y}, \tag{2}$$

$$\frac{N_y N_{ex}}{1 + \log_2(N_{ex})} \leq N_w \leq N_y\left(\frac{N_{ex}}{N_x} + 1\right)(N_x + N_y + 1) + N_y, \tag{3}$$

where N_y is the number of neurons in the output layer; N_{ex} is the number of training examples; N_w is the number of neural connections; N_x is the number of neurons in the input layer.

When replenishing the database of training examples, the structure of the neural network is adjusted. The choice of the number of neurons in the hidden layer of the autoencoder and perceptron affects the values of the quality indicators of the combined artificial neural network and the duration of training. Based on the experimental values in the structure optimization block of combined neural networks, the optimal number of neurons of the hidden layer is selected and the combined neural networks are adjusted.

6 Experiments

Experimental evaluation of the proposed neural network system for diagnosing computer incidents was carried out on the basis of the existing database of information security violations, compiled by experts. The database contains a description of two characteristics of a security breach, "the nature of the breach" and "the consequences". It consists of 276 records, which include sets of diagnostic indicators from event logs for 20 workstations. For a security breach "nature of the breach", 113 records correspond to an intentional violation and 163 correspond to an unintentional violation. For a security breach "consequences", 164 records correspond to major dam-age, and 112—to minor damage.

The experiments were carried out according to the following scheme. Initially, a database of training examples was formed, which includes many pairs of sets of diagnostic features and the corresponding values of the characteristics of a security violation. Then new training examples are saved to the same database for subsequent adjustment of neural networks. Learning is carried out by the method of back propagation of an error [8].

Further, the formation of the structure of combined neural networks was carried out. The size of the input layer for combined neural networks corresponds to the size of the vector of diagnostic features, which depends on the duration of the time interval for collecting features. In the course of the experiment, the size of the time interval Δt was chosen equal to the time interval between two consecutive violations of information security.

The time interval Δt was sequentially divided into uniform sections with different frequencies. In the course of the experiment, based on the previously obtained results in [2], the frequency k was selected in the range from 270 to 340. Accordingly, the dimension of the input layer of the autoencoder ranged from 29000 to 36000 neurons.

The output layer of each combined neural network is determined by the number of values of the secondary characteristic of the security violation.

To train and test the quality of the artificial neural network, the database was split into two parts. Initially 180 records were taken for training and 76 records for testing. The quality of the combined neural network was assessed according to the generally accepted F-measure, the value of which is calculated based on the accuracy Pr and completeness Rc:

$$F - measure = \frac{2 \cdot Pr \cdot Rc}{Pr + Rc}, \; Pr = \frac{TP}{TP + FP}, \; Rc = \frac{TP}{TP + FN}, \tag{4}$$

where TP is the number of records classified as a true characteristic value when it is true, FP is the number of records classified as a true characteristic value when it is actually false, FN is the number of records classified as false value while it is true.

With an increase in the frequency of collection of diagnostic features (a decrease in the interval Δt) and, as a consequence, an increase in the number of neurons in the input layer, the training time increases, and the values of the accuracy and completeness indicators decrease. This decline is due to a limited, relatively small set of training examples. With a decrease in the number of neurons in the input layer, along with a decrease in the training time, the values of the accuracy and completeness indicators decrease due to the impossibility of distinguishing the distribution of diagnostic features in time (since the features overlap).

7 Conclusion

The paper proposed a system for neural network diagnostics of computer incidents and solve the problem of optimizing the structure of combined neural networks in order to minimize the size of the input layer. The use of an autoencoder as part of a combined neural network made it possible to switch from an input set of high-dimensional diagnostic features to a compact representation due to compression by the method of principal components. The resulting set of group diagnostic features is classified into the set of values of the security violation characteristic. The classification problem is solved by a three-layer perceptron. Reducing the dimension of the set of diagnostic features allows training a combined neural network on a relatively small base of training examples.

Experimental evaluation of the proposed system on a test bench made it possible to identify the optimal number of neurons in the input and hidden layers of the combined neural network in order to minimize the training time. Given the simplicity of the proposed model, it seems possible to apply it in practice as part of the information security system of typical information and communication system.

Directions for further research are associated with the study of the application of combined neural networks to determine the values of secondary characteristics of information security violations that depend on other previously determined values of characteristics, including primary ones.

Acknowledgements This research is being supported by the grant of RSF No. 21-71-20078.

References

1. Alkasassbeh, M.: An empirical evaluation for the intrusion detection features based on machine learning and feature selection methods (2017). arXiv:1712.09623
2. Avramenko, V., Malikov, A., Kotenko, I., Saenko, I.: Combined neural network model for diagnosing computer incidents. In: 2020 CEUR Workshop Proceedings, pp. 280–294. CEUR (2020)
3. Baldi, P.: Autoencoders, unsupervised learning, and deep architectures. In: Proceedings of ICML Workshop on Unsupervised and Transfer Learning, pp. 37–49. JMLR Workshop and Conference Proceedings (2012)
4. Bose, R.J.C., Mans, R.S., van der Aalst, W.M.: Wanna improve process mining results? In: 2013 IEEE Symposium on Computational Intelligence and Data Mining (CIDM), pp. 127–134. IEEE (2013)
5. Cheng, H.-J., Kumar, A.: Process mining on noisy logs—can log sanitization help to improve performance? Decis. Support Syst. **79**, 138–149 (2015)
6. Fu, Q., Lou, J.-G., Wang, Y., Li, J.: Execution anomaly detection in distributed systems through unstructured log analysis. In: 2009 Ninth IEEE International Conference on Data Mining, pp. 149–158. IEEE (2009)
7. Hecht-Nielsen, R.: Kolmogorov's mapping neural network existence theorem. In: Proceedings of the International Conference on Neural Networks, vol. 3, pp. 11–14. IEEE Press, New York (1987)
8. Khaikin, S.: Neural networks: full course. M.: Williams, 1104 (2006)
9. Kotenko, I.: Multi-agent modelling and simulation of cyber-attacks and cyber-defense for homeland security. In: 2007 4th IEEE Workshop on Intelligent Data Acquisition and Advanced Computing Systems: Technology and Applications, pp. 614–619. IEEE (2007)
10. Kotenko, I., Polubelova, O., Saenko, I.: The ontological approach for siem data repository implementation. In: 2012 IEEE International Conference on Green Computing and Communications, pp. 761–766. IEEE (2012)
11. Kotenko, I., Saenko, I.: Creating new-generation cybersecurity monitoring and management systems. Herald Russ. Acad. Sci. **84**(6), 424–431 (2014)
12. Kurd, Z.: Artificial neural networks in safety-critical applications. PhD thesis, Citeseer (2005)
13. Larochelle, H., Erhan, D., Courville, A., Bergstra, J., Bengio, Y.: An empirical evaluation of deep architectures on problems with many factors of variation. In: Proceedings of the 24th International Conference on Machine Learning, pp. 473–480 (2007)
14. Lee, J., Kim, J., Kim, I., Han, K.: Cyber threat detection based on artificial neural networks using event profiles. IEEE Access **7**, 165607–165626 (2019)
15. Lv, F., Wen, C., Bao, Z., Liu, M.: Fault diagnosis based on deep learning. In: 2016 American Control Conference (ACC), pp. 6851–6856. IEEE (2016)
16. Nolle, T., Seeliger, A., Mühlhäuser, M.: Unsupervised anomaly detection in noisy business process event logs using denoising autoencoders. In: International Conference on Discovery Science, pp. 442–456. Springer (2016)
17. Sakurada, M., Yairi, T.: Anomaly detection using autoencoders with nonlinear dimensionality reduction. In: Proceedings of the MLSDA 2014 2nd Workshop on Machine Learning for Sensory Data Analysis, pp. 4–11 (2014)
18. Shone, N., Ngoc, T.N., Phai, V.D., Shi, Q.: A deep learning approach to network intrusion detection. IEEE Trans. Emerg. Top. Comput. Intell. **2**(1), 41–50 (2018)
19. Vaarandi, R.: A data clustering algorithm for mining patterns from event logs. In: Proceedings of the 3rd IEEE Workshop on IP Operations & *Management (IPOM 2003)(IEEE Cat. No. 03EX764)*, pp. 119–126. IEEE (2003)
20. Zou, D.-Q., Qin, H., Jin, H.: Uilog: Improving log-based fault diagnosis by log analysis. J. Comput. Sci. Technol. **31**(5), 1038–1052 (2016)

Approach to Anomaly Detection in Cyber-Physical Object Behavior

Anton Shulepov, Evgenia Novikova, and Ivan Murenin

Abstract Though the problem of anomaly detection in a cyber-physical object (CPO) behaviour is highly researched topic, there are still open challenges in this field that mainly relate to the lack of annotated real world data sets. Manual analysis of the anomalies and incidents in such data sets poses a certain difficulties as a typical cyber physical object is described by hundreds of heterogeneous parameters. This paper proposes an approach targeted to simplify the process of anomaly detection and investigation in CPO behaviour represented by multivariate time series. It is based on a novel metric that reflects a measure of the changes in CPO's behaviour. This metric incorporates data from a set of parameters defined by an analyst, and outputs a high level integral value that could be easily visualized using a standard timeline. The efficiency of the approach is demonstrated on real world data set describing the behaviour of aerospace object and consisting of almost 500 attributes.

1 Introduction

The cyber-physical systems (CPSs) are sensor-based autonomous or semi—autonomous systems with embedded communication and computing to monitor and control the physical environment, thus, the examples of cyber-physical objects (CPOs) are very diverse and could be found in almost all subject domains—aerospace, civil infrastructure, energy, healthcare, manufacturing, transportation etc. This fact makes a task of anomaly detection in such systems critical. The most widely used approaches to detect anomalies and possible failures in the CPS in practice are

A. Shulepov
Saint Petersburg Electrotechnical University "LETI", Prof. Popova str, 5, St. Petersburg 197376, Russia
e-mail: ao-shyleo@yandex.ru

E. Novikova (✉) · I. Murenin
St. Petersburg Federal Research Center of the Russian Academy of Sciences (SPC RAS), 14-th Liniya, 39, St. Petersburg 199178, Russia
e-mail: novikova@comsec.spb.ru

D. Camacho et al. (eds.), *Intelligent Distributed Computing XIV*, Studies in Computational Intelligence 1026, https://doi.org/10.1007/978-3-030-96627-0_38

rule-based and threshold-based ones due to the high level of determinism of CPO's behaviour and ability to construct its formal model of functioning. Application of supervised anomaly detection techniques are limited due to lack of the annotated real world data sets [1, 2] as such types of data sets represent sensitive and in many cases confidential data. Moreover, it is required to label data before training analysis models.

This paper focuses on the detection of anomalies in the industrial CPO's behavior based on the analysis of the historical data presented by multivariate time series, and the results of its application could be used for further data annotation. Currently, the analysis of the anomalies in CPO's functioning in many cases is implemented manually, the experts and technicians use simple visualization techniques such as line and point plots in combination with threshold based anomaly detection to investigate incidents. This task is quite resource consuming one, because the state of the CPO is represented by more then a hundred parameters with different update rate. The authors propose an approach targeted to simplify this process. It is based on an metric introduced by authors and named $Change\_Measure$ that reflects a measure of changes in CPO's behaviour. It is calculated on the basis of Delaunay triangulation of the data points' sequences describing state of the CPO during a given period of time. The $Change\_Measure$ metric incorporates data from a set of parameters defined by an analyst, and outputs an integral value that could be easily visualized using a standard timeline. The obtained plot reflects how the CPO's state changes over the time, and bursts and peaks in it could notify about possible deviations in the object state.

Thus, the authors' contribution is an approach that is used to detect anomalies in multivariate time series that describe the CPO's functioning, and is based on a novel $Change\_Measure$ metric that reflects how the object state changes over the time, incorporates data from a set of parameters, and could be visualized using traditional visualization techniques.

The novelty of the approach consists in the introduced novel metric $Change\_Measure$ and novel algorithms for its calculation and anomaly detection technique based on its analysis.

The paper is structure as follows. Section 2 studies the related works in anomaly detection in multivariate time series. Section 3 introduces the metric reflecting a measure of change in object state and approach for anomaly detection on its basis. Section 4 presents an use case of anomaly detection in aerospace object behaviour using proposed technique, and discusses the results obtained. Conclusions sums up the results and defines the direction of future research work.

2 Related Work

The task of anomaly detection in an industrial CPO's behaviour is a widely researched problem. There are solutions proposed for secure water treatment facilities [1, 3, 4], smart grids [2], smart buildings [5, 6]. For example, in [7] authors proposed a modular system that is based on Kohonen self-organizing map for the analysis of small spacecraft on-board equipment state.

In general case, it is possible to outline two major classes of approaches used to detect anomalies in CPO's functioning—classical and machine learning based ones [2]. Classical approaches include usage of statistical tests such as Grubbs's test, rule-based and threshold based techniques [8, 9]. For example, in [9] authors proposed an interactive tool MTSAD for study outliers in multivariate time series. It uses Tukey's method based on interquartile range with threshold anomaly filtering. The process of outlier detection is supported by visual interfaces that enable exploration of complex and large multidimensional data via filtering, ranking, and lensing.

Techniques for anomaly detection in CPO's behaviour based on machine learning (ML) techniques could be also divided into two groups: supervised and unsupervised techniques. For example, in [1] authors tested different classical "shallow" classifiers such as k-nearest neighbors and naive Bayes classifiers to detect anomalous deviations in secure water treatment facilities. However, application of supervised machine learning techniques requires annotated data sets, and the majority of ML-based approaches proposed recently are represented by a combination of unsupervised techniques, model specification and probabilistic filters such as Bayes filter [10]. As the functioning of CPOs could be represented as multivariate time series, many researchers investigate capabilities of recurrent neural networks and autoencoders to detect anomalies in such type of data. For example, Su et al. [11] adopted stochastic recurrent neural network for multivariate time series anomaly detection. Audibert et al., [12] proposed an encoder-decoder architecture within an adversarial training framework to detect anomalies in an unsupervised way. It consists of three elements: an encoder network and two decoder networks. First, the two AEs are trained to learn to reconstruct the normal input. Secondly, the two AEs are trained in an adversarial way, where the first AE tries to "fool" the second AE, and the second AE tries to learn when the data is real (coming directly from data source) or reconstructed (coming from the first AE). Data points with a high reconstruction error are considered to be anomalies and are define by threshold filtering. It was shown that such approach outperforms anomaly detection methods based on Isolation Forest, recurrent neural networks.

However, application of deep ML-based techniques for anomaly detection in CPO's behaviour requires large data sets to perform training step, moreover, some approaches have special requirements to perform model training. For example, Audibert et al. [12] needs to have training data set without anomalies. It is also should be noticed that such techniques lack transparency of analysis process that is necessary for analyst and experts when they analyze incidents. Thus, these techniques should be supplemented by additional explanatory techniques, for example, interactive visualizations.

The approach proposed by the authors is unsupervised threshold-based one. Inspired by visualization-driven approach presented in [5] it is more closer to [9], as it is targeted to provide means for exploration of multivariate time series. By introducing a novel metric *Change_Measure* we try to produce high-level representation of CPO's state in order to reduce the volume of input data and focus attention of an analyst on points of interest.

3 Approach for the Assessment of Changes in CPO's Behaviour

Formally, any complex technical object without loss of generality could be described by a set of parameters $P = \{p_i\}_{i=0}^{m}$ where each p_i could be either discrete (categorical) or analogue (continuous, numerical) parameter. In major cases, discrete parameters are represented by binary values, that take one or zero value. The analogue parameters have a range of possible values. At some moment of time t the state of the object is defined by the values of these parameters:

$$P^{(t)} = \{p_i^{(t)}\}_{i=0}^{m}.$$

The latter could be rewritten as $P(t) = \{p_0(t), p_1(t), ..., p_m(t)\}$. Thus, the system behaviour is represented by multivariate time series, that specifies the system state at some given moment of time as a point in m-dimensional space.

In [5] it was shown that application of data reduction techniques such as PCA or t-SNE for mapping multi-dimensional points to a plane allows producing graphical patterns different for normal and abnormal functioning of the cyber-physical system. Thus, the normal mode is usually characterized by a compact dense distribution of points with obviously seen transition periods, while the anomalous mode of functioning is characterized by a large scattering of the points. In [13] it was proposed to apply Delaunay triangulation to evaluate and compare the series of point distributions on the plane. Delaunay triangulation has a set of important properties, and one of which is that it is unique if no four points lie on the same circumference [14]. The conducted experiments showed that both parameters—mean square and total square of triangles—could be used to differentiate various patterns of points distribution. Moreover, the larger values of calculated squares notify about presence of the significant changes in the system behaviour. For example, Fig. 1 shows examples of Delaunay triangulation constructed for two time periods of aerospace object functioning, Fig. 1a corresponds to the period with anomalies in the system functioning, and Fig. 1b shows the normal mode of the object functioning. This led us to the idea to use the total square of triangles constructed for a sequence of n points as a measure of the changes in the system behavior during n timestamps. Let us denote it as $Change\_Measure(t, n)$, where t is a timestamp of the data point in two dimensional space, and n is a number of subsequent points used to construct Delaunay triangulation. Parameter n allows to consider the context of the data point and its impact on future state of the object. We propose to use metric $Change\_Measure(t, n)$ to analyze the changes in the system's behaviour and to detect suspicious deviations.

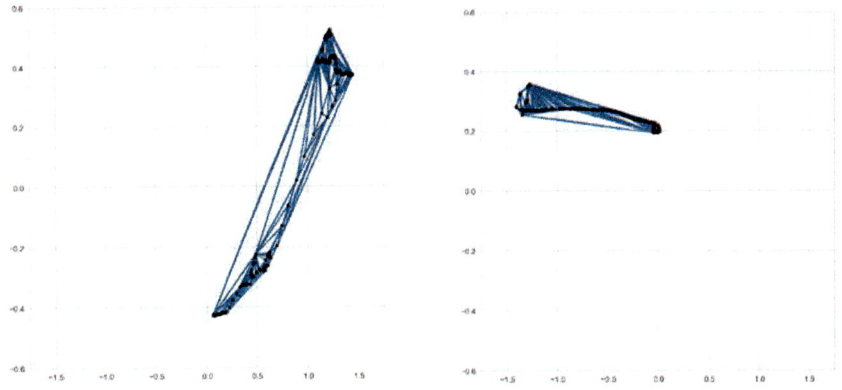

Fig. 1 Delaunay triangulation for two data sets that describe functioning of aerospace object during two different time periods generated by the PCA algorithm: **a** period with anomaly; **b** normal mode of object functioning

This metric could be easily visualized using a simple timelines that are the most frequently used visualization technique to analyze data.

Let S be a set of m-dimensional data points ordered in time or equipped with timestamps, $S(t, n)$ be a subset of S, consisting of tth data point and $n - 1$ data points subsequent to it. Parameter t denotes to a number of point in ordered set S or a timestamp. The algorithm for calculating $Change\_Measure(t, n)$ is as follows.

1. Apply data reduction technique, for example, PCA, to map original data set S to a two-dimensional space. Let us denote a data set with reduced dimensions as S'.
2. For each tth data point from the reduced data set S' select a subset $S'(t, n)$ and construct a Delaunay triangulation $DP(S'(t, n))$.
3. Calculate $Change\_Measure(t, n)$ as total area of obtained triangles of $DP(S'(t, n))$.

The complexity of proposed algorithm excluding data dimension technique is $O(N)$) where N is a number of data points is set S, the complexity Delaunay triangulation in the approach is constant and is defined by the parameter n. In the experiments we assessed the impact of the parameter n on the efficiency of the proposed approach. The complexity of data reduction step depends on the selected algorithm. Thus, the computational complexity of the algorithm depends linearly on the number of the input dataset, this means that for large volumes of data it is necessary to elaborate approaches that allow its distributed implementation. The authors believe that batching and parallel processing approach could be easily adopted to the offline processing of data and hybrid approaches for streaming data processing as the algorithm does not use aggregation procedures. Thorough analysis of this problem is included in the scope of future research tasks.

The anomaly detection based on $Change\_Measure(t, n)$ is done using threshold filtering. If $Change\_Measure(t, n)$ metric for some data point exceeds a given threshold the it considered as anomalous and labeled correspondingly. The threshold could be defined either based on statistical assessment or visually using the plot of the metric.

4 Experiments and Discussion

To evaluate the proposed approach, the authors used a real world data set that describes a behaviour of aerospace object. The state of the object during its functioning is monitored and controlled by almost 500 on-board measuring devices that generate and transmit data on regular basis. Thus, the state of the object at a given moment of time is presented by data point with almost 500 attributes. Based on consultations with experts from aerospace industry, we outlined following three sets of parameters:

- a set S_1 of the most significant parameters, it consists of 31 parameters mostly presented by continuous physical parameters;
- a set S_2 of significant parameters that consists of 89 parameters that contains both physical parameters and signs of failures;
- a set S_3 of insignificant parameters that consists of the rest 411 parameters, that were discarded from the analysis.

The data set contains information about three phases of object functioning—launch, flight and landing. The total functioning duration is 4848 s, during which 8606 data tuples were recorded. The duration of the launch phase is 1100 s (from 0 to 1100 s), the flight phase lasts for 3682 s (from 1100 to 4782 s), and the duration of landing phase is 66 s (from 4782 to 4848 s). The anomaly takes place on the 3438th second of the flight phase, and lasts for 4 s. During this period one of the most significant physical parameters exceeds the valid range of values.

The primary goal of the experiment was to understand if it is possible to identify the anomaly determined by experts, what the recommended value for the parameter n is. We also investigated how the visual patterns change when analyzing the most significant parameters only and all set of significant parameters.

Thus, in the experiments we used two data sets—S_1 and S_2. We selected PCA as multidimensional data projection technique, as it was shown that it highlights the significant deviations more clearly [13]. The application of PCA requires data prepossessing such as centering and normalization. We used mini-max normalization for each continuous parameter using known range of their values. Then we followed the steps of the algorithm presented in the Sect. 3. Figure 2 shows the projection of the data set S_1 that consists of 31 significant parameters and data set S_2 that consists of 89 parameters. It is clearly seen that two visualizations (projections) are generally alike, however, there are also observable distinctions. The projection of data points with the most significant attributes is more compact and occupies a smaller area, thus

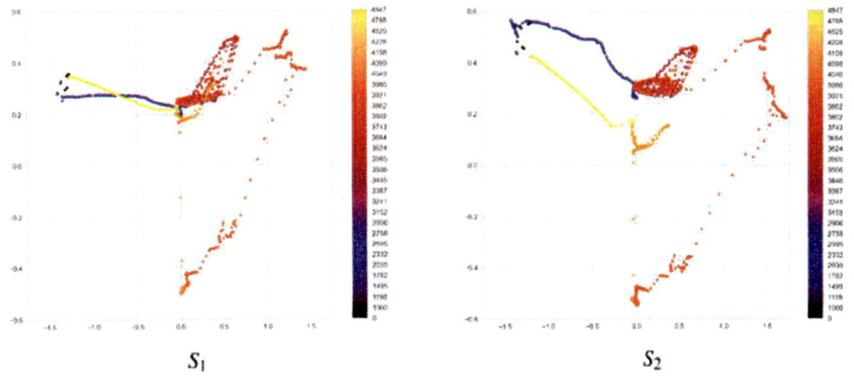

Fig. 2 PCA-based projection of two data sets S_1 and S_2 that describe functioning of aerospace object

it is possible to suggest that the values of $Change\_Measure(t, n)$ metric reflecting the changes in the state of the object in time will be less then the similar values of $Change\_Measure(t, n)$ metric calculated for a set of less significant but more numerous set of variables.

We evaluated different values of parameter n, and the experiments showed that the comparable results are obtained when parameter is set to 3, 4 and 5, while the plots of $Change\_Measure(t, n)$ with n greater then 5 do not show the outliers obviously. The increased value of n acts as smoothing filter, removing bursts in data and hiding thus the anomalies. Therefore, as the computation complexity of Delaunay triangulation depends on the number of points it is constructed for, the usage of value 3 for the parameter n is preferable.

The plot of $Change\_Measure(t, n)$ metric calculated for the set S_2 using sequential triangulation is shown in Fig. 3 by dotted timeline. The background color of the Fig. 3 shows the stages of object functioning. It is clearly seen at what time moments the behaviour of the analyzed object changes quite significantly. There are three significant peaks that takes place during the first and the second phases of object functioning. The experts recommended to ignore deviations registered during the first and third phase as these two periods are transition periods, and behaviour of the object is constantly changing. It is also possible to see small peaks when the object changes the phase of functioning. They could serve as markers of object's mode change. Thus, the most interesting peak is the one registered during flight phase. This peak is observed for the period 3400–3500 s, while the expect detected the anomaly at time 3438 s. Therefore, it is possible to conclude that this peak corresponds to the given anomaly. Having that there is the only one anomaly in the data set validated by the experts, we can conclude that the proposed method clearly highlights the indicated anomaly in the multivariate data.

We compared our approach with the widely used method for data annotating and anomaly detecting Isolation Forest [15]. This method assigns each data point from a

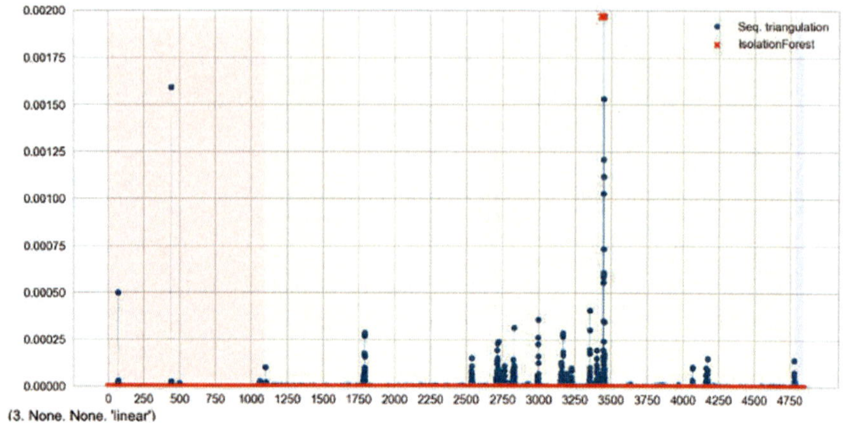

Fig. 3 Comparison of the results of sequential triangulation method and Isolation Forest with a contamination threshold of 0.01

given data set a value of 0 or 1, with 0 meaning no anomaly, 1—anomalous data point. Isolation Forest requires setting explicitly a ratio of points with abnormal activity to the total number of points. This parameter known as contamination is used to determine the threshold for assessing points' normality. In our experiments we set the contamination value to 0.01, meaning that no more than 1% of the points should be considered anomalous. We superimposed the obtained result on the plot obtained by the method of sequential triangulation for our data set S_2 marking anomalous data points revealed by Isolation Forest by red crosses (Fig. 3). The graph clearly shows that Isolation Forest method highlights the same time interval 3400–3500 s. We discovered that Isolation Forest marked exactly 87 points, which is 1% of all data. An increase in the contamination value also leads to an increase in the number of marked points. For example, setting the value of this threshold to 0.05 leads to marking 431 points as anomalous.

The certain benefit of the proposed approach is a possibility to establish a threshold for anomaly detection visually. Any point that falls above this threshold will be considered as a point at which the system was in an abnormal state. And setting a threshold value of 0.00098, we got only 6 points. We know that all these points, except for one located in the time interval 400–500 s, correspond to the known anomalous activity of the system, with a duration of 21 time counts. On the other hand, the Isolation Forest method mistakenly marked at least 66 points out of 87 due to the contamination threshold of 0.01.

The conducted experiments proved theoretical assessment of the algorithm complexity. The average time for calculating $Change\_Measure(t, n)$ metric including data reduction and plotting it is 200 ms for 8606 data points with 31 attributes. Isolation Forest has also linear complexity, however, its application to high dimensional data requires implementation of feature selection process that introduces additional computational expenses. In the conducted experiments on the similar data set its time performance was 750 ms that is 3,75 time higher then the running time of the proposed approach. This allows us to conclude that the proposed approach is comparable in accuracy with Isolation Forest and outperforms it in running time when detecting anomalies in multivariate time series.

5 Conclusion

Though the problem of anomaly detection in cyber-physical object (CPO) behaviour is highly researched topic, there are still open challenges in this field mainly relating to the lack of annotated real world data sets. This paper focuses on the detection of anomalies in the industrial CPO's behavior based on analysis of the historical data presented by multivariate time series, and the results of the proposed approach could be used for further data annotation. It is based on a novel metric $Change\_Measure$ that characterizes the measure of the changes in CPO's behaviour during a given period of time. This metric incorporates data from a set of parameters defined by an analyst, and outputs an high level integral value.

The conducted experiments showed that the threshold filtering of $Change\_Measure$ metric as anomaly detection technique in multivariate time series produces results comparable with Isolation Forest approach. Unlike to unsupervised deep learning techniques for anomaly detection it does not require training phase, and does not depend on volumes of input data. However, it is preferable to have representative data set describing object functioning including transition process to fine tune the threshold for anomaly detection.

The future research work includes analysis of additional filters such as simple sliding filters to improve visual identification of anomalies, thorough analysis of the approach efficiency on data sets describing different cyber security objects and comparison to other deep learning anomaly detection techniques for multivariate time series.

Another direction of future works relates to application of distributed and parallel computing techniques for accelerating the anomaly detection process when analyzing big streaming data.

References

1. Meleshko, A., Desnitsky, V., Kotenko, I.: Machine learning based approach to detection of anomalous data from sensors in cyber-physical water supply systems. In: The International Conference on Modern Trends in Manufacturing Technologies and Equipment 2019 (ICMTME 2019). Sevastopol, Russia. 9–13 September 2019. vol. 709. p. 033034 (2019). https://doi.org/10.1088/1757-899X/709/3/033034
2. Zhang J., Wu D., Boulet B.: Time Series Anomaly Detection for Smart Grids:A Survey. arXiv:2107.08835 [cs.LG]

3. Brentan, B., Rezende, P., Barros, D., Meirelles, G., Luvizotto, E., Izquierdo, J.: Cyber-Attack Detection in Water Distribution Systems Based on Blind Sources Separation Technique. Water. **13**(795) (2021). https://doi.org/10.3390/w13060795
4. Munir, M., Erkel, S., Dengel, A., Ahmed, S.: Pattern-based contextual anomaly detection in HVAC systems. IEEE Int. Conf. Data Min. Workshops (ICDMW) **2017**, 1066–1073 (2017). https://doi.org/10.1109/ICDMW.2017.150
5. Novikova, E. , Bestuzhev, M., Kotenko, I.: Anomaly Detection in the HVAC System Operation by a RadViz Based Visualization-Driven Approach. In: Katsikas, S. et al. (eds) Computer Security. CyberICPS 2019, SECPRE 2019, SPOSE 2019, ADIoT 2019. LNCS , vol. 11980. pp. 402–218. Springer, Cham (2020). https://doi.org/10.1007/978-3-030-42048-2_26
6. Puranik, T., Mavris, D.: Anomaly detection in general aviation operations using energy metrics and flight data records. J. Aerosp. Inf. Syst. Am. Inst. Aeronaut. Astronaut. **15**, 22–35 (2017). https://doi.org/10.2514/1.I010582
7. Skobtsov, V., Lapitskaja, N., Kim, D., Novoselova, N., Saksonov, R., Nikolaenya, E.: Analysis of reliability, survivability and telemetry data of on-board equipment of small satellites. In: 2019 IEEE 15th International Conference on the Experience of Designing and Application of CAD Systems (CADSM), pp. 1–5. IEEE (2019). https://doi.org/10.1109/CADSM.2019.8779289
8. Wei, Q., Ma, R., Wang, Y., Chen, M., Sun, Y., Liu, M., Lin, X.: Glad: a method of microgrid anomaly detection based on esd in smart power grid. In: 2020 IEEE International Conference on Power, Intelligent Computing and Systems (ICPICS), pp. 103–107 (2020)
9. Pham, V., Nguyen, N., Li, J., Hass, J., Chen, Y., Dang, T.: MTSAD: multivariate time series abnormality detection and visualization. In: 2019 IEEE International Conference on Big Data (2019). https://doi.org/10.1109/BigData47090.2019.9006559
10. Feng C., Tian P.: time series anomaly detection for cyber-physical systems via neural system identification and bayesian filtering. In: KDD 2021 (2021). https://doi.org/10.1145/3447548.3467137
11. Su, Y., Zhao, Y., Niu, C., Liu, R., Sun, W., Pei, D.: Robust anomaly detection for multivariate time series through stochastic recurrent neural network. In: Proceedings of the 25th ACM SIGKDD International Conference on Knowledge Discovery & Data Mining, pp. 2828–2837 (2019). https://doi.org/10.1145/3292500.3330672
12. Audibert, J., Michiardi, P., Guyard, F., Marti, S., Zuluaga, A.: USAD: unsupervised anomaly detection on multivariate time series. In: Proceedings of the 26th ACM SIGKDD International Conference on Knowledge Discovery & Data Mining, pp. 3395–3404 (2020). https://doi.org/10.1145/3394486.3403392
13. Shulepov, A., Novikova, E., Bestuzhev, M.: Approach to compare point distribution patterns produced by dimension reduction techniques. In: 2021 IEEE Conference of Russian Young Researchers in Electrical and Electronic Engineering (ElConRus), pp. 678–681. IEEE (2021). https://doi.org/10.1109/ElConRus51938.2021.9396692
14. de Berg, M., van Kreveld, M., Overmars, M., Schwarzkopf, O.C.: Delaunay Triangulations. In: Computational Geometry. Springer, Berlin, Heidelberg (2020)
15. Liu, F.T., Ting, K.M., Zhou, Z.-H.: Isolation-Based Anomaly Detection. ACM Trans. Knowl. Discov. Data **6**(1), 1–39. https://doi.org/10.1145/2133360.2133363

Security Analysis of Information Systems Based on Attack Sequences Generation and Testing

Roman Kryukov, Elena Doynikova, and Igor Kotenko

Abstract The paper considers an issue of analysis of security risks posed by the multistage cyber attacks using known tactics, techniques, and procedures. The technique for cyber attack sequences modeling and testing using source data from the MITRE ATT&CK database and the production model is proposed. Risk assessment is implemented for the generated cyber attack sequences, analysed information system components and an information system as a whole using risks decomposition and Common Vulnerability Scoring System metrics. The developed techniques for cyber attack sequences modeling and risks assessment are implemented within the software tool. The tool also allows recommending possible countermeasures for the successful attack sequences using MITRE ATT& CK data. The experiments for different infrastructures are conducted and application of the developed tool is demonstrated on the case study.

1 Introduction

Currently, the number of security incidents caused by complex multistage attacks including targeted attacks (so-called advanced persistent threat attacks) is constantly growing. In practice such attacks usually use already known tactics, techniques and tools, both malware and legitimate. Such known tactics and techniques are grouped together within database MITRE ATT&CK (Adversarial Tactics, Techniques & Com-

R. Kryukov
A.F. Mozhaysky Military-Space Academy, Zhdanovskaya str. 13, St. Petersburg 197198, Russia
e-mail: roman682@yandex.ru

E. Doynikova · I. Kotenko (✉)
Laboratory of Computer Security Problems, St. Petersburg Federal Research Center of the Russian Academy of Sciences, liniya 14-ya 39, St. Petersburg 197198, Russia
e-mail: ivkote@comsec.spb.ru

E. Doynikova
e-mail: doynikova@comsec.spb.ru

D. Camacho et al. (eds.), *Intelligent Distributed Computing XIV*, Studies in Computational Intelligence 1026, https://doi.org/10.1007/978-3-030-96627-0_39

mon Knowledge) [18]. They can be used to model attacks for penetration testing, security risk assessment, and security decision support.

As soon as modern information systems, including organizations' computer networks, cloud systems or internet of things systems, are usually large-scale and heterogeneous, manual modeling and analysis of multistage attacks against such systems based on known tactics and techniques is a rather laborious task. In this research we aimed to develop a technique and tool for automated attack modeling and testing, security risk assessment, and decision support. We considered application of MITRE ATT&CK database to achieve this goal.

Attack graphs are successfully used for multistage attack modeling. But existing techniques and tools usually use known vulnerabilities, weaknesses and exploits as source data and do not consider all stages of the multistage attacks, namely, the stages that do not require weaknesses exploitation.

There are researches that propose using the database of attack patterns CAPEC [17] as the source of data for attack modeling. But CAPEC database is focused on application security and describes techniques used by the attackers to exploit known weaknesses. MITRE ATT&CK is focused on the network defense, it describes exploitation phases of the attack's life cycle and details tactics and techniques used in multistage targeted attacks. Thus, MITRE ATT&CK is more suitable for the set goal. Besides, currently there is a number of projects based on MITRE ATT&CK. They include cyber security system CALDERA developed to automate attacker emulation and incidents response [3], CASCADE system developed to automate research of the scale and severity of suspicious behavior on the network using host data [4] and Cyber Analytics Repository (CAR) [6]. But they are not focused on modeling cyber attack sequences and security risks assessment considering these sequences, while these are the tasks considered in our research.

Thus the novelty of the proposed research consists in automation of multistage attack modeling and testing using known tactics and techniques described in MITRE ATT&CK and the proposed technique for security risks assessment based on the modeled attack sequences.

The contribution of this research is as follows: the technique for multistage attack modeling and testing using MITRE ATT&CK; the technique for security risks assessment based on the modeled attack sequences; the software tool for security assessment and decision support that implements the developed techniques.

The paper is organized as follows. Section 2 analyzes the related works. Section 3 describes the developed techniques for multistage attack modeling and security risks assessment. Section 4 describes the developed software tool, the case study, and experiments. Section 5 contains the conclusion and future research prospects.

2 Related Work

Attack graphs are widely used to model cyber attacks and assess the security risks [8, 10, 12, 13, 16, 19, 21, 23]. They represent possible attack paths that show how the vulnerabilities can be used in multistage attacks. There are different types of attack graphs including complete attack graph [7], predictive graph [15] and graph with multiple preconditions [11].

To assess security risks the probabilities are assigned to the attack graph nodes. Probabilities allow considering that attack action that corresponds to the attack graph node can fail. This probability usually depends on the attacker's skill level, attack action complexity, and the postconditions of the previous attack steps. The attack graphs can be classified on the probability attack graphs [22, 23] and Bayesian attack graphs [10, 12, 16, 19, 21] depending on the approach to the probabilities calculation. However, existing researches mainly use databases of known vulnerabilities, weaknesses or exploits while generating attack graphs, and do not take into account all stages of multistage attacks, namely, those stages that are not associated with the use of vulnerabilities or weaknesses. There are researches that propose using the database of attack patterns CAPEC [17] as the source of data for attack modeling [14]. But CAPEC database is focused on applications security and describes techniques used by the attackers to exploit known weaknesses as well.

In this research MITRE ATT&CK database is considered as soon as it is focused on the network defense, describes exploitation phases of the attack's life cycle, and details tactics and techniques used in multistage targeted attacks. Currently there is a number of projects based on MITRE ATT&CK. They include the CALDERA system developed to automate attacker emulation and incidents response [3], CASCADE system developed to automate research of the scale and severity of suspicious behavior on the network using host data [4] and CAR [6]. But they are not focused on modeling attack sequences while this is the task considered in our research.

In this research we propose the techniques for automation of multistage attack modeling and testing using known tactics and techniques described in MITRE ATT&CK and for the security risks assessment based on the modeled cyber attack sequences. The developed techniques differ by the possibility to test if the steps of multistage attacks (namely, techniques and procedures) can be really implemented against the analysed system and consider it while risks assessment. The tool implementing the proposed techniques can propose possible countermeasures to decrease the security risks based on data from MITRE ATT&CK.

3 The Technique for Information Security Analysis

To implement information security analysis we propose generating all possible known cyber attack sequences and assess risks based on the generated attack sequences. Thus, the proposed solution incorporates two techniques: the technique for attack sequences generation and testing and the technique for risks assessment. The techniques are described in detail in the subsections below.

3.1 Attack Sequences Generation and Testing

A multistage attack is represented as an attack sequence or the sequence of atomic attack actions. The technique for attack sequences generation takes as input data from MITRE ATT&CK. We take into account the following data from MITRE ATT&CK [18]: the sequence of tactics; the techniques that can be used to implement each tactic; and the procedures that can be used to implement each tactic if exist. The output of the technique is an attack tree incorporating all possible attack sequences for the information system (IS) under analysis.

The proposed technique incorporates the following stages: (1) specify common rules for the generation of multistage attacks based on the MITRE ATT&CK [18] tactics; (2) generate specific rules and multistage attacks for the IS under analysis considering its configuration.

On the first stage the sequence of tactics is used to determine the sequence of attack stages, i.e. for an enterprise there are reconnaissance, resource development, initial access, execution, persistence, privilege escalation, defense evasion, credential access, discovery, lateral movement, collection, command and control, exfiltration, and impact [5]. To model this process and generate attack sequences, namely, to connect attack stages within the attack sequences, we specify the common rules based on a production model. The production model uses IF-THEN rules to represent system operation and looks as follows [9, 20]: i; $sitA \rightarrow sitB$; P; N, where i—product (or rule) number; $sitA \rightarrow sitB$—product core, where $< sitA >$ is a current state of an object required to implement an action and transit to the next state $sitB$, $< sitB >$—state of the object after successful action implementation; $< P >$—logical predicate representing if the preconditions of product core are satisfied and product core is successfully activated and implemented (True if yes, otherwise—no); $< N >$—actions that should be implemented after successful product core activation and implementation.

On the first stage of the technique we developed 12 production rules corresponding to 12 MITRE ATT&CK tactics to generate cyber attack sequences. The general production rule looks as follows:

$$< Num >, < sitA > \stackrel{\text{A}}{\rightarrow} < sitB >, < 0, 1 >, < N_1, N_2 >,$$

where $< Num >$—rule number;

$< sitA >$—current state of the information system's object (preconditions);

A—atomic attacks (techniques or procedures) that can be implemented if the information system's object is in the state $< sitA >$ (i.e. preconditions are satisfied);

$< sitB >$—state of the information system's object after successful implementation of atomic attacks from A (postconditions);

$< 0, 1 >$—binary value that indicates if an atomic attack from A is successfully implemented. Value "1" (True) indicates that an atomic attack from A is successfully implemented and the information system's object is in the state $< sitB >$. Otherwise value "0" (False) indicates that an atomic attack from A can't be implemented;

$< N_1, N_2 >$—the element that specifies the next production rule, namely, if binary value is "1"—go to the production rule N_2, otherwise, go to the production rule N_1.

An example of production rule for the "Initial Access" tactics considering MITRE ATT&CK tactics for enterprise is as follows:

$$< 1 >< Sit(IP - address_1, type, vend, prod, vers) > \xrightarrow{A_1} < Sit(IP - address_1, type, vend, prod, vers, v) >< 0, 1 >< 1, 2 > .$$

Within this rule information on the software, such as type ($type$), vendor ($vend$), product ($prod$) and version ($vers$), installed on the IS object with ip-address $IP - address_1$ allows detecting the vulnerabilities v related to obtaining initial access to the object and their exploitation to get initial access using the connected techniques A_1. If the product core is successfully implemented, i.e. there are such vulnerabilities for the IS object with $IP - address_1$ and they can be successfully exploited, then we move to the next rule 2 (rule 2 specifies detection and exploitation of the vulnerabilities related with further IS object compromise after obtaining the initial access). If there are no such vulnerabilities or appropriate techniques can not be successfully implemented, we return to rule 1 for the next IS object with a different ip-address.

On stage 2 the available techniques and procedures within each tactic (one of 12 common product rules) are iterated. Thus, an attack sequence is represented as a sequence of techniques or procedures from MITRE ATT&CK. The techniques and procedures are used to represent atomic attacks within an attack sequence. The procedures, if exist for a technique, represent exploits for atomic attacks. Within an attack sequence, the techniques and procedures are connected based on the sequence of attack stages first (the sequence of tactics), and then based on the pre and post conditions of a technique or procedure implementation (namely, required platform and permissions, and obtained permissions). Successful implementation of an atomic attack action (technique or procedure from the set A) leads to changing of product's state and forming of preconditions for the next atomic attack.

To test if the vulnerabilities can be successfully exploited and appropriate techniques can be successfully implemented on stage 2 we use Atomic Red Team database with the tests that implement atomic attacks corresponding to the MITRE ATT&CK procedures [2].

Attack sequences generation proceeds till there are techniques or procedures with preconditions that correspond to the postconditions of the previously implemented technique or procedure, or till all information system's products are compromised.

3.2 Risk Assessment

Risk is usually considered as a combination of successful attack implementation probability and possible attack impact. We outline risks for the components of an information system (a host or product) and integral information security risk. Integral information security risk depends on the risks for all components of the information system. Risks for the components of information system depend on risks from the tactics, techniques, and procedures that are possible for these components. Risks from the tactics, techniques and procedures depend on the probability of their successful implementation and the possible impact from their implementation. Thus, the integral information security risk $Risk$ can be calculated as follows:

$$Risk = \sum_{j=1}^{n} (Risk_j \times C_j),$$

where $Risk, Risk_j \in [0, 10]$; n—number of risk metrics of the next level of the hierarchy (risks for IS hosts or products); $Risk_j$—jth risk metric of the next level of the hierarchy (for IS hosts or products); C_j—weight coefficients for the $Risk_j$ of each component.

Sum of the weight coefficients should be equal to 1: $\sum_{j=1}^{n} C_j = 1$. The weight coefficients should be set by the experts on the basis of corresponding risks severity for the system.

On the next level the components of information system (hosts and products) are used as risks decomposition factor. $Risk$ for each component is calculated as follows:

$$Risk_j = \frac{1}{\sum_{i=1}^{m} C_i} \times \sum_{i=1}^{m} (Risk_i \times C_i),$$

where $Risk_j, Risk_i \in [0, 10]$; m—number of risk metrics of the next level of the hierarchy (risks for the tactics, techniques and procedures); $Risk_i$—ith risk metric of the next level of the hierarchy (for the tactics, techniques and procedures); C_i—weight coefficients for the $Risk_i$.

While attacker elevates his/her privileges in the system under analysis it becomes more difficult to detect him/her and distinguish his/her actions from the actions of the legitimate users. It should be considered while determination of the weight coefficients C_i for each tactic. We propose to calculate the weight coefficient C_i for each tactic as follows (Table 1):

$$C_i = \frac{P_i + I_i + O_i}{3},$$

where P_i—implementation complexity; I_i—level of the potential negative impact; O_i—detection complexity.

Table 1 Metrics for determination of tactic severity

Value	Implementation complexity (P_i)	Impact (I_i)	Detection complexity (O_i)
Negligible	10	0,1	0,1
Low	6	3	3
Medium	3	6	6
High	0,1	10	10

On the third level the tactics from MITRE ATT&CK database (or attack stages) of the multistage cyber attacks are used as the second risks decomposition factor. On this level the $Risk$ for each tactic is calculated as follows:

$$Risk_i = \max(Risk_v),$$

where $Risk_v$ is calculated on the basis of Common Vulnerability Scoring System (CVSS) indexes for the vulnerabilities used to implement the specific technique or procedure [1].

According to CVSS 3.0 the vulnerability severity score is calculated as follows:

$$Risk_v = \begin{cases} Impact + Exploitability, & \text{if Scope is unchanged} \\ 1,08 \times (Impact + Exploitability), & \text{if Scope is changed}. \end{cases}$$

where $Exploitability$ represents complexity of vulnerability exploitation, $Impact$ represents potential negative impact from the vulnerability exploitation, $Scope$ is scope of the impact.

On the last level of the hierarchy the atomic attack actions (techniques or procedures from MITRE ATT&CK database) are used as the third risks decomposition factor. Risk [0; 4] is considered as low, (4; 7]—as average, and (7; 10]—as high.

4 Implementation and Experiments

The developed techniques are implemented within security analysis and decision support tool developed in C++. The complexity of the implemented algorithm is $O(logn)$. The common scheme of the tool operation is represented in Fig. 1. We used the stand for the experiments to evaluate the developed techniques. The stand incorporates several IS configurations from 1 to 30 hosts. Below the tool operation is demonstrated on the configuration that incorporates an administrator host with OS Windows 10 PRO 64 bit and three users' hosts with OS Windows 10 PRO 64 bit, Windows 7 64 bit and Windows 7 32 bit (Fig. 2). On the hosts there are various

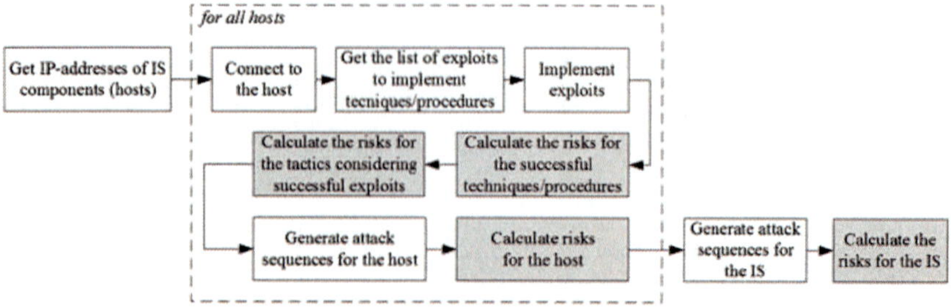

Fig. 1 Common scheme of the security analysis and decision support tool operation

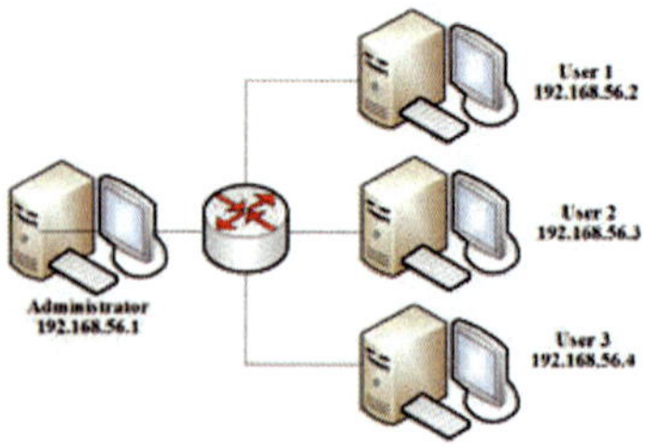

Host	OS Win 10 PRO 64 bit	OS Win 7 Ultimate	OS Win 7 Ultimate 32	Comodo FireWall	Windows FireWall	Kaspersky Free antivirus	COMODO Security	AVG AntiVirus Free	Avira Security antivirus	Weight coefficient
User 1 (192.168.56.2)	+	-	-	+	-	-	+	-	-	0.3
User 2 (192.168.56.3)	-	+	-	-	+	-	-	+	-	0.4
User 3 (192.168.56.4)	-	-	+	-	+	-	-	-	+	0.3

Fig. 2 The test configuration

security controls, including Comodo FireWall, Windows FireWall, Kaspersky Free antivirus, COMODO Security, AVG AntiVirus Free, and Avira Security antivirus.

Configurations and severities of the IS components (hosts) are represented in the table provided in Fig. 2.

The developed tool provides three functions: "This system", "DB" and "Network". Function "This system" implements risk assessment for the system where the tool is installed (usually it is the administrator's host) and represents recommendations to decrease the risks. Function "DB" allows selecting and managing the database used for the techniques/procedures implementation (for the penetration testing). This database can be supplemented with the owner's atomic attacks and their assigned vulnerability metrics in accordance with CVSS 3.0. Function "Network" allows assessing the risks. Risk can be calculated for each host separately, and for the entire IS.

The results of the tool operation for the test configuration in the Fig. 2 are provided in Fig. 3. Attack modeling and penetration testing were implemented using the developed tool from the Administrator host.

In Fig. 3 it is represented that for the test configuration 18 techniques were found in the database and were tested. 12 from these techniques were completed successfully and formed two attack sequences. These sequences are highlighted with red in Fig. 3. Thus, an attacker can compromise the host of User 1 and User 3 using these sequences with the risk values 8,39 and 8,14 accordingly. Existing security controls stopped possible attack sequences against the host of User 2. This sequence is highlighted

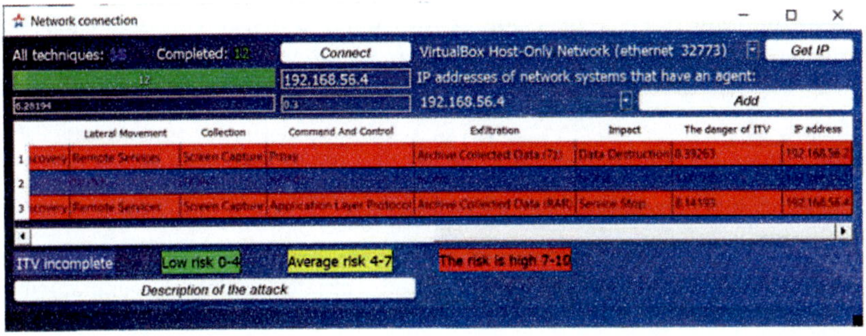

Fig. 3 Results of the tool operation for the test configuration

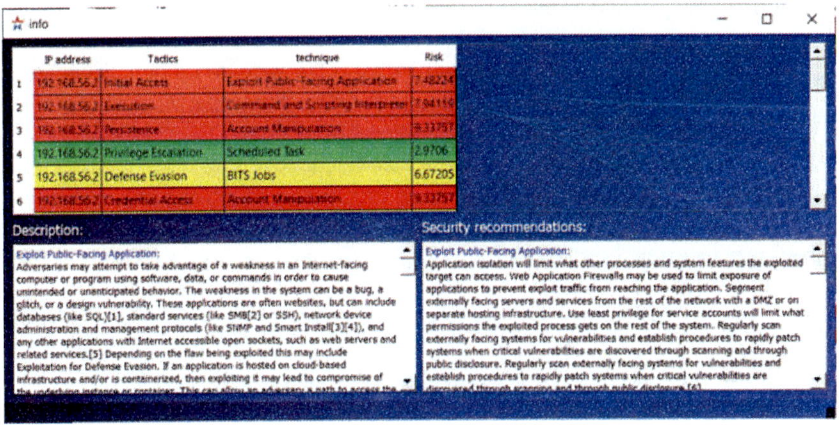

Fig. 4 Description of the attacker actions and possible countermeasures for the specific technique

with blue in Fig. 3. The integral security risk for the test environment is 6,28 that is an Average security risk (Fig. 3).

The "Description of the attack" function of the developed tool (Fig. 4) allows representing all techniques that were implemented within the successful attack sequences against the specific hosts. Besides, it allows representing the sequence of attacker actions to implement the technique and the possible countermeasures for the successful attack sequence (Fig. 4).

Finally, the described case study and conducted experiments for other environments (from 1 to 30 hosts with different configurations of products and security controls) demonstrate that the developed tool allows generating all possible known attack sequences considering known tactics, techniques and procedures, demonstrating if the generated sequences can be successfully implemented in the analysed environment and assessing the security risks ordering the generated attack sequences by the severity. Besides it provides security recommendations for the successful attack sequences. Required time and resources depend on the number of analysed hosts and products. In future research the algorithm for attack sequences generation and risks calculation will be optimised for application in real time.

5 Conclusion

In the paper the technique for assessing security risks of information systems based on the attack sequences generator was presented. The attack sequences generator uses known tactics, techniques and procedures to model and implement cyber attacks to test if they can be successfully implemented in the system under analysis. The software tool has been developed, which, based on the analysis of the security risks for the information systems, provides recommendations for preventing possible multistage cyber attacks. An efficiency of the developed tool for attack modeling, testing and risks assessment was tested on the experiments for various configurations. Optimisation of the proposed algorithms for application in real time will be considered in future research.

The implementation of the proposed recommendations will reduce the security risks for an information system. However, in order to select the optimal recommendations for preventing multistage cyber attacks, additional research is required that is the future research direction.

Acknowledgements This research is being supported by the grant of RFBR #19-07-01246 in SPC RAS.

References

1. First website. common vulnerability scoring system. https://www.first.org/cvss/
2. Github website. atomic red team tests. https://github.com/redcanaryco/atomic-red-team
3. Github website. MITRE CALDERA. https://github.com/mitre/caldera
4. Github website. MITRE CASCADE. https://github.com/mitre/cascade-server
5. MITRE website. MITRE ATT&CK enterprise matrix. https://attack.mitre.org/matrices/enterprise/
6. Mitre website. MITRE cyber analytics repository. https://car.mitre.org/
7. Artz, M.: Netspa, a network security planning architecture. Master's thesis (2002)
8. Dantu, R., Kolan, P., C. J.: Network risk management using attacker profiling. Sec. Commun. Netw. **2**(1), 83–96 (2009)
9. Feigenbaum, E.: The art of artificial intelligence: Themes and case studies of knowledge engineering. In: Proceedings of the International Joint Conference on Artificial Intelligence, pp. 1014–1029 (1977)
10. Frigault, M., Wang, L., Jajodia, S., Singhal, A.: Network Security Metrics (2017)
11. Ingols, K., Lippmann, R., Piwowarski, K.: Practical attack graph generation for network defense. In: Proceedings of 22nd Annual Conference on the Computer Security Applications, Miami Beach, FL, pp. 121–130. IEEE (2006)
12. Khosravi-Farmad, M., Ghaemi-Bafghi, A.: Bayesian decision network-based security risk management framework. J. Netw. Syst. Manag. 1–26 (2020)

13. Kotenko, I., Doynikova, E.: Security assessment of computer networks based on attack graphs and security events. In: L. et al. (ed.) ICT-EurAsia 2014, Lecture Notes in Computer Science (LNCS), vol. 8407, pp. 462–471. Springer (2014)
14. Kotenko, I., Doynikova, E.: The CAPEC based generator of attack scenarios for network security evaluation. In: Proceedings of the IEEE 8th International Conference on "Intelligent Data Acquisition and Advanced Computing Systems: Technology and Applications (IDAACS'2015), pp. 436–441 (2015)
15. Lippmann, R., Ingols, K. et al.: Validating and restoring defense in depth using attack graphs. In: Proceedings of the MILCOM 2006, Washington, DC, pp. 1014–1029 (2006)
16. Liu, Y., Man, H.: Network vulnerability assessment using bayesian networks. Proc. SPIE **5812**, 61–71 (2005)
17. MITRE. CAPEC database. https://capec.mitre.org/
18. MITRE. MITRE att&ck database. https://attack.mitre.org/
19. Muñoz-González, L., Sgandurra, D., Barrère, M., Lupu, E.: Dynamic security risk management using bayesian attack graphsexact inference techniques for the analysis of bayesian attack graphs. IEEE Trans. Depend. Sec. Comput. **16**, 231–244 (2019)
20. Newell, A.: Production Systems: Models of Control Structures. Academic, New York, NY (1973)
21. Poolsappasit, N., Dewri, R., R. I.: Dynamic security risk management using bayesian attack graphs. IEEE Trans. Depend. Sec. Comput. **9**(1), 61–74 (2012)
22. Singhal, A., Ou, X.: Security risk analysis of enterprise networks using probabilistic attack graphs. NIST interagency report (2011)
23. Stakhanova, N., Basu, S., W. J.: A cost-sensitive model for preemptive intrusion response systems. In: Proceedings of the 21st International Conference on Advanced Networking and Applications (2007)

TRIE (Special Session) & COMPUTING—Chair: Lidia Fotia

A Comparative Analysis of a Designed CNN and AlexNet for Image Classification on Small Datasets

Shahrukh Sultan and Yana Bekeneva

Abstract In recent years, the number of multimedia data sources has been steadily increasing, which leads to the need to develop tools for processing and analyzing generated images or video streams. Tasks related to pattern recognition are becoming relevant. Convolutional neural network (CNN) is the most suggested deep learning neural network for image classification and its efficiency mainly depends on the architecture of the convolutional layers and the size of data used to train the network. While training dataset is small it's quite challenging to yield high accuracy due to overfitting. Although data augmentation is a way to prevent overfitting but the process can't sufficiently overcome the issue. In such cases deployment of Alexnet which is already trained with large image dataset can give satisfactory result. This article demonstrates the comparative performance of a designed CNN and Alexnet for classification of animal faces from small dataset. During this work we designed a 15-layered CNN with 3 convolutional layers increasing the number of filters respectively. The experiments are performed in MATLAB varying the number of input images both for the designed CNN and Alexnet to compare the results in terms of accuracy and processing time.

1 Introduction

The problem of pattern recognition in images is becoming increasingly important in various fields of human activity, where there is a need to analyze homogeneous or heterogeneous data. There are various ways to process and analyse images, but the most widespread is the convolutional neural network (CNN) because of its complex learning architecture. However, performance of a CNN depends on the size of training data. When the dataset is large enough to train the network, it can extract and represent

S. Sultan (✉) · Y. Bekeneva
Saint Petersburg Electrotechnical University, St Petersburg, Russia
e-mail: shsultan@stud.etu.ru

Y. Bekeneva
e-mail: yana.barc@mail.ru

D. Camacho et al. (eds.), *Intelligent Distributed Computing XIV*, Studies in Computational Intelligence 1026, https://doi.org/10.1007/978-3-030-96627-0_40

all important features for classification. For small datasets, the network can't get sufficient features and there is a high risk of overfitting. Data augmentation is a method used to prevent overfitting but can't sufficiently overcome the issue in all cases [1]. When the architecture of a designed CNN is not efficient to deal with small datasets, here comes a suggestion to use a pre-trained network by transfer learning method [2]. AlexNet and GoogleNet are such pre-trained model and AlexNet has higher accuracy for sports video classification which primarily involves image classification [3].

Over the years researchers have developed various models for classifying different objects [4, 5], especially human faces with great accuracy. There are different approaches how to use artificial intelligence techniques to recognize objects on various shots in different conditions [6]. Some approaches assume to use large training sets to find patterns [7] but for classification of animal faces, those methods are not much accurate because of complex intra-class variability and inter-class similarity in animals [8]. So, classifying animal faces with better accuracy is still a motivation for researcher. AlexNet which is already trained with 1.2 million images to classify images into 1000 object categories including many common animals [9]. Thus, it is fascinating to observe the performance of Alexnet for animal image classification. In a similar study authors observed that Alexnet is comparatively less accurate because of high error rate, primarily due to the small number of layers in the network and the high selection of parameter values [10].

2 Convolutional Neural Networks

A convolutional neural network commonly known as CNN is a multilayer network architecture vastly used for image classification due to its ability to extract and learn features directly from input data, eliminating the need for manual feature extraction.

2.1 Architecture of CNN

A CNN has input layer, output layer and in between many hidden layers can be added as much as required for a specific task. The most common and effective layers of CNN are considered below.

Convolutional Layer. This layer applies convolution filters on the images. The network designer can set the size and number of filters accordingly. Number of filters specify the number of neurons that connect to the same region of the input image. Padding, stride are other attributes of a filter that can be applied by the designer.

Batch Normalization Layer. To make the training process easy and optimal this layer normalizes the activations and gradients propagating through a network.

ReLu Layer. Rectified Linear Unit layer is to apply an activation function. A CNN with ReLu layers trains several times faster than their equivalents.

Max pooling layer. This layer lies between convolution layers. Its main function is to perform a down sampling operation to reduce the spatial dimension of the feature map. This process increases the number of filters in deeper convolutional layers without increasing the required amount of computation per layer.

Fully Connected Layer. In this layer the neurons connect to all neurons of the next layer. This layer combines all the features learned by the previous layers across the image to identify the larger patterns for classification of the images.

Softmax Layer. This layer contains an activation function that normalizes the output of the fully connected layer in terms of positive numbers that summed to one and further used by output layer to calculate the loss and probabilities of classification.

2.2 *Pre-trained CNN: AlexNet*

AlexNet is a CNN architecture consists of five convolution layers and three fully connected layers. Some of the convolution layers are connected by max polling layer. It can reduce overfitting in the fully connected layers which makes it more effective [9]. Over the years it is widely used for image and video classification by method of transfer learning.

3 Experimental Setup and Results

3.1 *Dataset*

The dataset used for the experiment, Animal faces [11] consists of total 16130 (512 X 512) color images in three classes namely cats, dogs and wild animals. The images are in good resolution and machine readable format. With attribution to the licensor the dataset can be used for non-commercial purposes.

3.2 *Experimental Setup for Designed CNN*

In order to design a neural network in MATLAB we need Deep Learning Toolbox [12]. A CNN is designed with total 15 layers consisting of 3 convolutional layers as shown in Fig. 1 and specifications are mentioned below.

Input Image Size: 512 × 512, No. of convolution layers: 3, Filter Size: 3 × 3, No. of Filters in conv. layers: 8, 16, 32 respectively, Stride: 2, Padding: 1, Output Size in Fully Connected Layer: 3 (3 classes viz. cat, dog, wild).

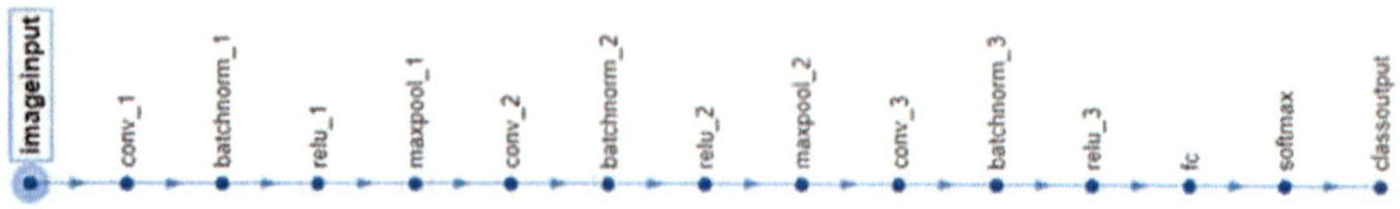

Fig. 1 Layers of the designed CNN

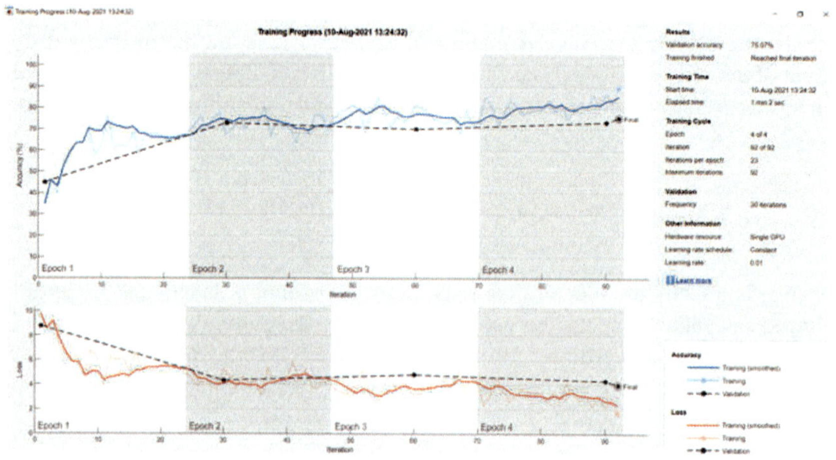

Fig. 2 Training progress of Alexnet

Training Options

Optimizer: Stochastic Gradient Descent with Momentum (SGDM), Hardware Source: Single CPU, Initial Learning rate:0.01, Learning rate schedule: Constant, Epoch: 4, Validation Frequency: 30.

3.3 Experimental Setup for AlexNet

With the help of AlexNet Network support package we can load the pre-trained Alexnet. The 25 layered AlexNet implemented in the experiment consists of one input image layer, a fully connected layer, a softmax layer, a classification layer and other 21 layers are loaded from pre-trained Alexnet by transfer learning method. The image input size and training options are kept same as in designed CNN. Figure 2 shows the training progress.

Table 1 Changes in accuracy as change in number of input images

Number of images	Accuracy: Designed CNN (%)	Accuracy: Alexnet (%)
200	35.68	60.84
400	34.74	65.91
600	35.92	68.89
800	40.58	68.50
1000	49.64	75.07

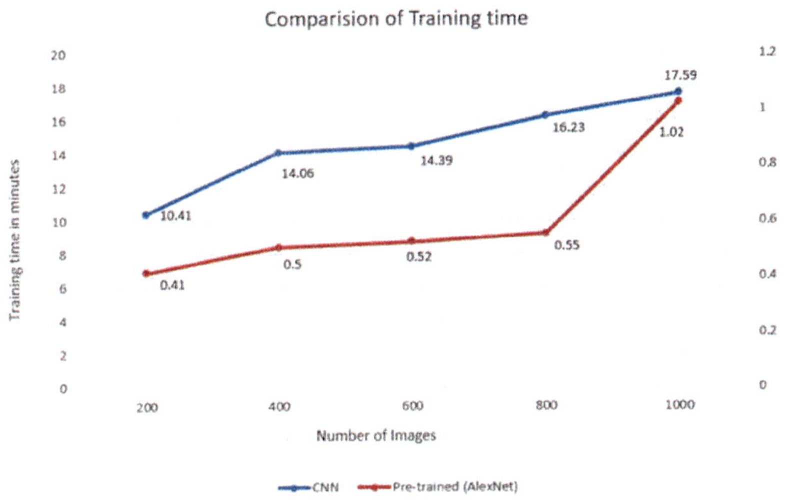

Fig. 3 Comparison of training time

3.4 *Results*

The program written for the experiment, randomly divides the dataset for training and testing by a ratio of 8:2. The experiment is performed repeatedly by varying the number of input images in order to observe the change in accuracy (Fig. 3 and Table 1).

4 Conclusions

Accuracy of the designed CNN increases as number of images increased, which confirms the low performance of the network in case of small training dataset. But accuracy of AlexNet is far better than the designed CNN as it is already trained with millions of images. Time taken by Alexnet can be counted in seconds while average training time of designed CNN is 14 times higher. This study is limited to experimental observations and suggests that is it better to use a pre-trained network like Alexnet when the architecture of the CNN is not appropriate to deal with small datasets. Our future plan aims to improve existing methods or develop new possible methods to increase the performance of a CNN even with small dataset from theoretical ground to practical experiments with different complex data sets.

Acknowledgements This paper was supported by the Russian Federation President Award SP-2581.2019.5.

References

1. Shorten, C., Khoshgoftaar, T. M: A survey on image data augmentation for deep learning. J Big Data **6**, 60 (2019)
2. Xingyi, Y., Xuehai, H., Yuxiao, L., Yue, Y.: Transfer Learning or Self-supervised Learning? A Tale of Two Pretraining Paradigms (2020). arXiv:2007.04234v1 [cs.CV]. Accessed from 19 Jun 2020
3. Ramesh, M., Mahesh, K.: A performance analysis of pre-trained neural network and design of CNN for sports video classification. Int. Conf. Commun. Signal Process. (ICCSP) **2020**, 0213–0216 (2020)
4. Hussain M., Bird J.J., Faria D.R., A Study on CNN Transfer Learning for Image Classification. Advances in Computational Intelligence Systems. UKCI: Advances in Intelligent Systems and Computing, vol. 840. Springer, Cham (2018)
5. Srivastava, S., Divekar, A.V., Anilkumar, C., et al.: Comparative analysis of deep learning image detection algorithms. J. Big Data **8**, 66 (2021)
6. Minyoung, H., Pulkit, A., Alexei, A.E.: Berkeley Artificial Intelligence Research (BAIR) Laboratory UC Berkeley. arXiv:1608.08614v2
7. Keyur, J., Thakkar, P., Sonal, P.: Rami: comparative analysis of variegated pre-trained models for discrete class-labels. In: International Journal of Latest Engineering and Management Research (IJLEMR), vol. 03, Issue 05, pp. 38–45 (2018)
8. Rafiul, H.K., Kyung-Won, K., Seon-Ja, L., Sung-Dae, Y., Oh-Jun, K.: Animal face classification using dual deep convolutional neural network. J. Korea Multimed. Soc. **23**(4), 525–538 (2020)
9. Krizhevsky, A., Sutskever, I., Hinton, G.E.: ImageNet Classification with Deep Convolutional Neural Networks. In: NeuroIPS (2012)
10. Ö-zkan, İ, Turan, B.: Classification of animals with different deep learning models. J. New Results in Sci. (JNRS), ISSN: 1304-7981
11. Animal Faces: https://www.kaggle.com/andrewmvd/animal-faces. Accessed 13 Aug 2021
12. MathWorks Documentation: https://in.mathworks.com/help/deeplearning/ug/create-simple-deep-learning-network-forclassification.html. Accessed 01 Aug 2021

A Clustering Reputation-Based Framework in Edge-Based IoT Environments

Giancarlo Fortino, Lidia Fotia, Fabrizio Messina, Domenico Rosaci, and Giuseppe M. L. Sarné

Abstract In this work we introduce a clustering reputation-based approach for edge-based IoT (Internet of Things) environments where it's possible to process and save data on edge servers. In particular, once in an edge domain two Smart IoT Objects (SOs) exchanged a service, supported by their associated agents, they will release a feedback about their respective counterparts to the edge server managing their current domain to update the counterparts' reputation score. However, even though in the proposed systems the reputation is managed and hosted in the Edge, the SOs' reputation scores are also periodically saved on the Cloud. In this way, when a SO moves itself from an edge domain to another edge domain its reputation score will be available on the Cloud. Moreover, each edge server provides to cluster its managed SOs on the basis of their reputation scores to strengthen the competitiveness of SOs into the domain. This paper describes the proposed distributed clustering reputation-based framework for heterogeneous SOs and its performance in a simulated scenario.

G. Fortino (✉) · L. Fotia
DIMES, University of Calabria, Rende (CS), Italy
e-mail: giancarlo.fortino@unical.it

L. Fotia
e-mail: lidia.fotia@dimes.unical.it

F. Messina
DMI, University of Catania, Catania, Italy
e-mail: messina@dmi.unict.it

D. Rosaci
DIIES, University "Mediterranea", Reggio Calabria, Italy
e-mail: domenico.rosaci@unirc.it

G. M. L. Sarné
Department of Psychology, University of Milano Bicocca, Milano (MI), Italy
e-mail: giuseppe.sarne@unimib.it

D. Camacho et al. (eds.), *Intelligent Distributed Computing XIV*, Studies in Computational Intelligence 1026, https://doi.org/10.1007/978-3-030-96627-0_41

1 Introduction

In the Internet of Things (IoT) [2], Smart IoT Objects (e.g. sensors and devices) can exchange data among them through network infrastructures potentially able to connect millions of IoT nodes and generate massive data.

In the past, many cloud-based environments have been proposed to ensure the access to IoT devices and supply them with communication, computational and storage resources [10]. However, approaches exploiting cloud computing solutions are subject to significant limitations, namely: (*i*) limited bandwidth and network resources are committed by massive data transmissions; (*ii*) communication can be exposed to significant latencies that is a critical aspect for time-sensitive applications; (*iii*) computation processes must be uploaded to the cloud. Furthermore, we have to consider another critical impact that cloud based solutions have on SOs in terms of power consumption, perhaps the most precious resource for SOs.

To solve these problems, the Edge Computing moves the computational and communication overhead from SOs to edge servers, provided with suitable resources and nearby the SOs. In this scenario, our goal is to promote the mutual cooperation among SOs making available the resources belonging to the more closed, equipped and performing SOs. Particularly in presence of competitive environments, trusting inappropriate counterparts can expose SOs to several potential threats due to malicious, fraudulent and/or disliked behaviors. To mitigate such risks a possible solution is the adoption of a reputation system to create a confidence atmosphere. More specifically, reputation systems consider indirect information about past behaviors or events [9, 12] to compute and provide a measure about the expectation that a trustor has to receive benefits from trustees.

In this paper, to manage the SOs cooperation, we present an IoT edged, agent-based framework implementing a clustering approach on the basis of the SOs reputation. More in detail, each SOs hosts a light tamper-proof software agent capable of basic interactions and social behaviors. The SO is supported by a reputation-system located on the edge server to require/accept of cooperating with a reliable partner placed in the proximity. Once two SOs exchanged a service, their associated agents will issue a feedback about their respective counterparts to an edge server (managing the IoT domain which SOs currently belong) for updating the counterparts' reputation score. Moreover, in our proposal the reputation systems is moved from the cloud to the edge servers managing the SOs affiliated with its domain but, however, reputation scores are periodically saved on the cloud from the edge servers. In such a way, whenever a SO moves from an edge domain to another one, its reputation score can be to retrieved on the Cloud.

Finally, each edge server is able to divide the SOs (i.e., agents) into clusters on the basis of their reputation scores. This clustering has been introduced to strengthen competitiveness into the SO environment because each SO is encouraged to gain a good reputation to take benefits from its cluster affiliation or, in other words, from its trustworthiness.

The paper is organized as follows. Section 2 presents an overview on the related literature. Section 3 discusses the proposed agent framework, while Sect. 4 describes the adopted reputation model. The experimental results are presented in Sect. 5. Finally, in Sect. 6 some conclusions are drawn.

2 Related Work

In this section, we first discuss edge computing issues in the IoT context and then we will introduce related technologies on a trust and reputation system in IoT.

IoT provides a different vision of big data in data storage, data processing and analytic activities. A proposed solution was that of displacing data into the cloud [1], but the nature of IoT data posed the question of choosing the best technology for performing such activities. Unlike the cloud infrastructure, the edge computing can manage the entire data management processes for its closeness to data sources (i.e., the IoT SOs). In [6], the authors explore the resource-efficient edge computing issues for intelligent IoT applications. They present the design of a resource-efficient computation offloading algorithm to permit an IoT device to use resources across the local device and the edge cloud in proximity. In particular, to reduce an intelligent IoT device's edge resource occupancy while satisfying the QoS requirement, this algorithm is composed of a delay-aware task graph partition algorithm and the optimal virtual machine selection method. Lin et al. [16] study an Intrusion-Detection System (IDS) architecture and resource allocation in edge computing. An IDS is used to mitigate security threats in edge computing. Their system facilitates multiple resources sharing and heterogeneous resource-demanding allocation. In particular, the authors present a general edge computing IDS architecture and use this as the basis for their model to allocate resources. Also, a Single-layer Dominant and Max-Min Fair (SDMMF) allocation is used and a multilayer resource allocation scheme is used to cope with the multiple resources fair allocation in multiple layers. In [19], the authors propose a resource allocation model for allocating computing resources in an edge computing platform, named Zenith. It uses a decoupled architecture where the infrastructure management at the Edge Computing Infrastructures (ECIs) is executed independent of the service provisioning and service management performed by the Service Providers (SPs). Then, they propose an auction-based mechanism for resource contract establishment and a latency-aware scheduling technique that maximizes the utility for both ECIPs and SPs.

Reputations systems, as well as group formation, play an important role in IoT contexts. Indeed, while cryptographic techniques protect privacy and authentication, trust and reputation systems estimate the trustworthiness of potential partners. To solve the problem of unreliable partners, several trust systems have been proposed in the past. For example, in [7, 8, 11, 13–15]. For instance, Bao et al. [3] presented a dynamic trust management protocol that exploits the social relationships among IoT objects to develop a trust-based service composition. The trust system introduced in [5] considers the evolution of social relationships over time and adapts itself to the trust swing. Hence, IoT SOs utilize direct experiences and available recommendations to evaluate the trustworthiness of their peers and to match services' demand and offer. In [18], the authors proposed an approach for machine-to-machine applications, named BETaaS, that supplements a trust model based on security, QoS, scalability, availability, and gateways reputation. This model is characterized by evaluating SOs reliability based on the monitoring of their behaviors. SOs can be arranged in groups/clusters formed by similar peers on the basis of their social interactions and mutual trust evaluations [17]. Finally, in [4], the authors studied the convergence among IoT, software agents and cloud computing to form agent groups/clusters. In particular, they introduced an algorithm to form agent groups/clusters through reliability and reputation gathered by the agents.

3 Our Scenario

Our scenario is composed by a set of local edge servers that manage some heterogeneous SOs acting into the associated local domain and working independently of each other. Also, we assume that SOs share resources for pay. We introduce two agents: (*i*) the *edge agent* (denoted by **e**) associated with the edge domain to assist SOs in their activities; (*ii*) the *trust agent*, a light tamper-proof component (denoted by **t**), handling reputation tasks, monitoring hardware resources and communication activities of its host, moreover it can also communicate with the edge agent of its current edge domain. In the remaining part of this paper we will refer at two different SOs, named SO_1 and SO_2, by means of their trust agents, denoted as t_1 and t_2, respectively.

Three different activities are carried out within the proposed framework, namely (*i*) edge tasks, (*ii*) trust task and (*iii*) SO task. In the edge task, when a SO acts within the agent framework, it must execute an *affiliation process* with **e** associated with one of the domains active into the framework. First, **e** verifies the "identity" and records it into the framework, then **t** starts to run on the SO. **e** associates **t** with the following parameters: (*i*) an initial reputation REP^0 initially set to 0.5, (*ii*) a pair of personal asymmetric cryptographic keys, (*iii*) the public key of the edge server and (*iv*) a certificate with expiration date and signed by e to witness device identity and reputation. During the *domain management*, **e** associates each SO with a cluster based on its trustworthiness. In particular, each **e**: (*i*) administers the reputation model; (*ii*) updates the SOs' reputation scores and certificates through the feedback of other SOs; (*iii*) forwards the updated certificates to **t** and a repository in the cloud. This repository is shared with the other edge agents and allows SO mobility between the edge domains.

The trust agent task consists of two phases: (*i*) *presentation* that carries out interactions with **t** associated with other SOs and sends information about their identities and reputation scores to its associated SO; (*ii*) *interaction* that manages the feedback and the certificate of its host SO by interacting with the trust agents associated with other SOs. In the SO task, each SO performs two phases: *Service Search* and *Ser-*

vice Provisioning. The former is executed when SO requests a service. In particular, SO requires to its **t** the certificates of SOs present in the same area to check their trustworthiness. Then it will choose SOs reflecting its individual *hazard threshold*[1] and which requires a resource r by sending them its descriptor d_r. The second phase starts once an SO_1 receives d_r of a resource r released by SO_2. If SO_2 satisfies the hazard threshold of SO_1, the tuple $\langle d_r, c_r \rangle$ is shipped to SO_2, where c_r is the cost that SO_1 is willing to pay for r. After agreeing the price together, SO_2 provides r to SO_1. At the end, the feedback about their counterparts are sent by both the SOs to **e**. The trust agents t_1 and t_2 receive by **e** the certificates with their updated reputation.

4 The Cluster Reputation Model

Let be $f^r_{1,2} \in [0, 1] \subset \mathbb{R}$ the feedback computed by SO_1 about SO_2 for the resource r, where 0/1 means the minimum/maximum appreciation that SO_1 has about SO_2 for their interaction (i.e., let be $f^r_{2,1}$ the feedback computed by SO_2 about SO_1, with $f^r_{2,1} \neq f^r_{1,2}$). Then t_1 (i.e., t_2) sends its feedback to the edge agent **e** in a secure way. **e** will update the SOs' reputation scores $REP \in [0, 1] \subset \mathbb{R}$ using the feedback of the trust agents t_1 and t_2. In this way, the reputation score of SO_1 (but analogously for SO_2) will be updated as follow:

$$REP_1 = \begin{cases} \beta \cdot {REP_1}^{old} + (1-\beta) \cdot \Theta^r_{1,2} & \Theta^r_{1,2} > 0 \vee {REP_1}^{old} \geq 0.5 \\ \\ REP^{old}_1 & \text{otherwise} \end{cases} \quad (1)$$

with $\Theta^r_{1,2} = \left(\alpha \cdot \Lambda_{2,1} + (1-\alpha) \cdot \Gamma_{2,1}\right) \cdot \Omega_2 \cdot {f_{2,1}}^r$ and where the parameter $\beta \in [0, 1] \subset \mathbb{R}$ weights the old reputation score ${REP_1}^{old}$ with the new contribution. REP is updated only if the condition $\Theta^r_{1,2} > 0 \vee REP^{old}_1 \geq 0.5$ is true in order to prevent the malicious activities. Finally, the edge agent **e** prepares and signs an updated certificate with the current reputation score (and identity) of SO_1 and sends it to the trust agent t_1. At the same time, the SO's reputation in the cloud repository is updated. To compute REP_1, **e** takes into account the reliability of SO_2 in providing honest feedback (i.e., its reputation score).

In particular, Ω is set to 1 if the SO's reputation is greater or equal than a threshold ϕ, where $\phi \in [0, 1] \subset \mathbb{R}$, or 0 otherwise. More formally, Ω is set to 1 if $REP \geq \phi$, while it is set to 0 otherwise and where Φ represents the relevance of r in term monetary cost c^r.

We suppose that the lower the cost, the lower the relevance and, accordingly, the effects on the reputation. In this way, we limit the benefits of malicious SOs in gaining a reputation for low relevant resources for then loosing it by cheating for high value resources. Also, the relevance of r is maximum when its cost is greater

[1] The hazard threshold is the probability of "failure" that the an entity accepts.

than a framework parameter c^r_{max} that indicates the maximum cost for r. In detail, Λ is equal to the ratio $c^r \;\; / c^r_{max}$ if $c^r < c^r_{max}$.

Finally, Γ is introduced to contrast the collusive behaviors where two or more SOs mutually and fictitiously increase their feedback with a high frequency, other wise it is set to 1. For this reason, Γ considers the time-frequency that a feedback was provided to the same SO (see the left part of Eq. 2).

$$\Gamma = \begin{cases} 1 & f^r < 0.5 \\ K(t_2)^{-1} & f^r \geq 0.5 \end{cases} \qquad K(t_2) = \begin{cases} 1 & t_1 = t_2 = 0 \\ K(t_l) + 1 & t_2 - t_l < T \\ max\left(1, K(t_l) - \left\lfloor \frac{t_2 - t_l}{T} \right\rfloor\right) & t_2 - t_l \geq T \end{cases} \quad (2)$$

When $f^r < 0.5$, the parameter Γ is set to 1, otherwise it is calculated as the inverse of K. When two SOs mutually interacting for the first time, the parameter $K \in [0, 1] \subset \mathbb{R}$ is set to 1 and then it is decreased/increased on the basis of the time elapsed between two consecutive feedback released by the same SOs and the time period T (see Sect. 5). We highlight that **e** can be always exploited to allow a correct representation of the malicious SO's reputation. Indeed, if a malicious behavior of SO is detected then its reputation is penalized as $REP_i = \sigma \cdot REP_i^{old}$, where $\sigma \in [0, 1[\subset \mathbb{R}$ is a system parameter.

Furthermore, the edge agent assigns SOs to clusters considering their reputation scores. The target of this process is to achieve a competitive environment to promote correct behaviors. To this aim, we assumed increasing benefits on the basis of the cluster affiliation, from one hand, in terms of resource costs for consumer SOs and, from the other hand, in terms of reliable payments for provider SOs. The number of clusters depends by the edge domain and, in each edge domain, it can further vary over time in order to be adaptive as more as possible to changes that could occur in that context. Each edge domain can set the clusters' number and the reputation score thresholds for the access to each cluster in an arbitrary way with the aim of maximizing the befits for honest SOs and maximizing the penalty for malicious SOs.

5 Experiments

Our framework has been tested in a simulated environment to verify its ability both to identify honest and malicious SOs and to cluster SOs based on their trustworthiness. We simulated one edge domain and two populations of SOs: 10,000 SOs acting as a consumer and 500 SOs playing the role of resource providers. Also, each SO's profile stores information about the: (*i*) honest or malicious SO's nature; (*ii*) type(s) of malicious behaviors of SO with the data characterizing each of such behaviors. The buying/selling of a resource will be possible only when the counterpart's reputation is at least equal to the own hazard threshold. Also, we assumed that the resource's price depends on the cluster to which the SO belongs (see Table 1).

Table 1 Clustering parameters of our scenario

Reputation range of the cluster
1—$REP \in [0.00; 0.45[$
2—$REP \in [0.45; 0.55[$
3—$REP \in [0.55; 0.85[$
4—$REP \in [0.85; 1.00]$

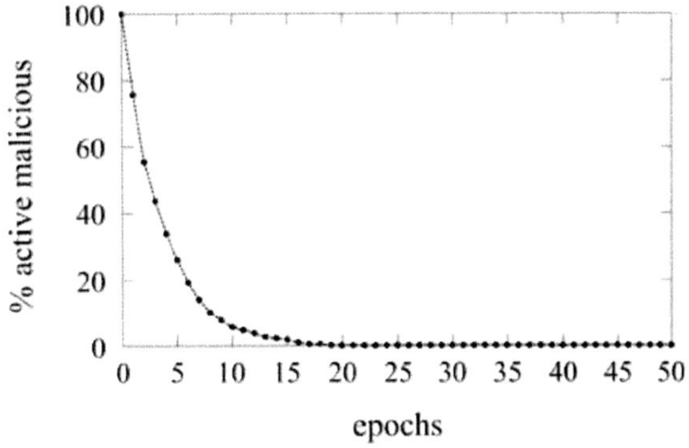

Fig. 1 Scenario A: Percentage of active malicious

Each experiment consisted of 50 epochs and for each epoch 2500 SOs purchase resources (i.e., the 25% of the consumer population) by interacting with all the providers. Two IoT scenarios have been simulated, called *A* and *B* (clustering parameters are shown in Table 1).

The scenarios are characterized by the presence of malicious SOs equal to 10% of the entire populations but they take on different behaviors in the two scenarios. In the scenario *A*, malicious actors trick and produce misleading feedback about their counterparts (e.g., 0 to reliable partners and, vice versa, 1 to unreliable partners), while in the scenario *B* malicious create positive reputations on low cost resources for tricking on high cost resources. $REP^0 = 0.5$ is the initial reputation value of all the SOs adopted to distinguish honest from cheaters. T of Eq. 2 is set to 60 min, the parameters β and α is both set to 0.5 and, c_r ranged in the domain [0.01; 1.00] $.

Our analysis was focused on malicious SOs to measure how many of them remain active based on their reputation scores (Figs. 1 and 2).

The results for scenario *A* (Fig. 1) show that about 10 epochs only less than ìthe 10% of malicious actors that are still active (i.e., SOs with $REP \geq 0.5$) in the simulation and their number decreases, as the number of simulated epochs increases, and characterize malicious SOs. The results of the scenario *B* (Fig. 2), show that the percentage of active malicious decreases a bit less rapidly than for the simulation *A* due to the alternate behaviors. As a consequence of these simulation we can argue as is resilient also in presence of both dynamic and alternate behaviors.

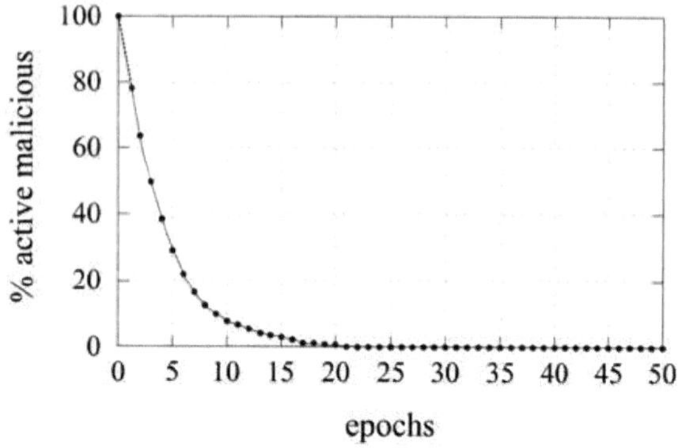

Fig. 2 Scenario B: Percentage of active malicious

6 Conclusions

In this paper, we proposed a reputation-based approach to mitigate the risks, for SOs, of interacting with unreliable partners in an IoT Edge-based environment. When two SOs interact to obtain a service, an edge server collects a feedback sent by each SO about its counterpart; starting from the latter, it calculate and update the reputation scores of the two SOs. The cloud is used as repository of the SOs' reputation scores; if a SO moves to another domain, our approach operates such that its reputation value will not be lost.

Furthermore, we designed a distributed framework based on trust where heterogeneous OSs exploit the assistance of associated tamper-proof software agents, called trust agents and implemented a clustering mechanism to enhance competitiveness in the framework. We highlighted this advantage by means of some experiments, aimed to verify the effectiveness of our approach to recognize honest by malicious SOs, to cluster IoT devices based on their trustworthiness. In our ongoing research we are introducing a further level of reputation in the cloud.

References

1. Andrade, E., Nogueira, B., de Farias Júnior, I., Araújo, D.: Performance and availability trade-offs in fog–cloud iot environments. J. Netw. Syst. Manag. **29**(1), 1–27 (2021)
2. Ashton, K.: That' internet of things' thing. rfid J. (2009). Accessed 22 June 2009
3. Bao, F., Chen, I.: Dynamic trust management for internet of things applications. In: Proceedings of the 2012 International Workshop on Self-aware Internet of Things, pp. 1–6. ACM (2012)
4. Chen, Min, Hao, Yixue, Lai, Chin-Feng., Di, Wu., Li, Yong, Hwang, Kai: Opportunistic task scheduling over co-located clouds in mobile environment. IEEE Trans. Ser. Comput. **11**(3), 549–561 (2016)
5. Chen, Ray, Bao, Fenye, Guo, Jia: Trust-based service management for social internet of things systems. IEEE Trans. Depend. Sec. Comput. **13**(6), 684–696 (2015)
6. Chen, Xu., Shi, Qian, Yang, Lei, Jie, Xu.: Thriftyedge: resource-efficient edge computing for intelligent iot applications. IEEE Netw. **32**(1), 61–65 (2018)

7. Comi, A., Fotia, L., Messina, F., Rosaci, D., Sarné, G.M.L.: Grouptrust: finding trust-based group structures in social communities. In: International Symposium on Intelligent and Distributed Computing, pp. 143–152. Springer (2016)
8. De Meo, P., Messina, F., Postorino, M.N., Rosaci, D., Sarné, G.M.L.: A reputation framework to share resources into iot-based environments. In: IEEE 14th International Conference on Networking, Sensing and Control, pp. 513–518. IEEE (2017)
9. Dumouchel, P.: Trust as an action. Eur. J. Sociol./Archives Européennes de Sociologie **46**(3) 417–428, (2005)
10. Fortino, G., Messina, F., Rosaci, D., Sarné, G.M.L.: Using trust and local reputation for group formation in the cloud of things. Future Gener. Comput. Syst. **89**, 804–815 (2018)
11. Fortino, G., Fotia, L., Messina, F., Rosaci, D., Sarné, G.M.L.: A meritocratic trust-based group formation in an iot environment for smart cities. Future Gener. Comput. Syst. **108**, 34–45 (2020)
12. Fortino, G., Fotia, L., Messina, F., Rosaci, D., Sarné, G.M.L.: Trust and reputation in the internet of things: state-of-the-art and research challenges. IEEE Access **8**, 60117–60125 (2020)
13. Fortino, G., Fotia, L., Messina, F., Rosaci, D., Sarné, G.M.L.: Using local trust measures to form agent cot groups 1. Intelligenza Artificiale **14**(1), 7–18 (2020)
14. Fortino, G., Messina, F., Rosaci, D., Sarné, G.M.L.: Using blockchain in a reputation-based model for grouping agents in the internet of things. IEEE Trans. Eng. Manag. **67**(4), 1231–1243 (2019)
15. Lingaraddi Hallappanavar, V., Birje, M.N.: A reliable trust computing mechanism in fog computing. Int. J. Cloud Appl. Comput. **11**(1), 1–20 (2021)
16. Lin, F., Zhou, Y., An, X., You, I., Raymond Choo, K-K.: Fair resource allocation in an intrusion-detection system for edge computing: ensuring the security of internet of things devices. IEEE Consum. Electron. Mag. **7**(6), 45–50 (2018)
17. Schooler, E.M., Zage, D., Sedayao, J., Moustafa, H., Brown, A., Ambrosin, M.: An architectural vision for a data-centric iot: rethinking things, trust and clouds. In: IEEE 37th International Conference on Distributed Computing Systems, pp. 1717–1728. IEEE (2017)
18. Vallati, Carlo, Mingozzi, Enzo, Tanganelli, Giacomo, Buonaccorsi, Novella, Valdambrini, Nicola, Zonidis, Nikolaos, Martínez, Belén, Mamelli, Alessandro, Sommacampagna, Davide, Anggorojati, Bayu, et al.: Betaas: a platform for development and execution of machine-to-machine applications in the internet of things. Wirel. Pers. Commun. **87**(3), 1071–1091 (2016)
19. Xu, J., Palanisamy, B., Ludwig, H., Wang., Q.: Zenith: utility-aware resource allocation for edge computing. In: 2017 IEEE international conference on edge computing (EDGE), pp. 47–54. IEEE (2017)

On Next-Generation Sequencing Compression via Multi-GPU

Pasquale De Luca, Annabella Di Mauro, and Stefano Fiscale

Abstract In the last decades, the human genoma analysis for addressing health-care problems, has widely grown. With the high throughput of biological data and, needing of represent them, the Next-Generation Sequencing was introduced. In order to store genomic features without losing information, different data format (such as FAST-A, FAST-Q, SAM, VCF) have been proposed. To overcome the storing process issues of these data, several genomic compressors have been presented. A specific VCF compressor is analyzed. Due to the restricted hardware resources limit of multi-core architecture when input size dimension data are given, large execution times are required. Thanks to the well-known parallel nature of the most recent Graphic Process Units, in this work we present a Multi-GPU based parallel implementation by exploiting CUDA framework. An ad-hoc memory approach combined with a suitable work decomposition strategy are able to give a strong increase in performance. To observe the benefits in terms of performance, tests and experiments complete our work.

1 Introduction

Nowadays, biological scientific field interest has widely grown. Strong genomic sequencing efforts lead us to introduce several computer science techniques for solving different problems. The increase in used memory and network traffic of

P. De Luca (✉) · S. Fiscale
Science and Technologies Department, University of Naples Parthenope, Centro Direzionale C4, 80043 Napoli, Italy
e-mail: deluca@ieee.org

S. Fiscale
e-mail: stefano.fiscale001@studenti.uniparthenope.it

A. Di Mauro
Department of Pathology, Istituto Nazionale Tumori, IRCCS-Fondazione "G. Pascale", 80131 Naples, Italy
e-mail: annabella.dimauro@istitutotumori.na.it

D. Camacho et al. (eds.), *Intelligent Distributed Computing XIV*, Studies in Computational Intelligence 1026, https://doi.org/10.1007/978-3-030-96627-0_42

new sequenced biological data has widely and deeply grown. Genomic projects such as Genome Aggregation Database and 1000 Genomes have contributed to a very large rise of databases and network traffic related to genomic data. The large-scale sequencing of samples of DNA has brought new attention by producing new works with related high interest. These kind of data need to be stored in a specific data format such as FAST-Q, FAST-A, SAM. Variant Call Format (VCF) contains meta-information lines, a header line, and then data lines each containing information about a position in the genome. The format also has the ability to contain genotype information on samples for each position. Nowadays, the constant growing of data dimension forces us to introduce specific genomic data compressor. In order to manage in simple and efficient way genomic data, several compressors are proposed in literature (i.e. Quip, DSRC, FQSqueezer). VCF Shark, is a novel genomic compressor framework for VCF files. Good compression rate results have been achieved from starting implementation [1]; while due to large input data dimension, the required execution time is very large. About this issue, High-Performance Computing architectures are mandatory for accelerating the main operation by reducing the execution times. More precisely, the Graphics Process Units (GPUs) was carried out to be very helpful for improving the performance, in time and reliability terms, of different research fields problems [2–9]. In this work, we propose a Multi-GPU software which exploits powerful of Compute Unified Device Architecture (CUDA) framework [10]. More precisely, ad-hoc parallel CUDA-kernels and efficient usage of parallel nature of the GPUs have been performed. Moreover, thanks to high programming flexibility of CUDA, an ad-hoc memory strategy combined with a specific work-load distribution operation are performed.

The paper is organized as follows: Sect. 2 gives some preliminary information about genomics by introducing the related file where are stored these kind of data. Section 3 describes the state of the art and focuses on some recent works in this field. Details about starting idea of VCF Shark are shown in Sect. 4. The Multi-GPU approach proposed, and the related ad-hoc work-load decomposition strategy, are discussed in Sect. 5. In order to confirm the efficiency and reliability of proposed implementation, Sect. 6 shows several experimental tests. Section 7 concludes the paper and proposes future works.

2 Genomics and VCF Format

In this section, we introduce the aims of genomics with related contributed given into medical and health-care fields combined with computer science techniques. Hence, we highlight the most recent genomic format used for representing biological data with related issues.

Genomics, with its technological innovations and authority as "big science," might be an optimal tool for solving different health-care problems. Despite the easy concept of building genomic libraries process, the genomic sequencing is a very hard challenge due to large kind of biological given information. Thanks to huge effort of most recent DNA sequencing methods, biomedical field offers different aims for solving hard health-care problems such as starting from common multi-factorial genetic disease to find new methodologies for building efficient medical treatments [11]. With the arrival of Next-Generation Sequencing (NGS), the interest for this scientific study field has exponential grown by introducing several challenges in numerous research areas [12–15]. Different formats have been introduced for representing in appreciable way the human genetic. VCF was introduced as a storing format standardized for rich annotations, structural variants and sequence variation as single nucleotide polymorphisms [17]. Composed of two distinct sections also called header and body, VCF files present in the first one the meta-information which describe annotations and tags; while the second one reports data of the samples. Starting with "##", meta-information lines highlight all the annotations, as data creation, software used, version of reference sequence and the related line number. This information could vary according to which information used in body section. Among there, both the first and last lines of this section are very important. While the first line starts with "##" and indicates the file format, the last one begins with single "#" and presents 8 different TAB separated fields: chromosome; position of variant start; identifier of the variant; reference allele; list of alternative non reference alleles comma separated; quality score; site filtering information; supplementary information.

3 Related Works

This section describes main techniques about VCF compression. In [18] the authors proposed *genozip*, a lossless compression tool for VCF files. The tool performs a main operation by splitting the input file into separate sections defined by data type and appropriately processing each section, before applying a general purpose data compressor, *bzip2*, to each section.

Deorowicz et al. in [19] try to answer the question how well a collection of genomes of the same species can be compressed by introducing the *Thousands Genomes Compressor (TGC)*. Their research goal concerns only genome collection compression, so TGC assumes that the input data are in Variant Call Format files restricted to their essential fields defined as VCFmin. Thus, the main idea of their algorithm is to transform the input data in a way to increase the locality and lengths of similar (identical) genome areas. In [20] the authors propose GTShark, a slightly parallelized tool able to shrink down the human genome to about 65 KB. A unique feature of GTShark is the ability to compress external, single-sample, VCF files with a use of the compressed collection as a reference. In terms of execution times, good results have been obtained by the aforementioned tools. Despite they are designed to fully leverage the hardware available, they all run on CPUs, and better performance could be obtained by providing a GPU based parallel implementation.

4 VCF Shark Details

Here we deal with a detailed description of the starting implementation of VCF Shark.

4.1 Streams

Each input file is computed in column order and, next is distributed in different streams. Moreover, we introduce the *fixed* streams whose are related to the first seven fields: CHROM, POS, ID, REF, ALT, QUAL, FILTER. The main property of these fields is closely to low number of contained information. According to previous data distribution, the INFO field could have different sub-fields whose are divided into different *info* streams, in recursively way. In order to offer a good support to VCF files composed by a non-uniform ploidy, an extended version of GTShark algorithm has been suitably performed.

4.2 Compression Schema and Computational Issues

Each stream is divided into chunk with related dimension equal to 8MB, while dimension set to 256 MB is fixed only for the genotypes chunks. In order to perform a good compression, BSC [21] compressor is used for compressing the starting header and the fixed streams. More in details, the usage of BSC is closely related to its high compression performance based on lossless and block-sorting algorithms [22]. A first exception encoding, so called *delta encoding* is applied for the *POS* stream before to execute the BSC compression. Hence, this kind of compression is applied also for the *info* and *format* stream if, for each chunk, their mean length is less than 64 bytes; the remaining related fields are called *wide* and are the main contribution about the large dimension of VCF file. Due to its nature, a pre-processing procedure is performed. Previous schema generate a pre-compression phase. Additional two compression operations are introduced:

1. in order to detect equal streams, a different sub-streams comparison is performed;
2. an identification of functional dependency among *info* and *format* streams is computed.

Hence, a multi-core implementation of the previous algorithm has been performed with appreciable performance results [1]. A large input sizes can be give worst performance in time terms. In second compression step previously discussed, the high number of operations can generate a large overhead due to huge dimension dataset. An accelerated VCF Shark implementation have to be mandatory.

5 Parallel Algorithm

Starting from interesting performance results obtained in [1], we observed for a huge dimension input size very large execution times are required.

5.1 *Work-Load Distribution*

Here, we deal with some details about the work-loading distribution approach used.

Let be:

$$G = \{g_1, g_2, \ldots, g_N\} \tag{1}$$

the ensemble of GPUs where N is the cardinality.

The local dimension D for each $g \in G$ is defined as follows:

$$D_{g_i} = \begin{cases} \lfloor \frac{M}{N} \rfloor & \text{if } \mathrm{mod}(M, N) = 0, \\ \lfloor \frac{M}{N} + 1 \rfloor & \text{otherwise,} \end{cases} \qquad i = 1, \ldots, N, \tag{2}$$

where M is the overall dimension of input dataset, and $\lfloor \cdot \rfloor$ is the floor operation. In order to perform a fair workload distribution, an additional sub decomposition has been applied. More precisely, we define a local chunk d related to columns distributed for each thread $j = \mathbf{0}, \mathbf{1}, \ldots, \tau - \mathbf{1}$ by following:

$$d_{g_i}^{(j)} = \begin{cases} \lfloor \frac{D_{g_i}}{\tau} \rfloor & \text{if } \mathrm{mod}(D_{g_i}, \tau) = 0, \\ \lfloor \frac{D_{g_i}}{\tau} + 1 \rfloor & \text{otherwise,} \end{cases} \tag{3}$$

where τ is the number of threads used for each CUDA block; $d_{g_i}^{(j)}$ is related to local dimension of thread j in $g \in G$. According to Eq. (3) a fair distribution of the work is performed. Moreover, each GPU works on the overall dataset in parallel and asynchronous way.

5.2 *GPU-parallel Implementation Details*

First we provide a specific memory strategy with relies onunify the global memory address of each GPU involved into the computation. An appreciable gain of performance is valued by merging several GPU global memory spaces. Hence, an Unified Virtual Addressing (UVA) technique has been performed [23]. One address space for all CPU and GPU memory is built. Previous equations combined with related issue to multicore version lead us to generate the following algorithm which provides the main operation computed on a single $g_i \in G$.

Algorithm 1 GPU parallel implementation

Input: input_data[], G **Output:** x_local[ ]

% set current GPU id = g_i, $i = 0, \ldots, N-1$
% STEP 0: Domain decomposition by following (2)
for each $g_i \in G$ **do**
  $d_{g_i}^{(j)}$:= (3) % build sub domain decomposition
  `index := threadIdx.x+(blockDim.x × blockIdx.x)` % define the thread's index
  % define the local chunk interval
  `chunk_idx :=` `(index` × $d_{g_i}^{(j)}$`) + ((index+1)` × $d_{g_i}^{(j)}$`)`
  % start parallel region
  for each thread **do**
    % STEP 1
    x_local`[index]:= input_data[chunk_idx + length]`
  end for
  % STEP 2: Performing compression schema
  for each thread ≤ chunk_id **do**
    x_local`[index]`$_{L2}$:= `#CHROM field into L2 cache`
    % % Dynamic library linking for run-time compression by means of GTShark
  end for
  % STEP 3: Compression of genotypes
  `len = x_local.length()`
  for each thread on **len do**
    % Move genotypes into global memory
    % Link dynamic library of BSC Compressor
    % Perform compression
  end for
  % end parallel region
end for

Algorithm 1 performs in STEP 0 the more important process of parallel algorithm which achieves a good performance. According to domain decomposition schema (3) a specific workload distribution has been performed. More precisely, the domain decomposition process provides to divide the starting input dataset into different chunk. The entire dataset is run-time loaded during the computation in order to avoid any overhead contention. Next, we combine a classical domain decomposition schema based on fine grain with a coarse grained parallelism. Let us:

$$S = \{s_0, s_1, \ldots, s_{k-1}\} \tag{4}$$

the ensemble of Symmetric Multiprocessors (SMP) where k_i is the number of SMP for each $g_i \in G$, $(i = 1, \ldots, N)$, which for every one a local L2 cache is present.

STEP 1 relies on loading of each stream into along $s_k \in S$. Thanks to redundant nature of genomic data, the starting streams have been loaded into local L2 cache. The property of L2 cache memory allows us to reduce the reading time. Here the fine grain parallelism is applied. While the starting dimension of genotypes is increased from 256 MB to 2048 MB by building a coarse grain parallelism. This strategy is very helpful since a parallel coarse grain is computed. We chose to move the data to global memory to achieve better access time. This idea is based on minor execution time observed on global memory with respect to shared memory. The mandatory synchronization operations in shared memory spent more time than of global memory computation. STEP 2 provides to main compression operation: here we execute a compression of initial fields with GTShark directly on GPU. This process reduces the execution time thanks to low access time into L2 cache memory. STEP 3 is related to genotypes compression computation which, as aforementioned, related block dimension is increased in order to distribute more data among each GPU thread. More in details, thanks to peculiarity of UNIX operating system we underline the run-time usage of BSC compressor by means of dynamic linking library. Hence, the GPU computation is performed in asynchronous way. According to the advanced-based adopted memory strategy, the CPU did not wait to GPU finishing work. At the end, the final output result is achieved. We highlight the parallel execution of Algorithm 1 on multiple GPUs concurrently. More in details, this procedure is computed on different GPUs performing related operations on different chunks of `input_data` dataset. Furthermore, two different of load balancing work load have been exploited. In a first part of computation a fine grain parallelism has been applied thanks to managing data into L2 cache. Hence, a coarse grain parallelism is performed on genotypes stream.

6 Results

This section deals with show some experimental tests in order to highlight the gain of performance achieved by our parallel implementation. Compression ratio and performance analysis experiments will be shown.

Tests are executed on a computer machine, so called MARCONI-100 of CINECA [24], with following hardware specifications: 980 Nodes; processors: 2 × 16 cores IBM POWER9 AC922 at 3.1 GHz, accelerators: 4 × NVIDIA Volta V100 GPUs, 5120 CUDA cores for each one, cores: 32 cores/node and RAM: 256 GB/node. In order to manage and distribute processes PBS scheduler has been used. Therefore, in order to increase the performance an additional memory-based operation has been employed presented in [25]. With respect to standard version, applying this approach a significant increasing of performance is obtained by exploiting the memory cache's property that can recover the most used data and instructions during the execution [25]. The read and write operation among GPUs is performed by the CUDA routine: `cudaDeviceEnablePeerAccess` combined with `cudaMemCpy` according to UVA memory strategy adopted. Each next shown result has been achieved by means of an average of 10 execution tests.

Data Compression Performance

Here we try to confirm about reliability of proposed implementation. We highlight the good rate of compression achieved due to potential error source could be introduced (i.e. round-off error):

Table 1 highlights the compression rate achieved by GPU-parallel implementation with respect to different canonical genomic compressor by varying the input

Table 1 Compression rate comparison, GPU version verses canonical compressors

		Compression rate		
Dataset	Size (GB)	7z	Genozip	VCF Shark GPU
Rice	27.80	21.15	80.35	81.39
ExAC	40.26	13.98	24.81	46.91
gnomADv2	63.19	28.93	50.90	66.54
gnomADv3.1	179.80	26.63	59.14	88.19
SGDP	180.56	8.52	13.54	15.92
1000GPp3	193.08	176.79	184.73	522.65

dataset ($N = 1$, $\tau = 64$). We denotes the low compression ratio obtained by *7z* so that is a generic compressor. We underline the good obtained reliability of GPU VCF Shark despite the advanced data distribution could generate losing of data. In case of the dataset 1000GPp3, when a large number of genotypes are present, our proposed implementation reaches excellent performance compare to other generic compressors. In addiction, we omitted the compression rate of the VCF Shark's CPU version because is the same of the obtained GPU results.

Time performance analysis. Here we deal with give a confirm about gain of performance achieved by our Multi-GPU parallel implementation.

Figure 1 denotes an appreciable improving of performance in terms of execution times. As comparison among different genomic compressors is shown. We observe the gain achieved by our Multi-GPU implementation. In particular, for each selected dataset, the GPU times are always lower than other multi-threads based compressors. A simple basic Multi-GPU configuration is setup, for this experimental test. More precisely, we used four GPUs with a low threads number set to 128. Moreover, the high dimension of input dataset allows us to apply the decomposition strategy (3) in

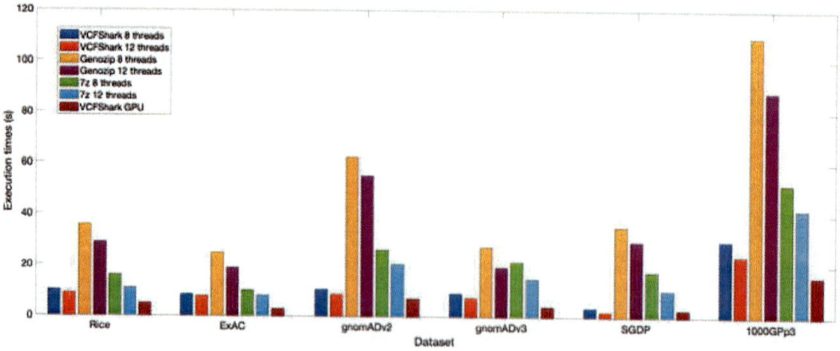

Fig. 1 Execution times for different dataset by varying the used genomic compressors combined with several threads configurations. 128 threads are fixed for GPU computations. $N = 4$, number of GPUs

Table 2 Times comparison by varying CUDA threads configuration ($\beta \times \tau$), $\beta = 1$

		GPU times (s)					
Dataset	Size (GB)	$\tau = 64$	$\tau = 128$	$\tau = 256$	$\tau = 512$	$\tau = 1024$	$\tau = 2048$
Rice	27.8	**2.71**	5.01	4.87	3.88	5.92	6.82
ExAC	40.26	4.85	3.09	2.83	**2.32**	3.91	4.34
gnomADv2	63.19	**4.99**	7.09	6.34	6.01	7.89	9.15
gnomADv3.1	179.80	**1.98**	4.14	3.52	3.12	5.03	6.81
SGDP	180.56	**0.74**	2.91	2.58	1.97	4.18	5.09
1000GPp3	193.08	**11.98**	15.94	14.05	13.76	17.94	21.83

order to compute a fair workload distribution among each GPU thread. Hence, we introduce the next test which deals with the optimal CUDA configuration.

Table 2 exhibits the optimal CUDA configuration in order to address next test setup. More precisely, here we find the best times achieved by a specific CUDA configuration (`block × threads`). We observe that, for the major number of executed tests, the 1×64 configuration relies on candidate to the best. Only for ExAc dataset the results are distinct due to particular chaotic nature of contained data.

7 Conclusions

In this work, a Multi-GPU based VCF Shark implementation, a novel genomic compression method, has been presented. A deep description of main issues related to this compressor has been addressed by highlighting the need of High-Performance Computing methods. A detailed parallel implementation by using GPUs has been performed. Experimental tests confirm the appreciable gain of performance achieved by our parallel implementation. Future works could see the introduction of a heterogeneous architecture, by expecting a strong increasing of performance.

Acknowledgements This paper has been supported for computational resources by project "Accelerated High Performance Methods for compressing Next-Generation sequencing data (AHNG20)" - CINECA.

References

1. Deorowicz, S., Danek, A.: VCF Shark: how to squeeze a VCF file. bioRxiv (2020)
2. De Luca, P., Formisano, A.: Haptic data accelerated prediction via multicore implementation. In: Science and Information Conference, pp. 110–121. Springer, Cham (2020)
3. Marcellino, L. et al.: Using GPGPU accelerated interpolation algorithms for marine bathymetry processing with on-premises and cloud based computational resources. In: International Conference on Parallel Processing and Applied Mathematics. Springer, Cham (2017)

4. De Luca, P., et al.: Accelerated Gaussian convolution in a data assimilation scenario. In: International Conference on Computational Science. Springer, Cham (2020)
5. De Luca, P., et al.: Recursive filter based GPU algorithms in a Data Assimilation scenario. J. Comput. Sci. **53**, 101339 (2021)
6. De Luca, P., et al.: Performance analysis of a multicore implementation for solving a two-dimensional inverse anomalous diffusion problem. In: International Conference on Numerical Computations: Theory and Algorithms. Springer, Cham (2019)
7. Fiscale, S., et al.: A GPU Algorithm for Outliers Detection in TESS Light Curves. In: International Conference on Computational Science. Springer, Cham (2021)
8. De Luca, P., Galletti, A., Marcellino, L.: Parallel solvers comparison for an inverse problem in fractional calculus. In: *2020 Proceeding of 9th International Conference on Theory and Practice in Modern Computing (TPMC 2020)* (2020, July)
9. Amich, M., et al.: Accelerated implementation of FQSqueezer novel genomic compression method. In: 2020 19th International Symposium on Parallel and Distributed Computing (ISPDC). pp. 158–163. IEEE (2020)
10. https://developer.nvidia.com/cuda-zone
11. Levy, S., et al.: PLoS Biol. **5**(10), e254
12. Shendure, J., et al.: Nat. Biotechnol. **26**(10), 1135–1145 (2008)
13. Ansorge, W.J.: New Biotechnol. **25**(4), 195–203 (2009)
14. De Luca, P., et al.: Distributed genomic compression in MapReduce paradigm. In: International Conference on Internet and Distributed Computing Systems. Springer, Cham (2019)
15. De Luca, P., Galletti, A., Ghehsareh, H.R., Marcellino, L., Raei, M.: A gpu-cuda framework for solving a two-dimensional inverse anomalous diffusion problem. In: Foster, I., Joubert, G.R., Kučera, L., Nagel, W.E., Peters, F. (eds) Parallel Computing: Technology Trends, Advances in Parallel Computing, vol 36, pp. 311–320. IOS Press (2020)
16. Zhang, X., et al.: Cancer Biol. Med. **16**(1), 189
17. Danecek, Petr, et al.: The variant call format and VCFtools. Bioinformatics **27**(15), 2156–2158 (2011)
18. Lan, D., et al.: Bioinformatics (Oxford, England) **36**(13), 4091–4092
19. Deorowicz, S., et al.: Genome compression: a novel approach for large collections. Bioinformatics **29**(20), 2572–2578 (2013)
20. Deorowicz, S., et al.: Bioinformatics **35**(22), 4791–4793 (2019)
21. Chandak, S., et al.: SPRING: a next-generation compressor for FASTQ data. Bioinformatics **35**(15), 2674–2676
22. Antonini, A., et al.: Low complexity algorithms for transmission of short blocks over the BSC with full feedback. In: (ISIT). pp. 2173–2178. IEEE
23. De Luca, P., Galletti, A., Marcellino, L.: A novel gpu implementation for image stripe noise removal. In: Intelligent Computing, pp. 232–243. Springer, Cham (2022)
24. https://www.hpc.cineca.it/hardware/marconi100 Accessed 30 June 30 2021
25. De Luca, P., Galletti, A., Marcellino, L.: A Gaussian recursive filter parallel implementation with overlapping. In: 2019 15th International Conference on Signal-Image Technology & Internet-Based Systems (SITIS), pp. 641–648. IEEE (2019)

Experimental Approach of the Quantum Volume on Different Quantum Computing Devices

Ilias P. Galanis, Ilias K. Savvas, and Georgia Garani

Abstract Quantum Computing is one of the most promising technology advancements of our time, promising to clarify problems considered unsolved for a classical computer. Real Quantum Computer Devices—once a science fiction concept—are now a reality. Many challenges still remain, on the way to achieve the so-called quantum supremacy. A phenomenon called Quantum Decoherence is the main cause of noise, making the available Quantum Computer Devices now limited to a few dozens of qubits at present. Despite that, most of the IT giants as IBM, Google and Rigetti are investing large amounts of money on deferent technological approaches to implement such devices and giving free access through the cloud not only to researchers but to common users also. So, it is important to evaluate and compare Quantum Computers of different philosophy, a process far more difficult than the equivalent for a classical computer. Quantum Volume represents a metric which measures the Quantum Computing Devices' reliability in terms of error rates. In this paper, quantum circuits implementing Hadamard and CNOT gates were examined on three dif-ferent IBM Quantum Computer Devices in order to experimentally investigate Quantum Volume and attempt to provide a meaning of this metric. Data of these experiments are presented and additionally, different metrics are used to compare the performance of the Quantum Computer Devices.

I. P. Galanis (✉) · I. K. Savvas · G. Garani
Department of Digital Systems, University of Thessaly, Gr Gaiopolis Campus, Larissa-Trikala Ring-Road, 415 00 Larissa, Greece
e-mail: ilgalanis@uth.gr

I. K. Savvas
e-mail: isavvas@uth.gr

G. Garani
e-mail: garani@uth.gr

D. Camacho et al. (eds.), *Intelligent Distributed Computing XIV*, Studies in Computational Intelligence 1026, https://doi.org/10.1007/978-3-030-96627-0_43

1 Introduction

Quantum computation and quantum information is the study of the information processing tasks that can be accomplished using quantum mechanical systems [15]. Quantum computing takes advantage of quantum phenomena such as super-position and entanglement. A device (quantum computer) capable of doing that, proposed in 1982 by Richard Feynman [9]. Quantum computing is expected to solve problems that are far from Classical Computers capabilities and help the development and research in areas such as chemistry, pharma-cology, machine learning and finance. As expected, such a promising tech-nology sparked great interest to some of the biggest IT companies. As a result, many of them have developed real Quantum Computer Devices (QCDs) and provide online access to not only to researchers but average users also.

Along with the QCDs, the essential platforms and APIs were developed. This software includes Python libraries such as IBM's Qiskit [2], Google's Cirq [1] and Rigetti's Pyquil [23], new programming languages, development kits (Microsoft's Q# and Azure Quantum) [6], fully managed quantum computing services (Amazon Braket) [4] and so on. While in classical computers information is stored in bits, on the contrary QCDs information is stored in qubits. A qubit can store a value of either 0 or 1 like in classic bit. What makes a qubit more powerful is that by taking advantage of the property of superposi-tion it can store 0 and 1 simultaneously. Quantum circuits and quantum com-puters consist of qubits and quantum gates. This can be considered as an ab-stract linear-algebraic generalization of a classical circuit. Physically a qubit can be implemented as a single atom, a single photon or a single electron. Al-gebraically the qubit can be represented by the ket notation (typically a column vector) [22] (Eq. 1).

$$|0\rangle = \begin{pmatrix} 1 \\ 0 \end{pmatrix}, |1\rangle = \begin{pmatrix} 0 \\ 1 \end{pmatrix}, |x\rangle = a\,|0\rangle + b\,|1\rangle\,, a^2 + b^2 = 1 \tag{1}$$

Quantum gates manipulate the state of the qubit and are represented as a matrix. An example of this manipulation can be found in Eqs. 2, and 3 (Hadamard gate, matrix representation: $H = \frac{1}{\sqrt{2}}\begin{pmatrix} 1 & 1 \\ 1 & -1 \end{pmatrix}$).

$$H\,|0\rangle = \frac{1}{\sqrt{2}}\begin{pmatrix} 1 & 1 \\ 1 & -1 \end{pmatrix}\begin{pmatrix} 1 \\ 0 \end{pmatrix} = \frac{1}{\sqrt{2}}\begin{pmatrix} 1 \\ 1 \end{pmatrix} = \frac{1}{\sqrt{2}}\begin{pmatrix} 1 \\ 0 \end{pmatrix} + \frac{1}{\sqrt{2}}\begin{pmatrix} 0 \\ 1 \end{pmatrix} = \frac{1}{\sqrt{2}}\,|0\rangle + \frac{1}{\sqrt{2}}\,|1\rangle \tag{2}$$

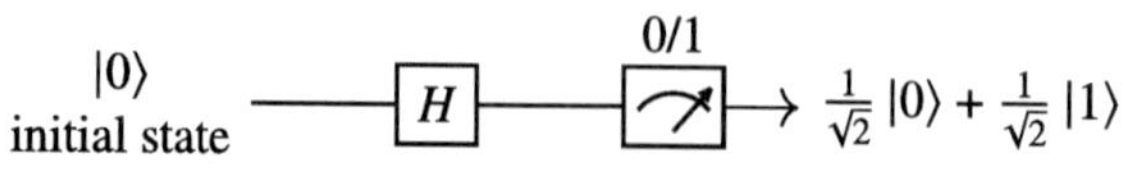

(3)

Quantum algorithms have been proposed to solve computational problems which remained unsolved by a classical computer through improving dramatically their time complexity. Such problems are integer factorization [21] (current RSA encryption relies on this fact), search in big databases [12], drugs development and so on.

All the above seem quite promising, but many issues arise on the effort to construct ideal QCDs. The QCDs now in use are not ideal and need a lot of technological advancement on hardware to be considered as efficient. An effect called Quantum Decoherence seems like an insurmountable obstacle for present. When this phenomenon occurs, information from a quantum system is lost towards the environment over time. Practically speaking, when a qubit is manipulated into a state $|1\rangle$, over time, it will return again into $|0\rangle$. One could think that maybe a total isolation of this system would solve this problem, but it is very difficult—maybe impossible—to isolate a quantum system from its external environment. Such an isolation would make it impossible our interaction with the system. Technological advancements made NISQ (Noisy Intermediate Scale Quantum) Computers with more than 50 qubits a reality.

Benchmarking and evaluating NISQ QCDs is an important step towards ef-forts on constructing more reliable QCDs and mitigate the errors occurring on these devices. In this work, a benchmarking metric, Quantum Volume (QV) is investigated [5]. Investigation—through experiments—focuses on what this metric, QV, means reliability-wise for the average user by running selected quantum circuits above the QV proposed by the manufacturer on three different IBM QCDs. For this purpose, experiments quantum circuits with Hada-mard and CNOT gates are implemented. The impact of the traspilation process (some QCDs don't have the capability to implement all gates) and the optimi-zation level of the device on QV is expected to have a great research interest. The code of the experiments can be found in [10].

The rest of the paper is organized as follows: In Sect. 2 the need for benchmarking of QCDs is being presented with and some main evaluation techniques are discussed. Section 3 is dedicated to a brief presentation of the benchmarking metric QV. Section 4 contains is the main part of this paper where the experimental data are presented. In Sect. 5 results of these experiments are discussed, while conclusions and future work suggestions are given in Sect. 6.

2 Benchmarking Quantum Computing Devices

NISQ quantum computers are far from considered reliable. Most of the IT giants develop quantum computers that differ from each other both on their hardware and their overall philosophy. There are quantum computers that use atoms, photons, or electrons. Numerous implementations of the physical real-ization of the quantum computers are presented, with supercondacting quan-tum computing and trapped ions be the most popular ones [25]. Implementa-tion and architecture (direct connection between qubits) in the QPU (quantum processing unit) are different between every quantum device in an attempt to find the optimal architecture for each a particular problem. So, a need arises for the benchmarking and the evaluation of a given QCD. In classical computers, common ways of evaluation are the measuring of operations per second (OPS) or floating-point operations per seconds (FLOPS). In real world applications however, these metrics may not represent the actual performance of a classical computer [19].

On the other hand, the main issue keeping QCDs in experimental state at present, is their reliability. Single qubits must be evaluated since there is high uncertainty of whether or not they function properly. Certainly, that is not necessary for a classical bit—it is expected that it will function perfectly. Another important question about QCDs concerns the size of quantum circuit (qubits and gates) can use a particular quantum computer for performing successfully. One step towards the evaluation of a quantum computer is the certification task. Certifications aims to establish whether a quantum device is functioning correctly or not. Certification schemes can be classified according to the effort required, the information gained, and the assumptions made about the device and its user.

The other aspect is benchmarking. **Randomized benchmarking** is a method for assessing the capabilities of quantum computing hardware platforms through estimating the average error rates that are measured under the implementation of long sequences of random quantum gate operations [26]. Randomized benchmarking (RB) refers to a collection of methods that aim to reliably estimate the magnitude of an average error of a quantum gate set in robust fashion against state preparation and measurement (SPAM) errors [8]. Briefly, the steps—logic behind RB [3] are:

1. Select l cliford gates from an n-qubit group randomly and calculate the in-version gate
2. Starting in $|0\rangle$ measure the polarization of each qubit after the application of the circuit
3. Average over many different random sequences (seeds)
4. Fit to $A * a^n + B$ (error is related to a).

Other methods of benchmarking are various **quantum tomography** methods. In quantum state tomography (QST), information about the state of the qubit is gained by performing multiple tomographic measurements to the system. Once many projective measurements are made, each probing a partic-ular aspect of the possible state, the complete quantum state is built up through a tomographic process [24]. QST and other relative techniques are of-ten small-scale and involve well characterized measurement devices probing a relatively less understood target device. The complexity of a technique will be less important and, while some assumptions regarding the measurement apparatus may be reasonable, they should be avoided as much as possible re-garding the device to be certified [8].

3 Quantum Volume

Despite all the evaluation works that had been carried out, i.e. the bench-marking techniques mentioned above, two critical questions remain: i) is there a way to benchmark deferent quantum computers (architecture-wise), ii) should QCD ceases to be considered an option when quantum decoherence makes it practically unusable? These two questions are addressed by a single number metric, the Quantum Volume (QV). The QV protocol is strongly linked to gate error rates and is influenced by underlying qubit connectivity and gate parallelism. Thus, it can be improved by moving toward the limit in which large numbers of well controlled, highly coherent, connected, and ge-nerically programmable qubits are manipulated within a state-of-the-art cir-cuit rewriting toolchain. High-fidelity state preparation and readout are also necessary [7]. QV tests the ability of the quantum device to implement SU(4) gates [27]. Two issues are taken into consideration: N the number of qubits and d the depth of the circuit (for this protocol the amount of continuous SU(4) gates, which may consist of a combination of one-qubit and two-qubit gates on the actual hardware). It is desirable to exclude extreme cases as such the smallest error rates and therefore the largest circuit depth will result from very few qubits; also the other extreme, where a device has many qubits but little co-herence, i.e. $d \approx 1$, resulting to uninteresting information about device [14]. Therefore, QV is defined as in Eq. 4.

$$V_Q = \tilde{min}(N, d(N))^2 \tag{4}$$

More clearly, e.g. if a quantum device is benchmarked as QV = 8, it means that it can properly function given a 3×3 quantum circuit ($2^3 = 8$). In order to determine the QV, the "heavy output generation" problem is used. The logic behind this goes as follows: Every QCD is expected to output a cor-rect bit string with an associated probability that can be measured. To achieve that, the quantum circuit must be ran on a simulator given that the execution would be ideal. A median of these probabilities can be calculated and the strings whose probability is higher than the median are the heavy outputs. When run on a real quantum device, the quantum circuit will output bit strings. The heavy output on these bit strings is then measured. If the probability of measuring any of the heavy outputs on the device is greater than two-thirds with a confidence interval greater than 97.725%, then the benchmark test is completed successfully and it can be proceeded to the next step, simply increasing by one both length and width of the circuit. When the benchmarking protocol fails, then the QV of the device has been determined.

4 Experimental Investigation of Quantum Volume

The goal of this research is the relevance and reliability of the QV in average usage conditions, where usually no calibration of the devices occurs and there is a probable ignorance of the transpilation process for quantum circuits close and above the depth (gates in a row) and width (number of qubits) suggested by the QV of the manufacturer. The QCDs can implement particular quantum gates. As a result, every quantum circuit is transpiled to an equal circuit that consists of these gates. A first concern for the selection of the circuits that will be used for the needs of the experiment was is using native Qiskit functions that return random quantum circuits with custom width and depth, though there is not an option of selecting particular gates for these circuits without altering the source code of this method at present. During the preliminary work, another approach is using only NOT and CNOT gates (all the quantum devices used for the experiments can implement these gates natively), since this would give us a deterministic system. Quantum physics and so quantum computers are not deterministic in their nature, therefore choosing these gates would be unfair for the purposes of this research. Finally, the idea behind choosing the quantum circuits was using Hadamard (frequently used gates in numerous quantum circuits and algorithms) and CNOT gates. Hadamard gate is transpiled to three quantum gates, while the CNOT gate is native to the system. So, all the circuits tested are of at least $3 + 1$ (measure gate) depth. Below, the experimental results of such circuits will be presented in detail, running on the following IBMQ QCDs:

1. ibmq_5_yorktown which is one of the Canary R1 processors (Optimized 2D lattice: all the qubits and readout resonators are on the same layer) with a QV of 8 (meaning it meets the requirements for a 3×3 quantum circuit),
2. ibmq_belem (QV of 16, 4×4 circuit), and
3. ibmq_lima (QV of 8) from the Falcon r4T processors (Multi-chip stack: the qubit chip is bump-bonded to an "interposer" chip that provides readout and signal delivery wiring) [13].

In Qiskit, the optimization level varies from 1 to 3 ($1 =$ lowest). Level 1 optimization uses light optimization by simple adjacent gate collapsing [16], while level 2 uses medium optimization by initial layout selection and gate cancelation using commutativity rules [17], and finally, level 3 uses heavy optimization by noise adaptive qubit mapping and gate cancellation using commutativity rules and unitary synthesis [18]. Each circuit was run 1024 times for each experiment. The following evaluation units are considered:

1. Metric I: The difference between the odds for the correct result when running on the simulator compared to when running on real devices (simulator probabilities—real device probabilities),
2. Noise: The noise levels on the real device (total probability for wrong results),
3. Metric II: The largest probability for a particular wrong bit string to appear in the results,
4. Metric III: The difference between the smallest probability for a correct bit string to appear minus the largest probability for a wrong bit string to appear,
5. How many times bigger is the lowest probability of the correct bit string to appear in the results compared to the bigger probability of the wrong bit string.

Bellow, a snapshot of the experiments is presented with detailed data. The snapshot refers to the results of the size 5×5 circuit when run on IBMQ_5_Yorktown QCD (Table 1). Comparison between QCDs using the metric IV can be found in Table 2. The traspilation impact on the quantum circuits' depth can be found in Table 3.

Table 1 BMQ_5_Yorktown 5 × 5 circuit results

bit-sequence	Probability on simulator	Probability on device	Metric I	Metric II	Metric III	Metric IV
Optimization level = 1						
01100	0,249	0,172	0,077	0,37	0,100	3,703
01111	0,255	0,137	0,118	0,37	0,100	3,703
10000	0,249	0,171	0,078	0,37	0,100	3,703
10011	0,247	0,177	0,070	0,37	0,100	3,703
Noise	0,000	0,343	–	–	–	–
Optimization level = 2						
01100	0,249	0,165	0,084	0,47	0,087	2,851
01111	0,255	0,144	0,111	0,47	0,087	2,851
10000	0,249	0,193	0,056	0,47	0,087	2,851
10011	0,247	0,134	0,113	0,47	0,087	2,851
Noise	0,000	0,364	–	–	–	–
Optimization level = 3						
01100	0,249	0,191	0,058	0,37	0,041	3,463
01111	0,255	0,142	0,113	0,37	0,041	3,463
10000	0,249	0,174	0,075	0,37	0,041	3,463
10011	0,247	0,151	0,096	0,37	0,041	3,463
Noise	0,000	0,342	–	–	–	–

5 Discussion

It can be assumed that if the metric suggested (IV) is ≥ 2 then it is quite clear what is noise in the results and what is not. Something also interesting is that despite the fact that the optimization level was indeed lowered the noise levels, the metric IV decreased in relation to lower optimization levels. That means that the probability of a particular wrong bit string in the results was getting bigger, making it more difficult to determine what is a correct bit string and what is noise. For 4 × 5 and 5 × 4 circuits the QCDs scored similar results with IBMQ_lima QCD been the best. As expected, the QCD IBMQ_Belem which is of QV = 16 compared to the other two QCDs (QV = 8) scored better for Quantum circuits of size 5 × 5 and 5 × 6, showing its edge over the other two QCDS tested. Even in most of such cases, the metric IV was consistently ≥ 2 (Table 2). Part of this is due to the fact that the traspilation process kept the depth of the circuits to manageable levels (Table 3). For the quantum circuit of size 5 × 6 CCNOT (one CCNOT gate was used) all three QCDs failed to score ≥ 2 on metric IV. The implementation of such gates increased the depth of the transpiled circuits from 5 × 6 to a range of 5 × 22 to 5 × 30 depending on the QCD. This level of depth made it difficult to draw safe conclusions about the results. Overall, the QV of a device represents its capability to implement and successfully run quantum

Table 2 Comparison between quantum devices using metric IV for various circuit sizes

Circuit size	Quantum device	Metric IV		
		Opt. level = 1	Opt. level = 2	Opt. level = 3
4 × 5	IBMQ_5_Yorktown	3,544	3,384	3,764
	IBMQ_Belem 5	5,468	7,929	4,000
	IBMQ_Lima	8,356	17,458	11,333
5 × 4	IBMQ_5_Yorktown	3,947	3,512	2,780
	IBMQ_Belem 5	4,742	5,355	3,921
	IBMQ_Lima	4,946	5,371	4,395
5 × 5	IBMQ_5_Yorktown	3,703	2,851	3,463
	IBMQ_Belem 5	4,676	3,349	5,059
	IBMQ_Lima	3,976	5,294	4,774
5 × 6	IBMQ_5_Yorktown	2,703	2,632	2,264
	IBMQ_Belem 5	3,514	2,115	3,432
	IBMQ_Lima	2,833	2,406	2,981
5 × 6 ccnot	IBMQ_5_Yorktown	0,786	0,957	1,127
	IBMQ_Belem 5	1,592	1,167	0,928
	IBMQ_Lima	1,325	1,175	1,157

circuits with Hadamard and CNOT gates, giving back meaningful results even for larger circuits both on width and depth and that is clear for an average user with little knowledge about the traspilation process with the exception of the Quantum circuits having a CCNOT gate.

6 Conclusion

Overall, using the metric IV, an indication is given on how much the results of a real QCD can be considered as reliable, based on QV, without need to take extra actions. It seems that even in cases where the circuits are slightly larger than the defined QV, one can draw safe conclusions about their reliability. Only in cases where the traspilation process highly increases the depth of the circuit, the number of the incorrect results increase also, making the estimation of the correct bit-sequences difficult to estimate. The increase of optimization levels, though improved the metric IV, most of the time did not alter the assumptions made prior to this action. In conclusion, it seems that NISQ quantum computers when dealing with circuits even above the QV proposed by the provider—given the fact that their depth remains reasonable after the traspilation—can be considered reliable and an unexperienced user can safely conclude which are the correct bit sequences. The observations made in this research strongly reinforce the observations made on prior work [20] and [11].

Table 3 The effect of the traspilation process on quantum circuits' size

Circuit size before transpile	Quantum device	Circuit size after transpile	Circuit size after transpile	Circuit size after transpile
		Opt. level = 1	Opt. level = 2	Opt. level = 3
4 × 5	IBMQ_5_Yorktown	4 × 7	4 × 7	4 × 7
	IBMQ_Belem 5	4 × 7	4 × 7	4 × 7
	IBMQ_Lima	4 × 7	4 × 7	4 × 7
5 × 4	IBMQ_5_Yorktown	5 × 6	5 × 6	5 × 6
	IBMQ_Belem 5	5 × 6	5 × 6	5 × 6
	IBMQ_Lima	5 × 6	5 × 6	5 × 6
5 × 5	IBMQ_5_Yorktown	5 × 7	5 × 7	5 × 7
	IBMQ_Belem 5	5 × 7	5 × 7	5 × 7
	IBMQ_Lima	5 × 7	5 × 7	5 × 7
5 × 6	IBMQ_5_Yorktown	5 × 9	5 × 8	5 × 8
	IBMQ_Belem 5	5 × 11	5 × 10	5 × 10
	IBMQ_Lima	5 × 11	5 × 11	5 × 13
5 × 6 ccnot	IBMQ_5_Yorktown	5 × 22	5 × 27	5 × 42
	IBMQ_Belem 5	5 × 26	5 × 29	5 × 27
	IBMQ_Lima	5 × 30	5 × 22	5 × 37

Future work direction will be two-fold. Firstly, machine learning algorithms will be examined in order to find connections and patterns between the fidelity of the quantum gates when implemented on QCDs related to their architecture and physical realization. Another direction is to implement the same quantum circuits used in this research on various QCDs from different providers like Rigetti, Google and Microsoft, etc. provided that access to them will be available.

References

1. C. developers: cirq. zenodo (2021). https://doi.org/10.5281/zenodo.5138274
2. Qiskit: An open-source framework for quantum computing (2021). https://doi.org/10.5281/zenodo.2573505
3. Randomized benchmarking: https://community.qiskit.org/textbook/ch-quantum-hardware/randomized-benchmarking.html (2021)
4. Amazon Web Services, I.: Amazon braket - amazon web services. https://aws.amazon.com/braket/
5. Bishop, L., Bravyi, S., Cross, A.W., Gambetta, J.M., Smolin, J.A.: March. Quantum volume (2017)
6. Bradben: Azure quantum documentation, qdk & q# api reference - azure quantum. https://docs.microsoft.com/en-us/azure/quantum/
7. Cross, A.W., Bishop, L.S., Sheldon, S., Nation, P.D., Gambetta, J.M.: Validating quantum computers using randomized model circuits. Phys. Rev. A **100**, 032328 (2019)

8. J. E. et al.: Quantum certification and benchmarking. Nat. Rev. Phys. **2**(7) (2020)
9. Feynman, R.P.: Simulating physics with computers. Theor. Phys. **21**(6), 467–488 (1982)
10. Galanis, I.: Investigate quantum volume (2021). https://github.com/iliasgalanis/Investigate-Quantum-Volume
11. Galanis, I.K., Savvas, I.P., Chernov, A.V., Butakova, M.A.: Reliability testing, noise and error correction of real quantum computing devices. TELFOR J. **13**(1), 41–46 08 (2021)
12. Grover, L.K.: A fast quantum mechanical algorithm for database search. In: Proceedings of the Twenty-Eighth Annual ACM Symposium on Theory of Computing, STOC '96, New York, NY, USA, pp. 212–219. Association for Computing Machinery (1996)
13. IBM: IBM quantum
14. Moll, N., Barkoutsos, P., Bishop, L., Chow, J., Cross, A., Egger, D., Filipp, S., Fuhrer, A., Gambetta, J., Ganzhorn, M., Kandala, A., Mezzacapo, A., Müller, P., Riess, W., Salis, G., Smolin, J., Tavernelli, I., Temme, K.: Quantum optimization using variational algorithms on near-term quantum devices. Quant. Sci. Technol. **3**, 10 (2017)
15. Nielsen, M.A., Chuang, I.L.: Quantum Computation and Quantum Information, 10 edn. Cambridge University Press, (2011)
16. Qiskit: Source code for qiskit.transpiler.preset_passmanagers.level1
17. Qiskit: Source code for qiskit.transpiler.preset_passmanagers.level2
18. Qiskit: Source code for qiskit.transpiler.preset_passmanagers.level3
19. Resch, S., Karpuzcu, U.R.: Benchmarking quantum computers and the impact of quantum noise (2021). arXiv:1912.00546
20. Savvas, I.K., Chernov, A.V., Butakova, M.A.: Experiments with IBM quantum devices for random number generation and string matching. In: 2020 28th Telecommunications Forum (TELFOR), pp. 1–4 (2020)
21. Shor, P.W.: Polynomial-time algorithms for prime factorization and discrete loga-rithms on a quantum compute. SIAM J. Comput. **26**(6) (1997)
22. Silva, V.: Practical quantum computing for developers: programming quantum rigs in the cloud using python, quantum assembly language and IBM experience (2018). https://doi.org/10.1007/978-1-4842-4218-6
23. Smith, R.S., Curtis, M.J., Zeng, W.J.: A practical quantum instruction set architecture (2016). abs/ arXiv:1608.03355
24. Toninelli, E., Ndagano, B., Vallés, A., Sephton, B., Nape, I., Ambrosio, A., Capasso, F., Padgett, M.J., Forbes, A.: Concepts in quantum state tomography and classical implementation with intense light: a tutorial. Adv. Opt. Photon. **11**(1), 67–134 (2019)
25. Wikipedia: Quantum computing
26. Wikipedia: Randomized benchmarking
27. Zulehner, A., Wille, R.: Compiling su(4) quantum circuits to IBM qx architectures. In: Proceedings of the 24th Asia and South Pacific Design Automation Conference, ASPDAC '19, New York, NY, USA, pp. 185–190. Association for Computing Machinery (2019)